Bubble and Drop Interfaces

Progress in Colloid and Interface Science

VOLUME 2

Bubble and Drop Interfaces

Edited by

R. Miller
L. Liggieri

Routledge
Taylor & Francis Group

LONDON AND NEW YORK

First published 2011 by Koninklijke Brill NV

Published 2018 by Routledge
2 Park Square, Milton Park, Abingdon, Oxon, OX14 4RN
52 Vanderbilt Avenue, New York, NY 10017

First issued in paperback 2018

Routledge is an imprint of the Taylor & Francis Group, an informa business

ISBN 13: 978-1-138-11787-7 (pbk)
ISBN 13: 978-90-04-17495-5 (hbk)
ISSN 1877-8569

Foreword

Progress in Colloid and Interface Science (PCIS)

This is the second volume in this book series. It is dedicated to monographs on all topics in the field of colloid and interface science. The main aim is to present the most recent research developments and progress made on particular topics. The different volumes will cover fundamental and/or applied subjects. The first volume entitled "Interfacial Rheology" was published in 2009 and represents the first overview on this subject in form of a book. The present second volume "Bubble and Drop Interfaces" summarizes the latest state of the art of interfacial studies based on the properties of bubbles and drops. It reviews routine methods like bubble pressure, drop profile or drop volume tensiometry, but is gives also insight into very specific processes such as rising bubbles in solutions and in flotation.

The PCIS series covers all topics of colloid and interface, including liquid/gas, liquid/liquid, solid/gas, solid/liquid interfaces, their theoretical description and experimental analysis, including adsorption processes, mechanical properties, catalysis reactions, emulsification, foam formation, nanoscience; the formation and characterization of all kinds of disperse systems, such as aerosols, foams, emulsions dispersions, and their stability; bulk properties of solutions, including micellar solutions and microemulsions; experimental methodologies, such as tensiometry, ellipsometry, contact angle, light scattering, electrochemical methods, bulk and interfacial rheology; self-assembling materials at interfaces, such as surfactants, polymers, proteins, (nano-)particles, and all their mixtures; applications in various fields, such as agriculture, biological systems, chemical engineering, coatings, cosmetics, detergency, flotation processes, food processing, household products, materials processing, environmental systems, oil recovery, pulp and paper technologies, pharmacy.

All colleagues are encouraged to contact the editors of this series in case they have ideas for books, plan to contribute chapters to single volumes or want to write even complete books on topics that fit into the described scientific fields.

Reinhard Miller and Libero Liggieri
Series Editors

Content

Introduction
Drops and Bubbles as Effective Tools for Interfacial Studies

Reinhard Miller [a] and Libero Liggieri [b]

[a] MPI of Colloids and Interfaces, 14424 Potsdam/Golm, Germany
[b] Istituto per l'Energetica e le Interfasi, IENI-CNR, UOS Genova, Via De Marini 6,
I-16149 Genoa, Italy

Contents

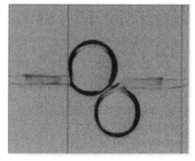

Two small drops approaching each other (studied with the DBMM of SINTERFACE Technologies).

A. General Introduction

Drops and bubbles are objects of everyday life and we meet them in various situations. Walking with open eyes through the nature allows us to see the beauty of liquid menisci, as demonstrated in Fig. 1.

Bubble and Drop Interfaces
© Koninklijke Brill NV, Leiden, 2011

Figure 1. Pendent and sessile water drops on leaves after rain show the beauty of drops in nature.

From a scientific point of view, however, we are not after the beauty but the information, drops and bubbles provide about their surface. Their shape is mainly determined by two forces. The surface tension tends to decreases the area to volume ratio, i.e., tries to make drops and bubbles spherical, while gravity tends to deform them according to their weight. As surface or interfacial tension (the tension between two immiscible liquids) are decreased by adsorption of surface active molecules, the determination of the drop/bubble profile or related quantities are a matter of scientific work in order to get information about the interfacial state. More generally tensiometric methods based on the utilisation of bubbles and drops are most efficient to investigate liquid interfaces and the kinetic processes involved with surface active species. This book aims at summarised the recent state of the art of the most important experimental techniques and gives examples of interesting applications.

A predecessor of this book was published in 1998 in the Elsevier book series "Studies in Interface Science" [1]. After a decade we can see that a tremendous amount of new work was done, which makes it necessary to describe this new advanced situation.

B. Basic Drop and Bubble Methods

While in the past ring and plate tensiometry were the methods of choice to measure the surface tension of liquids the drop profile analysis tensiometry is now the working horse in many laboratories. While this method was known and practiced form more than a century, only Neumann and his co-workers introduced it into surface science as a routine method [2]. This happened only about 25 years ago and became possible due to the availability of digital cameras and computers for data handling and for the solution of the Laplace equation, which by manual means is a burdensome work. Recent work was dedicated to improvements of calculation algorithms and efficiency of image analysis procedures

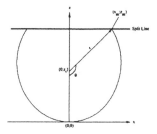

(cf. Chapters 2 and 3). Drop profile tensiometry, applied to drops covered with liquid films or membranes, is a suitable method, requires however a respective modification as shown in Chapter 4. Membrane covered drops serve as models for biological objects.

Again due to the progress in modern electronics, also the maximum bubble pressure tensiometry became a routine technique. The first who developed a breadboard of an electronic apparatus for using this methodology in an accurate and effective way was Fainerman [3]. He pioneered this technique over the last twenty years so that we can state now it is the only method providing reliable dynamic surface tensions in the time range of milliseconds and even below (Chapter 5).

The drop volume tensiometry belongs to the classical, maybe even historical techniques. Although still used in many laboratories, its efficiency is rather limited and therefore it is obviously a method which becomes less and less used in modern surface science (Chapter 6). An efficient technique for liquid analysis is the tensiograph platform (Chapter 16) a special mode of which corresponds also to the methodology of drop volume tensiometry.

The opposite is true for all capillary pressure methods, to which the maximum bubble pressure tensiometry belongs. The group of capillary pressure instruments includes mainly those dedicated to drops in gas or in a second liquids. The most important fact for their application is again the availability of short time data. In analogy to the bubble pressure method, this method provides the shortest adsorption times in particular for interfacial tensions between two liquids. So far adsorption times of few milliseconds are yet a target, but possibly it can be reached (Chapter 7). A trick is to practise this method under microgravity, which allows experiments in a broader time and frequency range due to the ideal spherical shape of all drops, independent of their size [4].

C. Drop Methodologies with Particular Applications

Due to its limited volume of small single drops, there is a depletion of surfactants when they adsorb at a very low concentration. This fact can be utilised for estimating the adsorbed amount of surfactants and even more of proteins, which belong to the strongest surface active molecules (Chapter 8). Single drops have also successfully be used to study the reversibility of adsorption, or to investigate successive adsorption processes of different molecules at the same interface. The features of this methodology, first proposed in 1999 by Cabrerizo-Vilchez *et al.* [5] (cf. Chapter 9).

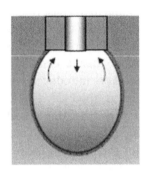

Studies with sessile drops are the most efficient investigations on the wetting of solid surfaces. First professional experiments go back to the so-called ADSA (Axisymmetric Drop Shape Analysis) technique of Neumann and co-workers [1]. All modern research of wetting, including practical questions like washing/cleaning of fabrics by surfactant solutions

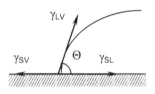

and super-spreading is performed by drop techniques (Chapter 10). Pendent and sessile drops represent also a suitable method for investigations of complex liquids such as polymers and liquid metals (Chapter 12).

An important application of bubble and drop tensiometry concerns the investigation of dilational rheology of adsorption layers. Modern technologies offer in fact the possibility to apply and accurately control variations of the interfacial area while measuring dynamic interfacial tensions. On that basis effective methods for the measurement of the dilational visco-elasticity [6] have been developed and applied in Drop Shape or Capillary Pressure tensiometers (Chapters 2, 3, 7).

D. Drop–Drop and Bubble–Bubble Interactions in Foams and Emulsions

Studies of rising bubbles are model experiments to understand the formation and response of bubble surface layers in a liquid flow field. Such experiments, combined with a fast video technique, have only recently demonstrated phenomena unknown so far, like bouncing bubbles upon rising and arriving at the free surface of a solution (cf. Chapter 11 and [7]).

Flotation is one of the famous applications of rising bubbles in technology (Chapter 14). The main principle is to form bubbles and let them rise in a slurry

in order to bind particles to their surface and carry them into a froth to be "harvested". The whole process is very complex and includes the dynamics of adsorption at bubble surfaces, the deformation of bubble surface layers due to their interaction with the liquid and formation of tension gradients along the bubble surface, attachment of particles in different way at the bubble surface, temporary stability of the froth before its wanted destructions etc.

A similarly complex technological process is the liquid mass transfer for extraction columns. The processes going on are adsorption from one or both liquid phases at the interface and the impact on mass transfer between the two liquids (Chapter 13). Due to the rising of droplets, also interfacial tension gradients are created which can affect the transfer process accordingly.

Elementary processes in emulsions and foams can be simulated with single drops or bubbles by using special micro-manipulators (Chapters 15 and 18). These new tools are not only to observe processes like coalescence but also allow direct measurements of the capillary pressure inside the drops/bubble and give also the opportunity to simulate external perturbations *via* the generation of oscillations.

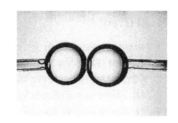

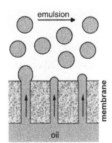

The new methodology of membrane emulsification allows producing emulsions with a very narrow droplet size distribution. It is a perfect example for the combination of fluid dynamics and surface science (Chapter 17). Single droplets are formed at the pores of a membrane under controlled liquid flow conditions. This opens the opportunity to quantitatively simulate the drop formation, and using capillary pressure methods, also to measure the relevant interfacial quantities.

Another combination of quantitative measurements of dynamic surface tension and correlation with stability parameters is the analysis of the dynamic processes in foams. This type of investigations represents the foundation for foam fractionation as a method to separate compounds from a solution or dispersion (Chapter 19).

E. Outlook

As mentioned above, the field of surface science requires an increasing number of specific tools with improved quality of data. In particular the quickly growing work in microfluidics requires quantitative tools in order to do more than only showing nice pictures of drop or bubble formation, transport of them, merging and coalescence of drops or bubbles etc. Therefore, further improvements of several methods, in particular of drop profile and capillary pressure techniques, will allow in the near future to be applied to such microfluidic devices for a controlled use and even completely new applications (Chapter 18). For example, a combination of fast video technique with drop–drop interactions provides direct access to very fast processes like coalescence. Another forthcoming development is concerned with the utilisation of capillary pressure tensiometers in the investigation of the dilational properties of spherical foam or emulsion films which have their own rheology due to the additional normal forces inside the liquid films (disjoining pressure components) [8].

The role of surfactants and/or particles will not only be analysed but also questions of very practical aims like particle attachment in flotation technology will experience a much better insight due to these new methodologies.

F. References

1. Drops and Bubbles in Interfacial Research, in "Studies in Interface Science", D. Möbius and R. Miller (Eds), Vol. 6, Elsevier, Amsterdam, 1998.
2. Y. Rotenberg, L. Boruvka and A.W. Neumann, J. Colloid Interface Sci., 93 (1983) 169.
3. V.B. Fainerman, Colloids Surfaces, 62 (1992) 333.
4. V.I. Kovalchuk, F. Ravera, L. Liggieri, G. Loglio, P. Pandolfini, A.V. Makievski, S. Vincent-Bonnieu, J. Krägel, A. Javadi and R. Miller, Adv. Colloid Interface Sci., 161 (2010) 102–114.
5. M.A. Cabrerizo-Vilchez, H.A. Wege, J.A. Holgado-Terriza and A.W. Neumann, Rev. Sci. Instr., 70 (1999) 2438.
6. R. Miller and L. Liggieri (Eds), "Interfacial Rheology". Progress in Colloid and Interface Science Series, Vol. 1, Brill Academic Pub., 2009.
7. K. Malysa, M. Krasowska and M. Krzan, Adv. Colloid Interface Sci., 114–115 (2005) 205.
8. V.I. Kovalchuk, A.V. Makievski, J. Krägel, P. Pandolfini, G. Loglio, L. Liggieri, F. Ravera and R. Miller, Colloids Surfaces A, 261 (2005) 115.

Determination of Interfacial Properties by the Pendant Drop Tensiometry: Optimisation of Experimental and Calculation Procedures

Giuseppe Loglio [a], **Piero Pandolfini** [a], **Libero Liggieri** [b], **Alexander V. Makievski** [c] and **Francesca Ravera** [b]

[a] University of Florence, Department of Organic Chemistry, Via della Lastruccia, 13, I-50019 Sesto Fiorentino (Firenze), Italy
[b] Istituto per l'Energetica e le Interfasi, IENI-CNR, UOS Genova, Via De Marini 6, I-16149 Genoa, Italy
[c] SINTERFACE Technologies, Volmerstrasse 5-7, D-12489 Berlin, Germany

Contents

Bubble and Drop Interfaces
© Koninklijke Brill NV, Leiden, 2011

A. Introduction

The Drop/Bubble Profile Analysis Tensiometry (PAT) is a well-established technique for determining static and dynamic properties of liquid-gas and liquid–liquid interfacial systems. Specifically, for the determination of interfacial tension, PAT is an accurate measurement method being metrologically based on a fundamental physical principle, that is, the Laplace equation.

This equation describes the mechanical equilibrium of two homogeneous fluids separated by an interface [1, 2] and relates the pressure difference across the interface to the surface tension and the curvature of the interface

$$\gamma \left(\frac{1}{R_1} + \frac{1}{R_2} \right) = \Delta P. \tag{1}$$

Here R_1 and R_2 are the two principal radii of interface curvature, γ is the interfacial tension and ΔP is the pressure difference across the interface.

In geometrical terms, for PAT the 3-dimensional interface (meniscus) belongs to the class of "rolled-up" axis-symmetric menisci (that is, the adjoining phases are bound by a surface of revolution, whose pivot-axis intersects the meniscus) [2]. Physically, this geometrical configuration involves an internal phase, confined in a small volume (drop or bubble), and an external phase with a volume usually some orders of magnitude greater than the drop/bubble volume.

The determination of interfacial tension by pendant drop tensiometry requires two constraining conditions, i.e., the two fluids possess a definite density difference and they are inside a gravitational field. Hence, here ΔP can be expressed as a linear function of the elevation

$$\Delta P = \Delta P_0 + (\Delta \rho) g z, \tag{2}$$

where ΔP_0 is the pressure difference at a reference plane, $\Delta \rho$ is the density difference, g is the local gravitational constant and z is the vertical height measured from the reference plane.

Substantially, the principal curvatures of the 3-D surface (i.e., the shape) of the drop (or bubble) are determined by a combination of surface tension and gravity effects. Surface forces tend to make drops and bubbles spherical whereas gravity tends to vertically elongate or squeeze them.

The advantages of this PAT method are numerous. Only very small amounts of the liquid are required, just enough to form one drop. It is suitable for both liquid–vapour and liquid–liquid interfaces, and applicable to materials ranging from organic liquids to molten metals and from pure solvents to concentrated solutions. There is also no limitation to the magnitude of surface or interfacial tension, accessible in a broad range of temperatures and pressures [3].

Operating at constant interfacial area, for a measurement run the time window ranges from parts of a second up to hours and even days so that even extremely slow processes can be easily followed. Under periodic perturbation of interfacial area, the frequency window spans some decades in the low frequency range, i.e., 10^{-5}–10^{-1} Hz. At faster oscillations, a frequency threshold appears when the interface no more is in mechanical equilibrium (no more Laplacian) as shape distortion occurs due to viscous forces and to triggering of drop/bubble normal oscillation modes [4–6].

In practice, in commercial or in laboratory instrumentation, the drop (or the bubble) is generated at the tip of a vertical or of a U-shaped capillary. According to the densities of the two adjoining phases, variant configurations are possible, namely pendant/emerging drops, sessile drops and captive/emerging bubbles.

In principle both pendant drop and emerging bubble are completely equivalent configurations, as in mathematical terms their shapes show a perfect mirror symmetry. The option to use a bubble instead of a drop appears to be unimportant.

On the other side, in some experimental studies, the two configurations have been utilized to discriminate between different transport phenomena related to adsorption processes [7, 8]. In fact, for extremely low bulk concentrations, while the adsorption at the surface of a drop can lead to a significant bulk depletion, the reservoir around a bubble supplies enough molecules so that such surfactant depletion is negligible.

The classical textbook by Rusanov [2] contains an exhaustive overview of the mathematics and the physical measurement principles relevant to pendant drop methods. In addition, this textbook describes all the milestone methods of measuring the surface tension based on studying the profile of the drop/bubble meridian-section, beginning from the initial efforts of Bashfort and Adams [9] and the earlier fixed parameter methods (geometrical properties, equators, diameters, heights, semi-empirical equations, tables), up to the recent regression variants of the profile analysis.

The advent of electronic computers allowed fast numerical integration of the Laplace equation. In this connection, a significant step ahead in axisymmetric drop shape analysis (ADSA technique) took place with the pioneering work by Neumann [10].

Essentially, Eqs (1), (2) can be reformulated as a set of three first-order differential equations expressed by the geometric parameters of the drop/bubble profile, i.e., the arc length, s, and the normal angle, ϕ, between the drop radius and the z-axis (i.e., the vertical direction) [11]

$$\frac{dx}{ds} = \cos(\phi), \tag{3}$$

$$\frac{dz}{ds} = \sin(\phi), \tag{4}$$

$$\frac{d\phi}{ds} = \pm\frac{\beta z}{b^2} + \frac{2}{b} - \frac{\sin(\phi)}{x}. \tag{5}$$

x is the abscissa of the profile point, z is the ordinate of the profile point, R is the radius of curvature at the point (x, z) on the meridian plane, b is the radius of curvature at the bubble apex $(0, 0)$, and β is the shape factor

$$\beta = \frac{\Delta\rho g b^2}{\gamma}. \tag{6}$$

The sign "plus" in Eq. (5) holds for sessile drops or captive bubbles and the sign "minus" holds for pendant drops or emerging bubbles.

Numerical integration of the Laplace equation, that is Eqs (3)–(5), and the fitting to the acquired profile coordinates of a drop/bubble (by a nonlinear regression procedure) results in the values of the apex radius and of the shape factor, and hence of the interfacial tension [2, 10–14].

Experimental aspects and numerical procedures of pendant drop tensiometry (drop/bubble image acquisition, edge detection, algorithms, digital image processing, profile acquisition errors and corrections, etc.) as well as various applications are extensively described in the excellent textbook by Neumann and Spelt [1].

As concerns the comprehensive treatment of the fundamental physical and mathematical principles for PAT, very shortly above summarised, the reader is directed to the referenced textbooks [1, 2]. It is also worth calling attention to recent progress on the determination of interfacial tension for drop/bubbles with nearly spherical shape, as described in Ref. [15] where all possible error sources are investigated by a systematic scrutiny of the software and hardware components of the measurement scheme. Moreover, in some experimental works specific advantageous features of the PAT by using a coaxial capillary have been exploited, opening a wide variety of experimental possibilities, as explained in a recent review [16].

This chapter addresses particular additional implementations in experimental and numerical procedures, taking into consideration the reliability and the analysis of the obtained results. With this aim, section B is dedicated to the improvement of the optical calibration for the acquired drop/bubble images. In sequence, section C highlights the importance of evaluating data consistency by the combination of complementary techniques, determining the interfacial tension of a given sample from distinct observed quantities.

Section D describes the fitting of a theoretical Laplacian surface, rather than a Laplacian profile, to the acquired image, advantageously applicable in strongly opalescent or dirty solutions. Section E presents particular details about the appropriate analysis of experimental results obtained for periodic oscillations. Specifically, the paragraph explains a possible approach for modelling and processing the measurement results of the real interfacial systems, which are more or less nonlinear. In this objective, the nonlinear system is represented and characterised by a consecutive combination of a linear part followed by an additional mathematical analysis parameterising the nonlinearity.

B. Fine Tuning of Optical Calibration

Calibration of scientific instrumentation is a mandatory step to be performed prior to the measurement of every specific physical quantity. A random or systematic error in the calibration parameters affects the accuracy of the measured value.

Actually, a measurement process can be thought of as a well-run process when the 'goodness' of measurement is quantified in terms of the errors that affect the measurement (bias, uncertainty, short and long term variability). The continuation of goodness is guaranteed by calibration tests, conducted on a regular basis, and by ongoing statistical control programs with the inherent remedial actions [17].

As concerns the determination of surface tension by PAT, the knowledge of the horizontal, c_x, and vertical, c_z, calibration factors (pertaining to the horizontal and the vertical axis of the drop meridian section) is required for the conversion of the CCD-camera-pixel unit to the length unit [1, 2, 13, 18–20]. Such conversion factors should be known with an accuracy better than 0.1%, as the resulting surface tension value is dramatically influenced by calibration factor errors. Specially, the error in the aspect ratio (i.e., the ratio between the values of c_x and c_z) results in fictitious surface tension changes, as a function of drop size, masking or emphasizing real square-pulse or harmonic oscillations [21, 22]. Thus, fine tuning (to the precision of four decimal digits) of the aspect ratio, c_x/c_z, is an imperative operation to obtain reliable responses to interfacial area disturbances.

The values of both horizontal, c_x, and vertical, c_z, calibration factors depend on the selected magnification for each particular experiment. In contrast, the ratio between c_x and c_z is independent of the operative magnification. Rather, this ratio is a constant property of the used instrument, exclusively depending on the characteristics of the digitising image-acquisition chain (that is, CCD camera, grabbing board, and software). The determination of accurate geometrical conversion factors involves the availability of an ultrapure reference liquid and the 'good' design of the following three issues (i) calibration procedure, (ii) diagnostic methods to ascertain the reliability of the obtained value, (iii) criteria for the adjustment of the c_x and c_z values and of their ratio, in case the outcome of the reliability test is not satisfactory.

1. Calibration Procedure

Various tensiometers for the measurement of surface tension, based on profile analysis, are commercially available. Such instruments exhibit different features, however substantially the same calibration design is applicable to the majority of the commercial realisations, with a general validity. The main parts of a profile analysis tensiometer are described elsewhere [14]. Specifically, the following-reported examples have been obtained by the PAT-1 model tensiometer (Sinterface Technologies, Berlin) [21, 23].

For the calibration process, like for regular interfacial tension measurements, a special care should be dedicated to the cleaning of the devices in contact with the samples: all the glass parts should be cleaned with concentrated sulfuric acid (ethanol for the other materials) and then washed several times with high-purity water.

Concerning the reference test liquid, high-purity water is often used which may be freshly prepared by double distillation on alkaline permanganate, using good quality drinking water for the starting material. Deionised water, obtained by exchange resins or by reverse osmosis, can also be utilised with special care to avoid possible contamination by polymer traces or bacteria.

Liquid-sample surface purity and cleanliness of glassware is assessed by the absence of any relaxation trend in the surface-tension raw-data, observing the required surface tension value from the very early generation of a drop up to several hour, at constant temperature.

The first stage of optical calibration is the proper selection and setting of the optical parameters, taking into account the experimental conditions and the variation range of the geometrical characteristics of the drops/bubbles. Thus, the established field of view (i.e., the estimated objective magnification), illumination intensity, vertical alignment and focus must be adjusted and remain fixed. Subsequently, the contrast enhancement of the acquired CCD-camera images is regulated and the grey-level distribution is optimised by the lookup table of the image processor. Also the same configuration of the foreseen experiments (i.e., pendant drop or emerging drop/bubble) should be used as the configuration of the calibration process.

An external length scale provides the value assignment for the conversion factors, from pixel units to length units. To this end, the experimental operation consists in the acquisition of a sequence of images for a reference object of known geometrical dimensions, suspended in air or immersed into water and placed in the same position as the established position of the drop/bubble. In classical calibration, the reference object is either a rod (a needle or the capillary itself) or a sphere [21, 22, 24]. For each observed reference-image, further processing operation involves detection of the limiting object-edge within a clean background, at a given threshold value of the 256 grey-levels, using a sub-pixel resolution (in case an internal reference rod image is picked within a turbid or noisy background, maximum grey-gradient or other robust algorithms must be properly adapted [25–28]).

Hence, each one of the two conversion factors, c_x or c_z, is determined by knowledge of the actual rod diameter (horizontally or vertically aligned). Alternatively, both c_x and c_z are simultaneously obtained from the fitting parameters of a circumference to the detected sphere profile, by knowing the diameter of the calibrated sphere.

The calibration procedure also includes a statistical analysis for a sequence of determinations of the conversion factors (value repeatability, standard deviation, etc.), as the calibration uncertainty is a significant component of the accuracy expression of the subsequent interfacial tension measurements.

The dimension of the reference object must be known at the best accuracy level (calibration rods or spheres with a diameter value known at one-micrometer level, or at very few micrometers, should be preferably used).

2. Diagnostic Methods for Assessing the Calibration Reliability

2.1. Test Experiments in Sequence to the Calibration Operations

The determination of surface tension by the PAT technique is based on a set of measured quantities which are correlated between them by the Laplace equation. Such a set is constituted by the coordinates of the drop profile, the density difference $\Delta\rho$ and the gravity constant g, as seen in Eqs (3)–(6). Since these physical quantities can be precisely and accurately measured, in principle the obtained results of surface tension are self-reliant.

However, in case the calibration object (rod or sphere) is not appropriate, the observed test results do not satisfactorily fulfil the two essential conditions, that is, (1) the determined surface tension value, for a reference liquid, must be in good agreement with the literature value, and (2) the determined surface tension value must be invariant with respect to a change of the geometrical size of the bubble (or the drop).

Thus, after the calibration operations, it is rational to conduct a few tests with a pure reference liquid, in order to ascertain the accuracy of c_x and c_z. The tests also verify as well the accuracy of the ratio c_x/c_z, usually referred as aspect ratio.

In practice, in sequence to the image acquisition of the rod or the sphere images, a time-series of images are acquired for a pendant drop of ultra pure water in air or alternatively for an air bubble in ultrapure water, taking into account the configuration of the foreseen experiments. The drop/bubble area is subjected to a succession of upward and downward trapezoidal pulses, as a function of time, spanning a large drop/bubble-size interval within the field of view. Then, the accuracy of the c_x and c_z is granted by the fit of the Laplace equation to the time-series of the different detected profiles, resulting in a constant value of surface tension (i.e., $\gamma = 72.8 \pm 0.1$ mN/m for pure water at temperature $T = 20°C$). Moreover, the determined surface tension value should be invariant with respect to a change of the geometrical size of the drop (or the bubble), as observed either during the trapezoidal pulses, or during sinusoidal oscillations as well as during a slow-growing (or decreasing) ramp, imposed to the drop area as a function of time [21, 22, 29].

Finally, on the basis of geometrical-optics considerations, it is worth noting that the values of c_x and c_z maintain their validity whatever is the external fluid around the drop or around the bubble, provided that the same focus sharpness is adjusted after a substitution of air or of water, respectively, with a liquid with a different refractive index.

2.2. Ongoing Diagnostic Tests for Assessing Short and Long Term Calibration Stability: Analysis of the Residual Population Distribution

Actually, in the course of running routine measurements of interfacial tension, data quality evaluation can be obtained by statistical inference from the population distribution of the Laplace-equation fitting residuals r_i [14, 15, 29], after convergence of the regression procedure:

$$r = \sqrt{(x_0 - x_c)^2 + (z_0 - z_c)^2},\qquad(7)$$

where r is the normal distance between each observed profile point $P(x_0, z_0)$ and the corresponding calculated $L(x_c, z_c)$ Laplacian point.

The moments about the mean of n-order quantitatively express the Gaussian characteristics of the data and the extent of discrepancy from the Gaussian distribution. Moreover, though qualitatively only, just the visual inspection of residual distributions in a sequence of acquired drop images allows the following sources of instrumental artefacts or of physical phenomena to be distinguished:

1. Distortion of acquired digital optical images, due to hardware defects such as the objective lenses, CCD and discretization [15].

2. A wrong ratio between calibration factors of the orthogonal coordinates for the meridian drop-section. In this case, rather than a random distribution around the zero-mean line, an S-shaped residual distribution becomes apparent for any drop/bubble in mechanical equilibrium (this circumstance is highlighted in Ref. [29] for water drops in hydrocarbon matrix. See also Figs 1, 2 and 3 where the residual distribution in the positive and negative sides is approximated by the difference $(x_0 - x_c)$ in Eq. (7), for each pair of corresponding points).

3. Viscosity effects of the matrix fluid. Drop shape is deformed to oblate while expanding and to prolate while contracting. Thus, an alternating left and right S-shaped residual distribution in concomitance with drop forced oscillation becomes apparent.

4. Excitation of drop-oscillation normal modes. The amplitude of the bubble oscillation resonance (if excited) may significantly deform the profile shape and, hence, result in erroneous values of surface tension, such a quantity being embedded in the shape parameter. In the case of drop resonance excitation, a particular

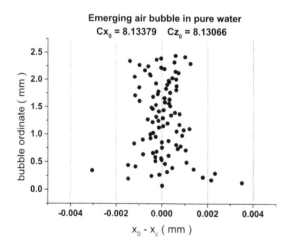

Figure 1. Random distribution of the residuals for a typical fit of an emerging-bubble profile to the Laplace equation. The distribution plot is approximated by the difference $(x_0 - x_c)$ in Eq. (7) at each vertical coordinate z_c. For this specific bubble, the standard deviation of the residual population is $\sigma = 0.78$ μm.

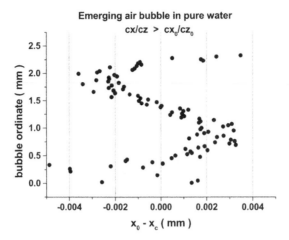

Figure 2. Same acquired profile as in Fig. 1, S-shaped residual distribution for a wrong calibration aspect ratio, in case of a value 3%-greater than the correct aspect ratio value. The standard deviation of the residual population is $\sigma = 2.3$ μm.

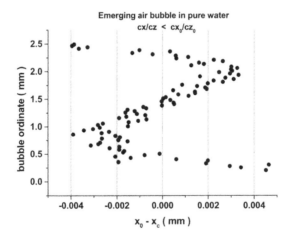

Figure 3. Same acquired profile as in Fig. 1. Reversed S-shaped residual distribution for a wrong calibration aspect ratio, in case of a value 3%-smaller than the correct aspect ratio value. The standard deviation of the residual population is $\sigma = 2.3$ μm.

pattern appears in the residual distribution for the interlaced image-acquisition of each drop.

5. Generation of a rigid surface skin, likely caused by a slow polymerisation of an adsorbed component at the surface, gives rise to an evolution of the extent of deformation in respect to the Laplacian shape. Hence, the sequence of residual distributions shows an initial random appearance followed by a continuously increasing S-shaped aspect.

In summary, the residual distribution manifests all physical and instrumental causes of profile deviation from the Laplacian equation. In this connection, note that the shape parameter, in the Laplacian first-order differential equation system, is much more sensitive to small profile disturbances than the apex-curvature parameter.

The scatter plot in Fig. 1 illustrates the random distribution of "well-behaved" fit residuals, to be pursued in accurate experimental runs.

The plots in Figs 2 and 3 show particular S-shaped and inverse-S-shaped distributions of the fit residuals, resulting from incorrect c_x/c_z-ratio of coordinate calibration factors.

For any regular experimental run, complementary to residual statistical analysis, short-term and long-term calibration variability can also be easily checked by knowledge of the capillary-diameter value. Actually, in order to continuously grant the data quality, capillary diameter is optionally saved as an internal reference device together with the coordinates of each observed drop. Invariance of the observed capillary diameter attests invariance of horizontal calibration factor, c_x.

3. Remedial Actions and Strategies

Different remedies are required to improve the accuracy, reducing constant errors and artefacts.

3.1. Tests with Ultrapure Water
Inaccuracies of calibration parameters, possibly manifested by trapezoidal pulses, sinusoidal oscillations or growing ramp test experiments with ultrapure water, can be corrected on the basis of the observed wrong values of surface tension.

3.1.1. Case when the Observed Surface Tension Value does not Match the Standard Value, but is Invariant with Respect to a Change in Bubble Size   In case of invariant but inaccurate values of surface tension obtained for a pure liquid, a quantitative study and the consequent remedial action is based on the calculation procedure adopted to solve the system of differential equations, i.e., Eqs (3)–(5), and the fitting to the acquired profile.

The value of the interfacial tension is obtained through β and b, provided as best fit values. According to the expression of β (see Eq. (6)), γ results proportional to b^2. Owing to their definition an error in both c_x and c_z affect b to a proportional extent. As the obtained value of γ is constant during the variation of the drop dimensions, we can assume the aspect ratio to be correct. Assuming also that $c_x \approx c_z$, which is true in most practical cases, we can derive a relation between the correct value of the interfacial tension γ and that derived from a wrong value of the calibration factors, that is

$$c_x/c_x^* = c_z/c_z^* = \sqrt{\gamma/\gamma^*}, \tag{8}$$

where the asterisk denotes the uncorrected values. Hence, after rearranging Eq. (8), the relationship between correct and wrong values is obtained

$$c_x = c_x^* \sqrt{\gamma/\gamma^*}, \qquad c_z = c_z^* \sqrt{\gamma/\gamma^*}. \qquad (9)$$

Thus, from Eq. (9) the corrected calibration values are computed from the reference and the obtained γ^* and from the used (inaccurate) c_x^* and c_z^* (see example illustration of Fig. 5 in Section 2.4).

3.1.2. Case when the Observed Surface Tension Value does not Match the Standard Value and is Dependent on a Change of the Bubble Size   The Laplacian shape is no longer maintained for the acquired image in case when the calibration factor c_x is affected by an error different from that affecting c_z. Provided that the error is very small, the least-squares algorithm may converge to a best-fit Laplacian profile, however in this case the value of surface tension determined by the fitting procedure is largely unreliable. Actually, while the drop/bubble is subjected to sinusoidal oscillations, with the aspect ratio lower than the correct one, $c_x^*/c_z^* < c_x/c_z$, a spurious in-phase alteration of surface tension, γ, becomes apparent; *vice versa* at $c_x^*/c_z^* > c_x/c_z$ we have a spurious out-of-phase alteration of γ.

In practice, in this circumstance, fine tuning of calibration is done iteratively by first varying the imposed value of c_x/c_z to minimise the apparent volume dependence of γ. Then the imposed values of c_x and c_z are varied at constant ratio until the reference value of γ is achieved [21, 24].

Substantially the dependence on the drop size of the apparent surface tension is useful in estimating the correction to be added or subtracted to c_x^*/c_z^*. Actually, the ratio between the value of γ obtained with a large drop and that obtained with a smaller drop, γ-large/γ-small are related to c_x^*/c_z^* through the following inequalities:

$$\gamma\text{-large}/\gamma\text{-small} > 1 \quad \text{at } c_x^*/c_z^* < c_x/c_z; \qquad (10)$$

$$\gamma\text{-large}/\gamma\text{-small} = 1 \quad \text{at } c_x^*/c_z^* = c_x/c_z; \qquad (11)$$

$$\gamma\text{-large}/\gamma\text{-small} < 1 \quad \text{at } c_x^*/c_z^* > c_x/c_z. \qquad (12)$$

Thus, by an iterative procedure, we can find a small number ε such as, added to c_x^*/c_z^*, allows for the invariance of γ^*, i.e.,

$$\gamma\text{-large}/\gamma\text{-small} = 1 \quad \text{at } c_x/c_z = c_x^*/c_z^* \pm \varepsilon. \qquad (13)$$

3.2. Routine Measurements

For routine measurements, the scatter plot of the residuals constitutes an ongoing control of the reliability (Ref. Section 2.2.2) and, in addition, it suggests a remedy treatment. An S-shaped distribution immediately indicates that the aspect ratio must be decreased, while it must be increased in relationship to an inverse-S-shaped distribution.

Moreover, a cross-correlation algorithm allows for a quantitative estimation of the correction factor, f_{corr}, to be used to find the correct aspect ratio, i.e., $c_x/c_z =$

$f_{corr}c_x^*/c_z^*$. In fact, in a pragmatic approach, the residual-population distribution-shape strongly suggests that a sinusoidal model might be appropriate for the correlation analysis. The basic sinusoidal model is:

$$y_i = \sin(\omega z_i + \phi), \quad 0 \leqslant z_i \leqslant z_{max}, \tag{14}$$

where z_i is the vertical ordinate of the drop/bubble, z_{max} is maximum ordinate value of the drop, $\omega = 2.75\pi/z_{max}$ and ϕ is the phase angle.

If the correlation (that is, specifically, the Pearson's index, m) between the residual and the model populations is high for all the sequence of acquired profiles, this implies it is worth to correct the aspect ratio. In contrast, if the correlation is weak, the correction action does not needs to be further pursued.

Quantitatively, trial and error correction strategy is expressed by the following inequalities:

$$m > 0.5 \quad \text{and} \quad \phi > 0, f_{corr} > 1, \tag{15}$$
$$m > 0.5 \quad \text{and} \quad \phi < 0, f_{corr} < 1, \tag{16}$$
$$m < -0.5 \quad \text{and} \quad \phi > 0, f_{corr} < 1, \tag{17}$$
$$m < -0.5 \quad \text{and} \quad \phi < 0, f_{corr} > 1, \tag{18}$$

where m is Pearson's index of linear correlation, the angle ϕ is the phase-angle of the sinusoidal curve that shows the maximum correlation with the residuals.

4. Example Illustrations

An example plot of the experimental results, which should be achieved for a satisfactory optical calibration, is reported in Fig. 4.

As seen, the measured average value of the surface tension compares favourably with the literature value. Moreover, the surface tension values appear invariant with

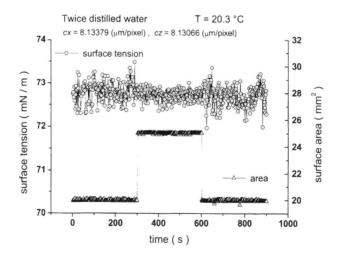

Figure 4. Surface tension of pure water, as a function of time, obtained by profile analysis technique on bubbles of two different sizes.

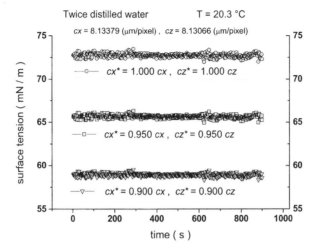

Figure 5. Plot of a sequence of simulated reprocessing results, with different fractional calibration factors (i.e., with different fractional errors in both calibration parameters): apparent surface tension of pure water, as a function of time, for the same acquired bubble profiles of Fig. 4.

respect to a change of the geometrical properties of the bubble; in particular, it is invariant in respect to a change of surface area (actually, a small change is within the experimental error).

Generally speaking, when a test experiment shows results similar to those reported in the example of Fig. 4, we can assert that the accuracy of the calibration parameters is guaranteed.

The effect of incorrect calibration factors, but correct aspect ratio c_x/c_z, is illustrated in Fig. 5, which shows the dependence of the apparent surface tension value for the case of an impartial change of both calibration factors.

Due to the mathematical properties of the Laplace equation, a fractional alteration of both c_x and c_z produces a similar geometrical Laplacian shape and, consequently, this variation maintains the fitting goodness unaltered. Only the issuing (apparent) surface tension value is changed.

The general case of inaccurate calibration parameters, when the observed surface tension value does not match the standard value and is dependent on a change of the bubble size, is simulated by a reprocessing of the acquired profile points with known inaccurate calibration parameters.

Figure 6 illustrates the plot of simulated numerical reprocessing results with a limited modification of the horizontal calibration factor only (namely, 0.01 and 0.02 fractional error). Inspection of this figure reveals that a small inaccuracy in the c_x/c_z ratio issues in a large inaccuracy in the surface tension value. Moreover, such inaccuracy becomes larger as the bubble size becomes smaller.

In Fig. 7 the comparison is presented between the effect of a ratio c_x^*/c_z^* greater than the correct value and the effect of a ratio c_x^*/c_z^* smaller than the correct value. As seen, at $c_x^*/c_z^* < c_x/c_z$, we have a spurious in-phase alteration of surface tension,

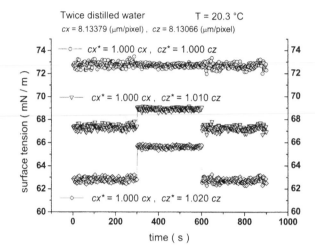

Figure 6. Same acquired bubble profiles of Fig. 4. Simulated numerical reprocessing results of surface tension, obtained by a limited modification of the vertical calibration factor only (namely, 0.01 and 0.02 fractional error).

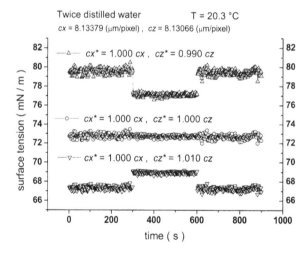

Figure 7. Simulated numerical reprocessing of the same acquired bubble profiles of Fig. 4. Comparison between the effect of a ratio c_x^*/c_z^* greater than the correct value (upper curve) and the effect of a ratio c_x^*/c_z^* smaller than the correct value (lower curve).

while at $c_x^*/c_z^* > c_x/c_z$ we have a spurious out-of-phase alteration in respect to the imposed interfacial area perturbation (Fig. 4).

Additional examples of unreal variations of surface tension of water, as measured with unadjusted aspect ratio c_x/c_z and cyclic (or step) variation of bubble volume, have been reported in Ref. [21] and more recently in Ref. [22], further emphasising the importance of aspect ratio in profile analysis tensiometry.

C. Combination of Profile-Analysis Tensiometry with Capillary-Pressure Tensiometry

A consistent data collection is a very useful means for demonstrating that the pursued improvements in the PAT measurement technique have been achieved, especially in instrument optical calibration. Furthermore, in addition to get reliance on the obtained results, data consistency also extends toward the range of decreasing drop/bubble sizes the PAT capability-limit in the determination of accurate values.

As a matter of fact, for micro-sized drops/bubbles, the optical resolution (and the inherent noise-shade) hides the deviation from the spherical shape. However, the apex radius still continue to be easily retrieved (substantially, in the fitting-regression procedure, the apex radius approaches the sphere radius and the shape factor approaches zero).

Internally-consistent data implicate the collection, from independent experimental observations, of two concomitant series of γ-values in good agreement. Essentially, the geometrical characteristics of the drop/bubble profile and the synchronic differential pressure value, observed on the same drop/bubble, allows the interfacial tension to be determined by the PAT technique and, simultaneously, by the capillary pressure technique (CPT) [29, 30]. Thus, CPT can be used here as an auxiliary tool and, in this concern, γ can be obtained by a relative calculation method [29], according to

$$\gamma = \gamma^0 \frac{R}{R^0} + \frac{R}{2}[(P - P^0) + \rho g(Z_{\max} - Z^0_{\max})], \tag{19}$$

where R is the apex radius, obtained from the profile analysis, P is the capillary pressure and $Z_{\max}$ is the apex level. The upper zero corresponds to a state of the drop where the interfacial tension is known. This state could be chosen where the profile is such that the determination of surface tension can be reliable by the PAT technique.

As an example, Fig. 8 illustrates the plots of the surface tension curve for pure water, at increasing surface area, obtained (a) from capillary pressure according to Eq. (19) and (b) from profile fitting to a Laplacian curve.

Inspection of Fig. 8 shows that γ-values, determined by capillary pressure and by profile analysis, compare favourably in a broad interval of drop dimensions. Upon decreasing the drop size, the drop shape turns into a spherical geometry, thus giving rise to unreliable data of surface tension obtained by profile fitting. At the same time, however, the pressure-signal amplitude increases, allowing consistent γ-values to be accessed by the capillary-pressure technique in an extended micrometer scale of drop size [29].

Figures 9 and 10 show the comparison of the surface tension responses, determined from profile fitting and from capillary pressure, to trapezoidal pulse and to sinusoidal perturbation of surface area, for aqueous solutions of *n*-dodecyl-dimethyl-phosphine-oxide ($DC_{12}PO$).

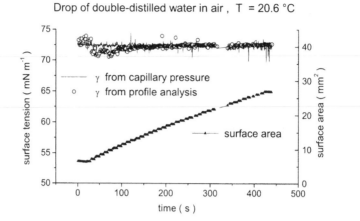

Figure 8. Comparison of γ-values for pure water, as determined by capillary pressure tensiometry (continuous line) and by pendant drop tensiometry (open dots).

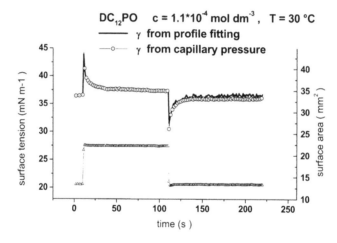

Figure 9. Comparison of γ-values for an aqueous solution of $DC_{12}PO$, as determined by pendant drop tensiometry and by capillary pressure tensiometry.

Actually, as reported in Figs 8–10, satisfactory agreement of γ-values, determined by capillary pressure and by profile fitting, grants consistency of the results with different experiment parameters and operating conditions.

D. Analysis of Optical Noisy Images

1. Bubble (or Drop) Profile Detection

The detection of the bubble (or drop) profile, from the acquired digitised images, is definitely an essential operation in PAT technique. The profile is represented by the coordinates of a finite number of points in an orthogonal Cartesian reference system, with the origin at the bubble/drop apex and the vertical axis aligned with

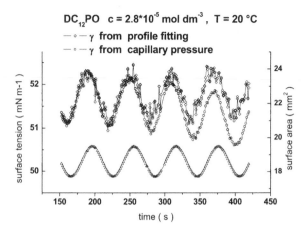

Figure 10. Comparison of γ-values for an aqueous solution of $DC_{12}PO$, as determined by pendant drop tensiometry and by capillary pressure tensiometry.

the bubble/drop symmetry axis. Precision and accuracy for the observed profile should be pursued at the maximum quality in the experimental procedure, as the resulting interfacial tension value is strongly affected by spread random errors in profile data.

In practice, the reliability of the profile coordinate values is granted by the primary condition of a good quality optical system (CCD-camera and objective), also provided that the experiment optical parameters (i.e., focus, light intensity and image contrast) are properly adjusted.

In most regular liquid/liquid or air/liquid experiments, the sequence of acquired images displays drop/bubble shapes inside a clean background. Thus, usually, the grey-level histogram of the image pixels shows a frequency distribution with two sharp peaks, one peak in proximity of the 255-level (white) and the other one in proximity of the 0-level (black). In this occurrence, the drop/bubble profile coordinates are extracted from the digitised image by a simple thresholding algorithm. By using a sub-pixel resolution, the values for the profile coordinates can be determined with a precision of one micro-meter or of a few micro-meters, depending on the optical magnification.

A big challenge for the profile detection appears when the drop/bubble is immersed in a turbid or opalescent liquid, as it can be guessed by bearing in mind the above described desirable conditions. Actually in turbid liquids, instead of a clean background, the acquired images show a noisy field with single- or many-pixel black spots, either isolated in the background or adjoining at the drop/bubble shape. Moreover, in opalescent liquids, the shape contours become rather blur and eventually indistinct.

Images with more or less optical noise are typical in experiments with biofluids, such as for instance the case of the measurement of surface tension with a captive bubble in extracted lung surfactants.

The determination of accurate and consistent values of interfacial tension for biofluids, with the captive bubble or the pendant drop configuration, has been tackled by Neumann and his co-workers by an improved analysis of the acquired noisy images. This critical task is successfully accomplished by two alternative strategies.

In particular, one of the two image-analysis procedures replaces the simple thresholding algorithm with a sophisticated edge detector which automatically discards isolated or adhering noise and retrieves a small number of smoothed profile points [27]. This algorithm is further upgraded by adding a preliminary component-labelling reduction of isolated noise [28].

2. Bubble-Image (or Drop-Image) Fitting Analysis

A second method of noisy-image analysis, eliminating the need of an independent edge-detector module, generates a theoretical image of the Laplacian meridian section for a drop/bubble and determines the interfacial tension by comparing pixel-by-pixel the theoretical image with the whole observed image [31–34]. Substantially, for calculating the theoretical image, this method adopts a similar numerical procedure as used for the profile analysis, where the basic mathematics is the system of three linear ordinary differential equations of Eqs (3)–(5), representing the Laplace physical principle.

2.1. Theoretical Laplacian Profile of the Drop/Bubble
Actually, the system of Eqs (3)–(5) can be numerically solved by a fourth-order Runge–Kutta routine, taking as initial values

$$x(0) = z(0) = \phi(0) = 0, \quad \text{at the apex } (s = 0), \tag{20}$$

$$\frac{d\phi}{ds} = \frac{1}{b}, \quad \text{at the apex } (s = 0). \tag{21}$$

Assuming that $\Delta\rho$ and g in Eq. (6) are known, the Laplacian shape just depends on the radius of curvature at the apex and on the interfacial tension.

2.2. Transformation of the Experimental Raw Image Into a Gradient Image
A digitised image is constituted by a certain set of pixels, $I(i, j)$. The grey-level of the pixels $I(i, j)$ in the acquired experimental image depends on the position of the drop/bubble and on the illumination intensity as well as on the image contrast. The effects of illumination and contrast are eliminated by transforming the original image into a gradient image $G(i, j)$, improving the reproducibility of the results, according to the following function [32]

$$G(i, j) = \max[|I(i + 1, j) - I(i - 1, j)|, |I(i, j + 1) - I(i, j - 1)|,$$
$$|I(i + 1, j + 1) - I(i - 1, j - 1)|,$$
$$|I(i + 1, j - 1) - I(i - 1, j + 1)|]. \tag{22}$$

After the transformation, the values of the pixels are high (white pixels) at the contour of the drop/bubble, where the grey-gradient is high, and low (dark pixels) in the rest of the image.

2.3. Determination of Interfacial Tension by Image Fitting

The theoretical counterpart of the gradient image $G(i, j)$ is generated from the Laplacian profile by defining the values of the pixels by the following rule

$$GT(i, j) = 255 \quad \forall(i, j) \in [(i_p, j_p)] \quad \text{and}$$
$$GT(i, j) = 0 \quad \forall(i, j) \notin [(i_p, j_p)], \quad (23)$$

where $[(i_p, j_p)]$ is the set of pixels pertaining to the profile points, as calculated in Section 4.2.1.

The interfacial tension is then determined by matching the observed and the theoretical gradient image.

The difference of the two gradient images is minimised by the Nelder–Mead simplex technique, the error function being defined as

$$\varepsilon = K - 255 \sum_{i_p}^{j_p} [2G(i, j) - 255], \quad (24)$$

where K is a constant, depending on the experimental conditions, which however does not influence the minimisation result. The algorithm, which fits the theoretical to the experimental image, searches for the values of four parameters, minimising Eq. (24), that is the parameters determining the drop/bubble shape (i.e., apex radius, b, and the interfacial tension, γ) and the parameters determining the drop/bubble position (i.e., the apex coordinates, x_a and z_a).

An important remark is that the image-fitting algorithm requires a satisfactory estimation of the initial values for the apex radius, b_0, and for the interfacial tension, γ_0. An effective approach for the initial estimation of the fitting parameters, b_0 and γ_0, is extensively described in Ref. [32].

In a subsequent article [33], the image-fitting analysis has been extended to various Laplacian objects, other than captive bubble or pendant drops, with and without apex (like for example sessile drops, liquid bridges and liquid lenses).

Experimental results of interfacial tension, obtained with a captive bubble inside very turbid liquids, also containing suspended particles, demonstrates the satisfactory reliability of the image-fitting procedure [32].

Substantially, such image-fitting technique, as well as the implemented generalised methodology [33], appears to be a robust and noise-resistant procedure and, hence, it can be advantageously applied when acquisition of experimental high-quality images is not possible.

E. Study of Interfacial Responses to Perturbations of Interfacial Area

It is now well-established that the profile analysis technique is most appropriate for investigating multi-component systems, containing surfactants, polymers or proteins which tend to adsorb into the interface. As a consequence of the adsorption process, the creation of a fresh interface or a perturbation of the interfacial area results in a change of the interfacial tension as a function of time.

Various commercially-available PAT-technique tensiometers, such as the instrument in Ref. [23], offer the option of controlling the interfacial area of a drop (or of a bubble) as a function of time, by a feed-back loop comparing the observed drop area-value with the set-value in a pre-defined time-line. This feature is an essential instrumental tool for studying the dynamic interfacial tension and the interfacial responses to controlled area perturbations. This kind of experimental studies, known as perturbation-response methods or drop/bubble oscillation methods, convey information for determining the interfacial visco-elastic properties of adsorption layers and, substantially, for measuring the interfacial dilational modulus, $\varepsilon(\omega)$, that is the quantity linking the issuing interfacial response to the imposed fractional-area perturbation.

1. Linear Visco-Elastic Regime of Interfacial Systems

In mathematical terms, the linear visco-elasticity can be defined either in the time domain or in the frequency domain, as shown in the following expressions,

$$\Delta\gamma(t) = \gamma(t) - \gamma_0 = \int_0^t F^{-1}\{\varepsilon(\omega),\tau\}\Delta A(t-\tau)/A_0\,d\tau, \qquad (25)$$

$$\Delta\gamma(\omega) = \varepsilon(\omega)\Delta A(\omega)/A_0, \qquad (26)$$

where $\Delta\gamma(\omega) = F\{\Delta\gamma(t)\}$, $\Delta A(\omega) = F\{\Delta A(t)\}$, F is the Fourier transform operator, τ is the dummy time in the convolution integral, ω is the angular frequency, and $\Delta A/A_0$ is the relative area change.

Limiting to appropriately-small disturbances of interfacial area, under not-far-from-equilibrium conditions, Eqs (25) and (26) allow to state that all interfacial transient or periodic processes can be represented by a linear time-invariant distributed-parameter dissipative system. Thus, adopting such a linear model, the basic conceptual proposition is now well-established that the interfacial dilational modulus, $\varepsilon(\omega)$ is an intrinsic constitutive property of a system which characterises the dynamics of adsorption layers, linking surface excitation, $\Delta A(t)/A_0$, forced with any functional form and surface response, $\Delta\gamma(t)$ [24, 35–50].

The definition of the surface dilational modulus relies on the assumption that a linearity range actually exists for small disturbances of surface equilibrium, i.e., on the assumption that the response amplitude is directly proportional to the disturbance amplitude [40, 41].

Within the linearity range, a harmonic multi-frequency $\Delta A(t)/A_0$ excitation generates an interfacial $\Delta\gamma(t)$ response with a similar harmonic multi-frequency composition. Since the Fourier-transform is a linear operator, the Boltzmann superposition principle is applicable, that is, the response to a composite excitation function is the summation of the particular responses to each elementary excitation function.

2. Nonlinear Dynamic Behaviour of Real Interfacial Systems

Real multi-component interfacial systems actually manifest a nonlinear dynamic (visco-elastic) response, with memory, consequent to an external forced change of interfacial area. Certainly, the nonlinear regime conveys additional information about the interfacial structure and dynamics, in respect to the linear regime.

The nonlinear regime becomes progressively more and more dominant on increasing the amplitude and the time scale (or the frequency) of the interfacial strain (the frequency dependence of the linear range can be denoted as dispersion of the linearity limit).

The extent of the nonlinearity behaviour strongly depends on the chemical nature and composition of each specific interfacial system. In particular, for adsorption layers constituted by polymers or by proteins the nonlinear regime already occurs at very small fractional area perturbations [51, 52].

Most technological operations involve violent hydrodynamic conditions, resulting in wide and fast area extension/compression of the interfacial layers, which are present inside the processed heterogeneous body. Hence the interested interface is characterised by a transient nonlinear visco-elasticity, rather than by a steady-state linear visco-elastic regime. The transient nonlinear visco-elastic behaviour is also likely occurring in drop–drop coalescence, where the thinning movement and rupture of the lamella, which intervenes between two adjoining droplets, definitely happens well-beyond the linearity-range limit.

Despite a given interfacial system may be nonlinear under particular conditions, however it exhibits a visco-elastic behaviour which may generate unexpected effects, impossible to be forecast by simple linear models. Thus, experimental and theoretical investigations of the nonlinear mechanisms pertaining to interfacial layers are relevant issues to understanding several phenomena and industrial processes.

From a general point of view, the behaviour of nonlinear systems with memory cannot easily be described in explicit theoretical terms. Recently, the nonlinear mechanical response of Langmuir lipid monolayers has been described by a polynomial expansion of the visco-elastic modulus [53]. Moreover the Volterra series appears as a good theoretical modelling tool, in the time domain or in the frequency domain, for (weakly) nonlinear dynamic systems [54]. In case of interfacial layers, in which $\Delta A(t)/A_0$ is the input perturbation and $\gamma(t)$ is the output response, the Volterra series reads in the time domain [55]:

$$\gamma(t) = \gamma_0 + \sum_{n=1}^{n=\infty} \frac{1}{n!} \int_0^t \cdots \int_0^t k_n(\tau_1, \tau_2, \ldots, \tau_n)$$

$$\times \prod_{i=1}^n \Delta A(t - \tau_i)/A_0 \, d\tau_1 \, d\tau_2 \cdots d\tau_n, \quad (27)$$

where γ_0 is the equilibrium surface tension at the instant $t = 0$ and the mathematical n-th order Volterra kernel k_n can be physically interpreted as the n-th higher-order impulse response of the system.

As seen in Eq. (27), the Volterra series includes the first-order impulse response of the interfacial system, which has been already introduced in Refs [35, 36] on the basis of the dilational visco-elastic modulus $\varepsilon(\omega)$, that is, the frequency-response function of the interfacial layers, earlier defined by Lucassen and Van den Temple [56].

The first-order impulse response, which characterises the linearity behaviour, is formulated as the inverse Fourier transformation F^{-1} of the frequency-response function of the system $k_1(t) = F^{-1}\{\varepsilon(\omega, t)\}$.

It is noticeable that Eq. (27) is a generalisation of the Taylor's series, because it includes the memory. Also, Eq. (27) is an extension of the previous linear treatment and Eq. (25), can be obtained by truncating the series at the zeroth-order plus the first-order terms. In addition, Eq. (27) defines the higher-order impulse responses k_n of the interfacial layers.

Thus Eq. (27) in principle is an appropriate mathematical resource for the representation and identification of the real interfacial systems which are more or less nonlinear.

However, the Volterra series has a serious drawback, because at present in the literature no efficient numerical algorithm, especially tailored for the interfacial layers, can be found. Also an hindrance, the lacking of experimental techniques for the kernel measurement, relevant to the nonlinear contribution of interfacial tension responses.

In practice, aiming at a simplifying treatment, various approaches are conceivable for describing the responses of interfacial layers subjected to area variation. Particularly, the following three situations have been envisaged.

- When the memory effect is negligible, i.e., when relaxation processes are considered as absent, the system behaviour can be described in the time domain, by determining a nonlinear response/excitation characteristic relationship which can be described by a Taylor's series expansion. This situation can be met in some insoluble interfacial layers.

- On the opposite side, if the memory effect plays a significant role, the system nonlinearity is neglected and the dynamic behaviour of the interfacial layers is described in the frequency domain, through the Volterra series truncated at the first order term as in Eq. (25).

- In an intermediate case, for a weakly nonlinear system with memory, the linear response of the system is given by $\varepsilon(\omega)$ as defined in Eq. (25), while the nonlinearity behaviour is phenomenologically described by giving the spectrum of the periodic response to a sinusoidal perturbation. Moreover, the nonlinearity extent is quantitatively expressed by the Total Harmonic Distortion (THD) parameter [46].

The possible approach for modelling the behaviour and processing the measurement results of the real interfacial systems for the above mentioned intermediate case is treated in the following paragraph 3.

3. Modelling of Weakly Nonlinear Interfacial Systems

3.1. Nonlinearity Evaluation by Fourier Analysis

The amplitude and phase of the higher-order harmonics convey information about the validity of the linearity assumption and, most important, they are a possible ancillary means, in concomitance with $\varepsilon(\omega)$, for characterising the dynamics of interfacial layers in real systems, like for example in crude-oil water interfaces [57] or in breathing processes [51].

The Total Harmonic Distortion (THD) parameter has been recently proposed in Refs [46, 58] as a suitable global index, whose value allows the assessment of the linearity-hypothesis validity for the interfacial excitation-response behaviour. Beyond the linearity threshold, the THD-value is important for the quantitative evaluation of the nonlinearity extent for an observed (weakly) nonlinear interfacial response [24, 52].

In a sequence of harmonic oscillation experiments, the THD-parameter can be used for defining a tolerability domain in the amplitude-frequency plane, fixing the limits where the nonlinearity of the interfacial system can be neglected.

In case the nonlinearity cannot be neglected and however the fundamental component is dominant in respect to the higher harmonics of the oscillating interfacial $\Delta\gamma$-response (that is, the interface behaviour is weakly nonlinear), nonlinearity effects may be circumvented by selecting just the amplitude and phase of the fundamental component in the determination of $\varepsilon(\omega)$. In all circumstances, when the oscillation is not purely sinusoidal, this Fourier-analysis procedure is more correct in respect to fitting a sinusoidal function to the $\Delta\gamma$-response.

In the following sub-paragraph the Fourier analysis algorithms are outlined, which are relevant both to the THD-parameter determination in the case of interfacial layers and to the determination of the linear visco-elastic modulus. Then, some representative example-results are illustrated for available data of surface tension responses to imposed periodic changes of surface area, at the air–water interface.

3.2. Basic Mathematical Expressions and Algorithms

Periodic phenomena are properly analysed by expansion into Fourier series.

In principle, a single cycle of a periodic phenomenon contains all the necessary physical information. In practice, in the harmonic oscillation experiments, the acquisition of data during a time interval longer than the single-oscillation period appears advantageous for a convenient experimental redundancy as well as for a verification of the transient or steady-state regime.

The Fourier analysis of the observed experimental data (namely, the interfacial tension and the interfacial area) is accomplished on the basis of the following mathematics.

The general analytical expression of a periodic function $g(t)$ expanded into a Fourier series is:

$$g(t) = \frac{a_0}{2} + \sum_{n=1}^{\infty}\left[a_n \cos\frac{2n\pi t}{T} + b_n \sin\frac{2n\pi t}{T}\right], \tag{28}$$

where:

$$a_n = \frac{2}{T} \int_0^T g(\tau) \cos \frac{2n\pi\tau}{T} d\tau, \tag{29}$$

$$b_n = \frac{2}{T} \int_0^T g(\tau) \sin \frac{2n\pi\tau}{T} d\tau. \tag{30}$$

Considering a sinusoidal signal (the interfacial tension or the interfacial area) at frequency ν and phase φ, only one term of the summation of Eq. (28) has to be taken, i.e.,

$$g(t) = g^0 + \tilde{g} \sin(2\pi\nu t + \varphi). \tag{31}$$

Considering the discretization of the acquired signal at the generic time t_j and using Eqs (28) to (30), one obtains

$$g_j = g^0 + a_1 \cos(2\pi\nu t_j) + b_1 \sin(2\pi\nu t_j), \tag{32}$$

where

$$a_1 = \frac{1}{N} \sum_{k=1}^N g_k 2 \cos(2\pi\nu t_k), \qquad b_1 = \frac{1}{N} \sum_{k=1}^N g_k 2 \sin(2\pi\nu t_k),$$

N is the number of experimental points and g_k is the value of g measured at the time t_k. Thus one obtains,

$$\tilde{g} = \sqrt{a_1^2 + b_1^2}, \tag{33}$$

$$\varphi = \arctan\left(\frac{b_1}{a_1}\right). \tag{34}$$

Here, the summation is performed over all points of all whole cycles included in a selected temporal interval.

Considering Eqs (29) and (30), we note that the numerical computation behaves as if the selected integer number of complete cycles is infinitely replicated as a function of time, giving rise to a periodical function.

The first term in Eq. (28), a_0, represents the mean value of the temporal sequence. Hence, in the present circumstances, the values of the mean interfacial tension and of the mean interfacial area, are obtained within the selected cycles.

We note that the amplitude value as determined by Eq. (33), coincides with the value obtained by fitting a sinusoidal function to the observed temporal sequence only in case of a purely harmonic response. However, when the experimental system contains also other frequency components, the least-squares fitting procedure may definitely ensue in meaningless amplitude and phase values.

Usually, in standard perturbation-response systems, the time-origin is arbitrarily chosen on the perturbation signal. What is significant is the phase difference between the response signal (interfacial tension) and the input signal (interfacial area).

According to the definition, the modulus of the dilational modulus is obtained from Eq. (26) by the amplitude ratio

$$|\varepsilon(\omega)| = \frac{\Delta\gamma(\omega)}{\Delta A(\omega)/A_0}, \tag{35}$$

where $\omega = 2\pi\nu$ and the amplitude of $\Delta\gamma$ and of ΔA are determined by Eq. (33).

The real and the imaginary part of the dilational modulus are obtained from

$$\varepsilon' = |\varepsilon(\omega)|\cos(\vartheta), \qquad \varepsilon'' = |\varepsilon(\omega)|\sin(\vartheta), \tag{36}$$

where θ is the phase difference between $\Delta\gamma$ and ΔA calculated using Eq. (34).

3.3. Determination of the Total Harmonic Distortion (THD)

The interfacial response to a given perturbation of interfacial area is often observed to be nonlinear. The amplitude spectrum of the oscillating interfacial properties is computed by Eq. (33).

An index of the non-harmonic contribution, occurring in the interfacial response, is expressed by the Total Harmonic Distortion (THD), that is, the ratio of the higher harmonics amplitude to the amplitude at the measured fundamental frequency. Alternatively, THD can be expressed as the ratio of the higher harmonics power to the power at the fundamental frequency [58].

In the following equation, a_1 is the amplitude value at the fundamental frequency and $a_2, a_3, \ldots, a_n$ are amplitude values of the higher harmonics:

$$\text{THD} = (a_2^2 + a_3^2 + \cdots + a_n^2)^{1/2}/a_1. \tag{37}$$

Thus, the THD index can easily be computed from the amplitude values of the fundamental frequency and of the higher order harmonic frequencies. THD is also expressed as a percentage of the fundamental-frequency amplitude.

For small-amplitude perturbations, the THD value becomes vanishingly small (linearity approximation). At constant-amplitude perturbation, THD depends on the excitation frequency (frequency dispersion of THD).

Due to a multitude of sources, the measurement of the interfacial response is affected by discretization and random errors, altogether denoted as noise.

The distinction between physically significant signal and aberrant noise usually demands the assumption of an appropriate model. Here, we assume that the interfacial response signal is completely constituted by the fundamental oscillation and by the subsequent higher harmonics, up the fifth order, while the remainder amount is noise.

The Total Harmonic Distortion with Noise (THD + N) is the ratio of the higher harmonics power, plus the noise power, to the power of the measured fundamental frequency. THD + N values will almost always be greater than the THD values for the same temporal sequence of data.

4. Example Results

Representative example results for experiments in the low frequency range, illustrating the application of the THD-parameter to the field of fluid interfacial systems, are reported in Refs [58–60], for very weakly nonlinear systems. In the numerical computation of the observed oscillating results, both the periodic area $\Delta A/A_0$ variation and the issuing surface tension $\Delta \gamma$ variation are expanded in Fourier series in selected time intervals and the averaged values, obtained by the algorithms of Eqs (34) and (35), are used for the determination of the dilational modulus as well as of the oscillation spectrum and the inherent THD-values.

The following sections report some examples for a variety of possible experimental periodic oscillatory deformations of interfacial area. These experimental examples were obtained on oscillating drops of aqueous solutions of the polymeric compound poly(sodium 4-styrenesulfonate) (shortly referred as PSS), adopting the pendant drop configuration [58].

4.1. Single-Frequency Oscillations at Different Amplitudes

Figure 11 illustrates the response of surface tension to sinusoidal oscillations of surface area with 3 different amplitudes (relative surface area $\Delta A/A_0 = \sim 5\%$, 10% and 15%).

Table 1 reports the relevant THD-values for the observed quantities, $\Delta \gamma$ and ΔA. In this concern, the importance of considering both $\Delta \gamma$-THD and ΔA-THD was remarked in Ref. [29], as certainly the response reflects the non-harmonicity of the excitation.

As seen in Table 1, at the lowest amplitude, the THD-value for ΔA is greater, in respect to the value for the other two amplitudes of ΔA. This occurrence can be attributed to the noise that may have a component at all frequencies. The effect is also propagated to the $\Delta \gamma$ response. The difference between $\Delta \gamma$-THD and ΔA-THD moderately increases with increasing amplitude, however remaining below THD = 5%, a value which can be assumed here as the tolerability limit for considering almost linear the perturbation-response relationship.

Also, Table 1 reports the real and imaginary part of the complex modulus ε^*, computed by selecting just the fundamental component of the oscillating re-

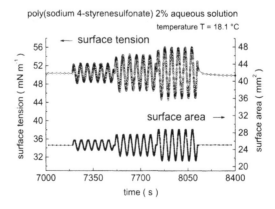

Figure 11. $\Delta \gamma$-responses to different monochromatic ΔA-perturbations ($\Delta A/A_0 = \sim 5\%$, 10% and 15%), at 0.02-Hz frequency.

Table 1.
Drop-oscillation experiments (PSS aqueous solution, concentration $c = 2$ wt%, $T = 18.1°C$, mean value of the surface area $A = 25.0$ mm^2; mean value of surface tension $\gamma = 50.5$ mN/m). Values of THD and of complex interfacial dilational modulus for different amplitudes of the external surface-area excitation ΔA

Freq. (Hz)	$\Delta\gamma$-THD (%)	ΔA-THD (%)	Re$\{\varepsilon^*\}$ (mN/m)	Im$\{\varepsilon^*\}$ (mN/m)
Oscillation amplitude of surface area: $\Delta A = 1.23$ mm^2				
(relative surface area: $\Delta A/A = 4.92\%$)				
0.02	4.5	3.2	33.13	17.54
Oscillation amplitude of surface area: $\Delta A = 2.50$ mm^2				
(relative surface area: $\Delta A/A = 10.00\%$)				
0.02	2.8	0.8	32.71	16.96
Oscillation amplitude of surface area: $\Delta A = 3.73$ mm^2				
(relative surface area: $\Delta A/A = 14.92\%$)				
0.02	3.3	0.8	31.77	16.58

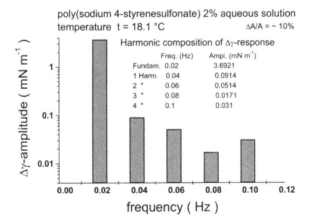

Figure 12. Oscillation spectrum of the $\Delta\gamma$-response the $\Delta A/A_0 = 10\%$ amplitude (same experiment of Fig. 11).

sponse/perturbation properties. Actually, the computed values of ε' and ε'' for the three different amplitudes compare favourably, within he experimental errors.

Inspection of Table 1 also suggests that the $\Delta A/A_0 = 10\%$ amplitude is to be preferred for the present system, due to very low THD-values.

The favourable linearity approximation for the amplitude $\Delta A/A_0 = 10\%$ is also visually represented by the oscillation spectrum of the $\Delta\gamma$-response as seen Fig. 12, where the fundamental-frequency magnitude-value is more than one order larger than the harmonics values.

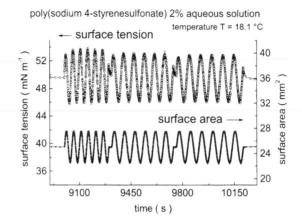

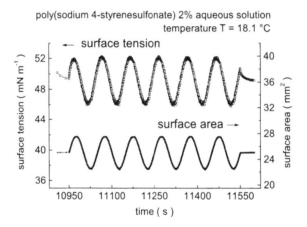

Figure 13. $\Delta\gamma$-responses to monochromatic ΔA-perturbations ($\Delta A/A_0 \approx 10\%$, at frequencies of 0.02, 0.01429 and 0.01333-Hz).

Figure 14. $\Delta\gamma$-response to monochromatic ΔA-perturbation ($\Delta A/A_0 \approx 10\%$, at frequency of 0.01-Hz).

4.2. Single-Amplitude Oscillations at Different Frequencies

Figures 13 and 14 by example display some sequences of $\Delta\gamma$-responses to monochromatic ΔA-oscillations with the same amplitude, at four different frequencies.

As shown in Table 2, the linearity approximation is satisfactory as the difference between $\Delta\gamma$-THD and ΔA-THD is below the 5%-limit. Moreover, the real and the imaginary part of ε^* exhibit a reasonable decreasing trend with decreasing frequency.

F. Summary

Pendant drop tensiometry is a reliable method for the measurement of static and dynamic interfacial tension in liquid–fluid multi-component systems.

Table 2.
Drop-oscillation experiments (PSS aqueous solution, concentration $c = 2$ w/w%, temperature $T = 18.1°C$, mean value of the surface area $A = 25.0$ mm^2; mean value of surface tension $\gamma = 49.7$ mN/m; oscillation amplitude of surface area $\Delta A = 2.50$ mm^2; relative surface area: $\Delta A/A_0 = 10\%$). Values of THD and of complex interfacial dilational modulus for different frequencies of the external surface-area excitation ΔA

Freq. (Hz)	$\Delta\gamma$-THD (%)	ΔA-THD (%)	Re{ε^*} (mN/m)	Im{ε^*} (mN/m)
0.02	2.92	0.95	32.97	17.33
0.01429	3.6	1.1	29.64	15.80
0.01333	3.4	1.2	28.89	15.51
0.01	3.7	0.80	26.47	14.38

Accuracy of calibration parameters is a priority concern. Experiments on a reference liquid, imposing a square pulse to the size of a bubble (or of a drop), allows the accuracy of the calibration parameters to be assessed by test analysis of the resulting interfacial tension value. The observed value must be in agreement with literature standard data and must be invariant in respect to the change of bubble (or drop) size. Discrepancies of the observed interfacial tension value in respect to the expected standard value are anyway effective for determining the corrected calibration parameters, by using (a) an analytical expression in case of drop-size invariance or (b) a numerical iterative procedure in case of drop-size dependence.

The population distribution (and its statistical analysis) of Laplace-equation fitting-residuals is a helpful diagnostic item for data quality evaluation. Such a distribution allows different error and artefacts sources to be ascertained: (a) distortion of acquired optical images; (b) wrong ratio between calibration factors (aspect ratio); (c) actual non-Laplacian shape of oscillating drops/bubbles due to viscosity, resonance effects or formation of rigid skins on the surface.

Consistency of observed values of static and dynamic interfacial tension is advantageously obtained by the combination of capillary-pressure and profile-analysis measurements, in a proper overlapping drop/size range, in case of agreement of the determined values from the independent physical quantities, that is, the shape factor and the capillary pressure.

Image-fitting analysis appears to be a robust and noise-resistant procedure which allows accurate determination of interfacial tension, when acquisition of experimental high-quality images is not possible and profile-fitting fails.

The interfacial dilational modulus is properly measured by taking the amplitude of a given frequency of the $\Delta\gamma$-responses, using Fourier series expansion. This can be effectively done even with a multi-frequency composition. Also the amplitude so obtained for the given frequency is more appropriate than the amplitude of the sinusoidal best-fit to the whole oscillation. In fact this procedure circumvents the amplitudes of the higher harmonics in the signal generated by nonlinearity effects as well as the component introduced by external disturbances.

The Fourier-series expansion of drop $\Delta\gamma$-responses, to forced interfacial-area periodic oscillations, appears the proper mathematical tool for quantitatively characterising system behaviour under weakly nonlinear conditions. In fact from the Fourier analysis, the total harmonic distortion (THD) parameter is defined. It is a suitable global index, whose value allows the assessment of the linearity hypothesis for the interfacial system. Moreover, beyond the linearity threshold, the THD-value provides a quantitative evaluation of the nonlinearity extent of the system, defining a tolerability domain in the amplitude-frequency plane and fixing the limits where the nonlinearity of the interfacial system can be neglected.

As the Volterra series is the general theoretical tool for modelling weakly nonlinear dynamic systems with memory, in case of interfacial layers, beyond the linearity limit, a possible simplifying treatment for characterising the nonlinear visco-elasticity is the determination of the first-order system function, i.e., the interfacial dilational modulus $\varepsilon(\omega)$, in conjunction with the analysis of the spectrum of the $\Delta\gamma$-response to a sinusoidal excitation. In other words, the nonlinear behaviour of the system is represented and characterised by a combination of a linear part and an additional mathematical analysis parameterizing the nonlinearity.

G. Acknowledgements

This work was performed within the framework of "MAP AO-99-052, Fundamental and Applied Studies of Emulsion Stability", FASES project (ESTEC Contract Number 14291/00/NL/SH).

H. References

1. A.W. Neumann and J.K. Spelt, Eds, "Applied Surface Thermodynamics", Surfactant Science Series, Vol. 63, Marcel Dekker Inc., New York, 1996.
2. A.I. Rusanov and V.A. Prokhorov, "Interfacial Tensiometry", D. Möbius and R. Miller (Eds), Studies in Interface Science Series, Vol. 3, Elsevier, Amsterdam, 1996.
3. R. Miller, V.B. Fainerman, A.V. Makievski, M. Ferrari and G. Loglio, Measuring Dynamic Surface Tensions, in "Handbook of Applied Colloid and Surface Science", K. Holmberg (Ed.), John Wiley & Sons, 2001, pp. 775–788.
4. Y.C. Liao, O.A. Basaran and E.I. Franses, Colloids Surfaces A, 250 (2004) 367–384.
5. E.M. Freer, H. Wong and C.J. Radke, J. Colloid Interface Sci., 282 (2005) 128–132.
6. M.E. Leser, S. Acquistapace, A. Cagna, A.V. Makievski and R. Miller, Colloids Surfaces A, 261 (2005) 25–28.
7. A.V. Makievski, G. Loglio, J. Krägel, R. Miller, V.B. Fainerman and A.W. Neumann, J. Phys. Chem. B, 103 (1999) 9557–9561.
8. L. Liggieri, M. Ferrari, A. Massa and F. Ravera, Colloids Surfaces A, 156 (1999) 455–463.
9. F. Bashfort and J.C. Adams, An Attempt to Test the Theory of Capillary Action, Cambridge University Press and Deighton Bell and Co., Cambridge, 1883.
10. Y. Rotenberg, L. Boruvka and A.W. Neumann, J. Colloid Interface Sci., 93 (1983) 169–183.

11. C. Maze and G. Burnet, Surface Science, 13 (1969) 451–470.
12. S. Lahooti, O.I. Del Rio, A.W. Neumann and P. Cheng, Axisymmetric Drop Shape Analysis (ADSA), in "Applied Surface Thermodynamics", A.W. Neumann and J.K. Spelt (Eds), Surfactant Science Series, Vol. 63, Marcel Dekker Inc., New York, 1996, pp. 486–487.
13. P. Chen, D.Y. Kwok, R.M. Prokop, O.I. del Rio, S.S. Susnar and A.W. Neumann, Axisymmetric drop shape analysis (ADSA) and its applications, in "Drops and Bubbles in Interfacial Research", Studies in Interface Science Series, D. Möbius and R. Miller (Eds), Vol. 6, Elsevier, Amsterdam, 1998, pp. 61–168.
14. G. Loglio, P. Pandolfini, R. Miller, A.V. Makievski, F. Ravera, M. Ferrari and L. Liggieri, Drop and Bubble Shape Analysis as Tool for Dilational Rheology Studies of Interfacial Layers, in "Novel methods to study interfacial layers", D. Möbius and R. Miller (Eds), Studies in Interface Science Series, Vol. 11, Elsevier, Amsterdam, 2001, pp. 439–485.
15. M. Hoorfar and A.W. Neumann, Adv. Colloid Interface Sci., 121 (2006) 25–49.
16. J.K. Ferri, C. Kotsmar and R. Miller, Adv. Colloid Interface Sci., in press.
17. NIST/SEMATECH e-Handbook of Statistical Methods, National Institute of Standards and Technology, U.S. Commerce Department's Technology Administration, 2006, http://www.itl.nist.gov/div898/handbook/.
18. G. Loglio, U. Tesei, P. Pandolfini and R. Cini, Colloids Surfaces A, 114 (1996) 23.
19. G. Loglio, P. Pandolfini, U. Tesei and B. Noskov, Colloids Surfaces A, 143 (1998) 301.
20. L. Liggieri and A. Passerone, High Temperature Technology, 2 (1989) 82.
21. G. Loglio, P. Pandolfini, A.V. Makievski and R. Miller, J. Colloid Interface Sci., 265 (2003) 161–165.
22. S.I. Karakashev and A.V. Nguyen, J. Colloid Interface Sci., 330 (2009) 501–504.
23. SINTERFACE Technologies, Berlin, Germany, http://www.sinterface.com.
24. M. Vrânceanu, K. Winkler, H. Nirschl and G. Leneweit, Colloids Surfaces A, 311 (2007) 140–153.
25. J. Canny, A Computational Approach to Edge Detection, IEEE Transactions on Pattern Analysis and Machine Intelligence, Vol. PAMI-8, No. 6, November, 1986, pp. 679–698.
26. V.S. Nalwa and T.O. Binford, On Detecting Edges, IEEE Transactions on Pattern Analysis and Machine Intelligence, Vol. PAMI-8, No. 6, November, 1986, pp. 699–714.
27. Y.Y. Zuo, M. Ding, A. Bateni, M. Hoorfar and A.W. Neumann, Colloids Surface A, 250 (2004) 233–246.
28. Y.Y. Zuo, Chau Do and A.W. Neumann, Colloids Surfaces A, 299 (2007) 109–116.
29. L. Del Gaudio, P. Pandolfini, F. Ravera, J. Krägel, E. Santini, A.V. Makievski, B.A. Noskov, L. Liggieri, R. Miller and G. Loglio, Colloids Surfaces A, 323 (2008) 3–11.
30. S.C. Russev, N. Alexandrov, K.G. Marinova, K.D. Danov, N.D. Denkov, L. Lyutov, V. Vulchev and C. Bilke-Krause, Rev. Sci. Instrum., 79 (2008) 104102–104110.
31. M.G. Cabezas, A. Bateni, J.M. Montanero and A.W. Neumann, Appl. Surface Sci., 238 (2004) 480–484.
32. M.G. Cabezas, A. Bateni, J.M. Montanero and A.W. Neumann, Colloids Surfaces A, 255 (2005) 193–200.
33. M.G. Cabezas, A. Bateni, J.M. Montanero and A.W. Neumann, Langmuir, 22 (2006) 10053–10060.
34. M.G. Cabezas, J.M. Montanero and C. Ferrera, Measurement Science Technology, 18 (2007) 1637–1650.
35. G. Loglio, U. Tesei and R. Cini, Ber. Bunsenges. Phys. Chem. Chem. Phys., 81 (1977) 1154–1156.
36. G. Loglio, U. Tesei and R. Cini, J. Colloid Interface Sci., 71 (1979) 316–320.
37. G. Loglio, U. Tesei and R. Cini, Colloid Polym. Sci., 264 (1986) 712–718.

38. G. Loglio, U. Tesei, R. Miller and R. Cini, Colloids Surfaces, 61 (1991) 219–226.
39. R. Miller, G. Loglio, U. Tesei and K.-H. Schano, Adv. Colloid Interface Sci., 37 (1991) 73–96.
40. G. Loglio, R. Miller, A. Stortini, U. Tesei, N. Degli Innocenti and R. Cini, Colloids Surfaces A, 90 (1994) 251–259.
41. G. Loglio, R. Miller, A. Stortini, U. Tesei, N. Degli Innocenti and R. Cini, Colloids Surfaces A, 95 (1995) 63–68.
42. P. Joos, Dynamic Surface Phenomena, VSP, Utrecht, The Netherlands, 1999, p. 247.
43. J. Lyklema, Fundamentals of Interface and Colloid Science, Vol. III, Liquid–Fluid Interfaces, Academic Press, San Diego, 2000, Chapter 4, p. 97.
44. F. Ravera, M. Ferrari, R. Miller and L. Liggieri, J. Phys. Chem. B, 105 (2001) 195–203.
45. L. Liggieri, V. Attolini, M. Ferrari and F. Ravera, J. Colloid Interface Sci., 255 (2002) 225–235.
46. G. Loglio, P. Pandolfini, R. Miller, A. Makievski, J. Krägel and F. Ravera, Phys. Chem. Chem. Phys., 6 (2004) 1375–1379.
47. A. Klebanau, N. Kliabanova, F. Ortega, F. Monroy, R.G. Rubio and V. Starov, J. Phys. Chem. B, 109 (2005), 18316–18323.
48. L. Zhang, X.C. Wang, Q.T. Gong, L. Zhang, L. Luo, S. Zhao and J.Y. Yu, J. Colloid Interface Sci., 327 (2008) 451–458.
49. C. Picard and L. Davoust, J. Colloid Interface Sci., 327 (2008) 412–425.
50. H. Hilles, A. Maestro, F. Monroy, F. Ortega, R.G. Rubio and M.G. Velarde, J. Chem. Phys., 126 (2007) 124904–124910.
51. R. Wüstneck, J. Perez-Gil, N. Wüstneck, A. Cruz, V.B. Fainerman and U. Pison, Adv. Colloid Interface Sci., 117 (2005) 33–58.
52. B.A. Noskov, D.O. Grigoriev, S.Y. Lin, G. Loglio and R. Miller, Langmuir, 23 (2007) 9641–9651.
53. L.R. Arriaga, I. López-Montero, R. Rodríguez-García and F. Monroy, Phys. Rev. E, 77 (2008) 061918-1–061918-10.
54. C. Bharathy, P. Sachdeva, H. Parthasarthy and A. Tayal, An Introduction to Volterra Series and Its Application on Mechanical Systems, in "Advanced Intelligent Computing Theories and Applications. With Aspects of Contemporary Intelligent Computing Techniques", De-S. Huang, D.C. Wunsh II, D.S. Levine and K.-H. Jo (Eds), Book Series Communication in Computing and Information Science, Vol. 15, Springer, Berlin, Heidelberg, 2008, ISBN 978-3-54085929-1, pp. 478–486.
55. G. Loglio *et al.*, paper in preparation.
56. J. Lucassen and M. Van den Temple, Chem. Eng. Sci., 27 (1972) 283–1291.
57. A. Hannisdal, R. Orr and J. Siöblom, J. Disp. Sci. Technol., 28 (2007) 361–369.
58. G. Loglio, P. Pandolfini, R. Miller, A.V. Makievski, J. Krägel, F. Ravera and B.A. Noskov, Colloids Surfaces A, 261 (2005) 57–63.
59. G. Loglio, P. Pandolfini, R. Miller, A. Makievski, J. Krägel, F. Ravera and L. Liggieri, Microgravity Sci. Technol., 16 (2005) 205–209.
60. G. Loglio, P. Pandolfini, R. Miller, A. Makievski, J. Krägel, L. Liggieri, F. Ravera, M. Ferrari and A. Passerone, Microgravity Sci. Technol., 18 (2006) 100–103.

Advances in Calculation Methods for the Determination of Surface Tensions in Drop Profile Analysis Tensiometry

S.A. Zholob [a], **A.V. Makievski** [b], **R. Miller** [c] **and V.B. Fainerman** [a]

[a] Donetsk Medical University, 16 Ilych Avenue, Donetsk 83003, Ukraine
[b] SINTERFACE Technologies, Volmerstr. 5-7, 12489 Berlin, Germany
[c] MPI for Colloids and Interfaces, 14424 Potsdam/Golm, Germany

Contents

A. Introduction

Dynamic surface tensions are extensively studied to gain properties of liquid adsorption layers [1]. The drop profile analysis tensiometry (PAT) is superior over other methods for the following advantages [2]:

PAT is a contactless method and therefore has a higher accuracy as compared to contact methods, for example ring or plate tensiometry.

PAT covers a very large range of surface formation times — from several seconds up to several hours and more, thus allowing to reach equilibrium states of adsorption layers.

PAT is based on the solution of the Gauss–Laplace equation, which describes the shape of axisymmetric profiles of bubbles and drops

$$\gamma \left(\frac{1}{R_1} + \frac{1}{R_2} \right) = \Delta P. \tag{1}$$

Bubble and Drop Interfaces
© Koninklijke Brill NV, Leiden, 2011

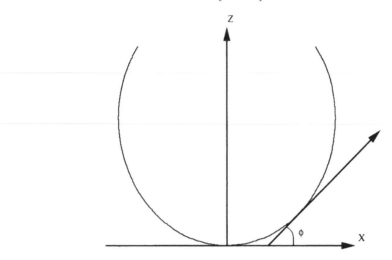

Figure 1. Definition of drop coordinates.

Here ΔP is the pressure difference across the interface, γ is the interfacial tension, and R_1 and R_2 are the principal radii of curvature. The drop profile is specified in a X, Z coordinate system according to Fig. 1.

The meridional curvature can be defined as

$$\frac{1}{R_1} = \frac{d\phi}{dS} \tag{2}$$

with ϕ — the angle of the tangent to the profile and S — the arc length along the profile. The azimuthal curvature is given by

$$\frac{1}{R_2} = \frac{\sin\phi}{X}. \tag{3}$$

In the gravitational field, the weight of the liquid column of height Z is added to the capillary pressure leading to

$$\Delta P = \frac{2\gamma}{b} + \Delta\rho g Z, \tag{4}$$

where b is the radius of curvature at the drop apex (hereafter apex radius), $\Delta\rho$ is the density difference between the drop and the surrounding medium, g is the gravitational acceleration constant. The parameter 2γ in the right-hand side of Eq. (4) appears due to the fact of equality of the two principal radii at the drop apex, i.e., the condition $R_1 = R_2 = b$ is met at $Z = 0$. So under the condition of constant density of both phases Eq. (1) reads

$$\gamma\left(\frac{d\phi}{dS} + \frac{\sin\phi}{X}\right) = \frac{2\gamma}{b} + \Delta\rho g Z. \tag{5}$$

By dividing both sides of Eq. (5) by γ/b we obtain the following equation, which is the starting point for many calculation algorithms

$$\frac{d\phi}{ds} = 2 + \beta z - \frac{\sin\phi}{x} \tag{6}$$

with the dimensionless arc length, horizontal and vertical coordinates, respectively,

$$s = \frac{S}{b}, \qquad x = \frac{X}{b}, \qquad z = \frac{Z}{b} \tag{7}$$

and the dimensionless Bond number

$$\beta = \frac{\Delta\rho g b^2}{\gamma}. \tag{8}$$

The parameter b is a scaling factor. For any curve the equations apply

$$\frac{dx}{ds} = \cos\phi, \tag{9}$$

$$\frac{dz}{ds} = \sin\phi. \tag{10}$$

As initial conditions for a solution of the given set of Eqs (6), (9) and (10) we can use

$$x(0) = 0; \qquad z(0) = 0; \qquad \phi(0) = 0. \tag{11}$$

The indeterminacy in Eq. (6) at $s = 0$ can be solved by

$$\left(\frac{\sin\phi}{x}\right)_{s=0} = 1. \tag{12}$$

B. Experimental Aspects of Drop Profile Analysis Tensiometry

Measurements of surface tension by drop profile analysis include the following four main steps [3]:

1. Acquisition of a drop/bubble image by using a video camera and transfer to a computer through a special digitizing board or the PC USB port [4].

2. Extraction of the profile coordinates of the drop/bubble from the image *via* specific edge detection algorithms. Different methods have been developed for edge detection, however, the algorithm proposed by Canny [5] is the most frequently used one.

3. Some "noise" can arise during the digitizing Step 1, and discontinuities or outliers appear in the extracted edge profile in Step 2. Therefore, some filtration and/or smoothing procedures have to be applied after Step 2.

4. The filtered and smoothed edge coordinates or some specific geometrical profile parameters are then used in the fitting algorithm to calculate the value of the surface tension.

In the general case, there is no direct analytical solution for the set of Eqs (6), (9) and (10), but there are approximate solutions for some special situations, e.g. for extremely large or small sessile drops [6]. Also an elliptical solution of Eq. (1) is available [7]. This type of solutions yields only approximate surface tension values. An integration of the drop volume along the vertical coordinate was suggested in [8], which provides a rather accurate determination of surface tension from sufficiently sharp images of drops far from a spherical shape.

In realistic investigations of dynamic surface tensions, where the experiments are performed under a wide range of conditions, an overall accuracy of better than ± 0.3 mN/m is required. The various sources of systematic errors, which influence the accuracy of surface tension determination from drop or bubble profiles, were summarised in [5]:

1. Software, including errors from the operating system and related software, and from the edge detection algorithms used in the image processing software;

2. Optical system, including lens distortion, blur, blooming, CCD sensor and random noise, aspect ratio and scaling;

3. Control of mechanical vibration;

4. Handling and position control, including levelling and vertical misalignment of the camera;

5. Precision of the frame grabber, if a frame grabber is applied;

6. Light system characteristics, i.e. uniformity of the light source.

Here we focus only on errors originated from the software subsystem, namely from edge detection algorithms and numerical solutions of the set of Eqs (6), (9) and (10), assuming that errors from all other subsystems are minimized.

As mentioned above, there are many algorithms for edge detection as well as for the numerical solution of a given set of equations suggested in literature. The most popular technique for determining interfacial tension from the profile of pendant or sessile drops is the so-called axisymmetric drop shape analysis of Rotenberg *et al.* [9], known as ADSA. ADSA acquires images of a drop (or bubble) and extracts the profile coordinates by using a respective edge detection technique. Then, a theoretical drop profile as defined by the system of Eqs (6), (9) and (10) is fitted to the extracted coordinates, taking the radius of curvature at the drop apex b and the capillary constant $c \equiv |\beta|/b^2$ as adjustable parameters. The numerical integration of ADSA is based on a 4th order Runge–Kutta method [10] and the target function required for the fitting procedure is defined by the sum of squares of differences between the theoretical and experimental drop/bubble profile. An optimization algorithm searches for the minimum of the target function using one of the of nonlinear optimization methods [11–14]. The minimum found corresponds to the correct value of the surface tension.

The accuracy of the profile coordinates can be improved by advanced edge detection methods, such as those proposed by Sobel [14] and Canny [5], and also by better image processing techniques such as the distortion correction proposed by Hoorfar and Neumann [15]. Under the assumption that images change only slowly from one to another a superposition of several subsequent images was suggested for example in [8], which gives rather good results even without subpixel resolution. However, such techniques are suitable only for sufficiently slow processes.

The accuracy of edge detection can be remarkably improved by using a subpixel resolution, such as cubic spline interpolation [16] or non-maxima suppression method [17]. A compensation of sub-pixel edge localization errors for any edge detection technique by a procedure of camera calibration by a specially designed test image was suggested in [19]. Moreover, the use some filtration technique in order to eliminate those detected points which lie too far from the theoretical profile, is very efficient. Two methods of this kind were compared in [4]: fifth order polynomial fitting (FOPF) and axisymmetric liquid-fluid interface-smoothing (ALFI-S). The FOPF method [19] is based on an approximation of the distances from a centre of the detected profile by a fifth order polynomial. The curve defined by this polynomial is used as a reference profile. Then any point that was more than threefold of standard deviation away from the fitted polynomial was rejected as an outlier. In ALFI-S the best matched Laplacian curve, defined by Eqs (6), (9) and (10), is used as a reference profile. Thus, a full optimization problem has to be solved in each iteration cycle. In both methods smoothing procedure is repeated iteratively until no outliers are found. A comparison demonstrated that ALFI-S is superior to FOPF with respect to accuracy and stability while FOPF requires less computation time [3]. A procedure, consisting of Canny edge detection followed by ALFI-S is claimed to be able to calculate the surface tension from captive bubble images under various conditions, including images with extensive noise, poor contrast, or non-uniform lighting conditions [4].

On the other hand, eliminating the need of an independent edge detection module would be useful when acquisition of sharp images is not possible due to experimental or optical limitations. As stated in [16] fuzzy or noisy images may not be successfully processed by standard edge detection techniques. Therefore, Neumann and co-workers suggested a new method called theoretical image fitting analysis (TIFA), which generates theoretical images of the drop and calculates an error function that describes the pixel-by-pixel deviation of the theoretical image from the experimental one. It then calculates the surface tension by matching the theoretical to the experimental image. This approach, however, instead of a suitable edge detection algorithm, requires knowledge on light intensity or gradient distribution near the edge profile, which was interpolated by third order spline functions and assuming the maximum of this function coincides with the position of the edge. Hence, even in the TIFA method an edge detection technique is included, but in a more sophisticated way.

Galerkin's finite element method [21, 22] was applied in [20] to integrate the dependence of drop volume on angle of rotation, which requires very much computation time and is based on drop shapes far from sphericity.

In [6] it was demonstrated that inexpensive CCD cameras with a maximum resolution of 640×480 pixels do not yield surface tensions with an accuracy better than ± 0.5 mN/m if the precision of edge detection is 1 pixel and one of the advanced edge detection algorithm with sub-pixel resolution is needed to achieve an acceptable surface tension accuracy.

We can conclude, that the drop profile analysis technique is most appropriate for the determination of interfacial tensions, and the present paper aims at describing some improvements in edge detection and numerical calculation algorithms.

C. Improvement in the Extraction of Profile Coordinates

A typical dependence of the brightness gradient along the direction normal to drop profile is shown in Fig. 2. This dependence is used in [16] to calculate the deviation of the theoretical drop profile from the experimental coordinates. The values of the brightness gradient between nodal points are interpolated with a 3rd order spline, and the position of the interpolated gradient maximum is assumed to be the drop edge coordinate. Due to experimental errors a wrong maximum can arise and the selection of the single true maximum is a complicated additional problem. Even if the true position is determined it can be significantly shifted due to errors. Furthermore, the interpolation by spline functions is advantageous only for sufficiently smooth curves with small changes between the points. Similar to the above discussion changes of the first derivative can be used as a measure of the curve's smoothness. However, it is evident from Fig. 2, that changes of the first derivative (dashed line) can be rather significant. Because of these two reasons the application of spline functions can be considered as not optimal, although the accuracy of reported results can be very good [16].

Regarding the disposition of points in Fig. 2 they can be guessed to lie rather close to a curve of a normal distribution density [23]

$$G(x) = G_{\mathrm{m}} \exp\left[-\left(\frac{x - x_0}{\sigma} \right)^2 \right].$$ (13)

By the previously defined edge positioning this algorithm should take not less than each 4 points inside of the drop and outside the drop. The selected points are then fitted according to Eq. (11) taking G_{m}, x_0 and σ as the parameters to be adjusted. G_{m} is considered as a parameter of the image sharpness and x_0 is accepted as the best edge position, while σ describes the gradient change rate. Such a quantity of at least 8 points appears to be quite sufficient as σ is less than 2 pixels in most cases. As can be shown with practical images, the proposed technique enhances the accuracy of γ for drops of small volume only when $G_{\mathrm{m}} \geqslant 200$.

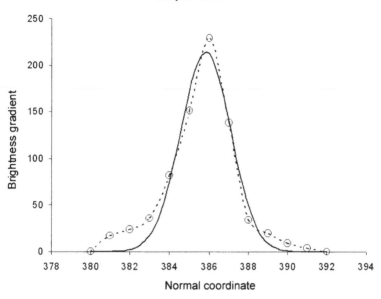

Figure 2. Normal brightness gradient: symbols — experimental points, solid line — fit curve according to Eq. (11), dotted line — 3rd order spline.

D. Initial Estimation of Adjustable Parameters

As it was stated above, calculation of surface tension by experimental drop/bubble images requires fitting of the theoretical profile, defined by Eqs (6), (9) and (10), to the detected coordinates by a suitable optimization procedure. Most frequently this procedure uses b and β (or some combination of them) as adjustable parameters. It is evident that real coordinates X and Z depend nonlinearly on b, which serves as a scaling factor, and β, which defines the drop shape. Thus, a fitting procedure must use certain algorithm of nonlinear optimization. Moreover, this algorithm should converge to an optimal solution even if starting far apart from an optimal set of parameters. Some of these algorithms using a least squares criterion are considered in [11–13]. Details of fitting procedures as applied to the calculation of b and β can be found elsewhere [24]. It is obvious that the convergence of the fitting procedure can be considerably improved with optimal starting parameters. Various attempts were made to obtain approximate values of b and β. The initial guess of value for β is possible in two ways, (i) based on analytical expressions or (ii) on a tabular interpolation.

The first approach was used by a number of authors. Most of them use the point on the drop profile at $\phi = \pi/2$ (called equator, with coordinates x_e and z_e) as characteristic point. As it follows from tables given in [25], coordinates of the equator point depend on β. A dependence of this kind for contact angles greater than $90°$

Table 1.
Coefficients a_j in Eq. (14), $0 \leqslant \beta \leqslant 100$ [24]

j	a_j
0	16.8621
1	−53.395889
2	26.5354
3	95.097412
4	−182.7348
5	146.43645
6	−60.568497
7	12.668255
8	−0.8999756

Table 2.
Coefficients a_j in Eq. (15), $0 \leqslant \beta \leqslant 2.0$

j	a_j
0	−16.44833611
1	50.43537305
2	−55.94313316
3	21.95477381

was proposed in [24], where the value of β can be determined *via* an eighth-order polynomial in the range of $0 \leqslant \beta \leqslant 100$

$$\beta = \sum_{j=0}^{8} a_j (x_e/z_e)^j. \tag{14}$$

The coefficients a_j are given in Table 1.

The average absolute deviation Δ_a is 0.003 [24]. The range of most frequent values of β is around 0.2. The numerical verification of Eq. (5) shows that in this given range $\Delta \leqslant 0.0007$. The resulting value β is only an approximation and has to be further improved in the fitting procedure Hence, the polynomial order can be considerably decreased. For the range $0 \leqslant \beta \leqslant 2$ a polynomial of third order can be applied

$$\beta = \sum_{j=0}^{3} a_j (x_e/z_e)^j, \tag{15}$$

with coefficients summarised in Table 2. The maximum absolute deviation is $\Delta_m = 0.00096$, the average absolute deviation $\Delta_a = 0.000274$.

The verification in real-time experiments with PAT [4] have shown that the approach based on a polynomial approximation is not optimum as the absolute error in determination of z_e is much larger than in x_e [26]. Therefore, many authors use another point on drop profile instead of z_e, having the coordinates $x_{(m)}$ and $z_{(m)} = mx_e$, where m is a number. This approach is used mainly in experiments with pendant drops, for example in [27–29] with $m = 2$.

Another approach uses the capillary constant $c \equiv |\beta|/b^2$ (also for $m = 2$) based on dependences of $T \equiv \frac{1}{4}cx_e^2$ as a function on $S \equiv x_{(m)}/x_e$ [30]. In [32] the range of m was extended to 1.25 up to 2.5 in steps of $\Delta m = 0.25$, and $P_e \equiv x_e\sqrt{c}$ was used as a resulting dimensionless parameter.

E. Transformation of the Gauss–Laplace Equation to Avoid Interpolation

1. Transformation to a Second Order Equation

A fitting procedure based on a direct numerical integration of the set of Eqs (6), (9) and (10) has the drawback of requiring interpolation between theoretical co-ordinates, which slows down calculation process and decreases the accuracy. This disadvantage can be bypassed using a transformation of Eqs (6), (9) and (10) into a 2nd order equation

$$\frac{d^2t}{dz^2} = -\beta - \frac{K}{1 - t^2} \tag{16}$$

with

$$t = \cos\phi, \tag{17}$$

$$K = tf\left(\frac{dt}{dz} + f\right), \tag{18}$$

$$f = 2 + \beta z + \frac{dt}{dz} \tag{19}$$

and the respective initial conditions

$$t(0) = 1; \qquad \frac{dt}{dz}(0) = -1. \tag{20}$$

The indeterminacy in Eq. (16) at $z = 0$ can be given by the limit

$$\lim_{z \to 0} \frac{d^2t}{dz^2} = -\frac{3}{4}\beta.$$

The elimination of s from the equation system gives

$$-\frac{d\cos\phi}{dz} = 2 + \beta z - \frac{\sin\phi}{x}. \tag{21}$$

Combining Eqs (17) and (21) at a given t, the coordinate x can be calculated from

$$-\frac{dt}{dz} = 2 + \beta z - \frac{\sin\phi}{x}. \tag{22}$$

2. Definition of the Volume Ratio

The use of Eq. (16) for the calculation of the interfacial tension leads to an improved accuracy, despite some remaining limitations. For their consideration, we need to first define V_d as the drop volume calculated from the drop apex to a position close to the capillary tip, called split line, avoiding any influence of irregularities of capillary wetting by the liquid (see below). The drop volume can be calculated *via* an approximation of the profile coordinates in the form $x^2 = f(z)$ [23]

$$V_d = \pi \int_0^{h_c} f(z)\, dz. \tag{23}$$

h_c is a distance of the split line from the drop apex.

We also define V_c as the volume of a sphere with the radius equal to the cross section at the level of the split line r_c

$$V_c = \frac{4}{3}\pi r_c^3 \tag{24}$$

and the dimensionless volume ratio

$$P_v = \frac{V_d}{V_c}. \tag{25}$$

P_v has a direct influence on the accuracy of the determined surface tension. Experiments with pendant drops of pure water with small volumes defined by

$$P_v < \frac{7}{4} \tag{26}$$

showed that some additional factors decrease the accuracy of interfacial tension calculations:

1. decreased number of profile points,

2. almost spherical drop shapes,

3. drop shape asymmetry due to irregularities of capillary wetting and deviation of the capillary axis from the vertical direction.

The pendant drop shapes for β-values of 0.2, 0.35 and 0.5 are shown in Fig. 3 in dimensionless coordinates x, z. It can be seen that the main differences between the drop shapes are observed above the equator, i.e., the largest diameter of the drop.

The higher the split line is located above the equator, the more accurate is the γ-value determined by fitting the theoretical profile to the experimental data. Due to the above mentioned point 3 the split line must not be placed too close to the capillary tip. Moreover, due to the factor 2, for drops of small volume the largest distance of the split line from the drop apex is yet not far enough from the equator, which also contributes to large scattering of calculated γ-values. The closer the split line is located to the equator the more appears the drop shape as spherical. A natural parameter that gives a measure of drop shape sphericity is β: at $\beta = 0$ the

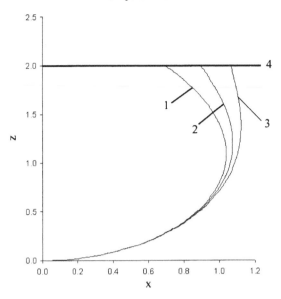

Figure 3. Drop shapes in dimensionless coordinates x, z: (1) $\beta = -0.2$; (2) $\beta = -0.35$; (3) $\beta = -0.5$; (4) split line.

drop is ideally spherical. However, the calculation of β is only part of the calculation routine to determine the surface tension. It appears therefore quite advantageous to use a dimensionless parameter that can be determined as criterion for the expected accuracy of surface tension calculation.

In [32] a so-called shape parameter was suggested to determine the sphericity of a drop shape

$$A_{\mathrm{pr}} = \int_0^{2\pi} \int_0^{r_c} R(\theta)\, \mathrm{d}R\, \mathrm{d}\theta, \tag{27}$$

$$P_{\mathrm{s}} = \frac{|A_{\mathrm{pr}} - \pi b^2|}{A_{\mathrm{pr}}}. \tag{28}$$

Here θ is the polar angle in respect to the vertical axis, R is the length of radius-vector (see below in Fig. 5) and A_{pr} is the projected area of the drop. For the calculation of A_{pr} (see below) another equation can be used

$$A_{\mathrm{pr}} = 2 \int_0^{h_c} x\, \mathrm{d}z \tag{27a}$$

which allows to calculate the drop shape independently of the projected area in one integration step. From a theoretical point of view it would be interesting to consider dependencies of P_{s} on the volume ratio P_{v} for different surface tensions [32]. Using the absolute value of the numerator in Eq. (28) negative values of P_{s} can arise for drops with $h_c \ll 2b$, which may cause a discontinuity in the dependence of P_{s}

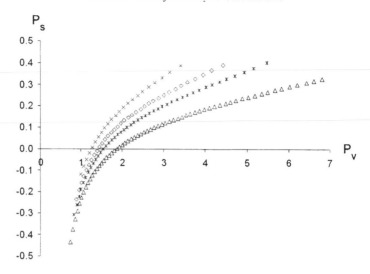

Figure 4. Dependences of P_s on P_v with varying γ and constant $r_c = 0.8$ mm. $\times - \gamma = 33$ mN/m; $\Diamond - \gamma = 42$ mN/m; $* - \gamma = 50$ mN/m; $\triangle - \gamma = 72$ N/m; according to [35].

(P_v). To avoid this case we can use Eq. (28) without taking the absolute value, so that Eqs (27a) and (28) provide

$$P_s = \frac{2\int_0^{h_c} x\,dz - \pi b^2}{2\int_0^{h_c} x\,dz}. \tag{28a}$$

The theoretical dependencies of P_s (P_v) can be computed as follows. A series of pendant drop shapes with the same contact diameter is calculated from Eq. (6) at constant γ and r_c and changing b from $b = r_c$ up to a maximum possible size for the given contact diameter. This procedure yields a dependence of V_d on b and finally P_s (P_v). Some of these dependencies of P_s on P_v are shown in Fig. 4.

The change in P_v is much larger than in P_s and therefore P_v can be applied instead of P_s to estimate a critical shape parameter which defines a minimum below which the deviation of the calculated surface tension from the correct and known value is less than a given error. In order to compare measured results for different surface tensions, a relative error was introduced in [32]:

$$\varepsilon_{rel} = \frac{\gamma_{meas} - \gamma_{true}}{\gamma_{true}}. \tag{29}$$

Here γ_{meas} and γ_{true} are the measured and the most probable surface tension values, respectively. Below some experimental dependences of P_s on ε_{rel} will be discussed.

3. Use of a Polar Coordinate System

The use of polar coordinates X, θ, as defined in Fig. 5, was suggested in [20] as optimum choice of coordinates:

$$R = \sqrt{X^2 + H^2} \quad \text{and} \quad \theta = \arctan\left(\frac{X}{H}\right). \tag{30}$$

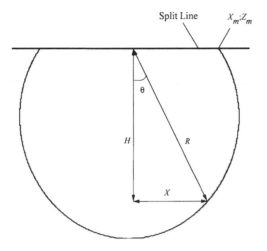

Figure 5. Definition of the spherical coordinate system [20].

$H = Z_m - Z$ and Z_m is the maximum value of Z, i.e. $Z_m = h_c$, and θ is the polar angle in respect to vertical axis. With these coordinates Eq. (5) turns into an implicit dependence $X(\theta)$ with the two boundary conditions $X(0) = 0$ and $X(\pi/2) = X_m$, where X_m is the X coordinate corresponding to Z_m. This transformation allows the solution of Eq. (5) by the Galerkin method [21, 22], which additionally increases the accuracy of calculation of γ, but it requires a rather large computation times. The combination of polar coordinates with a differential equation similar to Eq. (16), represented as a 2nd order equation with two initial conditions, appears to be optimum. Using the transformations (7) we can rewrite Eq. (30) in a dimensionless form

$$r = \sqrt{x^2 + h^2} \quad \text{and} \quad \theta = \arctan\left(\frac{x}{h}\right), \tag{31}$$

where $r = R/b$ and $h = H/b$ are the dimensionless radius and drop height, respectively. The dependence $r(\theta)$ in form of a general 2nd order differential equation reads [35]

$$\frac{d^2 r}{d\theta^2} = L(r), \tag{32}$$

where $L(r)$ is any differential operator. It is known from general considerations of this differential equation that their solution is more accurate the smoother the function $r(\theta)$ is.

If we consider the first derivative as a measure of smoothness of a function we can conclude for Eq. (32) that $dr/d\theta$ should be as constant as possible in order to get an accurate solution. In Fig. 6 the same dependencies as in Fig. 3 are shown but in coordinates of Eq. (16), i.e., z and $\cos(\phi)$, while in Fig. 7 in the coordinates of Eq. (32), i.e., θ and r, are used. The curves in Fig. 6 may be thought even smoother than those in Fig. 7, but the coordinates θ and r are superior over z and $\cos(\phi)$

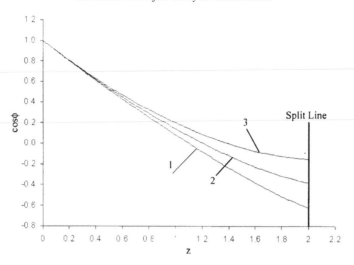

Figure 6. The same as in Fig. 3 but in coordinates z and $\cos\phi$.

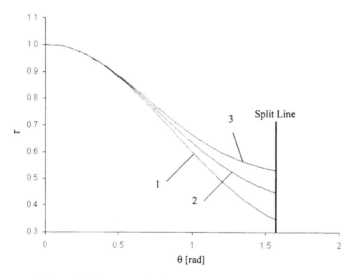

Figure 7. The same as in Fig. 3 but in coordinates θ and r.

because they are easier calculated from experimental drop profiles and thus can be easier matched to theoretical values, in contrast to z and $\cos(\phi)$.

The analysis of the origin of errors has shown that in the region close to drop apex the relative error of drop coordinates from an experimental image is comparatively high. The use of the radius r as dependent variable can remarkably decrease this relative error. For this, Eq. (5) has to be rewritten in coordinates defined by Eq. (31), which leads to

$$z = z_{\mathrm{m}}(1 - r\cos\theta) \quad \text{and} \quad x = z_{\mathrm{m}} r \sin\theta, \tag{33}$$

where $z_m = Z_m/b$. Introduction of the following notations

$$r' = \frac{dr}{d\theta}; \qquad r'' = \frac{d^2 r}{d\theta^2} \tag{34}$$

and $k = \frac{d\phi}{ds}$ as parameter for the curvature, we get

$$\sin\phi = \frac{1}{\sqrt{1 + P^2}}, \tag{35}$$

where

$$P = \frac{dx}{dz} = \frac{dx/d\theta}{dz/d\theta} = \frac{r' \sin\theta + r \cos\theta}{r \sin\theta - r' \cos\theta}. \tag{36}$$

Substitution of Eqs (33)–(35) into Eq. (6) leads to

$$k = 2 + \beta z_m (1 - r \cos\theta) - \frac{1}{z_m r \sin\theta \sqrt{1 + P^2}}. \tag{37}$$

In polar coordinates r, θ the curvature k can also be written as [23]

$$k = \frac{r^2 + 2r'^2 - rr''}{Q^{3/2}} \tag{38}$$

with $Q = r^2 + r'^2$. The solution of Eq. (38) in respect to r'' gives

$$r'' = r + \frac{1}{r}(2r'^2 - kQ^{3/2}). \tag{39}$$

The initial conditions for Eq. (39) are $r(0) = 1$; $r'(0) = 0$. A series expansion of the solution of Eq. (39) at $\theta = 0$ leads to

$$r(\theta) = 1 + (1 - z_m)\frac{\theta^2}{2} + O(\theta^4). \tag{40}$$

The indeterminacy for r'' at $\theta = 0$ is solved by the limit

$$\lim_{\theta \to 0} = 1 - z_m. \tag{41}$$

The calculation accuracy can be increased again by extending the range of independent variables. This can be achieved by shifting the origin of the coordinate system to the drop centre, which can be defined as the point in a distance equal to that from the drop apex $(x_m; z_m)$ (see Fig. 8).

This defines the position of the centre z_c by the equation

$$(z_m - z_c)^2 + x_m^2 = z_c^2 \tag{42}$$

which gives

$$z_c = \frac{1}{2}\left(\frac{x_m^2}{z_m} + z_m\right). \tag{43}$$

In Eqs (33), (37), (40)–(43) z_m and z_c can be interchanged.

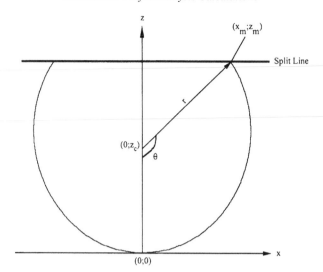

Figure 8. Definition of the drop centre z_c: the distance from $(0; z_c)$ to $(x_m; z_m)$ is equal to z_c.

F. Experimental Results

In order to determine the range of independence of the measured surface tension on drop volume two special experiments were performed using the drop/bubble profile analysis tensiometer PAT (SINTERFACE Technologies, Berlin) [4]. For calculation of the surface tension an optimization algorithm given in [12] was used. Detailed of its implementation can be found elsewhere [34].

In the first experiment Milli-Q water and in the second a 1 mmol/l solution of $C_{14}EO_8$ (approximately 100 CMC) were used [35]. The determined surface tension of pure water at different drop volume is shown in Fig. 9. The calculation of the initial adjustable parameters as proposed in Section 4 was not used in these experiments (see results including this feature later). For a better accuracy five subsequent drop images were superimposed as discussed in [8, 35]. The experiments were performed with a stainless steel capillary of 2 mm in diameter. A value of $r_c = 1.05$ mm was taken for the calculation of P_v according to Eqs (24) and (25). The dependencies in Fig. 9 look very similar to those discussed in [32]. The dependence of ε_{rel} on P_v is shown in Fig. 10 ($\gamma_{true} = 71.9$ mN/m) from which one can see that stable measurements with constant error smaller than 0.7% are available for $P_v \geqslant 4.4$.

Due to Fig. 4 for $\gamma_{true} = 72$ mN/m this critical value of P_v corresponds to $P_s = 0.21$, which is very close to the value defined as critical in [32]: $P_{s(critical)} = 0.19$. For the 1 mmol/l solution of $C_{14}EO_8$ a similar dependence of ε_{rel} on P_v is shown in Fig. 11.

Thus, stable measurements with an error less than 1.5% are possible for $P_v \geqslant 3.0$. According to Fig. 4 for $\gamma_{true} = 33$ mN/m this value of P_v corresponds to $P_s = 0.33$, which is almost the same value as the critical one given in [32] for experiments with a Teflon capillary: $P_{s(critical)} = 0.29$. In the experiments of Fig. 11 steel capillary

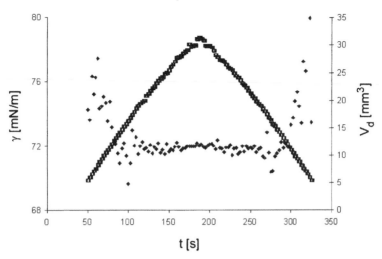

Figure 9. Dependence of surface tension and drop volume of pure water on time; according to [35].

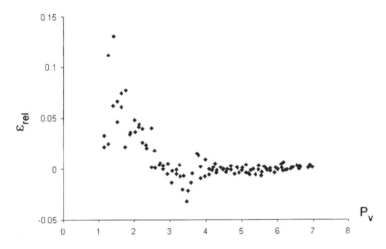

Figure 10. Dependence of ε_{rel} on P_v for pure water, according to [35].

with a diameter of 2 mm was used instead of 3 mm in [33]. Thus, the presented data on $P_{v(critical)}$ discussed recently in [35] coincide very well with those on $P_{s(critical)}$ as defined in [33].

In order to estimate the effect of initial estimations for b and β, as proposed in Section 4, the results of some additional pendant drop and buoyant bubble experiments are analysed below. The fitting procedure was started with initial values of b and β estimated using specific tables, as discussed above. Experimental results for buoyant bubbles in pure water are shown in Fig. 12 and those with a 5 mmol/l mixed SDS/Triton-X100 solution (molar ratio is 2:1) in Fig. 13. Results of pendant drop experiments with pure water are shown in Fig. 14 and those with a mixed SDS/Triton-X100 in Fig. 15.

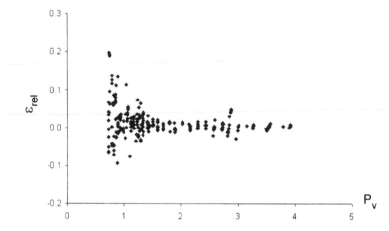

Figure 11. Dependence of ε_{rel} on P_v for 1 mmol/l $C_{14}EO_8$ solution, according to [35].

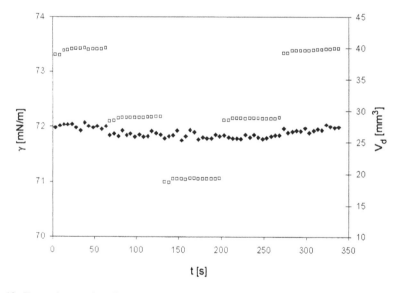

Figure 12. Dependence of surface tension and drop volume of pure water on time, estimation of initial b and β, buoyant bubble configuration; ($\blacklozenge$) surface tension, ($\square$) bubble volume.

One can see that the estimation of initial values for b and β significantly improves the final results of surface tension calculation, especially for the buoyant bubble configuration.

The relative error decreases, and the decrease of the bubble volume to half leads to a surface tension change no more than 0.2% for water and 0.3% for SDS/Triton-X100 solution, respectively. This result can be explained hydrodynamically by the mass of liquid surrounding the bubble, which is about six orders of magnitude larger than that of the bubble. This factor greatly reduces any hydrodynamic instability due to the fast damping of any accidental vibrations. In the pendant drop configuration the situation is opposite: pendant drops tend to wobble due to its large mass

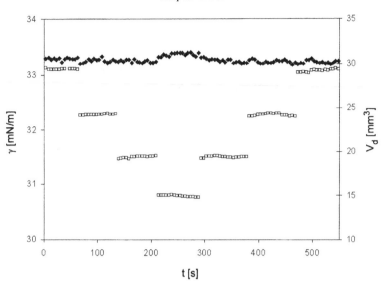

Figure 13. The same as in Fig. 12 for mixed SDS/Triton-X100 solution (molar ratio 2:1, concentration 5 mmol/l); (◆) surface tension, (□) bubble volume.

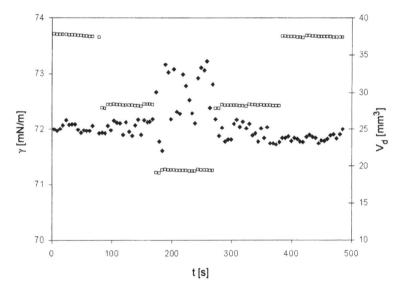

Figure 14. The same as in Fig. 12 but using pendant drop configuration; (◆) surface tension, (□) drop volume.

as compared to that of the surrounding air. Moreover, buoyant bubbles are more convenient, as compared to pendant drops for studies of highly surface active surfactants at low total bulk concentrations, because depletion of the bulk surfactant concentration due to adsorption of surfactant molecules at the bubble surface is negligible. This depletion effect is considerable in pendant drop experiments [36–39]

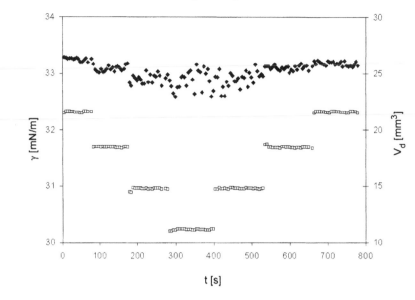

Figure 15. The same as in Fig. 14 for mixed SDS/Triton-X100 solution (molar ratio is 2:1, concentration 5 mmol/l); (◆) surface tension, (□) drop volume.

and can be used for a direct estimation of the adsorbed amounts of surfactants. This is demonstrated in Chapter Eight of this book.

G. Conclusion

There are various possibilities to improve the standard technique for measuring surface or interfacial tension by drop profile analysis. One possibility to increase the accuracy is the fitting of experimental grey level gradients obtained from experimental drop or bubble images to normal distribution functions and locate the edge into maximum of the distribution. During integration of the Gauss–Laplace equation many algorithms require interpolation between nodal points to calculate the target function. The transformation of the Gauss–Laplace into polar coordinates allows avoiding this inaccuracy. The extension of the range of angle of rotation as independent variable by setting the origin of the coordinate system into the drop centre also increases the accuracy of calculations. Proposed initial estimation of parameters characterizing drop shape and size by previously calculated table considerably improves convergence of optimization procedure used in the main calculation algorithm. For droplets close to a spherical shape the relative error of surface tension determination increases. A dimensionless criterion allows the indication of how large the droplet should be so that the relative error in the resulting surface tension does not exceed a given value.

H. Acknowledgements

The work was financially supported by DLR 50WM0941 and the DFG SPP 1273 (Mi418/16-2).

I. References

1. A.I. Rusanov and V.A. Prokhorov, Interfacial Tensiometry, in "Studies in Interface Science", Vol. 1, D. Möbius and R. Miller (Eds), Elsevier, Amsterdam, 1996.

2. D. Möbius and R. Miller (Eds), Drops and Bubbles in Interfacial Research, in "Studies in Interface Science", Vol. 6, Elsevier, Amsterdam, 1998.

3. Y.Y. Zuo, M. Ding, A. Bateni, M. Hoorfar and A.W. Neumann, Colloids and Surfaces A, 250 (2004) 233.

4. R. Miller, C. Olak and A.V. Makievski, SÖFW-Journal, 130 (2004) 2.

5. J. Canny, IEEE Transactions on Pattern Analysis and Machine Intelligence, 8 (1986) 679.

6. Y.Z. Zhou and J. Gaydos, J. Adhesion, 80 (2004) 1017.

7. E. Hernández-Baltazar and J. Gracia-Fadrique, J. Colloid Interface Sci., 287 (2005) 213.

8. G. Faour, M. Grimaldi, J. Richou and A. Bois, J. Colloid Interface Sci., 181 (1996) 385.

9. Y. Rotenberg, L. Boruvka and A.W. Neumann, J. Colloid Interface Sci., 93 (1983) 169.

10. E. Kamke, Differentialgleichungen. Lösungsmethoden und Lösungen, Leipzig, 1959.

11. K. Levenberg, The Quarterly of Applied Mathematics, 2 (1944) 164.

12. D. Marquardt, SIAM Journal on Applied Mathematics, 11 (1963) 431.

13. P. E. Gill and W. Murray, SIAM Journal on Numerical Analysis, 15 (1978) 977.

14. Source code for the Sobel method is available at: http://www.pages.drexel.edu/~weg22/tutorials.html.

15. M. Hoorfar and A.W. Neumann, J. Adhesion, 80 (2004) 727.

16. M.G. Cabezas, A. Bateni, J.M. Montanero and A.W. Neumann, Colloids Surfaces A, 255 (2005) 193.

17. F. Devernay, A non-maxima suppression method for edge detection with sub-pixel accuracy, Technical report RR 2724, INRIA, 1995.

18. F. Pedersini, A. Sarti and S. Tubaro, IEEE Trans. Pattern Analysis and Machine Intelligence, 19 (1997) 1278.

19. R.M. Prokop, A. Jyoti, M. Eslamian, A. Garg, M. Mihaila, O.I. del Río, S.S. Susnar, Z. Policova and A.W. Neumann, Colloids Surfaces A, 131 (1998) 231.

20. N.M. Dingle, K. Tjiptowidjojo, O.A. Basaran and M.T. Harris, J. Colloid Interface Sci., 286 (2005) 647.

21. B.A. Finlayson, The Method of Weighted Residuals and Variational Principles, Academic Press, New York, 1972.

22. D. Zwillinger, Handbook of Differential Equations, Academic Press, San Diego, 1989.

23. A.G. Korn and T.M. Korn, Mathematical Handbook for Scientists and Engineers, McGrow-Hill, New York, 1961.

24. C. Maze and G. Burnet, Surface Sci., 13 (1969) 451.

25. F. Bashforth and J.C. Adams, An Attempt to Test the Theories of Capillary Action, University Press, Cambridge, UK, 1883.

26. E.B. Dismukes, J. Phys. Chem., 63 (1959) 313.

27. S. Fordham, Proc. Roy. Soc. Ser. A, 194 (1948) 1.

28. O.S. Mills, Brit. J. Appl. Phys., 4 (1953) 247.

29. C.E. Stauffer, J. Phys. Chem, 69 (1965) 1933.
30. M.D. Misak, J. Colloid Interface Sci., 27 (1968) 141.
31. S. Ramakrishnan, J.F. Princz and S. Hartland, Indian J. Pure Appl. Phys., 15 (1977) 228.
32. M. Hoorfar, M.A. Kurz and A.W. Neumann, Colloids Surfaces A, 260 (2005) 277.
33. M. Hoorfar and A.W. Neumann, Adv. Colloid Interface Sci., 121 (2006) 25.
34. S. Brandt, Data Analysis, 3rd edn, Springer, New York, 1999.
35. S.A. Zholob, A.V. Makievski, R. Miller and V.B. Fainerman, Adv. Colloid Interface Sci., 134–135 (2007) 322.
36. A.V. Makievski, G. Loglio, J. Krägel, R. Miller, V.B. Fainerman and A.W. Neumann, J. Phys. Chem. B, 103 (1999) 9557.
37. V.B. Fainerman, S.V. Lylyk, A.V. Makievski and R. Miller, J. Colloid Interface Sci., 275 (2004) 305.
38. R. Miller, V.B. Fainerman, A.V. Makievski, M. Leser, M. Michel and E.V. Aksenenko, Colloids Surfaces B, 36 (2004) 123.
39. V.B. Fainerman, S.A. Zholob, J.T. Petkov and R. Miller, Colloids Surfaces A, 323 (2008) 56.

Axisymmetric Drop Shape Analysis with Anisotropic Interfacial Stresses: Deviations from the Young–Laplace Equation

James K. Ferri[a] **and Paulo A.L. Fernandes**[b]

[a] Departmental of Chemical and Biomolecular Engineering, Lafayette College, 18042 Easton, Pennsylvania, United States of America
[b] Max-Planck-Institut für Kolloid- und Grenzflächenforschung, Am Mühlenberg 1, 14424 Potsdam, Germany

Contents

A. Introduction

The interface between two immiscible fluids can be described as a two dimensional continuum; in the Gibbs convention, an interface possesses negligible thickness and therefore cannot support bending moments. The principal curvatures K_1 and K_2 in the two surface directions characterize the shape of the interface, and the forces within the interface can be expressed as a surface stress tensor which has units of force per unit length. For a pure fluid, the in-plane stresses can be described in terms of a single isotropic tension γ, the thermodynamic surface tension, which can be related to attractive intermolecular interactions in the bulk phase and the loss of coordination associated with molecules at the interface. The conditions of mechanical equilibrium can be used to derive the classical Young–Laplace relationship, shown in Eq. (1). Equation (1) describes the shape of a fluid interface by relating

Bubble and Drop Interfaces
© Koninklijke Brill NV, Leiden, 2011

the isotropic interfacial tension, the curvature of the interface, and the pressure jump [p] across it:

$$\gamma(K_1 + K_2) = [p]. \tag{1}$$

When amphiphilic molecules adsorb at a fluid interface, the surface excess concentration Γ increases and the thermodynamic surface tension $\gamma(\Gamma)$ decreases from the pure fluid surface tension $\gamma(\Gamma = 0)$ according to the Gibbs adsorption equation [1]. However, in the absence of deformation, the surface tension remains isotropic. As an interface is stretched, interfacial stresses can develop in response to the applied strain. When surface molecules do not interact, i.e., an ideal surface gas, the interfacial stress remains isotropic. The existence of interactions between adsorbed molecules necessitates additional considerations, see for example, Refs [2, 3]. One common way to assess the interactions between surface molecules is the use of interfacial dilational rheology [4]. In these experiments, the surface of a pendant drop or bubble is expanded or contracted, and the interfacial stress response as a function of deformation is measured. When surface molecules interact to form an elastic interfacial network, the inflation of a pendant drop can result in anisotropic surface stresses. Here we discuss a mechanical framework for describing the inflation of a pendant drop or bubble and the associated interfacial strains and tensions for fluid interfaces that can support anisotropic stresses. We specifically treat purely elastic interfaces, although extension to viscoelasticity is straightforward. Purely dissipative interfaces can be described by Eq. (1).

The formation of elastic surface networks by adsorbed molecules has received increased interest in recent years. This is because templating at fluid/fluid boundaries has become a widely used method for fabrication nanostructured membranes with asymmetric functionality because of the access afforded by each fluid phase [5–7]. Progress in the understanding of the mechanical performance of nanomembranes fabricated at mesoscopic sizes is has been limited due to the lack of suitable instrumentation and methods; particularly because the metrology of nanomaterials at intermediate ($10^{-4} < h < 10^{-2}$ m) length scales is difficult to access. Therefore, the elaboration of both experimental methodology and theoretical frameworks for their interpretation are of significant importance in advancing both the science and technology of mesoscopic materials transversely constrained to molecular and supramolecular dimensions. There is also fundamental interest in the mechanics of complex interfaces resulting from adsorption of surfactants, macromolecules, and nanoparticles, and their mixtures.

This chapter is restricted to the development of the theoretical framework and a parametric study of the effect of surface shear modulus on the deformation of a two-dimensional membrane of a purely elastic continuum constrained to a fluid interface of axisymmetric curvature under tensile loading from a uniform field such as hydrostatic pressure. In Section B, a description of the theory is provided. Section C described results and implications and provides signatures of elastic network formation in experiments such as error profiles resulting from the use of the Young–

Laplace equation in axisymmetric drop shape analysis. It is shown that systematic deviation of experimentally measured drop profiles from the shape predicted by the Young–Laplace equation can be used to identify the onset of elastic network formation. Section D provides conclusions and future prospectus, and Acknowledgements and References are given in Sections E and F, respectively.

B. Theory

Consider the shape of a pendant drop covered by a purely elastic network of surface molecules. The coordinates describing this shape are shown schematically in Fig. 1.

This surface can be treated as an interface having both an isotropic thermodynamic surface tension γ and elastic membrane tensions T_1 and T_2 in the two surface directions presenting the possibility of an anisotropic stress distribution. A point on the membrane in the undistorted state is characterized by the cylindrical coordinates (r_0, ϕ_0, z_0) or the surface coordinates (ξ_0, ϕ_0); as shown in Fig. 1.

Because the elastic membrane tensions only result from deformation, $T_1 = T_2 = 0$ in the absence of deformation. In this case, the Young–Laplace equation describes the shape of the undeformed interface in terms of the equilibrium isotropic surface tension γ. There is extensive literature [8–10] documenting the recasting of equation (1) into:

$$\theta' = \frac{2}{R_0} - \frac{(\Delta\rho)gz_0}{\gamma} - \frac{\sin\theta}{r_0},\qquad(2)$$

where R_0 is the radius of curvature at the origin, $\Delta\rho$ is the density difference across the interface, g is the gravitational constant, and γ is the equilibrium thermodynamic surface tension.

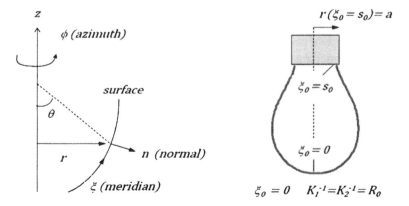

Figure 1. Axisymmetric surface generated by revolving the meridonal curve (ξ) around the axis of symmetry (z) and associated coordinates. The surface coordinates ξ_0, ξ describe the arc length along the meridian; the undeformed arc length (ξ_0) is $0 < \xi_0 < s_0$; ξ is the deformed arc length. At the drop apex, the principal curvatures K_1, K_2, defined in equations (9), are equal.

The solution of Equation (2) together with the geometric relations:

$$r_0' = \cos\theta, \tag{3a}$$

$$z_0' = \sin\theta \tag{3b}$$

yields the drop shape. In Eqs (2), (3) and subsequently, the prime denotes differentiation with respect to the undeformed arc length ξ_0. The initial conditions needed for the integration of Eqs (2) and (3) are $r_0(0) = z_0(0) = \theta_0(0) = 0$. The drop shape is given by $r_0(\xi_0)$ and $z_0(\xi_0)$. The reference configuration $r_0(z_0)$ is characterized by an undeformed area:

$$A_0 = \int_{s_0} 2\pi r_0 \, d\xi_0, \tag{4}$$

where s_0 is the total curvilinear length of the undeformed membrane.

When subjected to an internal inflationary load such as hydrostatic pressure, this deforms to a shape described by the coordinates (r, ϕ, z) or (ξ, ϕ). The expectation of rotational symmetry implies $\phi_0 = \phi$, therefore motion of a point during inflation can be expressed as $r(\xi_0)$ and $z(\xi_0)$ or $\xi(\xi_0)$; knowledge of these functions specifies the shape of the membrane in the deformed state.

The deformation of the interface is characterized by the eigenvalues of the strain tensor; the principal stretches in the directions of the surface coordinate lines associated with the surface strain are:

$$\lambda_1 = \frac{d\xi}{d\xi_0} = \sqrt{r'^2 + z'^2}, \tag{5a}$$

$$\lambda_2 = \frac{r}{r_0}, \tag{5b}$$

where the subscripts 1 and 2 denote the meridional (ξ) and azimuthal (ϕ) surface directions, respectively.

As the dilation of the interface proceeds, surface stresses develop in response to its stretching. For isotropic surface materials, the stretches and stresses are collinear. The physical components for the interfacial stress tensor $S_{11} = S_1$ describing the shape of the deformed interface are assumed to be of the form:

$$S_1 = \gamma(\Gamma) + T_1(\lambda_1, \lambda_2). \tag{6}$$

The surface stress in the orthogonal surface direction 2 is obtained by exchanging indices 1 and 2. Equation (6) states that the total surface stress S is the sum of the thermodynamic surface tension γ and the membrane stress T resulting from deformation. This approach was used in Ref. [11] to separate surface tension effects from membrane contributions in Pickering emulsion droplet deformation. The simplest form of membrane tensions assumes that the surface stress is linear in the surface deformation; this is the surface equivalent of Hooke's law. Linear elasticity is typically restricted to small deformations. In this case, the membrane tension is:

$$T_1 = \frac{G_s}{1 - \nu_s} [\lambda_1^2 - 1 + \nu_s(\lambda_2^2 - 1)], \tag{7a}$$

where G_s (usually expressed in mN/m) and ν_s (a dimensionless quantity $-1 < \nu_s <$ 1) are the surface shear modulus and the surface Poisson ratio of the membrane. For an incompressible Hookean surface $\nu_s = 1$. To describe large deformation behavior, non-linear constitutive laws are usually required. For example, the Mooney–Rivlin relationship for an incompressible elastic surface solid is:

$$T_1 = 2C_1 h \left(\frac{1}{\lambda_1 \lambda_2} - \frac{1}{(\lambda_1 \lambda_2)^3} \right)(1 + \alpha_{MR}\lambda_2^2), \tag{7b}$$

where $C_1 h$ is proportional to the surface shear modulus G_s of a Hookean membrane, and α_{MR} is a dimensionless constant. For further discussion of surface constitutive behavior, see, for example, Ref. [12].

The equations of equilibrium are the result of the balance of linear momentum and the geometric relations. The local force balance in the membrane requires that the divergence of the stress in the membrane equals the jump of pressure across it. The force balance has two tangential components and one normal component. The normal and the tangential (ξ) components of the force balance are:

$$\frac{\partial S_1}{\partial r} + \frac{1}{r}(S_1 - S_2) = 0, \tag{8a}$$

$$K_1 S_1 + K_2 S_2 = [p], \tag{8b}$$

where the principal curvatures can be expressed as:

$$K_1 = \frac{\theta'}{\lambda_1}, \tag{9a}$$

$$K_2 = \frac{\sin\theta}{r}. \tag{9b}$$

The tangential (ϕ) component is identically satisfied by axisymmetry.

The normal pressure jump $[p]$ is given by the hydrostatic pressure, i.e., $[p] = [p]_0 - \Delta\rho g z$. Equations (8) can be rearranged using the relations:

$$f_1 = \frac{\partial S_1}{\partial \lambda_1}; \quad f_2 = \frac{\partial S_1}{\partial \lambda_2}; \quad f_3 = S_2 - S_1; \quad \lambda_2' = \frac{r'}{r_0} - \frac{r r_0'}{r_0^2} \tag{10}$$

to yield:

$$\lambda_1' = \frac{1}{f_1}\left(\frac{r'}{r} f_3 - \lambda_2' f_2 \right), \tag{11a}$$

$$\theta' = \frac{\lambda_1}{S_1}\left([p]_0 - \Delta\rho g z - S_2\frac{\sin\theta}{r} \right). \tag{11b}$$

Equations (11) together with the geometric relations:

$$r' = \lambda_1 \cos\theta, \tag{12a}$$

$$z' = \lambda_1 \sin\theta \tag{12b}$$

can be numerically solved, subject to the initial conditions $r(0) = z(0) = \theta(0) = 0$, to find the deformed shape. Additionally, geometry at the apex requires $z'(0) = 0$, therefore $r'(0) = \lambda_1(0)$. It should be noted that at $\xi_0 = 0$, Eq. (11) are ill-defined and can be replaced by:

$$\lambda_1'(\xi_0 = 0) = 0, \tag{13a}$$

$$\theta'(\xi_0 = 0) = \frac{[p]_0}{2R_0}. \tag{13b}$$

The solution is constrained by the boundary condition at $\xi_0 = s_0$, i.e., the membrane remains pinned to the capillary during inflation. Therefore, $\lambda_2(s_0) = 1$. The solution procedure is as follows: (1) a guess is made for $\lambda_1(0)$, (2) Eqs (11) and (12) are solved, and the pressure at the apex $[p]_0$ is adjusted iteratively to satisfy the condition $\lambda_2(s_0) = 1$, and (3) $\lambda_1(0)$ is adjusted iteratively to satisfy the specified inflation:

$$\alpha = \frac{A - A_0}{A_0}, \tag{14}$$

where the reference area A_0 is defined in Eq. (4) and the deformed area A is given by:

$$A = \int_{s_0} 2\pi r \lambda_1 \, d\xi_0. \tag{15}$$

Solution of these equations yields the field load $p(z)$, the principal stretches $\lambda_1(\xi_0)$ and $\lambda_2(\xi_0)$, the principal stresses $S_1(\xi_0)$ and $S_2(\xi_0)$, and most importantly the deformed configuration $r(z)$ as a function of the constitutive parameters of the membrane, cf. (G_s, ν_s) or $(C_1 h, \alpha_{MR})$ as in equations (7). Comparison of $r(z)$ to measured drop or bubble inflation profiles form the basis of an inverse method to determine the constitutive parameters of the nanomembrane.

C. Results and Discussion

The balance of linear momentum constrains the shape of a surface described by both thermodynamic interfacial tension and purely elastic membrane tensions. The total surface stresses are assumed to be a linear combination of thermodynamic and membrane tensions. Before these shapes are discussed different limiting cases are described.

1. Limiting Cases: Isotropic Interfacial Tension and Flat Disc Membrane

The shape of a pendant drop governed by isotropic interfacial tension alone can be recovered for two different cases: (1) zero deformation, i.e., the principal stretches $\lambda_1 = \lambda_2 = 1$, and (2) zero membrane elasticity, i.e., $G_s = 0$. It should also be noted that under microgravity conditions, the internal pressure is constant, and an isotropic tension is capable of describing the shape of this interface. This is because in this situation $\lambda_1 = \lambda_2 = \lambda$ corresponds to a spherically symmetric inflation. For isotropic materials, when the stretches are equal, the stresses are equal.

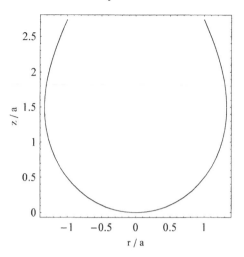

Figure 2. No deformation; linear elastic interface Eq. (7a). $\lambda_1 = \lambda_2 = 1$; $G_s = 100$ mN/m; $\nu_s = 0$; $R_0 = 13.2$ cm^{-1}; $\Delta\rho = 997$ kg m^{-3}; $\gamma(\Gamma) = 50$ mN/m.

In case 1, equations (7) show that the membrane stresses are identically zero when $\lambda_1 = \lambda_2 = 1$. Equation (6) becomes $S_1 = S_2 = \gamma(\Gamma)$, and equation (8a) is identically satisfied. Equation (8b) becomes Equation (1). There is only one inflation from equation (14) that satisfies this criteria; $\alpha = 0$. Figure 2 compares the shapes calculated by equations (2) and (3) — the Young–Laplace equation and Eqs (11)–(13) — the anisotropic surface stress equations. The shapes are identical. The same occurs for all deformations, i.e., for all α, when $G_s = 0$.

Case 2 is a purely elastic membrane inflation. Consider a flat sheet of Mooney–Rivlin material bound to a flat rigid plate with a circular orifice. When a uniform pressure $[p]$ is applied to one side, the sheet is inflated to an axisymmetric surface. For this situation, $s_0 = a$; $\gamma = 0$ (there is no interfacial tension); and $\Delta\rho = 0$. The deformed shape computed using the Mooney–Rivlin constitutive law, i.e., Eq. (7b) with $\alpha_{MR} = 0.1$, together with the momentum balance and the geometric relations, i.e., Eq. (11), can be compared with previous results [13]. Results are presented in Fig. 3 and show the deformed profiles, stretch, and stress resultants are identical to previously reported calculations.

2. Linear Elastic Nanomembranes at Axisymmetric Fluid Interfaces: Effect of Surface Shear Modulus on Drop Shape, Surface Deformation, and Surface Stresses

The central question addressed in this chapter is: what are the differences between the inflation of a pendant drop described only by isotropic surface tension and a surface having an anisotropic distribution of surface stresses caused by the presence of an elastic membrane?

For this purpose, all remaining calculations will utilize the assumption of linear elasticity, i.e., Eq. (7a), to describe surface constitutive behavior. Extension to

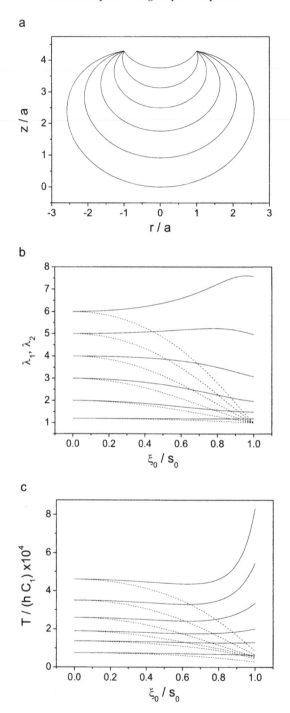

Figure 3. Flat disc inflation of Mooney–Rivlin, $\alpha_{MR} = 0.1$, membrane for inflations of $\alpha = 0.12, 0.2, 0.3, 0.4, 0.5$ and 0.6: (a) deformed shape $r(z)$, (b) stretch λ_1 (–) and λ_2 (– –) and (c) total interfacial stress S_1 (–) and S_2 (– –).

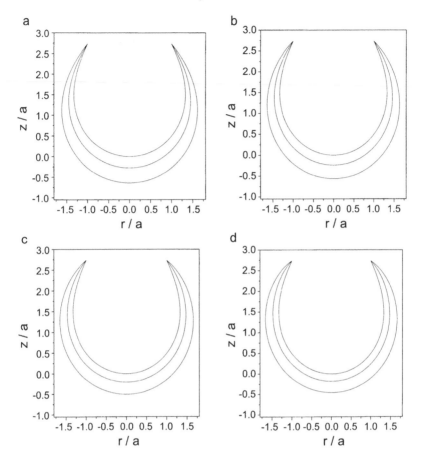

Figure 4. Effect of surface shear modulus on deformed configuration; $G_s = $ (a) 50, (b) 100, (c) 200 and (d) 500 mN/m; $\nu_s = 0$; for all calculations $\alpha = 0$, 0.2 and 0.5; $R_0 = 13.2$ cm^{-1}, $\Delta\rho = 997$ kg m^{-3}, $\gamma(\Gamma) = 50$ mN/m.

non-linear elasticity, for example using Eq. (7b) to describe membrane stresses, is straightforward.

The isotropic tension $\gamma(\Gamma)$ used in these calculations that appears in Eq. (6) is 50 mN/m. This is approximately the interfacial tension of an oil/water interface. Therefore, the physical situation corresponds to a pendant drop of water in continuous oil phase having a cross-linked macromolecular network (e.g., a protein) at the interface. As the cross-link density increases, the surface shear modulus also increases. The effect of the surface shear modulus on deformed shape is shown in Fig. 4 for $G_s = 50$, 100, 200 and 500 mN/m with $\nu_s = 0$ for inflations of $\alpha = 0, 0.2$ and 0.5. The axes of the deformed coordinates (r, z) are scaled using the radius of the capillary $a = r_0$ ($\xi_0 = s_0$) from the undeformed shape. Results show that the deformed shape is sensitive to the value of the surface shear modulus; as the surface shear modulus increases, the deformed profile tends toward the spherical shape.

This behavior can be described by considering the forces which compete during inflation: the gravitational force acts to extend the drop from its spherical shape and the surface elastic membrane tensions act to oppose elongation. The ratio of gravitational to surface tension forces is described by the Bond number, $Bo = \Delta\rho g a^2 / S$, where the surface tension force S is can be represented using a characteristic value of the total surface stress.

When Bo is zero, the deformed shape is spherical; this occurs in the absence of gravity (or microgravity), i.e., $g = 0$; for isodense fluids, $\Delta\rho = 0$; small experimental geometries, i.e., $a = 0$; and for deformations which result in sufficiently large surface stresses, either because of large surface shear moduli or large deformation. This last case is most important. Therefore, regardless of G_s, as inflation increases, the Bond number decreases asymptotically toward zero. For the same inflation, the rate of approach to zero increases with increasing surface shear modulus. This suggests that shape differences between surfaces of different moduli decrease as inflation proceeds. However, inflation of any elastic surface will eventually tend toward the spherical shape. This was previously discussed by Müller *et al.* [14].

Figures 5 and 6 show the surface stretch and total surface stress resultants for the deformed configurations under the same conditions as in Fig. 4. In both Figs 5 and 6, the undeformed arc length coordinate ξ_0 is scaled by the total undeformed arc length, s_0.

From Fig. 5, it can be seen that meridional stretching (λ_1) is maximum at the drop apex and that as the surface shear modulus increases, non-monotonic behavior develops in λ_2. However, the magnitude of the stretches is similar in all cases. This is because the undeformed Laplacian shape is similar to a sphere ($\lambda_1 \approx \lambda_2$). And for a specified inflation, the product of the stretches is constrained by:

$$\alpha = \lambda_1\lambda_2 - 1. \tag{16}$$

It is then expected that the magnitude of the stretches be similar regardless of G_s.

More dramatic are the differences in the principal stresses S_1 and S_2 for each case as shown in Fig. 6. This is because the magnitude of the stress scales more directly with the magnitude of the surface shear modulus. However, it should be noted that although there is an order-of-magnitude increase in G_s between Fig. 7a and d, there is only a roughly five-fold increase in the maximum value of the total surface stress S_1. It should also be noted that the surface stress is maximum at the contact line with the capillary $\xi_0 = s_0$; this is intuitive because the weight of the drop is supported by the membrane at this point. This suggests an intrinsic vulnerability to membrane failure at the boundary.

3. Shape Differences Between Surfaces with Anisotropic Stresses and the Young–Laplace Equation

As discussed in the previous section, the inflation of any purely elastic surface will eventually tend toward the spherical shape therefore, it is expected that the difference between the shapes will both systematically differ and increase with increasing inflation.

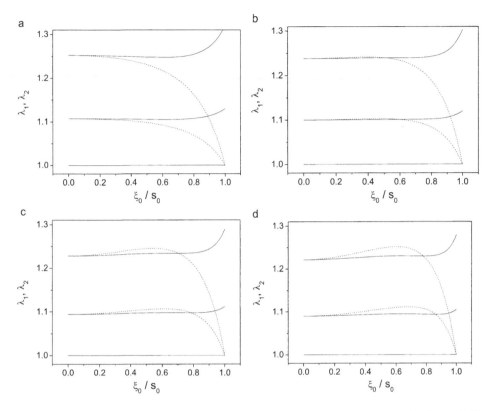

Figure 5. Effect of surface shear modulus on principal stretches λ_1 (–), λ_2 (– –): $G_s =$ (a) 50, (b) 100, (c) 200 and (d) 500 mN/m; $\nu_s = 0$; for all calculations $\alpha = 0, 0.2$ and 0.5; $R_0 = 13.2$ cm^{-1}, $\Delta\rho = 997$ kg m^{-3}, $\gamma(\Gamma) = 50$ mN/m.

The question now becomes: are the differences between drop shapes calculated using Eqs (6), (7a) and (11)–(13) and the Young–Laplace equation experimentally observable?

To address this question, a drop shape was calculated using the assumption of anisotropic stress distribution. Then, the Young–Laplace equation was fit to this shape using the isotropic surface tension γ as the sole fitting parameter, hereafter γ_{fit}. Figure 7 compares these shapes.

In Fig. 7, a systematic deviation between the two shapes can be observed. This can be quantified in terms of the difference between the radii of the two shapes for the same value of the axial coordinate z. For reference, this will be referred to as the Young–Laplace or r-error. Therefore, the deviation from Young–Laplace is expected to increase with increasing deformation (α) and increasing surface shear modulus (G_s). This is because, as discussed in Section C.2, all inflated profiles tend toward the spherical shape. Figure 8 shows the deviation from Young–Laplace for the anisotropic shapes as a function of deformation and surface shear modulus.

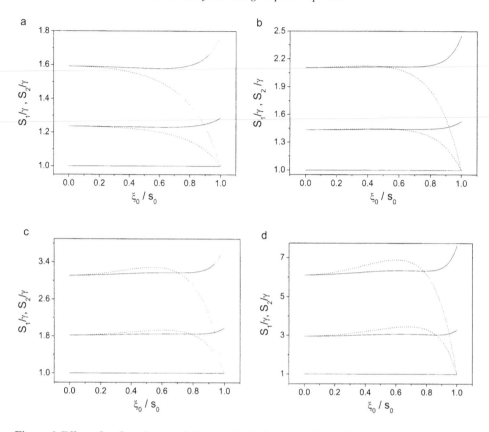

Figure 6. Effect of surface shear modulus on principal stresses S_1 (–), S_2 (– –) $G_s =$ (a) 50, (b) 100, (c) 200 and (d) 500 mN/m; $\nu_s = 0$; for all calculations $\alpha = 0$, 0.2 and 0.5; $R_0 = 13.2$ cm^{-1}, $\Delta\rho = 997$ kg m^{-3}, $\gamma(\Gamma) = 50$ mN/m.

As expected the deviations increase with increasing inflation and surface elasticity. These calculations are provided to demonstrate this deviation. However, here it is important to note that this error profile can be used to provide an experimental signature of elastic network formation without integrating equations (11)–(13). Examination of the Young–Laplace error profile that results from fitting experimentally measured shapes with the Young–Laplace equation, i.e. Equation (2), can yield important clues about the nature of the surface stress distribution and therefore the intermolecular surface interactions.

D. Conclusions and Prospectus

In this chapter a framework for evaluation surface rheology data is presented for experiments which study purely elastic interfacial networks. It should be noted that because all materials are viscoelastic, the macroscopic behavior depends on the ratio of the timescale for internal relaxation τ_R and the timescale of an experimental observation τ_O known as the Deborah number, $De = \tau_R/\tau_O$. For $De \gg 1$ macro-

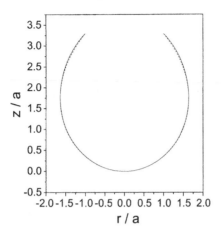

Figure 7. Drop shape calculated using anisotropic stress distribution, equations (6), (7a), and (11)–(13), and best fit to the Young–Laplace equation (– –), i.e., γ_{fit} as sole fitting parameter; for the anisotropic stress-drop shape (–): $G_s = 100$ mN/m; $\nu_s = 0$; $\alpha = 0.5$; $R_0 = 13.2$ cm^{-1}, $\Delta\rho = 997$ kg m^{-3}, $\gamma(\Gamma) = 50$ mN/m.

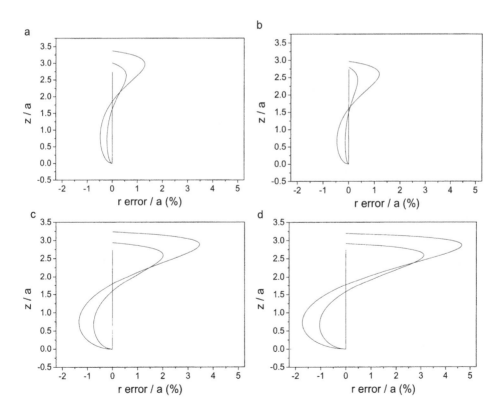

Figure 8. Effect of surface shear modulus on Young–Laplace error; $G_s =$ (a) 50, (b) 100, (c) 200 and (d) 500 mN/m; $\nu_s = 0$; for all calculations $\alpha = 0, 0.2$ and 0.5; $R_0 = 13.2$ cm^{-1}, $\Delta\rho = 997$ kg m^{-3}, $\gamma(\Gamma) = 50$ mN/m.

scopic behavior is elastic, $De \ll 1$ viscous and $De \sim 1$ viscoelastic. Therefore, the difference of applicability between Eqs (7), (11)–(13) or Eqs (2) and (3) to describe the shape and state of stress of the interface arises not from the difference between solids and liquids, but rather the capacity of the interface to support a non-zero deviatoric stress $\gamma' = S_1 - S_2$. Future publications focus on elucidating the nature of the error arising from using the Young–Laplace equation alone to describe surface stresses in elastic interfacial systems and on the development of simplified metrics to extract information regarding surface shear modulus from experimental data.

E. Acknowledgements

This work was financially supported in part by the United States through the National Science Foundation (NSF) Award 0729403 and the National Aeronautics and Space Agency (NASA) cooperation with the European Space Administration (ESA) through opportunity AO-2009-0813. We also acknowledge the Max Planck Society for support for research stays for both JKF and PALF provided through the Max Planck Institute for Colloids and Interfaces.

F. References

1. C.H. Chang and E.I. Franses, Colloids and Surfaces A, 100 (1995) 1.
2. A. Yeung, T. Dabros and J. Masliyah, Langmuir, 13 (1997) 6597.
3. A. Yeung and L.C. Zhang, Langmuir, 22 (2006) 693.
4. R. Miller, R. Wüstneck, J. Krägel and G. Kretzschmar, Colloids Surfaces A, 111 (1996) 75.
5. J.B. He, Z.W. Niu, R. Tangirala, J.Y. Wan, X.Y. Wei, G. Kaur, Q. Wang, G. Jutz, A. Boker, B. Lee, S.V. Pingali, P. Thiyagarajan, T. Emrick and T.P. Russell, Langmuir, 25 (2009) 4979.
6. S. Kutuzov, J. He, R. Tangirala, T. Emrick, T.P. Russell and A. Boker, Physical Chemistry Chemical Physics, 9 (2007) 6351.
7. Y. Lin, A. Boker, H. Skaff, D. Cookson, A.D. Dinsmore, T. Emrick and T.P. Russell, Langmuir, 21 (2005) 191.
8. P. Cheng, D. Li, L. Boruvka, Y. Rotenberg and A.W. Neumann, Colloids Surfaces, 43 (1990) 151.
9. O.I. del Rio and A.W. Neumann, J. Colloid Interface Sci., 196 (1997) 136.
10. F.K. Skinner, Y. Rotenberg and A.W. Neumann, J. Colloid Interface Sci., 130 (1989) 25.
11. J.K. Ferri, P. Carl, N. Gorevski, T.P. Russell, Q. Wang, A. Boker and A. Fery, Soft Matter, 4 (2008) 2259.
12. D. Barthes-Biesel, A. Diaz and E. Dhenin, J. Fluid Mechanics, 460 (2002) 211.
13. W.H. Yang and W.W. Feng, J. Applied Mechanics-Transactions of the ASME, 41 (1970) 1002.
14. I. Müller and P. Strehlow, Rubber and Rubber Balloons, Paradigms of Thermodynamics, Springer-Verlag, 2004.

Maximum Bubble Pressure Tensiometry: Theory, Analysis of Experimental Constrains and Applications

V.B. Fainerman [a] and R. Miller [b]

[a] Donetsk Medical University, 16 Ilych Avenue, Donetsk 83003, Ukraine
[b] Max-Planck-Institut für Kolloid- und Grenzflächenforschung, 14424 Potsdam/Golm, Germany

Contents

A. Introduction

The maximum bubble pressure technique is a classical method in interfacial science. Due to the fast development of new technique and the great interest in experiments at very short adsorption times in recent years, efficient instruments were made commercially available for a large number of researchers. In 1851, more than 150 years ago, Simon [1] proposed the maximum bubble pressure method (MBPM) for measuring the surface tension of liquids. In recent reviews the various aspects of this method were discussed [2–8]. The historical development of the method was reviewed in detail by Mysels [4]. The physical processes taking place during the growth at and separation of a bubble from the tip of a capillary, the problems of measuring bubble pressure, lifetime and the so-called dead time were considered in many papers [7–29]. Theoretical considerations and experimental implementation become both complicated when the method is used for studies in the short time range, i.e., in the millisecond and sub-millisecond time range. However, this is exactly the time range which attracted recently the great attention to this methodology as it promises new important physico-chemical results. Using the MBPM, significant results have been obtained in many fields of application where dynamic surface tensions at short adsorption times are required, including industrial and biological applications [30–58].

After the development of a suitable theoretical basis and the solution of respective technical problems of the MBPM a number of bubble pressure tensiometers appeared on the market. The most important facts for these developments are the possibility of automation of measurement and calculation procedures, and in particular the increasing demand for studies of dynamic surface phenomena at very short times. The various available tensiometers on the market use capillaries of different size, material and design. Also the technical principles employed for measuring the surface lifetime are different. This leads to dynamic surface tensions which deviate significantly from each other when measured with different instruments [59], although the surface tension is a characteristics of the studied solution and must not depend on the devise used for the measurement. Hence, data obtained only from some of the instruments available on the market are reliable, in particular at short adsorption times.

In some instruments, the surface tension is directly determined from the maximum pressure in the instrument's gas system connected to the capillary. The lifetime of the bubble (t_l) and the dead time (t_d) are estimated from changes in the measured pressure: t_l from the increasing part, and t_d from the decreasing part. This procedure is self-contradictory: precise measurements of surface tension based on the pressure in the system require a system volume much larger than the volume of a single separating bubble [2, 7, 29]. At the same time, precise measurements of the lifetime can be performed only if the system volume is relatively small.

In few tensiometers the precision of surface tension measurements is high due to the large volume of the measurement system (up to 40 ml). The optimum procedure to be employed for the determination of the lifetime was described in [2, 17,

28]: first, the critical pressure (or critical flow rate) is to be localised in the pressure — flow rate curve $P(L)$. The interval between bubbles in the point of the critical flow rate (the transition point between the bubble and jet regimes) is measured. This time interval can be assumed to be equal to the dead time, which for any arbitrary flow rates can be recalculated *via* the measured pressure using the Poiseuille law. This method has the drawback, that the consideration of hydrodynamic effects in the calculation of dead time and lifetime is quite cumbersome [7, 24–27], and its implementation requires a preliminary measurement of the entire $P(L)$ characteristic and determination of the critical point. This restricts the applicability of the method and makes it rather slow.

In [8, 60] an essentially new method for the measurement of the surface lifetime was proposed, which is free of the deficiencies mentioned above. The method employs explicit measurements of the dead time and lifetime from the oscillations of the gas flow from the measurement system into the capillary. The method is based on the fact that during the period of lifetime the air flow from the measuring system to the capillary is very small, and finally the growth of the bubble becomes almost zero. During the rapid bubble growth and final separation from the capillary, the dead time period, the air flow from the measuring system to the capillary increases strongly and attains its maximum in the moment of bubble separation. Therefore, the air flow from the external system to the measuring system changes systematically and can be used to determine the interval between two bubbles, and to determine the bubble dead time and lifetime exactly. In contrast to a direct determination of t_l from pressure oscillations, this new method does not require a small volume of the measuring system, so additional errors in the surface tension measurement do not arise.

This chapter gives an overview of the main theoretical and experimental aspects of bubble pressure tensiometry. An analysis of the influence of the geometry of the measuring system, and the design of the capillary on the measured surface tension data is given. Finally, selected experimental examples show the capacity of the presented methodology for understanding very fast surface processes, such as the impact of the micelle kinetics on the formation of adsorption layers.

B. The Optimum Design of Bubble Pressure Tensiometers and Measuring Procedure

The designs of bubble pressure tensiometers are various. Fig. 1 illustrates the scheme of one of them as described in [8, 60]. It is equipped with a gas flow sensor, with which the above mentioned new method for the determination of bubble surface lifetime is accessible. The air or any other gas is pressed by a compressor through the flow capillary and gas volume, which smoothes the gas flow and hence the pressure at the inlet into the measuring system. The air flow is determined from the pressure difference along the flow capillary *via* a differential pressure sensor. The optimum internal gas volume of the instrument and selection of the right cap-

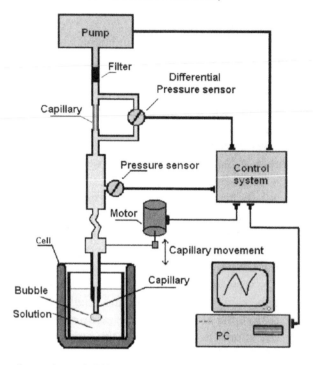

Figure 1. Scheme of a maximum bubble pressure tensiometer with a gas flow oscillation analyser (BPA).

illary are two important points in the design of a device and will be discussed in detail further below.

The measurement procedure is usually quite simple – the pressure is measured as a function of bubble formation rate and further analysed depending on the algorithms implemented in the respective instrument. In the device discussed here (BPA-1S designed by SINTERFACE Technologies, Berlin) a more complex procedure is implemented which finally makes it suitable for measurements at very short adsorption times [8, 60]. The pressure in the internal gas volume P_s and the gas flow rate L are measured as a function of time and recorded in 0.5 ms time intervals. In Figs 2 and 3 examples of the dependencies $L(t)$ and $P_s(t)$ are shown. Arrows indicate the various phases of bubble growth. The time intervals between successive maxima or minima in P_s and L constitute the total bubble time, necessary for the formation and separation of each bubble: $t_b = t_l + t_d$. The time interval of increasing L (Fig. 2) and decreasing P_s (Fig. 3), respectively, corresponds to the dead time t_d. The lifetime t_l can be calculated in two ways. At first it can be calculated from the relation $t_l = t_b - t_d$ with t_b and t_d determined as explained above. Secondly, the lifetime t_l can be obtain directly from the time intervals of decrease in L and increase in P_s, respectively. Averaged values of the results obtained from the two methods are compared and the value with the lower standard deviation is accepted [60].

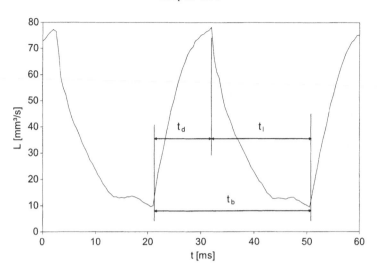

Figure 2. The example of the $L(t)$ dependence. Arrows indicate the duration of corresponding stages of bubble growth.

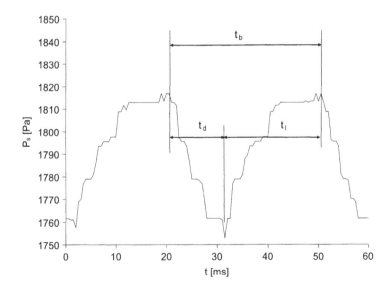

Figure 3. The example of the $P_s(t)$ dependence. Arrows indicate the duration of corresponding stages of bubble growth.

The maximum pressure in the measuring system is calculated from the peaks in the dependence $P_s(t)$ (cf. Fig. 3) also averaged over a certain number of measured oscillations (10–400 cycles depending on the given bubble formation frequency). All calibration and test routines of modern bubble pressure tensiometers (also of the BPA-1S discussed in detail here), the measurement and calculation procedures are fully automated.

To reach bubble surface lifetimes below 10 ms, a special procedure has to be used [53] (see below), including special capillaries coupled with a bubble deflector. The volume of the separating bubble and, therefore, the dead time can be adjusted by variation of the distance between the capillary tip and the deflector located opposite to it. A minimum dead time t_d of 5 ms is obtained for a bubble volume V_b of approximately 2 mm^2. It was shown in [2] that such low dead times are needed to measure the dynamic surface tension in the surface lifetime range of sub-milliseconds. In this way, bubble pressure tensiometers can provide data in a surface lifetime range from 100 μs to 50–100 s, i.e., a huge time interval spanning over 6 orders of magnitude [53].

C. Surface Tension Determination

1. Effect of Non-spherical Bubble Shape

The surface tension γ can be calculated from the measured maximum capillary pressure P and the known capillary radius r_0 via the Laplace equation

$$\gamma = f \frac{r_0 \cdot P}{2}.$$ (1)

f is a correction factor which accounts for the deviation of the bubble from a spherical shape. This correction factor f was tabulated by Sugden [62] as a function of the ratio r_0/a, where a is the capillary constant:

$$a = (2\gamma/\Delta\rho g)^{1/2}.$$ (2)

Sugden's tables were later transformed into a polynomial form by a number of authors (see [2]), for example by Bendure [34]:

$$f = \sum a_i \left(\frac{r_0}{a}\right)^i$$ (3)

with the polynomial coefficients $a_0 = 0.99951$, $a_1 = 0.01359$, $a_2 = -0.69498$, $a_3 = -0.11133$, $a_4 = 0.56447$ and $a_5 = -0.20156$. Mysels analysed in [4] the various corrections to the Laplace equation.

It follows from Eq. (3) that the correction factor f is needed for capillaries with a radius $r_0 > 0.1$ mm, which is the case for all commercial instruments. For capillaries of radii $r_0 < 0.1$ mm the bubbles are sufficiently spherical and the error in the calculation of γ via Eq. (1) with $f = 1$ is less than 0.5% and a calibration with a know liquid can reduce this error significantly. For wide capillaries, for example, 1 mm as used in some tensiometers [63], the error in the surface tension of water can amount up to 10%. When surfactant solutions are studied, this error increases and is roughly proportional to the ratio between the surface tensions of water and this surfactant solution. This means that for surfactant solutions a calibration with respect to water produces errors of 10% and more. Note, that all known corrections are valid under static conditions. The fast bubble formation is a highly dynamic process and errors arising from Eq. (1) due to dynamic effects can be even higher. Especially for surfactant solutions this leads to increased radii of the growing bubbles [7, 27, 29].

The measured capillary pressure P can be expressed *via* the excess maximum pressure in the measuring system P_s, the hydrostatic liquid pressure $P_H = \Delta \rho g H$, and the excess pressure P_d (caused by dynamic effects, such as aerodynamic resistance of the capillary, viscous and inertia effects in the liquid etc. [7]), while $\Delta \rho$ is the difference between the densities of liquid and gas, g the gravity constant, and H the immersion depth of the capillary into the liquid. Hence the capillary pressure is given by [2, 7]

$$P = P_s - P_H - P_d, \tag{4}$$

and from (1) and (4) we obtain

$$\gamma = f \frac{r_0 (P_s - P_H)}{2} - \Delta \gamma_a - \Delta \gamma_v. \tag{5}$$

The term $\Delta \gamma_a$ represents the effect caused by the aerodynamic resistance of the capillary, $\Delta \gamma_v$ reflects the effect of viscous and inertia effects in the liquid.

2. Influence of Bulk and Dilational Viscosities

The expansion of the bubble surface and the displacement of the meniscus into the liquid bulk during its lifetime can contribute to the excess pressure P_d. Keen and Blake [64] have considered the effect of bulk and surface dilational viscosity on the growth dynamics of a bubble at a capillary tip immersed into a liquid. It was shown, in particular, that the effect of dilational viscosity k_s is significant for very large values of $k_s / r\mu > 1000$ (μ is the bulk viscosity of the liquid). The excess pressure in a growing bubble arising from the bulk viscosity of the liquid can be expressed as [65]:

$$P_d = \frac{\mu L}{\pi r^3}, \tag{6}$$

where r is the current bubble radius, and L the gas flow rate. At the final lifetime stage it can be assumed that $L = \pi r^3 / t_1$, and Eq. (6) transforms into an approximate expression [66]: $P_d \approx \mu / t_1$. This relationship agrees qualitatively with experiments performed with higher viscous liquids. However, a logarithmic dependence of P_d on μ fits the experimental data better [66]. The contribution of the dilational surface viscosity to P_d was estimated in [65], and for the final lifetime stage in the MBPM we have:

$$P_d = \frac{k_s}{r_0 t_1}. \tag{7}$$

The values of surface dilational viscosity for sodium dodecyl sulphate and octanoic acid presented in [65] ($k_s = 10^{-5} - 10^{-4}$ mN s m^{-1}) show that its contribution to P_d in the MBPM is negligible in the millisecond and sub-millisecond time range.

Dukhin *et al.* estimated the liquid inertia contribution to P_s and showed that the inertia effects are most significant in the final lifetime stage and do not amount to more than one percent of P. The liquid inertia contribution can be neglected if

$$\frac{\rho_L \gamma r_0^3 \delta}{(32\eta l)^2} \ll 1, \tag{8}$$

where ρ_L is the liquid density, η is the dynamic viscosity of the gas, l is the capillary length, δ is a small factor ($\delta \ll 1$). The inequality (8) is only violated for very wide and short capillaries.

3. Gas Flow Regime and Influence of Resistance of the Capillary

The present state of the art of the hydrodynamic theory for MBPM was developed in [7, 20–26]. The bubble lifetime can be split into the following main stages [7]. At the moment when the bubble separates, the meniscus curvature radius is approximately equal to the radius of the separating bubble $r_b \gg r_0$. At this time moment the next bubble starts to grow. During the time interval t_{l1} the radius of curvature of the meniscus approaches the capillary radius, and the meniscus itself moves into the capillary by some depth h (forward meniscus motion), depending on the properties of the internal capillary surface. At the next stage during the time t_{l2} the excess pressure in the system moves the meniscus to the end of the capillary (reverse meniscus motion). During the following stage of the growing bubble evolution the bubble radius decreases and becomes equal to the capillary radius t_{l3}. The time interval between bubble separation and the moment when $r = r_0$ is the bubble lifetime $t_l = t_{l1} + t_{l2} + t_{l3}$. Forward meniscus movement into the capillary after bubble separation and its reverse movement is a periodical process with a frequency equal to that of bubble formation (slow oscillation). It should be noted that also a fast oscillation exists.

After the gas bridge collapses (bubble separates) the gas excess pressure in the capillary at the orifice P_b is significantly lower than the excess pressure at the opposite capillary side P_s. In this moment the gas velocity in the capillary reaches its maximum. The gas from the measuring system flows into the capillary driven by the pressure difference between the capillary ends and by the gas inertia. This leads to a smoothing of the pressure profile along the capillary.

The excess pressure in the capillary $P(x, t)$ was expressed analytically as a function of time t starting from the moment of gas bridge collapse and the coordinate x measured along the capillary axis. The following expression was obtained in [7, 24]

$$P(x, t) = P_s - \frac{2P_0}{\pi^2} e^{-t/\tau} \sum_{k=0}^{\infty} \frac{\cos[\pi(k + 1/2)(l - x)/l]}{(k + 1/2)^2}$$

$$\times \left[\cos(t\beta_k) + \frac{\sin(t\beta_k)}{\tau\beta_k} \right], \tag{9}$$

where $\tau = \frac{2N}{c^2} = \frac{r_0^2}{3v}$ is the damping time, $\beta_k = \omega_k(1 - 1/(\omega_k^2\tau^2))^{1/2}$, v is the gas kinematic viscosity, c is the velocity of a disturbance propagation, P_0 is the pressure drop over the capillary at $t = 0$, and N is the parameter which describes the damping. Hence, the pressure variation in the capillary can be approximately expressed in the form of a superposition of direct waves and that reflected from the meniscus. All frequencies $\omega_k = 2\omega_0(k + 1/2)$ can be expressed *via* the smallest frequency $\omega_0 = \pi c/2l$.

The process behaviour turns out to depend strongly on the value of $\omega_0\tau$. For $\omega_0\tau < 1$ Eq. (9) predicts a smooth damping process of the pressure restoration in the absence of oscillations, with the characteristic time $t_a = 2/(\omega_0^2\tau)$. For this regime Eq. (9) transforms into

$$\bar{P}(l,t) = 1 - \frac{2\bar{P}_0}{\pi^2}\sum_{k=0}^{\infty}\frac{1}{(k+1/2)^2}\exp\left[-N\left(\frac{\pi}{l}(k+1/2)\right)^2 t\right], \qquad (10)$$

where $\bar{P} = P(x)/P_0$ and $\bar{P}_0 = P(x = 0)/P_0$. For $\omega_0\tau > 1$ the gas flow into the capillary oscillates with the frequency ω_0 (fast oscillations with the characteristic decay time of τ [24]). The value of $\omega_0\tau$ is defined primarily by the capillary dimensions:

$$\omega_0\tau = \frac{\pi c r_0^2}{6vl}. \qquad (11)$$

According to Eq. (11) the process of pressure restoration in the capillary after bubble detachment is of oscillatory character for small capillary length and/or large diameter, while this process is an aperiodic process for long and/or narrow capillaries. Thus, the theory predicts that the change of either capillary length or radius should lead to the transition between the two modes at certain values of some characteristic capillary parameters.

The main conclusion that can be drawn from the given discussion is that there is a qualitative difference between phenomena taking place in systems with long narrow capillaries and those characteristic for wide and short capillaries. A criterion for the two different types of capillaries is the following ratio [7]:

$$\frac{r_0^2}{l} \approx \frac{6v}{\pi c}. \qquad (12)$$

For air this corresponds to capillaries with $r_0^2/l \approx 0.1$ μm. Inertia effects in gas are significant for wide short capillaries for which the relation $r_0^2/l \geqslant 0.1$ μm holds, whereas they can be neglected for long narrow capillaries with $r_0^2/l \ll 0.1$ μm.

The restoration of pressure in wide short capillaries after bubble detachment is very fast. Moreover in such capillaries, gas flow oscillations arise when the bubble detaches: as the initial state of the whole system is in non-equilibrium, these oscillations develop due to the inertia and elasticity of the gas. In contrast, in long narrow capillaries the pressure restores during a much longer time without oscillations in a slow aperiodic regime.

The time during which the meniscus grows up to a hemisphere is called the bubble lifetime t_1. For short wide capillaries the time required for pressure smoothening along the capillary is much lower than the lifetime. Therefore the maximum pressure overcome in the bubble takes place when the pressure along the capillary has already become uniform, and the pressure in the reservoir is almost equal to that within the bubble. Therefore, for wide short capillaries even at small lifetimes only minor excess pressure in the reservoir is required.

For long narrow capillaries the pressure smoothening time is high as compared to the lifetime: it increases proportionally with the ratio l^2/r_0^2:

$$t_{11} \cong \frac{8l^2}{\pi^2 N} = \frac{64\eta}{\pi^2 P_0} \left(\frac{l}{r_0} \right)^2. \tag{13}$$

Therefore, in the short lifetime range (<10 ms) the overcoming of maximum pressure takes place before the pressure within the capillary smoothens. At this moment there still exists a significant pressure difference between the capillary ends, and the pressure in the reservoir exceeds the pressure in the bubble. Therefore, for long and narrow capillaries a significant excess pressure in the reservoir is necessary to ensure the formation of bubbles of very short lifetimes. The increase of the excess pressure in the reservoir explains the increase of apparent surface tension observed for long narrow capillaries at short lifetimes [2, 7]. This shows that the use of long narrow capillaries leads to incorrect data at low lifetimes when the dynamic surface tension is calculated directly from the pressure in the reservoir without consideration of the pressure drop over the capillary length.

The effect of gas compressibility depends on the capillary length l and radius r_0 and bulk elasticity modulus B. For short and wide capillaries the restoration of gas pressure after a bubble separation is very rapid, which corresponds to an almost incompressible gas behaviour. It was shown by Kovalchuk and Dukhin [7] that the approximation for incompressible media can be used under the condition

$$\frac{r_0^2}{l} \gg 4b/\kappa, \tag{14}$$

where $b = \gamma/P_{atm}$, P_{atm} is the atmospheric pressure, and $\kappa = 1.4$ is the adiabatic constant. For water we have $b = 0.73$ µm and $r_0^2/l \gg 2$ µm. The condition (14) is satisfied only for very short and wide capillaries, for example, a capillary of radius $r_0 = 150$ µm must be shorter than 1 cm. A criterion for the transition between the aperiodic regime of pressure variation within the capillary after bubble separation and the oscillation regime is Eq. (12). Combining it with the condition (14) we can see that the capillaries can be divided into three groups depending on the ratio l^2/r_0^2 [7]. When this ratio is larger than 2 µm one can neglect the gas compressibility but needs to consider the non-stationarity and inertia effects. In turn, when the ratio is smaller than 0.1 µm it is possible to neglect the non-stationarity and inertia but it is necessary to consider the gas compressibility. For capillaries with a ratio in between all effects influence the gas flow.

The available hydrodynamic theory for the MBPM enables us to determine the conditions which minimise the aerodynamic component of the excess pressure P_d. Then, as proposed in [2, 5], using an empirical correction of the measured P_s obtained for pure liquids, one can exclude P_d and gets the relation $P = P_s - P_H$.

D. Bubble Time and Its Constituents

The advantage of many tensiometers are its ability to directly measure the lifetime and dead time for lifetimes longer then 10 ms. However, this time range is insufficient for measurements in concentrated surfactant solutions, and in particular micellar solutions, because the surface tension decreases significantly during very short times, such as 1 ms and less. For lifetimes <10 ms a special procedure for the accurate determination of the dead time is needed.

1. Dead Time Determination

The first theoretical calculations of the dead time t_d were performed in the Poiseuille approximation for a gas flow through a capillary of length l. The differential equation for the rate of bubble growth due to the pressure difference between the two capillary ends can be expressed then by [67]

$$\frac{dr}{dt} = \frac{r_0^4(P_s - P_H - 2\gamma/r)}{32l\eta r^2}.$$ (15)

Integration leads to the dead time, i.e., the time during which the bubble is growing rapidly so that its radius increases from r_0 to the radius r_b of the separating bubble

$$t_d = \frac{32l\eta}{r_0(P_s - P_H)}\left[\frac{1}{3}\left(\frac{r_b}{r_0}\right)^3 + \frac{\gamma^*}{r_0(P_s - P_H)}\left(\frac{r_b}{r_0}\right)^2\right].$$ (16)

The first term on the right hand side of Eq. (16) describes the gas expansion into the infinite space, while the second term corresponds to the capillary pressure in the growing bubble. The surface tension for a growing bubble γ^* during the dead time, which enters the second term, is in fact unknown for surfactant solutions. The analysis performed in [68] has shown that for solutions the value of γ^* in Eq. (16) lies between the equilibrium value γ_∞ and the dynamic value of γ at $t = t_1$. Another important conclusion of this analysis is that the variation of γ^* in the range $\gamma \geqslant \gamma^* \geqslant \gamma_\infty$ does not affect the t_d value. Thus γ^* can be substituted by γ in Eq. (16) and for the Poiseuille approximation one obtains [68]

$$t_d = t_b \cdot \frac{L}{k_p P}\left(1 + \frac{3}{2}\frac{r_0}{r_b}\right),$$ (17)

where k_p is the Poiseuille equation constant for a capillary not immersed into the liquid ($L = k_p P$), L is the gas flow rate, $P = P_s - P_H$, and t_b the time interval between successive bubbles. A more rigorous dead time theory developed in [20–26] shows that the corrections related to the non-stationarity of the gas flow through the capillary and to the effect arising at the initial section of the capillary, does not exceed a few per cent of the t_d value calculated from Eq. (17), leading to slightly larger t_d values, which is in agreement with experimental findings [68].

2. Bubble Lifetime

Kloubek proposed that for high bubble formation frequencies, the dead time interval becomes equal to the time interval between two successive bubbles [69]. Indeed, the dependence of P_s on t_b, shows a sharp pressure increase at the transition from the regime $t_d < t_b$ to the regime $t_d = t_b$ [2, 70]. Hence, in the moment when the pressure starts increasing significantly we have $t_b = t_d$. More precise is a combined experimental/theoretic procedure based on Eq. (17). This procedure assumes that t_d depends on t_b and P. The dependencies of excess pressure in the measuring system on the gas flow rate L for capillaries of various lengths are shown in Fig. 4.

All curves show a transition point, named critical point at P_c and L_c. The linear sections of the curves at $L > L_c$ can be described by the Poiseuille equation and correspond to the jet regime of gas expansion from the capillary. For $L < L_c$ the injection of the gas into the liquid results in the formation and separation of individual bubbles with $t_l > 0$. In the transition point ($L = L_c$) the lifetime vanishes and the time interval between two successive bubbles becomes equal to the dead time.

For constant r_b one can combine Eq. (17) for the two cases $L = L_c$ and $L < L_c$ and obtains

$$t_d = t_b \frac{L P_c}{L_c P} \tag{18}$$

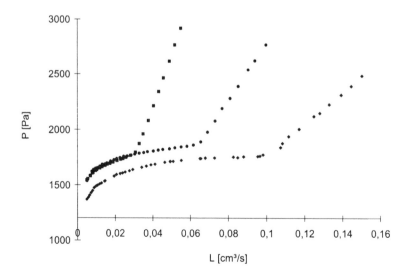

Figure 4. Dependence of pressure in the measuring system on the air flow rate for a 0.2% Triton X-100 solution, $r_0 = 0.0084$ cm, $l = 6$ cm (■), 3 cm (●) and 1.5 cm (◆).

and

$$t_1 = t_b - t_d = t_b \left(1 - \frac{L P_c}{L_c P}\right). \tag{19}$$

Thus, the procedure developed in [6, 70] provides a calculation routine of the lifetime using the experimental parameters t_b, $P = P_s - P_H$ and L. The co-ordinates of the transition point in the $P(L)$ dependence can be easily determined. Equation (19) accounts for the dependence of t_1 on t_b (or, more precisely, on P) which is predicted by Eq. (17). The value of t_d usually increases with t_b, because in this case P decreases (for surfactant solutions). Equation (19) was derived assuming constant r_b or V_b. The necessary control of the volume of separating bubbles can be arranged by a deflector located opposite to the capillary at a definite distance, as it was first described in detail elsewhere [71].

The determination of the lifetime *via* the difference $t_1 = t_b - t_d$ yields good results for large lifetimes, i.e., when errors in the dead time are negligible [7]. However, for $t_1 \ll t_d$ even a small error in the dead time can substantially influence the resulting lifetime. In this case the gas flows through the capillary very quickly and the Poiseuille law does not apply due to the inertia of the gas flow. The necessity of considering the gas inertia leads to the conclusion that the time intervals of lifetime and dead time cannot be described independently because the inertia effects in both time intervals are interconnected through the initial conditions [7, 22].

When surfactant solutions are studied, the bubble surface expands during the last (third) stage of the lifetime period. Various types of surface deformation are characteristic also for other dynamic surface tension methods, such as the drop volume or oscillating jet methods. For comparison of the various methods, and also to make use of a unique standard independent of specific features of a method, the results of dynamic surface tension measurements are usually presented as a function of the so-called effective adsorption time t_{ef}. Within a sufficient accuracy the effective time can be expressed by equations presented in [2, 5]. For surfactant solutions with $\gamma/\gamma_0 < 0.8$ the result for the MBPM is similar to that obtained for a growing drop, $t_{ef} = (3/7)t_1$.

3. Hydrodynamic Relaxation Time of the Liquid Meniscus

The hydrodynamic relaxation time t_h represents the sum of the first two components of the lifetime: $t_h = t_{11} + t_{12}$, i.e., the times of the forward and reverse meniscus motion. For short capillaries with a hydrophobic internal surface the liquid penetration depth h into the capillary is small or even zero, while for a hydrophilic internal surface the value of h can be of the order of the capillary radius. Therefore for hydrophilic capillaries the time interval t_h can contribute significantly to the value of t_1. Values of t_{11} for hydrophilic capillaries were first estimated in [23]. For long and narrow capillaries t_{11} can amount to 1 ms, while for short and wide capillaries the forward meniscus motion time is very short, $t_{11} < 100$ μs.

The penetration depth of the meniscus into the hydrophilic capillary can be expressed as [7, 24]:

$$h_{max}^2 \approx \frac{128\eta\gamma}{\pi^4 \mu \kappa P_{atm} r_0} l^2, \tag{20}$$

where η and μ are the dynamic viscosities of gas and liquid, respectively. Introducing numerical values for air and water, $\eta = 0.018 \cdot 10^{-3}$ Pa s and $\mu = 1 \cdot 10^{-3}$ Pa s we obtain:

$$h_{max}^2 \approx \frac{0.0124}{r_0} l^2, \tag{21}$$

where r_0 is given in μm. Therefore for a capillary radius of 100 μm the rise height is of the order of 1% of the capillary length l. From this it follows that the values of t_{l1} and t_{l2} for long hydrophilic capillary can achieve even several milliseconds. Thus, the employment of long and hydrophilic capillaries is not suitable for measurements in the millisecond time range.

A relationship for the hydrodynamic relaxation time of the liquid meniscus in short and wide capillaries was derived in [21]. This relationship holds for the case when the liquid does not penetrate into the capillary after bubble separation, i.e., the case of short hydrophilic or hydrophobic capillaries. It was shown that

$$t_h = \frac{4\nu\rho l}{\gamma} \left\{ \frac{4m^2 + 11m + 19}{6(m+1)^3} + \frac{2}{(m+1)^4} \log \frac{2m}{m+1} \right.$$
$$+ 2\frac{1 - m^2 - 2m}{(m+1)^4 \sqrt{m(m+2)}}$$
$$\left. \times \left[\arctan \sqrt{\frac{m}{m+2}} + \arctan \frac{1}{\sqrt{m(m+2)}} \right] \right\}, \tag{22}$$

where ρ is the gas density, $m = P_d/P$. Equation (22) leads to very small values of t_h. The results of calculations from Eq. (22) for capillaries of 1 cm length are presented in [2, 21]. It is shown that even for very low values of excess pressure, t_h does not exceed 1 ms. For capillaries short enough so that the ratio $P_d/P = (1-2)\%$, the value of $t_h < 0.1$ ms. It means that short and wide capillaries are recommended to be used for the MBPM in the milli- and sub-millisecond range of bubble lifetime.

4. Surface Lifetime in the Millisecond and Sub-millisecond Range

For the calculation of the so called critical gas flow rate L_c (the gas flow measured at the transition from the bubble regime to the gas jet regime) the following equation following from Eqs (17), (18) can be used [72]:

$$L_c = \frac{V_b}{t_d(1 + K^0)}, \tag{23}$$

where K^0 is a coefficient depending on surface tension, capillary radius and radius of the separating bubble. For capillaries with a respective bubble deflector, the bubble volume is almost independent of the gas flow rate L [2]. However, to improve the accuracy of the calculation of L_c from Eq. (23), the dependencies $V_b(1/L)$ and $t_d(1/L)$ are measured experimentally in the time range between 10 and 50 ms. Then the obtained dependencies are approximated by a polynomial, and the values of L for higher gas flow rates (up to the critical value) were calculated from the extrapolation using Eq. (23). This procedure was proved to be reliable by comparing the L_c values found in this way with those determined by the 'classical' method, i.e., *via* the kink point in the dependence of pressure P on gas flow rate L [2, 5, 68]. The values of L_c calculated from Eq. (23) coincide to within 1–2% with those determined from the kink point procedure.

The bubble surface lifetime t_l in the gas flow range from the minimum value L_{min} (corresponding to a time of 10 ms) to the maximum critical value L_c at $t_l = 0$ (because at the critical flow rate the bubble lifetime is zero [2, 5]) is determined, for example by the BPA-1S, in the following way. The experimental dependence of $L(t_l)$ in the surface lifetime range below 50 ms (including the zero time value for the critical flow L_c) is approximated by a polynomial. The BPA-1S control routine then calculates and adjusts the gas flow for any given surface lifetime in the range between 0 and 10 ms. In this procedure for the measurements and calculations in the time range below 10 ms the critical point is determined before the measurement run. Therefore, it becomes possible to calculate and adjust automatically the necessary gas flow corresponding to a respective surface lifetime. Also, for this method the accuracy of determination of L_c for concentrated surfactant solutions is higher, because the slope of the experimental dependence $P(L)$ in the jet and bubble regime, respectively, is quite similar, and therefore the intersection point of the corresponding lines may have a large error [61].

E. Analysis of Experimental Constrains

1. Comparison of Two Methods for the Determination of the Bubble Lifetime

For the experiments described in [60] the tensiometer BPA-1S was applied. The special capillary used for the technical studies had a radius of 0.085 mm and a length of 10 mm, and the deflector was located in a distance of 1 mm from the capillary tip. As mentioned above, the deflector ensures the formation of separating bubbles with constant volume, almost independent of the bubble frequency. The respective system volume V_s was changed from 1.5, to 4.5 and 20.5 ml by a special design. For a bubble volume of $V_b = 2$ mm^3, the ratios of $V_s/V_b = 750$, 2250 and 10250, respectively, result.

The dependencies of the relative change of air flow $\Delta L/L_{max}$ and pressure $\Delta P/P_{max}$ in the measuring system on the system volume V_s are shown in Fig. 5. Here ΔL and ΔP are differences between the maxima and minima of the air flow L and pressure P, respectively (cf. Figs 2 and 3).

The analysis of the dependencies shown in Figs 2, 3 and 5 yields a number of significant differences:

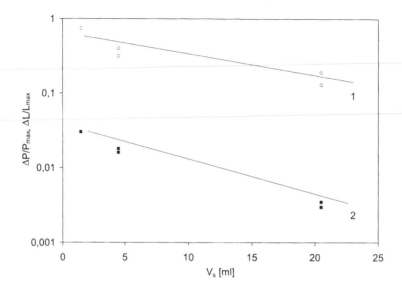

Figure 5. The dependence of $\Delta L/L_{max}$ (1) and $\Delta P/P_{max}$ (2) on V_s.

(i) the relative amplitude of measured flow rate changes is more than one order of magnitude higher than the relative amplitude of system pressure changes;

(ii) the values of t_l and t_d can be precisely determined from the flow pressure changes irrespective of the measuring system volume;

(iii) the bubble lifetime can be accurately determined from pressure changes only for a sufficiently small system volume.

The expected pressure decrease in the measuring system due to the growth and separation of a bubble during the dead time interval can be estimated from the expression [7]:

$$\Delta P_s/P_s = (P_{max} - P_{min})_s/P_s \cong \frac{V_b}{V_s}\frac{P_{atm} + P_s}{P_s}. \tag{24}$$

As the excess pressure in the system P_s is small as compared to the atmospheric pressure P_{atm}, one can take only the atmospheric pressure $P_{atm} = 10^5$ Pa for estimations with Eq. (24). For example, for the compared system volumes the obtained $\Delta P_s/P_s$ values are 0.048, 0.016 and 0.004, respectively, which are quite close to the experimental values shown in [8, 60].

Note that the results discussed here were obtained for narrow capillaries. For wider capillaries (with radius $r_0 \geqslant 0.5$ mm), and small measuring system volumes (as it is the case for example in the SITA tensiometers [63]), the course of pressure minima in the $P_s(t)$ curve is different. In this case, the growth of the bubble during the dead time leads to a significant drop of the excess pressure in the system so that for large separating bubbles the pressure minima can be comparable to the hydrostatic pressure at the bubble tip. For conditions of the SITA tensiometers the difference ΔP_s is relatively high (close to P_{max}), due to the very low capillary pressure and the low ratio V_s/V_b. It will be shown below that alone a small ratio V_s/V_b causes large errors in the measured maximum pressure, which finally leads to large errors in the surface tension of surfactant solutions. P_{min} cannot be lower than the hydrostatic pressure, otherwise a formation and separation of bubbles were impossible. Thus, for large separating bubbles and a small ratio V_s/V_b we can get $P_{min} \approx P_H$. In this case, however, the separation of a bubble is followed by a penetration of liquid into the capillary even when the internal surface of the capillary is hydrophobized [7, 28]. This fact, in addition to the necessary account for the non-spherical bubble shape in Eq. (1), makes an analysis of such tensiometry data in the framework of common theoretical adsorption models quite cumbersome [2, 28].

When we compare the standard measurement errors for the time values measured by the two determination methods one can see that the absolute error for t_l and t_d from flow rate in the range between 5 and 20 ms does not exceed 1 ms (i.e., is close to the instrumental error) and is independent of the measuring system volume. In contrast, the calculations based on pressure changes for the minimum volume system (1.5 cm³) have a standard error in the determined t_l values of 1.5–3 ms, for $V_s = 4.5$ cm³ the standard error is 4–5 ms, and for $V_s = 20.5$ cm³ the standard error becomes as high as 10 ms. Therefore, the measurement of bubble lifetimes based on pressure changes is possible only for very small system volumes, while for larger volumes this method becomes unacceptable for surface lifetimes in the range of milliseconds. The absolute standard error in the determination of t_l from flow changes at $t_l > 20$ ms becomes higher because of certain instabilities in the bubble generation process, however, the relative standard error in the range from 20 to 1000 ms does not exceed 4%.

2. Determination of the Dead Time

Measurements of bubble time $t_b = t_d + t_l$ and dead time t_d based on the flow rate method [8, 60] were performed for water and a dodecyl dimethyl phosphine oxide (C_{12}DMPO) solution with measurement systems of different gas volume (Fig. 6). For a given liquid, the dependence t_d (t_b) is almost independent of V_s, in agreement with the maximum bubble pressure theory and supporting the importance of the bubble deflector [2]. The deflector guarantees a stable bubble size irrespective of the formation frequency, and, therefore, the dead time remains approximately constant at a fixed system pressure P_s. This is impressively shown by the data for water in Fig. 6. The variation of P_s due to the variation in the surface tension of the studied surfactant solution or additional dynamic effects in the short time range is taken into account *via* the Poiseuille equation. The two lines 1 and 2 were calculated from the system pressure according to the relationship

$$t_d = t_d^* \frac{P_s^*}{P_s},$$
(25)

where the asterisks denote the corresponding parameters in the critical point of the $P_s(L)$ dependence. Therefore, the dead time in the critical point t_d^* (shown by

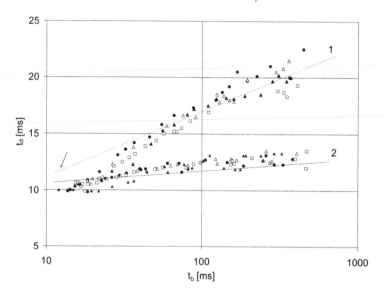

Figure 6. The dependence of the dead time t_d on the bubble time $t_b = t_d + t_1$ for water (below) and $C_{12}DMPO$ solution (above) for the capillary radius 0.85 mm and measuring system volume $V_s = 1.5$ ($\bullet$), 3.7 ($\square$), 4.5 ($\triangle$) and 20.5 ml ($\blacktriangle$). Straight lines 1 ($C_{12}DMPO$ solution) and 2 (water) were calculated from Eq. (3), according to [60].

an arrow in Fig. 6) for the capillary employed is 11.0 and 11.5 ms for water and $C_{12}DMPO$ solution, respectively. The lines, calculated from Eq. (25), are satisfactory agreement with the measured values. In addition, the results indicate that the minimum dead time measured from flow rate changes in the BPA (i.e., the t_d^* value in the critical point of the $P_s(L)$ dependence) is about 1 ms lower than that calculated from the assumption $V_s = $ const. For the two liquids (water and $C_{12}DMPO$ solution) this value is 10 ms. This difference results from the hydrodynamic effects occurring at high bubble formation frequencies and is considered in the measurement procedure of the BPA-1P [60]. As the lifetime in the critical point is taken to be zero, this error in the determination of the dead time would lead to a larger relative error in t_1 and amounts to approximately 10% in the millisecond time range [8]. The relative error in the determination of the time t_1 in the range 20 to 100 ms would not exceed 10%, and decreases with increasing t_b down to 5%. Thus, the explicit lifetime measurement *via* flow rate changes decreases the standard error of t_1 measurements to half as compared to other procedures.

There is also another procedure for the life time determinations below 10 ms (see Section 4.3): the experimental dependence of $L(t_1)$ in the surface lifetime range 10–50 ms is approximated by a polynomial and extrapolated up to zero time value. This procedure (also implemented in the software of the BPA-1S tensiometer) strongly decreases the standard error of t_1 measurements (maximum 5–10%).

The critical flow rate L_c is determined by the capillary parameters, i.e., the longer the capillary and the smaller its diameter, the lower is the gas flow in the critical point. Therefore, for a fixed capillary radius one can control the critical gas flow rate L_c (and, therefore, the dead time) by varying the capillary length l [61]. However, for wide capillaries the respective capillary length can be quite long, so that this procedure becomes impracticable for the control of the critical flow. This obstacle could be overcome by using combined narrow and wide capillaries, arranged by a wide capillary immersed into the liquid and connected to a narrow capillary, or by decreasing the cross-section of a wide capillary in its upper part by a special insert.

The critical gas flow L_c can be readily calculated from the aerodynamic resistance of the capillary. The following relationship for a capillary with arbitrary geometry follows from the Poiseuille law (see Eq. (17)):

$$\frac{P_c}{L_c k_P} = \text{const} = K. \tag{26}$$

The constant $k_P = 8\eta l / \pi r_0^4$ is the Poiseuille coefficient which depends on the gas viscosity η, the capillary length l and radius r_0. The coefficient k_P can be determined experimentally when the capillary is not immersed into the liquid, and is given by $k_P = dP/dL$. The theoretical value of the dimensionless constant $K > 1$ is in agreement with experimental findings ($K \cong 1.6$) [61].

As the dead time is determined by V_b and L_c, it is possible to control t_d for any capillary diameter. This can be achieved by varying the critical flow rate L_c which depends on the aerodynamic resistance of the capillary, and also by the value of V_b which in turn depends on the position of the bubble deflector. Note that the shorter the dead time, the lower is the initial adsorption at the bubble surface, and consequently the lower is the surface lifetimes available for the studies [2, 61].

It was demonstrated experimentally [61] that, under identical conditions (capillary diameter, surface tension), the elongation of the capillary, introduction of an additional resistance and formation of smaller bubble volumes V_b lead to smaller the value for L_c. In contrast to static conditions, i.e., when bubbles separate due to buoyancy forces, under dynamic conditions the pressure exerted onto the bubble by the gas inflow from the capillary plays a significant role. In this case, the higher the pressure created by the gas flow, the smaller is the bubble size. However, if the gas flow velocity is too high, the bubble can exceed a critical size, determined by the balance between buoyancy and capillary forces. For water and a capillary diameter of 0.25 mm this critical size is 5.7 mm^3. This is caused by the continuous gas inflow into the bubble from the measuring system during the separation of the bubble from the capillary [61]. In summary, the variation of the capillary geometry can help to achieve optimum characteristics of the bubble formation process, such as critical gas flow, bubble volume and dead time.

3. *Effect of the System Volume on Measured Dynamic Surface Tensions*

The MBPM usually involves the supply of gas into the capillary through a reservoir of respective volume V_s which is much larger than the volume of a single separating bubble V_b. The ratio V_s/V_b varies between 10^2 and 10^4 for different commercial instruments. If $V_s \gg V_b$, the resulting pressure variations are very small, but even small pressure variations can affect significantly the bubble formation regime. When a small volume of gas ΔV is supplied to or withdrawn from the reservoir, the small change in pressure in the reservoir for an adiabatic regime is approximately given by

$$\Delta P_s = \kappa \frac{P_s \Delta V}{V_s}. \tag{27}$$

After bubble detachment the pressure in the reservoir drops and is then restored due to further gas inflow. The pressure in the reservoir is restored simultaneously with the pressure equilibration between the bubble and the reservoir which is determined by hydrodynamic and aerodynamic processes. At a small gas supply rate the equilibrium between the bubble and the reservoir establishes earlier than the pressure restores to the maximum capillary pressure. In this regime the bubble lifetime is determined by the pressure restoration. At a large gas supply rate the pressure restores earlier than the equilibrium between the bubble and the reservoir establishes. The bubble can detach already when a pressure difference still exists between the capillary and the reservoir. This leads to an increased apparent dynamic surface tension.

The apparent dynamic surface tensions of water at 20°C for various volumes of the measuring system are shown in Fig. 7 (data obtained by the tensiometer BPA-1P [8]). To eliminate errors caused by an incorrect capillary radius and bubble non-sphericity, a calibration with respect to water can be performed, using the known reference surface tension (72.75 mN/m at 20°C).

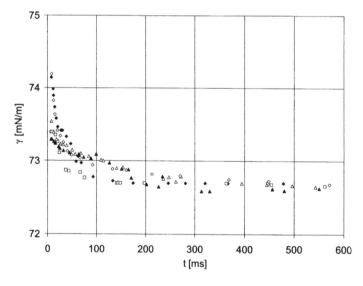

Figure 7. The dependence of apparent dynamic surface tension of water at 20°C and measuring system volume $V_s = 1.5$ (◆), 3.7 (□), 4.5 (△) and 20.5 ml (▲). (◇) the results obtained with capillary radius 0.125 mm and $V_s = 4$ ml.

One can see that in the surface lifetime beyond 300 ms, the measurement results made for different system volumes reproduce the reference value for water within 0.1 mN/m. In the BPA, the calibration is performed in a time range of 0.5 to 1 s. To eliminate aerodynamic resistance effects, the dependence of the excess surface tension correction term $\Delta\gamma_a(t_1)$ (see Eq. (5)) in the range $t_1 < 1$ s can be approximated by a fourth order polynomial.

In the short time range $t_1 < 100$ ms the apparent surface tension is higher than the reference value. This is due to the hydrodynamic effects caused by the rapid bubble growth and has to be corrected [2, 7]. It should be noted that for a small system volume ($V_s = 1.5$ cm^3) the increase in the apparent surface tension value is significantly higher. This additional effect is predicted by the theory [7] where the use of small volume reservoirs is considered.

The effect of the system volume V_s on the dynamic surface tension of C$_{12}$DMPO solutions is illustrated in Fig. 8. The results are in agreement with the predictions of the theoretical model [7, 29]: the lower the system volume, the higher is the apparent dynamic surface tension in the millisecond time range, because an excess pressure in the system is needed to generate bubbles at a given frequency. Only at very long times (10 s and more; not shown in Fig. 8), i.e., in the vicinity of equilibrium, one can expect overlap of all dependencies for different system volumes. We can see that for a system volume $V_s = 1.5$ cm^3 the error in the measured γ values, as compared with the value for $V_s = 20.5$ cm^3, is between 5 and 10%. At the same time, the error in the dynamic surface tensions measurement for $V_s = 20.5$ cm^3 are only 1–2% lower than those for $V_s = 4.5$ cm^3.

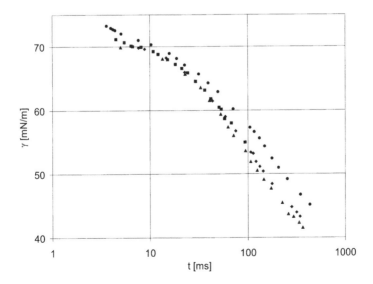

Figure 8. The dependence of measured dynamic surface tension of C$_{12}$DMPO solution at 20°C and measuring system volume $V_s = 1.5$ (●), 3.7 (■), 4.5 (◆) and 20.5 ml (▲).

If the reservoir volume is very large then the pressure decrease after bubble detachment is very small (Eq. (27)). The pressure deficit can be so small that it is compensated by the inertia of gas and liquid, i.e., due to inertia the bubble can overcome the maximum capillary pressure when the pressure is not yet restored completely. So, a bubble can detach a short time interval after the previous bubble. Analogously the next bubbles can also detach by this mechanism. However, after separation of each bubble, the pressure within the reservoir decreases by ΔP_s because of a non-complete pressure restoration. Finally the moment is reached when the pressure decrease in the reservoir becomes significant enough, preventing the next bubble from overcoming the maximum bubble pressure barrier. The next bubble grows slowly with a pressure restoration to the value which will be sufficient for overcoming the maximum capillary pressure without the inertia effect.

On the basis of these results we can make a rational choice of the measuring system volume which ensures precise measurement of both the lifetime and surface tension. Clearly, systems with a volume less than 4 ml (i.e., $V_s/V_b < 2000$) cannot ensure a precise measurement of the maximum pressure (i.e., surface tension) sufficient for scientific investigations and most technologic purposes. Therefore, to increase the measurement accuracy one should employ systems with a larger gas volume. On the other hand, if the system volume is very high (say, $V_s/V_b > 10^4$), undesirable hydrodynamic effects and technical problems arise. One of the major negative effects for large V_s is the formation of bubble series [7]. This formation of bubble series is known to introduce significant errors in the determination of surface tension and lifetime. The results of measurements for a capillary of radius 0.125 mm, immersed into water, and a system volume of about 40 cm^3, are shown in Fig. 9. The formation of series of two subsequent bubbles is clearly visible. An-

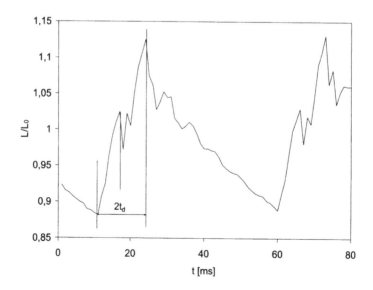

Figure 9. Example of the series consisting of two bubbles detected from the gas flow changes.

other drawback related to systems of very large volume is the increased pressure relaxation time in the system upon changing the gas flow rate. For example, a system of 40 cm^3 volume requires 15–20 s for pressure stabilisation after a change of the flow rate. This makes the measurement procedure rather time consuming. Also much more foam could be formed in such cases, which prevents the use of the method for small sample volumes, e.g., in medical applications [48].

Therefore, the optimum system volume, if all factors are taken into account, is that yielding a ratio of V_s/V_b in the range between 2000 and 5000. As for such volumes the use of pressure changes for the determination of lifetime has large errors, the use of gas flow changes are superior and should be analysed instead [60].

4. Influence of the Liquid Viscosity on Surface Tension Measurements

Some tensiometers (for example MPT2 and BPA) have the option to correct for the viscosity effect of the studied liquid. The viscosity is considered to be known for the studied system. In this case, the measuring program employs a procedure which accounts for an additional resistance at short bubble lifetimes (for example BPA, see Eq. (5)) as [2]

$$\Delta \gamma_v = (-\Psi - \phi \ln t_l) \frac{P}{P_{cal}} \ln \mu, \qquad (28)$$

where Ψ and ϕ are constants which depend on the capillary radius and their values decrease with increasing radius. P_{cal} is the calibration excess pressure for water, and μ is the viscosity of the liquid. As shown experimentally the dead time t_d measured here by the BPA tensiometer decreases almost linearly with the non-corrected dynamic surface tension γ_m (or the pressure in the measuring system) [73]. The additional procedure implemented in BPA tensiometer, is based on a correlation between the viscosity and dead time values [73]

$$t_d = A - \alpha \gamma, \qquad (29)$$

where A and α are constants characteristic for each liquid. It is important that the slope of the t_d vs surface tension curve depends on the viscosity of the liquid. Some examples of such dependencies are shown in Fig. 10. The higher the liquid's viscosity, the larger is the slope of the dependence $t_d(\gamma)$. This fact can be explained by an increasing bubble volume separating from the capillary, because the increase in viscosity leads to a retardation of the bubble separation process.

For viscosities 2 mm^2/s $< \mu <$ 100 mm^2/s, the dependence of α on μ is almost linear [73]. The dependencies of α on the viscosity of the liquid for solutions of various surfactants (water and aqueous glycerine solutions used as solvents) are shown in Fig. 11. Therefore, we obtain

$$\alpha = B + \beta \mu, \qquad (30)$$

where B and β are constants. If the viscosity is in the given range, the user does not need to input the viscosity value along with other parameters of the studied

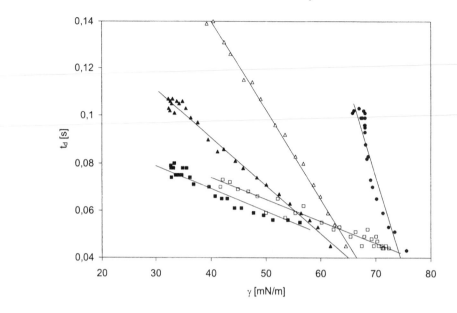

Figure 10. Dead time as a function of surface tension for the Triton X-100 (2 mmol/l) and $C_{12}EO_6$ (0.3 mmol/l) aqueous solutions (■ and □, respectively); the Triton X-100 (2 mmol/l) solutions in 50% and 70% solutions of glycerine in water (▲ and △, respectively), and 80% glycerine aqueous solution (●); the viscosity values were 1 mm^2/s (■ and □), 5, 23 and 52 mm^2/s (▲, △ and ●).

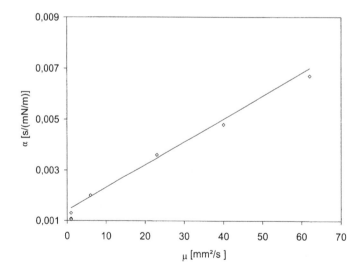

Figure 11. Dependence of the coefficient α on the liquid viscosity μ; points are experimental data; line corresponds to Eq. (30).

system because the software determines it automatically (BPA-1S). As an example, non-corrected (open circles) and corrected (filled circles) surface tensions for a 2 mmol/l Triton X100 in a 50% glycerine solution are shown in Fig. 12. The vis-

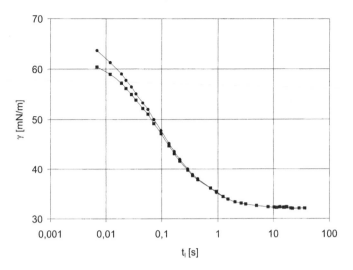

Figure 12. Non-corrected (●) and corrected (■) dependence of dynamic surface tension of 2 mmol/l Triton X-100 in a 50% glycerine solution.

cosity obtained from Eq. (30) is $\mu = 6.1$ mm^2/s, instead of 6.3 mm^2/s as measured directly for the water/glycerine mixture. Both effects, $\Delta\gamma_a$ and $\Delta\gamma_v$ (see Eq. (5)), tend to zero at lifetimes $t_1 > 0.5$ s.

5. Correction of the Hydrodynamic Pressure

In his experiments Simon used a single capillary and hence the immersion depth of the capillary into the liquid had to be measured accurately to take the hydrostatic pressure into consideration [1]. This immersion depth does not need to be measured if two capillaries (narrow and wide) are used. This idea, proposed by Sugden [9], was further developed by other authors leading to various modifications of the MBPM, based on the measurement of the immersion depth difference for two capillaries with different diameters in the same liquid, or two identical capillaries in different liquids, etc. [10–14]. Due to our opinion, set-ups with two capillaries can only be used successfully for pure liquids or highly concentrated solutions (cf., for example, [15, 16]). For typical surfactant solutions, however, due to the dynamic character of the surface tension, significant errors can occur due to the fact that surface tensions in bubbles of different size are different. It is very difficult to fulfil the condition $\gamma = $ const, as it is insufficient to ensure identical lifetimes for the two compared bubbles [72].

For the Sugden method with two capillaries a set of two equations of type (4) has to be considered simultaneously, which after some rearrangements yields:

$$P_{s1} - P_{s2} - (P_{d1} - P_{d2}) = 2\gamma/f_1 r_1 - 2\gamma/f_2 r_2. \qquad (31)$$

The subscripts 1 and 2 refer to one or the other capillary ($r_2 > r_1$). Equation (31) can be transformed into a principal equation for the Sugden method [9] only if the aerodynamic and viscous resistances of the two capillaries are identical (i.e.,

$P_{d1} - P_{d2} = 0$, which is true only if the bubble formation is very slow), and also if $f_1 = f_2 = 1$ (i.e., for spherical bubbles):

$$\gamma = \frac{P_{s1} - P_{s2}}{2(1/r_1 - 1/r_2)}. \tag{32}$$

As the sphericity factors f_i are present in Eq. (3), it is impossible to express explicitly the surface tension *via* the difference of maximum pressures and the difference of the inverse capillary radii even under static experimental conditions. As the radius of the wider capillary is 1 mm or more, the error caused by neglecting the factor f_2 in Eq. (32) becomes too high. Therefore, it was already proposed by Sugden [62] that a calibration should be performed with respect to a liquid of known surface tension, which introduces the calibration factor C and an additional empiric coefficient k_2 (which depends on the radius r_2 and approximately accounts for the sphericity factor f_2) in Eq. (32). The Sugden equation which involves these empiric corrections reads [62]:

$$\gamma = C(P_{s1} - P_{s2})\left[1 + \frac{k_2}{P_{s1} - P_{s2}}\right]. \tag{33}$$

It was argued in [74] that the error in the surface tension values introduced by Eq. (33) for pure liquids measured under static experimental conditions is less than 0.5%. At high bubble formation frequencies the dynamic effects were shown to be quite significant. Nevertheless, it seems to be possible to accurately correct the dynamic effects, i.e., to include the dependence of $P_{d1} - P_{d2}$ on the liquid viscosity, capillary radii and bubble lifetime into Eq. (31).

Another factor which affects significantly the experimental accuracy of the Sugden method is the difference between the dead times for the two capillaries of different diameters. Two measurement options have been proposed for using Sugden's method (see, for example, [75]): a fixed bubble formation frequency in both capillaries, and a fixed surface lifetime in both capillaries. However, such experimental protocols cannot simultaneously control the dead time, i.e., the influence of dead time on the results is neglected.

The dependence of the dynamic surface tension on the bubble lifetime t_1 is illustrated in Fig. 13 for aqueous solutions of Triton X-100 (Sigma) at two concentrations (0.48 and 1.2 mmol/l) and measured with two capillaries at different dead times [72]. For the concentrated Triton X-100 solution one can obtain equal dynamic surface tensions only if the dead times of the two capillaries are identical. A significant difference between the dynamic surface tension values in the short time range was also observed when the bubbles formation time t_b were equal for the both capillaries (i.e., for identical bubble frequencies). The results shown in Fig. 13 are re-plotted in Fig. 14 as a function of bubble time t_b. It can be seen that the surface tensions for the wider capillary are higher than those for the narrow capillary by up to 5 mN/m. As the dynamic surface tensions in the short time range for the two capillaries with different dead time values differ significantly, it is impossible to simplify Eq. (31) into Eq. (32) or (33), because the condition

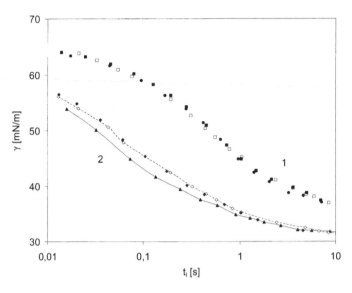

Figure 13. Dynamic surface tension for Triton X-100 solutions as a function of bubble lifetime t_l for the concentrations of: (1) — 0.48 mmol/l; and (2) — 1.2 mmol/l. Open points: experimental results for the capillary 0.24 mm in diameter; filled points: experimental results for the capillary 0.46 mm in diameter. The dead time values were: 40 ms ($\square$, $\diamondsuit$, $\blacksquare$, $\blacklozenge$ and dotted line); and 110 ms ($\blacktriangle$, $\bullet$ and solid line).

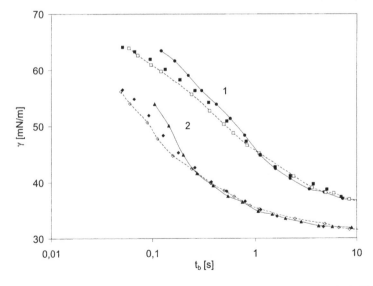

Figure 14. Dynamic surface tension for Triton X-100 solutions as a function of the bubbles time t_b; notations as in Fig. 13.

$\gamma = $ const does not hold. To obtain reliable results for surfactant solution in the short time range or in the regime of high bubble formation frequencies, one has also to ensure that the dead time values for the two capillaries are equal to each other. This has to be fulfilled as extra requirement to the experimental procedure of the Sugden method, despite the usual condition of identical bubble lifetimes or bubble formation frequencies in the two capillaries. It seems rather unrealistic that all conditions discussed above can be fulfilled at the same time, hence, the Sugden methods appears not suitable for studies of surfactant solutions.

An approximation of the hydrostatic pressure, without any measurements of the capillary immersion depth, can be performed using a single capillary. Such a procedure is implemented for example in the SITA tensiometer T60 [63]. The low V_s / V_b ratio and the large radius of separating bubbles ensure that the pressure in the minimum point of the $P_s(t)$ curve is approximately equal to the hydrostatic pressure.

In the LAUDA tensiometer MPT2 the existing procedure of determination of the critical point P_c in the $P_s(L)$ dependence can be used for the correction of the hydrostatic pressure. Examples of such dependencies for water and a $C_{12}DMPO$ solution, obtained for various values of the hydrostatic pressure (100–500 Pa) are shown in [8].

It was demonstrated that P_c for any given hydrostatic pressure is exactly equal to the pressure in the measuring system throughout the whole range of the bubble regime ($P < P_c$) for solvents with the same hydrostatic pressure. Obviously, the accuracy of the calculation of P_h with this method depends on the accuracy of the determination of P_c. For diluted solutions, where the slope of the $P_s(L)$ curve at $P < P_c$ changes significantly, such problem does not exist. For higher concentrated solutions one can improve the accuracy of the determination of P_c *via* the analysis of the gas flow changes, as described above. When the critical pressure (i.e., transfer from single bubble to the gas jet regime) is passed, a sharp decrease is observed in the flow rate amplitude, and the flow rate changes become irregular. Consequently the standard deviation of the determination of t_b becomes 100% and more, instead of 10–15%. In addition to the correction of the hydrostatic pressure, this method allows to compensate small excess pressure caused by other reasons.

Precise measurements of dynamic surface tension require accurate measurements of the hydrostatic pressure, i.e., of the capillary immersion depth. It should be noted that for narrow capillaries an even rather low accuracy of this procedure (say, to within 0.3 mm) results in surface tension errors less than 0.5%. In the BPA single capillaries are employed. To exactly account for the hydrostatic pressure P_H, the liquid/gas interface is automatically detected *via* the pressure jump appearing when the capillary touches the liquid surface. Subsequently, the capillary is immersed into the liquid to a depth chosen by the user (2 to 20 mm), with an accuracy of ±0.05 mm.

6. Limits of Applicability of the Maximum Bubble Pressure Method

The time range available by the MBPM for measuring dynamic surface tensions is mainly determined by the capillary parameters and the size of the separating bubble. As discussed above, the dependence of the excess pressure P_s on the gas flow L splits into two regimes (cf. Fig. 15) — the gas jet flow regime for flow

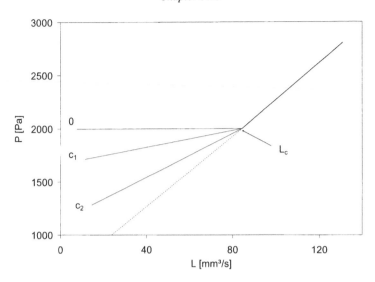

Figure 15. Schematic dependence of pressure in the measuring system on the gas flow through the capillary for pure solvent (0) and for surfactant solutions of different concentrations, $c_1 < c_2$.

rates above the critical point $L > L_c$ (indicated by an arrow), and the single bubble regime for $L < L_c$. In the critical point bubbles are formed continuously and the interval between bubbles t_b is exactly equal to the dead time t_d.

The slope of the line in the jet regime range is determined by the aerodynamic resistance of the capillary k_P (26):

$$\frac{dP_s}{dL} = k_P \equiv \frac{8\eta l}{\pi r_0^4}. \tag{34}$$

The dependence of P_s on L for any solvent in the bubble regime is parallel to the abscissa axis. The higher is the surfactant concentration in the solution ($c_2 > c_1$), the higher is the slope of the dependence $P_s(L)$ in the bubble region $L < L_c$. Clearly the slope cannot exceed k_P, because the gas flow velocity along the capillary cannot exceed that predicted by the Poiseuille law (34). Therefore, expressing the kinetics of adsorption at a growing bubble *via* the derivative dP_s/dL, one can determine the values of the capillary parameters for which the surface tension γ of a given solution can be studied. The excess pressure in the system can be approximated by $P_s = 2\gamma/r_0$, while the air flow in the capillary is given by $L = V_b/t_b$. The derivative dP_s/dL ($V_b = \text{const}$ and $t_d = \text{const}$) caused by adsorption (index a) is given by

$$\left(\frac{dP_s}{dL}\right)_a = -\frac{2t_b^2}{r_0 V_b}\left(\frac{d\gamma}{dt_1}\right) = \frac{2t_b^2}{r_0 V_b}\left(\frac{d\Pi}{dt_1}\right), \tag{35}$$

where $\Pi = \gamma_0 - \gamma$ is the surface pressure of the surfactant solution, and γ_0 is the surface tension of the pure solvent. As mentioned above, the relation between any derivatives dP_s/dL and $(dP_s/dL)_a$ holds

$$\frac{dP_s}{dL} \geqslant \left(\frac{dP_s}{dL}\right)_a \tag{36}$$

which, together with Eq. (35), yields the requirement for $(d\Pi/dt_1)_a$ for a studied solution:

$$\left(\frac{d\Pi}{dt_1}\right)_a \leqslant \frac{k_P r_0 V_b}{2t_b^2}. \tag{37}$$

Therefore, only capillaries with high enough aerodynamic resistance (the constant k_p) are suitable for the analysis of adsorption processes characterised by high values of the derivative $(d\Pi/dt_1)_a$. Eq. (17) can be specified if we assume a diffusion controlled adsorption kinetics for the surfactant. For very short lifetimes, $t_d \gg t_1$, a criterion was derived in [61] which allows to estimate the potential capacity of instruments worker with the principle of MBPM:

$$\left(\frac{d\Pi}{dt_1}\right)_a \leqslant \frac{\gamma_0 t_d}{1.6(t_1 + t_d)^2}. \tag{38}$$

For a diffusion controlled adsorption mechanism and a capillary with $t_d = 100$ ms solutions with concentrations $c < 1$ mmol/l can be studied in a time interval $t_1 > 3$ ms. If a capillary with $t_d = 10$ ms is used, we can investigate solutions up to concentration of 5 mmol/l [61]. Clearly, these estimates are only the lower concentration limits, especially when the adsorption is not diffusion controlled. For micellar solutions and solutions of ionic surfactants the adsorption rate is lower than predicted by the diffusion model [5]. Therefore, the MBPM can usually be applied here to higher concentrations.

F. Examples of Experimental Results

1. Comparison of MBPM with Other Methods

There are several publications which show the comparison of results of dynamic surface tension measurements for surfactant solutions obtained by MBPM with those obtained by other methods, such as Wilhelmy plate, dynamic capillary, inclined plate, strip, drop volume and oscillating jet (see, for example, [2, 76]). Good agreement is found when the data are represented as the function of the effective life time. In a number of studies an implicit comparison of data can be performed by comparing the diffusion coefficients or adsorption–desorption rate constants calculated from the experimental data. In these cases a satisfactory agreement was also obtained [2]. The dynamic surface tensions for Triton X-100 solutions of various concentrations measured by MBP, oscillating jet and inclined plate methods were summarised in [76]. The last two methods have been described in detail elsewhere

[77, 78] and have partly a common time window with the MBPM. The capillary used in the MBPM studies was wide and short enough ($r_0 = 0.0088$ cm, $l = 1.5$ cm) to prevent liquid inflow after bubble separation ($r_0^2/l > 5 \cdot 10^{-5}$ m) and fast pressure oscillation in the capillary. It was shown in [76] that good agreement exists between the experimental methods. Recently, it was demonstrated the in particular drop and bubble profile analysis and maximum bubble pressure tensiometry provide experimental data that nicely complement each other and represent therefore a perfect combination of experimental methods which cover finally a large time interval of more than eight orders of magnitude [79].

2. Comparison of Dynamic Surface Tensions Measured by Different MBP Tensiometers

There are various commercial instruments on the market and some of them have been tested with the same surfactant solutions in order to compare the results and analyse the differences observed [59]. In Fig. 16 the dynamic surface tensions are shown for a 10^{-4} mol/l $C_{12}EO_6$ solution. The values obtained with the tensiometer T60/2 are lower than those measured by the BP2, and both sets of data are essentially lower than those measured by the MPT2. We believe that the main cause of the inconsistency between the results obtained by various devices is the difference between the geometric characteristics of the capillaries and measuring systems, which leads to the differences in the dead time (and hence, the bubble volume) values characteristic for each device.

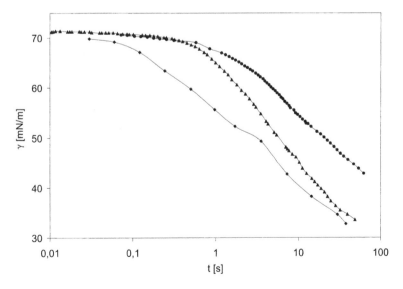

Figure 16. Dependence of dynamic surface tension in the $C_{12}EO_6$ solution with concentration 10^{-4} mol/l on the life time measured by the T60 (▲), BP2 (◆) and MPT2 (●) tensiometers; according to [59].

Of all the tensiometers studied in [59], only the MPT2 instrument fulfilled the condition $V_b \ll V_s(P_s/P_{atm})$, which corresponds to the situation when the gas mass excess in the system is large enough to enable a spontaneous (controlled only by the capillary resistance, nd not by the gas supply system, such as a compressor) formation of the bubble during the dead time period. Therefore, the bubble growth during the dead time stage in MPT2 takes place at almost constant pressure in the measuring system, i.e., both at the beginning and the end of the bubble life the pressure in the system is almost constant, which prevents the liquid to penetrate into the capillary. Possible reasons of these discrepancies are the differences in the bubble formation mechanisms for different devices, which are determined by the ratio of the gas mass excess in the measuring system to the mass of the gas contained in the bubble, $M = (V_s/V_b)(P_s/P_a)$. Both the increase in V_b and the decrease in P_s lead to a decrease of M. This is just the case for the BP2 and, especially, for the T60, with $M = 7$–10, and $M \cong 1.0$, respectively [59]. This ratio should be larger than 10, and for scientific applications in the short time range at least 40–50.

Figure 17 presents the dynamic surface tensions measured for various Triton X-100 concentrations using MPT2 (detection of t_b and recalculation of t_l *via* the critical point in the $P(L)$ dependence) and BPA-1P (direct detection of t_l *via* the gas flow oscillations $L(t)$) [8]. The γ values obtained with the MPT2 having a system volume of about 40 ml and a capillary of radius $r_0 = 0.125$ mm, are in good agreement with those obtained by the BPA having a system volume $V_s = 4.5$ ml and the same capillary radius $r_0 = 0.125$ mm. The data measured with the BPA tensiometer and the wider capillary ($r_0 = 0.25$ mm) are essentially lower [59]. For

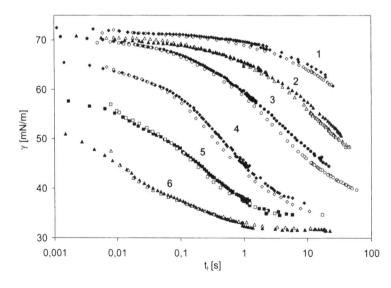

Figure 17. The dependence of dynamic surface tension of Triton X-100 solution at temperature 22–23°C for different concentration: 0.05 (1), 0.1 (2), 0.2 (3), 0.5 (4), 1.0 (5) and 2.0 mmol/l (6); open symbols — BPA, closed symbols — MPT2.

the BPA tensiometer and the narrow capillary the value of $M = (V_s/V_b)(P_s/P_a)$ is between 30 and 40, while for the wider capillary this ratio is about 10.

3. Dynamic Surface Tension of Surfactants Mixtures

In practice surfactants are typically mixtures of components of different surface activity. To reach special effects mixtures are even often used on purpose. This makes a deeper understanding of the adsorption process from mixed surfactant solutions necessary. Figure 18 shows the dynamic surface tensions of mixtured solutions of the alkyl dimethyl phosphine oxides $C_{10}DMPO$ and $C_{14}DMPO$ (for three mixtures of the two surfactants) measured with the maximum bubble pressure tensiometer BPA and profile analysis tensiometer PAT1 (both SINTERFACE Technologies, Germany) [51].

As expected the data of the two experimental methods complement each other adequately. The shown theoretical curves were calculated for a diffusion controlled adsorption kinetics for surfactant mixtures discussed in [51] with diffusion coefficients D between $1 \cdot 10^{-10}$ m^2/s and $3 \cdot 10^{-10}$ m^2/s (see legend to the figure). The experiments are quantitatively described by the diffusion model. It was stated in [51] that the thermodynamic model and the diffusion controlled adsorption kinetics quantitatively describes the adsorption layers of surfactant mixtures of the two homologous decyl and tetradecyl dimethyl phosphine oxides. This represents the basis for understanding practical surfactants, which due to their degree of purity are typically surfactant mixtures.

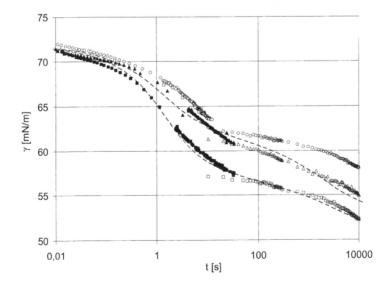

Figure 18. Dynamic surface tensions of a mixture of $C_{10}DMPO$ and $C_{14}DMPO$; concentration ratio $c_{10}/c_{14} = 10^{-7}$ mol/cm^3/10^{-9} mol/cm^3 (O, ●), 10^{-7} mol/cm^3/$3 \cdot 10^{-9}$ mol/cm^3 (△, ▲), $2 \cdot 10^{-7}$ mol/cm^3/$3 \cdot 10^{-9}$ mol/cm^3 (□, ■); dotted lines — calculated for $D = 1 \cdot 10^{-10}$ (1), $3 \cdot 10^{-10}$ (2), $2 \cdot 10^{-10}$ (3) m^2/s using the surfactant parameters given in [51], open symbols — BPA1, closed symbols — PAT1, according to [51].

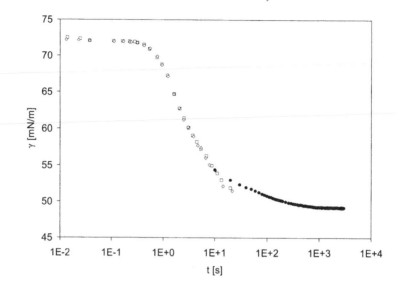

Figure 19. Dynamic surface tension of β-casein solutions with the concentration $5 \cdot 10^{-6}$ mol/l, measured by the drop shape method (●) and maximum bubble pressure method ($\Diamond$, $\Box$).

4. Dynamic Surface Tension of β-Casein at the Solution/Air Interface

The direct measurement of protein adsorption kinetics at liquid interfaces is a quite difficult experimental problem. Measurements of the dynamic surface tension are technically most simple to perform. A number of results obtained by various tensiometric measurements made with protein (and, in particular, β-casein) solutions are referred to in a recent review [80].

The dynamic surface tension in the time range of 10 to 30 000 s as measured by PAT1 for a $5 \cdot 10^{-6}$ mol/l β-casein solution are shown in Fig. 19. The drops were formed at the tip of a PTFE capillary immersed into a cuvette filled with a water-saturated atmosphere. The drop volume was kept constant over the experimental time (about 16 mm^3). The dynamic surface tension in the complementary time range of 10 ms to 50 s was measured by the BPA equipped with a steel capillary of 0.5 mm internal diameter. One can see that the γ values obtained with the PAT1 tensiometer are in a good agreement with the data obtained by the BPA tensiometer.

The diffusion controlled adsorption kinetics of proteins at liquid interface as proposed in [81], is based on the simultaneous solution of the Ward–Tordai equation and the respective set of equations (adsorption isotherm, surface layer equation of state, function of adsorption distribution over the states with different molar areas, etc.) which follows from the theory of protein adsorption [82]. The effective diffusion coefficient, obtained from a best fit of the experimental data, is in good agreement with the values obtained from direct studies of the adsorption dynamics using radiotracer techniques and ellipsometry.

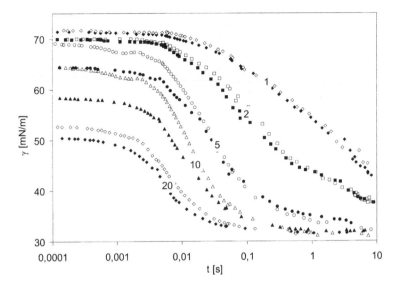

Figure 20. Dynamic surface tension for Triton X-100 solutions as a function of the effective surface lifetime; open symbols — experimental results for dead time 4–5 ms; filled symbols — experimental results for dead time 8–9 ms; the labels correspond to the surfactant concentrations given as multiples of the CMC.

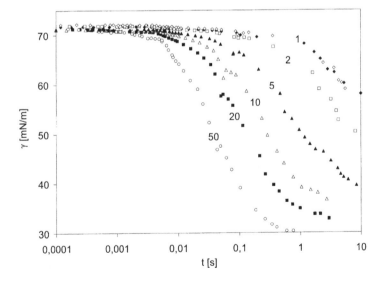

Figure 21. Dynamic surface tension for Triton X-45 solutions as a function of the effective surface lifetime; labels correspond to the surfactant concentrations given as multiples of the CMC.

5. Dynamic Surface Tension of Micellar Solutions

The Figs 20–22 show the dynamic surface tensions (measured with the BPA-1S tensiometer) of micellar solutions of Triton X-100, Triton X-45 and $C_{14}EO_8$ as

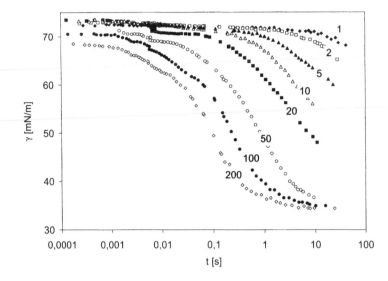

Figure 22. Dynamic surface tension for $C_{14}EO_8$ solutions as a function of the effective surface lifetime; labels correspond to the surfactant concentrations given as multiples of the CMC.

functions of the effective lifetime [53]. The CMC values of these surfactants are 0.24, 0.1 and 0.011 mmol/l (at 25°C), respectively [83]. The maximum studied concentrations of these surfactants are 4.8, 5.0 and 2.2 mmol/l for Triton X-100, Triton X-45 and $C_{14}EO_8$, respectively.

The dynamic surface tension of Triton X-100 solutions was measured at two dead time values: 8–9 ms (filled points in Fig. 20) and 4–5 ms (open points). For Triton X-45 and $C_{14}EO_8$ solutions the dead time was 6–8 ms. It is seen from Fig. 20 that the lower the dead time is, the higher is the measured surface tension in the short time range. The difference is especially evident for concentrated solutions.

The effective dead time was estimated and compared with the effective bubble surface lifetime in [53]. The deformation (expansion) of the bubble during the experiment affects the adsorption dynamics and hence the dynamic surface tensions. To account for this effect, the so called effective dead time is introduced, which corresponds to an identical bubble surface load at a non-deformed surface. For spherical bubbles, before they come into contact with the deflector, it holds $t_{d(ef)} = (3/7)t_d$ [2, 5]. The presence of the deflector leads to a flattening of the bubble, for which at constant system pressure we get $t_{d(ef)} = 0.2t_d$. The asymmetric shape of the bubble during its separation (cf. video images shown in [2]) leads to an additional deformation of the bubble. The expansion is especially pronounced in the area of the bubble that provided the initial surface of the subsequent bubble. When we approximately these asymmetry effect we finally get $t_{d(ef)} = 0.1t_d$. Therefore, the effective dead time in the BPA-1S device is approximately one order of magnitude lower than the physical dead time, i.e., the dead time of 5–8 ms corresponds to an effective dead time of about 0.5–0.8 ms. Hence, the condition $t_{ef} < t_{d(ef)}$ to be obeyed in order to obtain reliable results of surface tension measurements, is satisfied in the BPA-1S rigorously only for $t = 1$ ms. Results obtained at shorter times are accurate only for low-concentrated solutions, when the surface tension decrease is comparatively small. For higher surfactant concentrations the dynamic surface tensions could be underestimated due to a significant surfactant adsorption during the dead time stage. Thus, these results are of interest only for comparative studies.

The results shown in Figs 20–22 are in agreement with the analysis given above for the effective and real dead times. It is seen that for solutions of Triton X-100, which has the highest CMC among the three studied surfactants here.

The surface tension at 10 CMC and 20 CMC and at $t < 1$ ms is essentially lower than the surface tension of the pure solvent (water), and depends only slightly on the surface lifetime, caused by the significant adsorption of Triton X 100 during the dead time period. For lower Triton X 100 concentrations, and for the other surfactants studied (characterised by significantly lower CMCs), also for $t < 1$ ms a decrease in surface tension is observed.

The effect of micelle dissolution on the adsorption dynamics of micellar solutions was theoretically analysed in a various papers [84–92]. All relevant theories involve at least two stages: a fast process governed by the separation of monomers from the micelle, and a slow process corresponding to the complete dissolution of micelles.

For the analyse of the results of Figs 20–22, we used a linearised theoretical model proposed in [92]. The general expression for the dynamic adsorption $\Gamma(t)$ for a fast micelle dissolution at short times and under the assumption $k_f \cdot t \ll 1$ reads:

$$\Gamma = 2c_k \left(\frac{Dt}{\pi}\right)^{1/2} \left[1 + \frac{2}{3}\left(\frac{c_0 - c_k}{c_k}\right)k_f t\right]. \tag{39}$$

In this equation c_0 is the total concentration of the surfactant, D is the diffusion coefficient of surfactant monomers, $c_k = $ CMC and k_f is the fast micelle dissolution constant. Assuming that the adsorbed surface layer is not fully packed, i.e., the dynamic surface pressure is proportional to the adsorption Γ, $\Pi = \gamma_0 - \gamma = RT\Gamma$, Eq. (39) yields:

$$\Pi = 2RTc_k \left(\frac{Dt}{\pi}\right)^{1/2} \left[1 + \frac{2}{3}\left(\frac{c_0 - c_k}{c_k}\right)k_f t\right], \tag{40}$$

where T is the temperature and R is the gas law constant. For the fast micelle dissolution ($k_f \cdot t > 1$) a limiting equation for the adsorption dynamics and dynamic surface pressure was derived in [92]:

$$\Pi = RT\Gamma = 2T Rc_0 \left(\frac{Dt}{\pi}\right)^{1/2}. \tag{41}$$

Calculations with Eq. (41) require only one parameter — the diffusion coefficient, the value of which was obtained by best fit of the equation to experimental dynamic surface tensions [53]. To perform calculations with Eq. (40), one extra parameter is necessary — the rate constant of fast micelle dissolution k_f. Studies of the fast relaxation process performed by various authors using ultrasonic and thermal relaxation, stopped flow and other techniques were summarised in [93]. The expression used to calculate the relaxation time of the fast micelles dissolution process has the form:

$$\tau_1^{-1} = k_f\left[q + p\frac{c_0 - c_k}{c_k}\right]. \tag{42}$$

The parameters p and q are constants specific for each surfactant. A value of $p = 1$ was assumed in [93] and the respective values for q were found in the range between 0 and 1 for all surfactants studied, while the constant k_f increased with increasing CMC. More than 30 surfactants were studied in [93] (cf. open points in Fig. 23) and for all these substances the dependence of k_f on CMC plotted on a logarithmic scale can be well described by a linear dependence.

To obtain the best agreement between Eq. (40) and the experimental results, the optimum values for k_f for Triton X-100, Triton X-45 and $C_{14}EO_8$ were 525, 85 and 20 s^{-1}, respectively (cf. closed symbols in Fig. 23). The agreement with the results obtained earlier for other surfactants is rather good. The time range in which Eq. (40) is applicable for Triton X-100, Triton X-45 and $C_{14}EO_8$ is below 0.002, 0.012 and 0.05 s, respectively. At the same time, Eq. (41) is valid in the time range above $t > 1/k_f$. It was shown in [53] that the theoretical calculations oincide quite well for t $\approx$ $1/k_f$, i.e., the curves obtained from Eq. (41) agree reasonably well with experimental data of Figs 20–22 in the time range for $t > 1/k_f$, and in turn

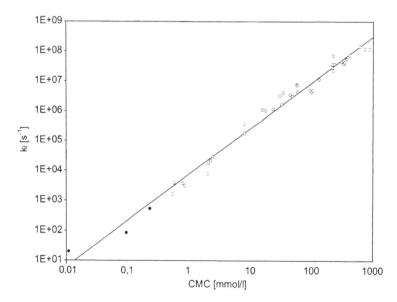

Figure 23. Fast micelle dissolution rate constant k_f on the critical micellization concentration (CMC) for various surfactants; open symbols — data from [93]; filled symbols — data from [53].

the values calculated from Eq. (40) agree with experimental data in the short time range.

The maximum bubble pressure method was recently used in [94] to measure the dynamic surface tensions of Tritons X-45, X-100, X-165 and X-405 micellar solutions. The experimental data were analysed using the theoretical model based on the numerical solution of Fick's equations for surfactant adsorption [95], and the expression proposed by Joos [87] for the calculation of an effective surfactant diffusion coefficient D^* of monomers in micellar solutions:

$$D^* = D(1 + \beta)(1 + \alpha\beta), \qquad (43)$$

where $\alpha = D_m/D \approx 0.25$ [84], $\beta = (c_0 - c_k)/c_k$, D_m is the diffusion coefficients of surfactant micelles. For micellar solutions we have to assume that $c_0 = \text{CMC}$ for all $c_0 > \text{CMC}$, and instead of the value of D for monomers we should use in the theoretical model [95] an effective diffusion coefficient D^* for the monomers.

As it was shown by Danov *et al.* [91], Eq. (43) (similar to Eq. (41)) describes the diffusion processes in micellar solutions for a time range exceeding the relaxation time of fast micelles dissociation. In [94] it was found that the theoretical calculations for micellar Tritons X-45 and X-100 solutions agree very well with the experimental data. At the same time, for micellar Tritons X-165 and X-405 solutions only a qualitative agreement exists between the theoretical values and experimental dependencies, because the Tritons are mixtures of homologues with different oxyethylation degree [94].

G. Conclusions

Significant progress was made in MBPM in the millisecond time range. This progress is based on the analysis of gas flow rather than pressure oscillations in the measuring system applied for the quantitative determination of all phases of bubble formation. The design and operational principles of different tensiometers are described here, and an analysis of the influence of the main parameters of the measuring system (capillary radius, measuring system volume and V_s/V_b ratio) on the measurement precision of the bubble lifetime and dynamic surface tension is presented. In all cases, the optimum procedure found is to employ sufficiently narrow capillaries (with a radius <0.15 mm) and measuring systems with a ratio V_s/V_b in the range between 2000 and 5000. The errors arising for wide capillaries (radius up to 1 mm) and by lifetime measurements based on the analysis of pressure oscillations are examined and it is shown that wide capillaries are not acceptable due to uncontrolled errors in the dynamic surface tension measurements (of the order of 10% and more). Time detection *via* pressure oscillations, for narrow capillaries, provides quite acceptable surface tension results for technical applications (accuracy within 5%) in a surface lifetime range $t_l \geqslant 10$ ms. The analysis of various methods used to account for the hydrostatic pressure (two capillaries method, minimum system pressure method or jet regime critical pressure shift method) show that for studies of surfactant solutions with narrow capillaries best results can be obtained when the capillary immersion depth is directly measured.

The procedures used in most recently developed maximum bubble pressure tensiometers (BPA series) allow to significantly decrease the minimum analysed bubble surface lifetime from 10 ms down to 0.1 ms. The influence of the physical and effective dead time during bubble formation on the measured surface tension is analysed. It is shown that for physical dead times of about 5–10 ms, the effective dead time in the BPA-1S does not exceed 1 ms, which makes studies in the millisecond and sub-millisecond time range possible. The dynamic surface tensions of aqueous solutions of Triton X-100, Triton X-45 and $C_{14}EO_8$ in the concentration range between CMC up to 200 times CMC were measured at surface lifetimes between 0.1 ms and about 50 s. The results of theoretical calculations made by using the fast micelle dissolution model agree quite well with the experimental data.

H. List of Symbols

$a = (2\gamma/\Delta\rho g)^{1/2}$ — capillary constant

c_0 — bulk concentration

c_k — critical micelle concentration (CMC)

D — diffusion coefficient

f — correction factor in the Laplace equation

g — gravity constant

H — capillary immersion depth

h — penetration depth of liquid into the capillary

k_P — Poiseuille equation constant

k_s — dilational surface viscosity

k_f — fast micelle dissociation constant

L — gas flow rate

l — capillary length

P — capillary pressure

P_{atm} — atmospheric pressure

P_b — excess pressure in the bubble

P_d — excess dynamic pressure

P_H — hydrostatic pressure

P_s — excess pressure in the measuring system

R — gas constant

r — current value of bubble radius

r_b — radius of a separating bubble

r_0 — capillary radius

T — temperature

t — time

t_b — time interval between successive bubbles

t_d — dead time

t_{ef} — effective adsorption time

t_h — hydrodynamic relaxation time

t_l — lifetime

V_b — bubble volume

V_s — volume of measuring system

Γ — adsorption

γ — surface tension

γ_0 — surface tension of the solvent

γ_∞ — equilibrium surface tension of a solution

η — gas dynamic viscosity

κ — the adiabatic constant

μ — dynamic viscosity of liquid

ν — kinematic viscosity of gas

Π — surface pressure

ρ — gas density

ρ_L — liquid density

τ — characteristic time of pressure oscillations

ω_0 — characteristic frequency of pressure oscillations

I. References

1. M. Simon, Ann. Chim. Phys., 32 (1851) 5.
2. V.B. Fainerman and R. Miller, The maximum bubble pressure technique, monograph in "Drops and Bubbles in Interfacial Science", in "Studies of Interface Science", Vol. 6, D. Möbius and R. Miller (Eds), Elsevier, Amsterdam, 1998, p. 279–326.
3. J. Eastoe and J.S. Dalton, Adv. Colloid Interface Sci., 85 (2000) 103.
4. K.J. Mysels, Colloid Surfaces, 43 (1990) 241.
5. R. Miller, P. Joos and V.B. Fainerman, Adv. Colloid Interface Sci., 49 (1994) 249.
6. A.I. Rusanov and V.A. Prokhorov, Interfacial Tensiometry, in "Studies in Interface Science", Vol. 3, D. Möbius and R. Miller (Eds), Elsevier, Amsterdam, 1996.
7. V.I. Kovalchuk and S.S. Dukhin, Colloids Surfaces A, 192 (2001) 131.
8. V.B. Fainerman and R. Miller, Adv. Colloid Interface Sci., 108–109 (2004) 287–301.
9. S. Sugden, J. Chem. Soc., 121 (1922) 858.
10. F.M. Jaeger, K. Ned. Akad. Wet. Versl. Gewone Vergad. Atd. Natuurkd., 23 (1914) 330; Z. Anorg. Chem., 101 (1917) 1.
11. E.L. Warren, Philos. Mag., 4 (1927) 358.
12. P.P. Pugachevich, Zh. Fiz. Khim., 38 (1964) 758.
13. P.T. Belov, Zh. Fiz. Khim., 55 (1981) 302.
14. K. Lunkenheimer, R. Miller and J. Becht, Colloid Polymer Sci., 260 (1982) 1145.
15. R. Razouk and D. Walmsley, J. Colloid Interface Sci., 47 (1974) 515.
16. J.L. Ross, W.D. Bruce and W.S. Janna, Langmuir, 8 (1992) 2644.
17. R. Miller, V.B. Fainerman, K.-H. Schano, W. Heyer, A. Hofmann and R. Hartmann, Labor Praxis, N8 (1994).
18. E.N. Stasiuk and L.L. Schramm, Colloid Polymer Sci., 278 (2000) 1172.
19. G. Biesmans, L. Colman and R. Vandensande, J. Colloid Interface Sci., 199 (1998) 140.
20. S.S. Dukhin, V.B. Fainerman and R. Miller, Colloids Surfaces A, 114 (1996) 61.
21. V.I. Kovalchuk, S.S. Dukhin, V.B. Fainerman and R. Miller, J. Colloid Interface Sci., 197 (1998) 383.
22. V.I. Kovalchuk, S.S. Dukhin, A.V. Makievski, V.B. Fainerman and R. Miller, J. Colloid Interface Sci., 198 (1998) 191.
23. S.S. Dukhin, N.A. Mishchuk, V.B. Fainerman and R. Miller, Colloids Surfaces A, 138 (1998) 51.
24. S.S. Dukhin, V.I. Kovalchuk, V.B. Fainerman and R. Miller, Colloids Surfaces A, 141 (1998) 253.
25. V.I. Kovalchuk, V.B. Fainerman, R. Miller and S.S. Dukhin, Colloids Surfaces A, 143 (1998) 381.
26. V.I. Kovalchuk, S.S. Dukhin, V.B. Fainerman and R. Miller, Colloids Surfaces A, 151 (1999) 525.
27. N.A. Mishchuk, V.B. Fainerman, V.I. Kovalchuk, R. Miller and S.S. Dukhin, Colloids Surfaces A, 175 (2000) 207.
28. S.V. Lylyk, A.V. Makievski, V.I. Kovalchuk, K.-H. Schano, V.B. Fainerman and R. Miller, Colloids Surfaces A, 135 (1998) 27.
29. N.A. Mishchuk, S.S. Dukhin, V.B. Fainerman, V.I. Kovalchuk and R. Miller, Colloids Surfaces A, 192 (2001) 157.
30. R. Kuffner, M.T. Bush and L.J. Bircher, J. Am. Chem. Soc., 79 (1957)
31. A.M. Kragh, Trans. Faraday Soc., 60 (1964) 225.
32. M. Austin, B.B. Bright and E.A. Simpson, J. Colloid Interface Sci., 23 (1967) 108.
33. J. Kloubek, Tenside, 5 (1968) 317.
34. R.L. Bendure, J. Colloid Interface Sci., 35 (1971) 238.
35. T.E. Miller and W.C. Meyer, American Laboratory, (1984) 91.

36. X.Y. Hua and M.J. Rosen, J. Colloid Interface Sci., 124 (1988) 652.
37. S.G. Woolfrey, G.M. Banzon and M.J. Groves, J. Colloid Interface Sci., 112 (1986) 583.
38. P.R. Garrett and D.R. Ward, J. Colloid Interface Sci., 132 (1989) 475.
39. D.E. Hirt, R.K. Prud'homme, B. Miller and L. Rebenfeld, Colloids Surfaces, 44 (1990) 101.
40. K.J. Mysels, Langmuir, 2 (1986) 428; 5 (1989) 442.
41. C.D. Dushkin, I.B. Ivanov and P.A. Kralchevsky, Colloids Surfaces, 60 (1991) 235.
42. C.P. Hallowell and D.E. Hirt, J. Colloid Interface Sci., 168 (1994) 281.
43. T.H. Iliev and C.D. Dushkin, Colloid Polymer Sci., 270 (1992) 370.
44. B.V. Zhmud, F. Tilberg and J. Kizling, Langmuir, 16 (2000) 7685.
45. J. Eastoe, J.S. Dalton and R.K. Heenan, Langmuir, 14 (1998) 5719.
46. E. Alami and K. Holmberg, J. Colloid Interface Sci., 239 (2001) 230.
47. K. Theander and R.J. Pugh, J. Colloid Interface Sci., 239 (2001) 209.
48. V.N. Kazakov, O.V. Sinyachenko, V.B. Fainerman, U. Pison and R. Miller, Dynamic Surface Tensiometry in Medicine, in "Studies in Interface Science", Vol. 8, D. Möbius and R. Miller (Eds), Elsevier, Amsterdam, 2000.
49. R. Crooks, J. Cooper-Whitez and D.V. Boger, Chem. Eng. Sci., 56 (2001) 5575.
50. Ch. Frese, S. Ruppert, M. Sugár, H. Schmidt-Lewerkühne, K.P. Wittern, V.B. Fainerman, R. Eggers and R. Miller, J. Colloid Interface Sci., 267 (2003) 475–482.
51. R. Miller, A.V. Makievski, C. Frese, J. Krägel, E.V. Aksenenko and V.B. Fainerman, Tenside Surfactants Detergents, 40 (2003) 256–259.
52. Ch. Frese, S. Ruppert, H. Schmidt-Lewerkühne, K.P. Wittern, R. Eggers, V.B. Fainerman and R. Miller, Colloids Surfaces A, 239 (2004) 33–40.
53. V.B. Fainerman, V.D. Mys, A.V. Makievski, J.T. Petkov and R. Miller, J. Colloid Interface Sci., 302 (2006) 40–46.
54. R. Jiang, J. Zhao and Y. Ma, Colloids Surfaces A, 289 (2006) 233.
55. R. Jiang, Y. Ma and J. Zhao, J. Colloid Interface Sci., 297 (2006) 412.
56. M.S. Kalekar and S.S. Bhagwat, J. Dispersion Science and Technology, 27 (2006) 1027.
57. J. Gao, J. Chai, J. Xu, G. Li and G. Zhang, J. Dispersion Science and Technology, 27 (2006) 1059.
58. E.P. Calogianni, E.M. Varka, T.D. Karapantsios and S. Pegiadou, Langmuir, 22 (2006) 46.
59. J. Meissner, J. Krägel, C. Frese, S. Rupert, V.B. Fainerman, A.V. Makievski and R. Miller, SÖFW-Journal (English Version), 130 (2004) 41–46.
60. V.B. Fainerman, A.V. Makievski and R. Miller, Rev. Sci. Instruments, 75 (2004) 213–221.
61. V.B. Fainerman, V.N. Kazakov, S.V. Lylyk, A.V. Makievski and R. Miller, Colloids Surfaces A, 250 (2004) 97–102.
62. S. Sugden, J. Chem. Soc., 125 (1924) 27.
63. Patent DE 197 55 291 C1, 1997; Patent DE 199 33 631 A1, 1999.
64. G.S. Keen and J.R. Blake, J. Colloid Interface Sci., 180 (1996) 625.
65. R.L. Kao, D.A. Edwards, D.T. Wasan and E. Chen, J. Colloid Interface Sci., 148 (1992) 247.
66. V.B. Fainerman, A.V. Makievski and R. Miller, Colloids Surfaces A, 75 (1993) 229.
67. V.B. Fainerman, Kolloidn. Zh., 41 (1979) 111.
68. V.B. Fainerman, Kolloidn. Zh., 52 (1990) 921.
69. J. Kloubek, J. Colloid Interface Sci., 41 (1972) 7.
70. V.B. Fainerman, Colloids Surfaces, 62 (1992) 333.
71. V.B. Fainerman and R. Miller, J. Colloid Interface Sci., 175 (1995) 118–121.
72. V.B. Fainerman, V.D. Mys, A.V. Makievski and R. Miller, J. Colloid Interface Sci., 304 (2006) 222–225.
73. V.B. Fainerman, V.D. Mys, A.V. Makievski and R. Miller, Langmuir, 20 (2004) 1721–1723.

74. J.L. Ross, W.D. Bruce and W.S. Janna, Langmuir, 8 (1992) 2644.

75. Patent USA, number 6085577, Jul. 11, 2000.

76. V.B. Fainerman, R. Miller and P. Joos, Colloid Polymer Sci., 272 (1994) 731.

77. P. van den Bogaert and P. Joos, J. Phys. Chem., 83 (1979) 2244.

78. R. Defay and G. Petré, in "Surface and Colloid Science", Vol. 3, E. Matijevic (Ed.), Wiley, New York, 1971.

79. R. Miller, C. Olak and A.V. Makievski, SÖFW (English version), (2004) 2–10.

80. R. Miller, V.B. Fainerman, A.V. Makievski, J. Krägel, D.O. Grigoriev, V.N. Kazakov and O.V. Sinyachenko, Adv. Colloid Interface Sci., 86 (2000) 39.

81. R. Miller, V.B. Fainerman, E.V. Aksenenko, M.E. Leser and M. Michel, Langmuir, 20 (2004) 771–777.

82. V.B. Fainerman, E.H. Lucassen-Reynders and R. Miller, Adv. Colloid Interface Sci., 106 (2003) 237–259.

83. V.B. Fainerman, R. Miller, E.V. Aksenenko and A.V. Makievski, in "Surfactants — Chemistry, Interfacial Properties and Application", Studies in Interface Science, Vol. 13, V.B. Fainerman, D. Möbius and R. Miller (Eds), Elsevier, 2001, pp. 189–286.

84. J. Lucassen, Faraday Discuss. Chem. Soc., 59 (1975) 76–87.

85. K.D. Danov, P.M. Vlahovska, T. Horosov, C.D. Dushkin, P.A. Kralchevsky, A. Mehreteab and G. Brose, J. Colloid Interface Sci., 183 (1996) 223–235.

86. K.D. Danov, D.S. Valkovska and P.A. Kralchevsky, J. Colloid Interface Sci., 251 (2002) 18–25.

87. P. Joos, Dynamic Surface Phenomena, VSP, Utrecht, 1999.

88. B.A. Noskov, Adv. Colloid Interface Sci., 95 (2002) 237–293.

89. B.A. Noskov and D.O. Grigoriev, in "Surfactants — Chemistry, Interfacial Properties and Application", Studies in Interface Science, Vol. 13, V.B. Fainerman, D. Möbius and R. Miller (Eds), Elsevier, 2001, p. 401–509.

90. K.D. Danov, P.A. Kralchevsky, N.D. Denkov, K.P. Ananthapadmanabhan and A. Lips, Adv. Colloid Interface Sci., 119 (2006) 1–16.

91. K.D. Danov, P.A. Kralchevsky, N.D. Denkov, K.P. Ananthapadmanabhan and A. Lips, Adv. Colloid Interface Sci., 119 (2006) 17–33.

92. V.B. Fainerman, Kolloidn. Zh., 43 (1981) 94–100.

93. V.B. Fainerman, Kolloidn. Zh., 43 (1981) 926–933.

94. V.B. Fainerman, A.V. Mys, E.V. Aksenenko, A.V. Makievski, J.T. Petkov, J. Yorke and R. Miller, Colloids Surfaces A, 334 (2009) 22–27.

95. V.B. Fainerman, S.V. Lylyk, E.V. Aksenenko, L. Liggieri, A.V. Makievski, J.T. Petkov, J. Yorke and R. Miller, Colloids Surfaces A, 334 (2009) 8–15.

Drop Volume Tensiometry

A. Javadi [a], V.B. Fainerman [b] and R. Miller [a]

[a] Max-Planck-Institute of Colloids and Interfaces, 14424 Potsdam/Golm, Germany
[b] Donetsk Medical University, 83003 Donetsk, Ukraine

Contents

A. Historical Survey

There are numerous methods for measuring the surface tension of a liquid or the interfacial tension between two liquids. The drop volume or weight method is one of the standard technique [1] due to its simplicity and the advantage of being applicable to both liquid/gas and liquid/liquid interfaces. One of the strongest limitations is the short time range in which it provides experimental data.

    This method was derived from the so-called stalagmometer method, where only the number of drops is counted and compared to a known liquid. Traube proposed this method for determining the composition of liquids, for example the contents of

Bubble and Drop Interfaces
© Koninklijke Brill NV, Leiden, 2011

ethanol in alcoholic beverages [2]. The fundamentals of the drop volume method were developed at the beginning of the last century by Theodor Lohnstein [3, 4]. Since this time several instruments were developed [5–11] and some of them are fully automatic commercial ones available on the market.

The drop volume or drop weight method was possibly first used by the pharmacist Tate in 1864 [12], who used the number of drops as a measure of a certain liquid volume in order to dose liquid medicine. As result he formulated several laws, among which the proportionality between drop volume V, the capillary radius r_{cap} and the capillary constant is most important for our target here. Tate postulated that the weight W of a drop detaching from a capillary of given size is proportional to the product of capillary radius r_{cap} and surface tension γ, known now as the law of Tate:

$$W = 2\pi r_{cap} \gamma. \tag{1}$$

This law has been used for a long time to determine the surface tension of liquids despite the fact that it is only a rough estimation. Especially Lohnstein [3, 4] criticised it and made a series of calculations to establish a basis for an accurate theory.

Lord Rayleigh [13] has already suggested that the basic Eq. (1) has to be correction by a factor f in order to obtain a more accurate relationship

$$W = 2\pi r_{cap} \gamma f. \tag{2}$$

The coefficient f is a function of r_{cap} and the capillary constant given by

$$a = \sqrt{2\gamma / \Delta\rho g}. \tag{3}$$

Starting from here, Lohnstein developed a theory which consequently leads to a systematic improvement of the experimental technique.

The stalagmometer, still used in some laboratories, represents the simplest first version of drop volume tensiometry. Its principle consists in counting the number n of drops formed from a definite volume of a liquid at the tip of a capillary as compared to the number of drops n_s of a calibration liquid of known surface tension γ_s. The surface tension γ of the liquid under study is then obtained from a simple relationship

$$\gamma = \frac{n_s \gamma_s \rho_s}{n \rho}. \tag{4}$$

where ρ and ρ_s are the respective densities.

The drop volume tensiometry, in contrast, does not work in a relative way but determined accurately the volume of a drop formed at the tip of a given capillary. The measuring procedure is realised by means of a precise dosing system which allows a continuous formation of drops at a capillary. Figure 1 shows the subsequent stages of this process. At stage 6 the drop becomes unstable and detaches. Once the detachment process is completed, the entire process of drop formation starts again. Due to the force balance between the acceleration due to gravity and interfacial

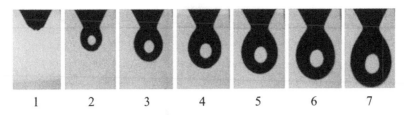

1   2   3   4   5   6   7

Figure 1. Images of subsequent stages of a drop formation and detachment.

tension, the critical drop volume correlates directly with the interfacial tension and the density difference $\Delta\rho$ of the two adjacent phases

$$2\pi r_{cap}\gamma \sim V\Delta\rho g. \tag{5}$$

The term $2\pi r_{cap}$ represents the circumference of the capillary where the interfacial tension γ acts and counterbalances the gravitational force $V\Delta\rho g = W_{cr}$.

As the drop does not detach directly from the capillary tip but necks and breaks off such that a remnant drop remains, Eq. (5) requires this correction factor f:

$$\gamma = \frac{\Delta\rho g V}{2\pi r_{cap} f}. \tag{6}$$

The development from a simple stalagmometer to the drop volume tensiometer of today was made more than 100 years ago by Lohnstein [3]. He performed calculations of the volume of detaching and residual drops as a function of the capillary radius and the capillary constant and improved the law of Tate. First attempts of automation applied a light barrier for counting the drops, while now even the special developments in individual labs are computer driven.

Further improvement of the theoretical foundation of the drop volume method has been proposed recently [14]. In addition to the development of the two-stage drop formation model, computational fluid dynamics simulations have been considered as new tool to explain the complexity of this method. This work was also supported by the availability of fast video technique, as demonstrated in [15].

B. The Drop Volume Experiment

A simple drop volume instrument can be set up in any laboratory easily. The only thing needed is an accurate syringe equipped with a needle of a certain diameter and design. The syringe as dosing system has to be mounted vertically, and the experiment can be started right away (Fig. 2).

From knowledge of the volume of a drop detaching from the tip of the needle, i.e. from reading the syringe the surface tension of the liquid under study can be calculated. Commercial instruments do the reading of the drop volume and calculated the respective surface tension *via* Eq. (6) automatically.

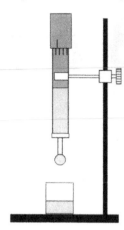

Figure 2. Design of a manual drop volume method.

1. Experimental Scheme

There is a wide variety of drop volume tensiometers described in literature. The important part of any instrument is an accurate volume determination, typically arranged by a dosing system of constant liquid flow and an accurate timer. From the dosing rate Q and the time of drop formation t_d, the drop volume $V = Qt$ can be calculated accurately.

Figure 3 shows the principle design of a drop volume tensiometer. The dosing system equipped with the syringe allows for a constant and accurate delivery of the liquid to form continuously drops at the tip of a capillary. A light barrier detects the detachment of each drop, and in this way determines the drop time, i.e. the time between two drops. Drop time multiplied by the dosing rate leads to the drop volumes. When the drop volume is measured for different dosing rates, the surface tension is obtained as a function of drop time. Further below we will discuss the recalculation of drop time into effective adsorption time in order to get the dynamic surface tension as a function of effective surface age.

The scheme shown in Fig. 3 corresponds to the DVA1 of SINTERFACE Technologies, however, the designs of other companies are similar and all are driven by a PC *via* a respective electronic interface. Although various commercial instruments of high quality are available on the market, new laboratory set-ups for special purposes are still constructed [11, 15–17].

2. Measuring Procedures

Three different measurement modes can be employed with the drop volume method which can yield different data. However, when all peculiarities of each measuring procedure are considered, the different types of results should finally be transferred into the same dynamic surface tensions.

The dynamic version of the drop volume method is the classical procedure for measuring the surface and interfacial tensions of pure liquids. It consists of continu-

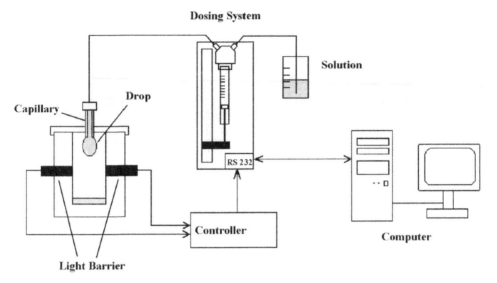

Figure 3. Principle of an automated drop volume instrument; according to the DVA1 of SINTERFACE Technologies, Germany.

ously growing drops at the tip of a capillary by means of an accurate dosing system, which eventually detach and a new drops starts to grow. The interfacial tension is calculated from the averaged volume measured for several subsequent drops formed at the same dosing rate. Note, the first two drops should be discarded in the calculation of the average volume as their volumes could be affected by uncontrolled conditions at the start of the measurement.

For pure liquids, the measured interfacial tension should be independent of the drop formation time. Nevertheless, at small drop times a systematic increase in the obtained interfacial tensions is observed, caused by the dynamic conditions of the measuring procedure. This hydrodynamic effect must be corrected, as will be shown further below.

Some drop volume tensiometers provide also so-called static and quasi-static measuring mode. Due to our knowledge the quasi-static measuring procedure was first applied by Addison and co-workers [18] and later used by other authors [5, 19, 20] to determine the adsorption kinetics of surfactants without the need of considering an area change. While the static mode is suitable only for studies of surfactant solutions, the quasi-static mode can also be applied to pure solvents. The principle of both modes is based on the adsorption process of surfactants at the interface leading to a decrease in interfacial tension. In both modes a drop is formed such that its volume is small enough to be kept stable with the actual surface tension. While keeping now the drop volume constant (static mode) the interfacial tension decreases due to increasing adsorption. After a certain period of time the drop volume will reach a critical value at which it detaches. If the pre-set initial drop volume was too small it will never detach. Thus, a certain strategy and experience is needed

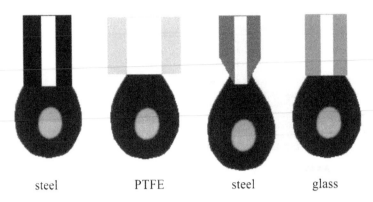

steel   PTFE   steel   glass

Figure 4. Wetting conditions at the tip of capillaries with different shape.

to perform such experiments manually, however some instrument software has this procedure implemented. In the quasi-static mode, after a certain time interval has passed and the drop is not yet detached, the drop is increased until detachment.

In both modes the drop is initially quickly formed and therefore the lifetime of the drop corresponds approximately to the effective age of the interface.

The major part in drop volume experiments is the capillary. The capillaries of cylindrical shape proved to be most suitable. Depending on the material the liquid will wet the capillary such that the drop sits either at the inner or outer circumference of the capillary. In these two cases the outer or the inner diameter, respectively, is taken for as parameter $2r_{cap}$ (Fig. 4).

For unclear wetting conditions, as shown in Fig. 4 with the two left capillaries, an uncontrolled formation of the drop can happen and the effective radius r_{cap} for the drop detachment is unknown. Therefore, capillaries with a conical shape should better be avoided.

3. Correction Factors and Optimisation of the Experiments

It was shown in Eq. (6) that the shape and size of a drop is controlled by the capillary radius r_{cap}, the drop volume V, the interfacial tension γ, the difference $\Delta\rho$ between the densities of the two fluids, and the gravitational constant g. For the calculation of the interfacial tension from a measured drop volume an additional correction factor f is needed. From the Gauss–Laplace-Equation of capillarity it becomes evident that this correction factor depends on the capillary constant a given in Eq. (3).

The first who calculated the correction factors f as a function of a and r_{cap} was Lohnstein [3, 4]. He considered that the residual drop after detachment has the same radius of curvature as the drop before detachment. Later Freud and Harkins [21], Hartland and Srinivasan [22] and Harkins and Brown [23] improved the accuracy of Lohnstein's calculations.

The different approaches were discussed then by Wilkinson [24] who published the correction factors in the form $r_{cap}/a = f(r_{cap}/V^{1/3})$ by the following polynomial:

$$r_{cap}/a = z(A + z(B + z(C + zD))) + E \tag{7}$$

with $z = r_{cap}/V^{1/3}$, and the coefficients A to E with the following values: $A = 0.50832$, $B = 1.5257$, $C = -1.2462$, $D = 0.60642$, $E = -0.0115$. From Eqs (3) and (7) the interfacial tension can be calculated:

$$\gamma = a^2 \Delta \rho g / 2 \tag{8}$$

as typically done by the software of commercial drop volume instruments: from the averaged drop volume V first the value $z = r_{cap}/V^{1/3}$ is calculated and then *via* Eq. (7) the value of r_{cap}/a. As r_{cap} is known the value for a and finally *via* Eq. (8) the value of γ is obtained.

Regarding the accuracy of drop volume measurements Earnshaw et al. [25] presented an error analysis, taking into account various uncertainties while neglecting the covalence terms,

$$\left(\frac{\sigma_\gamma}{\gamma}\right)^2 = \left(\frac{\sigma_{\Delta\rho}}{\Delta\rho}\right)^2 + \left(\frac{\sigma_{r_{cap}}}{r_{cap}}\right)^2 \left[1 + \frac{r_{cap}}{V^{1/3}}\frac{f'}{f}\right]^2$$
$$+ \left(\frac{\sigma_V}{V}\right)^2 \left[1 + \frac{r_{cap}}{3V^{1/3}}\frac{f'}{f}\right]^2, \tag{9}$$

where σ_i are the uncertainties on the parameters $\gamma, \Delta\rho, r_{cap}$ and V, respectively. The experimental uncertainties can be given easily, however, the second and third terms on the right-hand side contain the correction factor f and its derivative f'. Both quantities are functions of $r_{cap}/V^{1/3}$ and exist only as experimental data or numerical values calculated from theories. Thus the overall accuracy mainly depends on the right choice of the radius r_{cap} such that the total relative uncertainty is minimum. Earnshaw's was based on the experimental data of Harkins and Brown [23] and Wilkinson [24] and the theoretical calculations of Lohnstein [3, 4] and Hartland and Srinivasan [22]. From Eq. (9) we can estimate that the highest accuracy can be reached for $\frac{r_{cap}}{V^{1/3}}\frac{f'}{f} < 0$, i.e. for $\frac{r_{cap}}{V^{1/3}} < 0.85$. This result is different from the one of Harkins and Brown, who suggested $0.6 < \frac{r_{cap}}{V^{1/3}} < 1.2$. The optimal range of the method can be established by using very small tip radii, however, a limit given by the accuracy of volume measurements.

4. Peculiarities

The drop volume method has a lot of advantages in comparison with other surface tension methods, most of all due to its easy handling and the reliable temperature control. Its accuracy is comparatively high, especially when applied to liquid/liquid interfaces. Disturbing wetting effects, as it is observed in the ring or plate tensiometry, are also much less important.

As a large disadvantage we have to note that equilibrium interfacial tension values cannot be measured for low surfactant concentrations when the adsorption slow and takes hours [26]. Also, at small drop times the so-called hydrodynamic effects have to be taken into consideration, which requires additional experiments. This procedure is explained in detail in the subsequent paragraph. Another disadvantage are the irregularities in drop formation at very high drop formation rates, as shown further below.

C. Effect of Hydrodynamics on Measured Drop Volumes

The drop volume method is mainly used due to its dynamic character which gives access to rather short adsorption times. If used, however, at drop formation times of less than 10 s, the so-called hydrodynamic effects come into play. The larger the dosing rates, the more important are these effects and therefore have to be considered in the determination of correct dynamic surface tensions [26].

As the first who analysed these effects systematically, Davies and Rideal [27] defined two main effects influencing the drop formation and drop detachment: the so-called "blow-up" and "circular current" effects. While the first effect increases the detaching drop volume and simulates a higher surface tension, the second effect leads to an earlier break-off of the drop and therefore decreases the obtained value of γ. The two processes are effective in different time windows [28]. While the "circular current" leads to an early drop break-off at drop times less than 1 second, the blow-up effect is effective up to 10 seconds of drop formation time (for typical capillaries with a diameter of about 1 mm).

It is clear that the surface or interfacial tension of pure liquids is independent of time, and deviations can be expected only at extremely short times in the order of molecular rearrangements, well below 1 µs. Such short times are not available by any experimental set-up. Thus, drop volume experiments have to yield drop volumes for water constant at any drop formation time. This however is not the case and significant deviations occur as demonstrated in Fig. 5. There is obviously an increase in measured drop volumes and consequently in the calculated apparent surface tension γ_{app} at shorter drop times. This apparent increase is much larger than the experimental accuracy and is more pronounced for larger capillaries.

1. Experimental Observations of Hydrodynamic Effects

To quantify the hydrodynamic effect Kloubek et al. [29] measured the drop volume of pure liquids at small drop formation times and presented an empirical relationship for $V(t)$

$$V(t) = V_e + k/t. \tag{10}$$

The un-effected drop volume V_e can be obtained from fitting experimental data with this equation while K is an empirical constant. As an example data for pure water are given in Fig. 6. The intersection of the line with the ordinate yields the drop

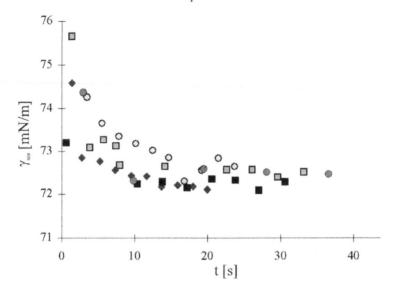

Figure 5. Measured apparent surface tensions γ_{app} of pure water using capillaries with different tip radii; $r_{cap} = 0.254$ mm (■), 0.504 mm (▨), 0.633 mm (◆), 1.051 mm (●), 1.322 mm ( ), according to [48].

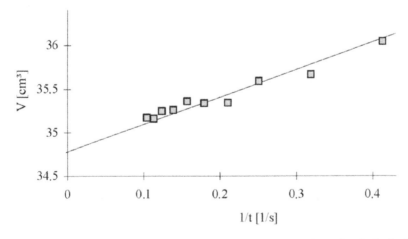

Figure 6. Plot V as a function of $1/t$ according to Kloubek *et al.* [4] to determine the hydrodynamic correction for the drop volume of pure water; $r_{cap} = 1.051$ mm.

volume V_e which corresponds to a volume not effected by the hydrodynamics and k is the slope of the line.

Another result was obtained by Jho and Burke [30]. Based on a large number of experiments with pure liquids they found a dependence of the drop volume on the drop formation time in the following form:

$$V(t) = V_e + kt^{-3/4}, \tag{11}$$

where V_e is again the un-effected drop volume and k is a regression coefficient obtained by a linear regression of the experimental data. They found that k depends on surface tension, density difference, and tip radius, but an effect of the viscosity of the liquid was not obtained. The results were confirmed by similar experiments performed by van Hunsel [31]. In all the works no physical model has been presented to explain the statistically proved relationship.

2. A Simple Correction Model Based on a Drop Detachment Time Concept

Possibly Rao *et al.* [32] were the first who proposed the strategy of a two-stage process for the formation of a drop at the tip of a capillary and its detachment once it has reached a critical size. During the first stage (static stage), the drop is assumed to expand until the drop weight is balanced by the interfacial tension force. During the second stage, when the drop is detaching from the capillary, the drop continues to grow and the volume of a detaching drop was assumed to be always larger than the one proposed by the equilibrium force balance. Scheele and Meister [33–35] and Heertjes *et al.* [36, 37], improved the theoretical basis and modelled the drop formation at low liquid velocities. These new models incorporated essential variables in the second stage, which are the forces acting upon a drop, the way the dispersed phase enters the drop, the necking of the drop as a function of time, and finally the velocity of neck formation. Although, the comparison with a number of experimental data showed good agreement, no efforts for a direct utilizing in drop volume tensiometry was reported. Recently Barhate *et al.* [38] stated that the models proposed in literature for the drop volume of liquid–liquid systems are successful only under restricted conditions, due to the complexity of the drop formation process. They were able to develop a new estimation of the drop detachment time by introducing an analysis of the drag force with internal circulations.

A quantitative analysis of the droplet formation process should actually be based on a solution of the full Navier–Stokes equation at proper boundary conditions. However, the mathematical complexities did not allow quick progress of this problem [39–42].

In their most recent work Yildirim and Basaran [14] considered the effects of flow rate and viscosity, in addition to surface tension, density, and capillary radius, on the drop volume method. They reached a quantitative agreement between experimental and their computational predictions obtained from solving the full Navier–Stokes equations for low to moderate liquid flow rates. The efforts for understanding the different stages and aspects of drop formation process are of continuous interest [43, 44] and modelling of the surface tension forces and the computational reconstruction of liquid interfaces with high accuracy for different conditions is still needed and under investigation [45].

The complete analysis of the formation and detachment process shows that a drop, after it has reached its critical volume and starts to detach, is for a certain time still connected with the liquid flow through a liquid bridge. During this time, necessary for the act of detachment itself, a certain amount of liquid flows addi-

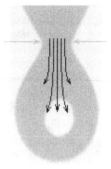

Figure 7. Drop in the moment of detachment (left) and inflow of additional liquid (right).

tionally into the drop. A schematic of the flow pattern of a detaching drop from a capillary is shown in Fig. 7. From this picture it becomes evident that the volume of a detaching drop is always larger than the critical volume. Under the conditions that the process of detachment happens under laminar conditions and neglecting any "circular current" effects, the additional volume should be proportional to the dosing rate. This is also reflected by the empirical formula given in [30].

From the given physical model the drop volume V_a pumped additionally into the detaching drop during the time t_o (detachment time) can be calculated from

$$V_a = Q \times t_o, \tag{12}$$

where Q is the liquid flow rate. From the experimental conditions we obtain

$$Q = V_e/(t - t_o) = V(t)/t. \tag{13}$$

Here $V(t)$ is the drop volume measured at the drop formation time t, and V_e is the undisturbed or corrected volume which has to be used for the calculation of interfacial tensions. Equations (12) and (13) lead to the following final expression [48]

$$V(t) = V_e(1 + t_o/(t - t_o)) = V_e t/(t - t_o). \tag{14}$$

The presented physical picture also holds for surfactant solutions, and consequently the hydrodynamic effect has to be corrected in the same way.

Further above in Fig. 5, the effect of the capillary radius r_{cap} on the measurements of surface tensions of pure water was shown: for a bigger capillary radius a more pronounced apparent increase in surface tension was observed. Hence, the drop detachment time t_o increases with the capillary radius r_{cap}. By fitting the data of Fig. 5 by the relation given in Eq. (14), the drop detachment time is obtained as a function of r_{cap} and a linear regression of these data yields [46]

$$t_o = \alpha + \beta r_{cap}. \tag{15}$$

There is also a small surface tension effect on the parameter t_o, however, this is negligible and for liquids with a viscosity $\mu < 5$ mm^2/s we obtain the following values: $\alpha = 0.008$ s and $\beta = 0.041$ s/cm. Note, the correction for the hydrodynamic

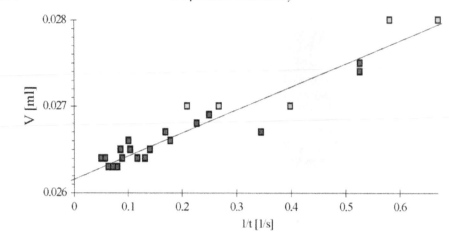

Figure 8. Drop volumes as a function of $1/t$ measured for 75% glycerol in water; symbols refer to two different instruments.

effect on surface tension has not been implemented in any software commercial instruments yet.

The effects of liquid's viscosity on the drop formation process and consequently drop volume tensiometry has been investigated by several authors [47–50], using the two-stage-approach of drop formation. Miller *et al.* proposed an empirical relationships to correct the drop volume for the effect of viscosity. Kaully *et al.* [50] presented a simple easily applicable experimental method for evaluation of surface tension and shape factor of liquids for a wide range of viscosities. Yildirim and Basaran [51] computed the dynamics of formation of drops even for non-Newtonian liquids and indicated that the rheology is described by a six-parameter rate-thinning/rate-thickening constitutive model which is shown to profoundly affect the dynamics of dripping.

As an example the drop volumes measured for water–glycerol mixtures as a function of $1/t$ are depicted in Fig. 8 [47]. At low viscosities the dependence of drop volume on drop time is quite regular and yields drop detachment times comparable to those for water.

With increasing viscosity the hydrodynamic effect seems to oscillate passing through regions of stronger and weaker influence. This tendency was also observed with capillaries of larger and smaller tip radii [47]. This behaviour is not understood so far and requires further detailed investigations.

For liquid/liquid interfaces, the principle situation is the same, however, drop detachment times can be much different.

3. Irregularities in Drop Formation

At drop times shorter than the range of the "blow-up" effect, which in general increases the volume of detaching drops, irregularities can be observed. Careful studies showed that these irregularities are rather regular and highly reproducible.

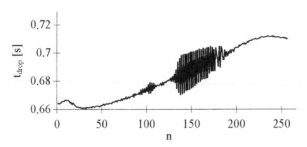

Figure 9. Drop times measured as a function of the drop number n obtained for a continuously decreasing liquid flow rate; $5.10-6$ mol/cm^3 SDS, tip diameter 7.8 mm; according to [52].

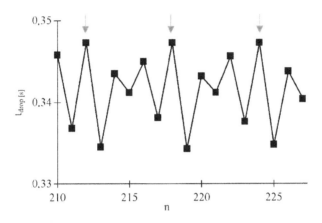

Figure 10. Periodic pattern of measured drop times as a function of drop number; 0.5 g/l Triton X-165, container diameter 60 mm, tip diameter 7.8 mm; according to [52].

These effects could not be studied by any of the available commercial instruments as the volume of single drops is typically not provided but only averaged values. For this reason Fainerman and Miller [52] designed a special apparatus with which single drops at very small drop formation times could be investigated.

In Fig. 9 experiments are shown for a solution of sodium dodecyl sulphate, the possibly most frequently studied surfactant worldwide [53]. On a first glance the results look chaotic, as emphasised above. However, there are ranges of comparatively regular drop time changes between ranges of high "scatter". The analysis of these ranges of high scatter shows that there is a kind of alternation of larger and smaller drops, in some cases even complex pattern can be observed. Such case is shown in Fig. 10, obtained for a Triton X-165 solution. One can see that there is a sequence of smaller and larger drop from number 212 to 217, which is exactly repeated from number 218 to 223, and the same pattern starts again from 224 on. Then it stops and drop volumes are measured following another law. A physical reason for the drop volume bifurcations, as discussed in [52], can be capillary waves generated at the remnant drop and are not sufficiently damped out before the subsequent drop starts to detach. It is important to note here that such regions of drop formation instabilities cannot easily be detected with existing commercial instruments.

D. Theory of Surfactant Adsorption at the Surface of Growing Drops

From drop volume experiments the adsorption kinetics of surfactant solutions can be studied in order to determine their adsorption mechanisms. A quantitative analysis of drop volume tensiometry data requires quantitative theories of the adsorption process at a constant drop surface (quasi-static version) or at the surface of a growing drop. Similar problems are to be solved also for the data interpretation of experiments from bubble and drop pressure tensiometry [28].

The adsorption kinetics of surfactants at the aqueous solution/air or solution/organic solvent interface can be described by the model first derived by Ward and Tordai [54]. This model is based on the assumption that the time dependent interfacial concentration Γ of the adsorbing molecules is governed mainly by their transport from the bulk to the interface. The result of this so-called diffusion controlled adsorption kinetics is given by the following integral equation [54]:

$$\Gamma(t) = \sqrt{\frac{Dt}{\pi}} \left[c_0 \sqrt{t} - \int_0^{\sqrt{t}} c(0, t - \tau) \, d\sqrt{\tau} \right]. \tag{20}$$

Here D is the diffusion coefficient, c_0 is the surfactant bulk concentration, and $c(0, t)$ is the bulk concentration close to the surface, the so-called subsurface concentration. The application of Eq. (20) to dynamic surface tensions $\gamma(t)$ requires numerical calculations [55]. A simpler approximate equation was derived by Sutherland [56], but its range of application is rather restricted [57].

The adsorption kinetics models by Ward and Tordai, and the approximation by Sutherland are only valid for interfaces of constant interfacial area. This condition is given only in the quasi-static mode of drop volume tensiometry after the drop has been formed. The classical measuring procedure is the continuously growing drop until its detachment. Pearson and Whittaker [58] were the first to formulate the physical picture considering a radial flow inside the drop. In analogy to the equation of Ward and Tordai Eq. (20) the following integral equation was be derived in [59]:

$$\frac{d\Gamma}{dt} = -(R(t))^2 \sqrt{\frac{D}{\pi}} \int_0^t \frac{(\frac{dc(0,t)}{dt}) t_0}{\int_{t_0}^t ((R(\xi))^4 \, d\xi)^{1/2}} \, dt_0 - \Gamma \frac{d \ln A}{dt} \tag{21}$$

and numerically analysed. It was shown that the rate of adsorption at the surface of a growing drop at linear volume change $V(t)$ is about $1/3$ of that at a surface of constant area [60], and the same approximation can be used for other kinetic models.

1. Approximation for the Diffusion Model

Equation (21) is a rather complex equation and cannot be simply applied to experimental data. Ilkovic was the first to describe the adsorption at the surface of a growing drop by some simple approximation [61]. He defined the boundary conditions such that the model corresponded to a mercury drop in a polarography experiment but not to the adsorption of surfactants at a liquid droplet. Later Addison *et al.* [19] proposed to consider the area change during the drop growth, which ended up in a semi-empirical model consisting in a step-wise calculation of $\Gamma(t)$ from the equation (20) in a certain time interval, and a stepwise correction of the surface coverage inversely proportional to the area increase. Such model, however, does not take into account any flow in the bulk phase and hence is only an estimate. A similar procedure was later elaborated by Kloubek [62], but the neglected flow inside the drop lead to an overestimation of the effect of drop area expansion [28]. In 1957 Delahay and Trachtenberg [63], based on the theory of Ilkovic, derived an approximation for the initial period of the adsorption process at the surface of a growing drop

$$\Gamma(t) = 2c_0\sqrt{\frac{3Dt}{7\pi}},\qquad(22)$$

which indicates the ration between the adsorption time at a growing drop surface and a constant interface, given by the factor $3/7$, which is close to $1/3$ as discussed in [57].

2. General Diffusion Theory

Pierson and Whittaker formulated the first complete model for the adsorption process at a growing drop surface [58] assuming a diffusion as well as mixed diffusion-kinetic-controlled adsorption mechanism. The equation to describe the transport by diffusion inside or outside a spherical drop or bubble has the following form:

$$\frac{\partial c}{\partial t} + v_r \frac{\partial c}{\partial r} = D\left(\frac{\partial^2 c}{\partial r^2} + \frac{2}{r}\frac{\partial c}{\partial r}\right).\qquad(23)$$

For drops of a surfactant solution, this diffusion equation is valid in the interval $0 < r < R$, where R is the drop radius. In contrast, when the diffusion process happens outside a drop or bubble, Eq. (23) applies to the interval $R < r < \infty$ [28].

During the growth of a drop the flow inside the drop can be assumed radial. The drop surface area expansion is not ideally isotropic, however, no significant flow within the interface is expected. Nonetheless, the surface expansion leads to a thinning of the adjacent liquid layers and hence the diffusion concentration profile is compressed (cf. [28], Chapter 4), so that diffusion layer compression due to drop surface expansion, and diffusion layer expansion due to the drop growth are superimposed. The radial flow is given by [64]

$$v_r = \frac{R^2}{r^2}\frac{dR}{dt}.\qquad(24)$$

Instead of a radial a turbulent flow in the drop was considered in [65] and used for the discussion of experimental dynamic drop volume data.

In analogy to the Ward and Tordai equation (20) a non-linear integral equation was derived in [60]:

$$\Gamma(t) = 2c_0 \sqrt{\frac{3Dt}{7\pi}} - \sqrt{\frac{D}{\pi}} t^{-2/3} \int_0^{-3/7t^{7/3}} \frac{c(0, 7/3\tau^{3/7})}{3/7t^{7/3} - \tau} \, dt. \tag{25}$$

A significantly advanced theoretical analysis of the given problem was made by MacLeod and Radke [59], starting from a finite drop size instead of a point source at the beginning of the adsorption process. In addition to transport equation in spherical coordinates Eq. (23) the following boundary condition was proposed:

$$\frac{\partial \Gamma}{\partial t} = \pm D \frac{\partial c}{\partial r} - \Gamma \frac{d \ln A}{dt}, \quad r = R(t), \tag{26}$$

where the sign of the first term on the right-hand side depends on whether the diffusion takes place inside or outside the drop. As final result Eq. (21) was obtained, which can take into consideration any change in drop radius $R(t)$, and consequently $A(t)$. For a spherical geometry we have

$$A(t) = 2\pi r_{cap} h \quad \text{and} \quad R(t) = \frac{r_{cap}^2}{2h} + \frac{h}{2}, \tag{27}$$

where the drop height h as a function of time t is given by

$$h = \sum_{i=1}^{2} \left(\frac{3}{\pi}(Qt + V_0) + (-1)^i \sqrt{r_{cap}^6 + \frac{9}{\pi^2}(Qt + V_0)^2} \right)^{1/3}. \tag{28}$$

(Q is the liquid volume flow rate, and V_0 is the volume of the remnant drop.)

Equation (21) can be compared with experimental data only *via* numerical calculations. MacLeod and Radke [59] performed some model calculations which demonstrate the suitability of the theory for drop volume as well as growing drop experiments.

For fast adsorption processes, the initial load at the drop surface can be a significant parameter, as it was observed in dynamic interfacial tension data for some Triton solutions [16, 66]. The theoretical model for the diffusion controlled adsorption at a constant surface with initial load was proposed first by Cini *et al.* [67] and later generalised to the case of a growing bubble surface for any values of initial adsorption by Joos [68]. The result has the general form of the equation of Ward and Tordai Eq. (20)

$$\Gamma(\tau) f(t) = \Gamma_d + 2 \sqrt{\frac{D}{\pi}} \int_0^{\sqrt{\tau}} [c_0 - c_s(\tau - \lambda)] \, d(\sqrt{\lambda}). \tag{29}$$

However, the initial adsorption values Γ_d at the drop surface has been considered, and also the area change $f(t) = (1 + \alpha t)^n$, n is a constant, λ is the integration variable, and $\tau = \int_0^t [f(t)]^2 \, dt$. Still, we can conclude that the theory derived in [59] is the most general and allows a quantitative analysis of experimental data.

E. Experimental Results

For pure liquids the drop volume technique can easily provide surface and inter-facial tensions. This method can also be applied to solutions of surfactants and polymers. One of the most important advantages is the fact that for measurements of the interfacial tension between two immiscible liquids no modifications of the instrument are needed.

Although not as efficient as the drop profile tensiometry, the drop volume ten-siometry is of still continuous interest in many applications, and there are various examples that describe its advantages. In addition to the comparatively low price of respective instruments, the drop volume method is also easily temperature con-trolled and provides reliable data, as it was demonstrated for example for the tem-perature dependence of surface tension of water in the interval 15–50°C [69].

To demonstrate the interpretation of adsorption kinetics data the results for two decyl diethyl phosphine oxide solutions, obtained from dynamic and quasi-static drop volume experiments, are presented in Fig. 11.

The data are plotted in coordinates $\gamma(1/\sqrt{t})$ following the diffusion theory given above. According to Eqs (20) and (25), corresponding to the quasi-static and con-tinuously growing drop mode, the final slopes of the curves should differ by a factor of $\sqrt{7/3} \approx 1.5$. This is what we observe in Fig. 11. The extrapolation of the experimental data meet in the same intersection point on the ordinate defining the expected equilibrium surface tension. A comprehensive quantitative analysis of dynamic surface tension data requires, however, additional information on the adsorption isotherm of the studied surfactant, which will not be discussed here.

The dynamic interfacial tensions of different dietary oils are shown in Figs 12 to 14. The measurements were done in two ways:

1. direct measurement between the respective oil and pure water

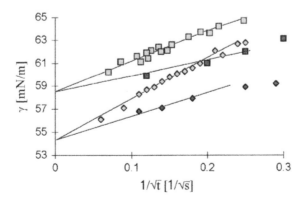

Figure 11. Dynamic surface tension of a decyl diethyl phosphine oxide solution of concentration $c_0 = 10^{-8}$ mol/cm^3 (□ ■), $c_0 = 10^{-7}$ mol/cm^3 (◇ ◆); dynamic mode (□ ◇), quasi-static mode (■ ◆), data from [16].

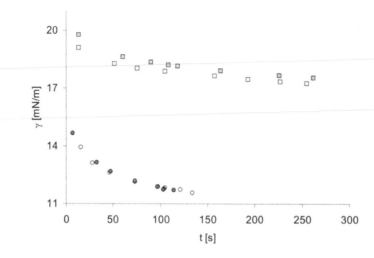

Figure 12. Dynamic interfacial tension of dietary oil against water; Pumpkinseed oil (○●), Sunflower oil 1 (■□); filled symbols — original oil against pure water; open symbols — mutually saturated oil and water; data from [73].

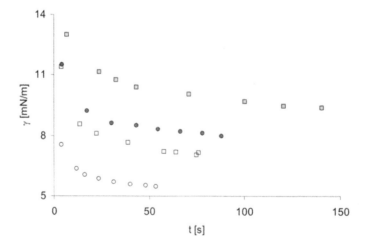

Figure 13. Dynamic interfacial tension of dietary oils against water; Linseed oil (○●), Walnut oil (■□); filled symbols — original oil against pure water; open symbols – mutually saturated oil and water phases; data from [73].

2. mutual mixing of equal amounts of oil and water, subsequent separation of the two liquids, and then measurement of the interfacial tension between the two separated liquids.

The dynamic interfacial tensions for Pumpkinseed and Sunflower oils are shown in Fig. 12. The prior mutual saturation of oil and water against each other does not influence the results remarkably.

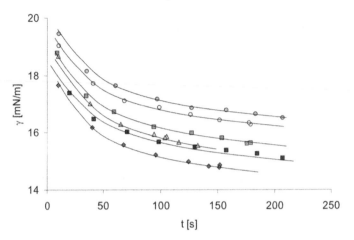

Figure 14. Dynamic interfacial tension of six extra virgin olive oils against water; data from [73].

This is in contrast to what was measured for Linseed and Walnut oils. As we can see in Fig. 13, the mutual saturation of these two oils with water prior to the measurement washes out water-soluble surface-active compounds from the oil, which then also adsorb at the interface from the water side. Hence, the interfacial tension decreases as compared to the measurements without prior mixing. This behaviour is very common for surface active compounds soluble in both liquid phases [70].

In Fig. 14 the interfacial tensions are shown of six extra-virgin olive oils of different origin. There are several mN/m difference in the interfacial tensions for these samples, obviously caused by the different composition of the oils. Although the studied olive oils belong to the same category "extra-virgin", the measured differences underline significant differences in the composition due to the presence of minor compounds. Extra-virgin olive oil is solely obtained *via* physical processing and is therefore rich in minor compounds originating from the olive fruits. The composition of virgin olive oils depends on the production methods and conditions [71], and several compounds adsorb at the oil/water interface [72]. The kinetics of the measured interfacial tensions $\gamma(t)$ indicate in an excellent way the presence of significant amounts of these bio-surfactants preset in these studied oil samples.

F. Summary and Conclusions

In this chapter we described the drop volume method in detail and demonstrate that it is a rather powerful method to investigate the dynamics of liquid interfacial layers. Despite of a number of advantages, such as easy handling and temperature control, and its straightforward application to liquid/liquid interfaces, it has also drawbacks. First, the measured data have to be interpreted correctly, i.e. the growth of the drops during the measurement process has to be considered. Moreover, the method is restricted to a rather narrow interval of drop formation times. When drops are formed too fast the measured drop volumes are no longer directly proportional to the surface tension and additional hydrodynamic effects have to be considered to correct the obtained results. For too slow drop formation rates, the experiments become extremely slow and inefficient.

Considering the pros and cons for this experimental method, we can conclude that it still is of important value in surface science and will surely be practiced also in future.

G. Acknowledgements

The work was financially supported by projects of COST D43 and P21, the DFG (Mi418/18-1) and the German Space Agency (DLR 50WM0941).

H. List of Symbols

A area of the interface $[\text{cm}^2]$

A_o initial interfacial area $[\text{cm}^2]$

a capillary constant as defined in Eq. (3)

D diffusion coefficient $[\text{cm}^2/\text{s}]$

D_{eff} effective diffusion coefficient $[\text{cm}^2/\text{s}]$

f correction factor

g gravitational constant $[9.81 \text{ m/s}^2]$

h drop height $[\text{cm}]$

k slope of the dependence $V(1/t^n)$

Q liquid volume flow rate $[\text{cm}^3/\text{s}]$

R gas law constant $[8.314 \text{ g cm}^2/(\text{s}^2 \text{ mol K})]$

$R(t)$ drop radius $[\text{cm}]$

r_{cap} radius of capillary tip

T absolute temperature $[\text{K}]$

t time $[\text{s}]$

t_o drop detachment time

V volume $[\text{cm}^3]$

V_o initial drop volume $[\text{cm}^3]$

V_e corrected drop volume $[\text{cm}^3]$

W drop weight

v_r radial flow velocity [cm/s]

α coefficient for hydrodynamic correction

β coefficient for hydrodynamic correction

$\Delta\rho$ density difference between two liquid phases

$\Gamma(t)$ adsorption as a function of time

γ interfacial tension [mN/m]

σ_i uncertainty on the parameter i

I. References

1. F.C. Goodrich, in "Surface and Colloid Science", E. Matijevic (Ed.), Vol. 1, Wiley-Interscience, New York–London–Sydney–Toronto, 1969, pp. 1.
2. J. Traube, Fresenius' J. Anal. Chem., 27 (1888) 655.
3. T. Lohnstein, Ann. Physik, 20 (1906) 237; 20 (1906) 606; 21 (1907) 1030.
4. T. Lohnstein, Z. phys. Chem., 64 (1908) 686; 84 (1913) 410.
5. E. Tornberg, J. Colloid Interface Sci., 60 (1977) 50.
6. P. Joos and E. Rillaerts, J. Colloid Interface Sci., 79 (1981) 96.
7. B.J. Carroll and P.J. Doyle, J. Chem. Soc., Faraday Trans. 1, 81 (1985) 2975.
8. R. Miller and K.-H. Schano, Colloid Polymer Sci., 264 (1986) 277.
9. P.J. Doyle and B.J. Carroll, J. Phys. E: Sci. Instrum., 22 (1989) 431.
10. R. Miller, A. Hofmann, R. Hartmann, K.-H. Schano and A. Halbig, Advanced Materials, 4 (1992) 370.
11. R. Gunde, A. Kumar, S. Lehnert-Batar, R. Mäder and E.J. Windhab, J. Colloid Interface Sci., 244 (2001) 113.
12. T. Tate, Phil. Mag., 27 (1864) 176.
13. Lord Rayleigh, Phil. Mag., 48 (1899) 321.
14. O.E. Yildirim, Q. Xu and O.A. Basaran, Phys. Fluids, 17 (2005) 062107.
15. W. Wang, K.H. Ngan, J. Gong and P. Angeli, Colloids Surfaces A, 334 (2009) 197.
16. V.B. Fainerman, S.A. Zholob and R. Miller, Langmuir, 13 (1997) 283.
17. B.-B. Lee, P. Ravindra and E.-S. Chan, Colloids Surfaces A, 332 (2009) 112.
18. C.C. Addison, J. Chem. Soc. (1946) 579.
19. C.C. Addison and S.K. Hutchinson, J. Chem. Soc. (1948) 943; (1949) 3387; (1949) 3406.
20. R. Miller and K.-H. Schano, Tenside Detergents 27 (1990) 238.
21. B.B. Freud and W.D. Harkins, J. Phys. Chem. 33 (1929) 8.
22. S. Hartland and P.S. Srinivasan, J. Colloid Interface Sci., 49 (1974) 318.
23. W.D. Harkins and F.E. Brown, J. Amer. Chem. Soc., 41 (1919) 499.
24. M.C. Wilkinson, J. Colloid Interface Sci., 40 (1972) 14.
25. J.C. Earnshaw, E.G. Johnson, B.J. Carroll and P.J. Doyle, J. Colloid Interface Sci., 177 (1996) 150.
26. G. Kretzschmar and R. Miller, Adv. Colloid Interface Sci., 36 (1991) 65.
27. J.T. Davies and E.K. Rideal, Interfacial Phenomena, Academic Press, New York, 1969.
28. S.S. Dukhin, G. Kretzschmar and R. Miller, Dynamics of Adsorption at Liquid Interfaces, in "Studies in Interface Science", D. Möbius and R. Miller (Eds), Vol. 1, Elsevier, Amsterdam, 1995.

29. J. Kloubek, K. Friml and F. Krejci, Czech. Chem. Comm., 41 (1976) 1845.

30. C. Jho and R. Burke, J. Colloid Interface Sci., 95 (1983) 61.

31. J. Van Hunsel, "Dynamic Interfacial Tension at Oil Water Interfaces", Thesis, 1987, University of Antwerp.

32. E.V.L.N. Rao, R. Kumar and N.R. Kuloor, Chem. Eng. Sci., 21 (1966) 867.

33. G.F. Scheele and B.J. Meister, AIChE J., 14 (1968) 9.

34. B.J. Meister and G. F. Scheele, AIChE J., 15 (1969) 689.

35. B.J. Meister and G. F. Scheele, AIChE J., 15 (1969) 700.

36. P.M. Heertjes and L.H. De-Nie, Chem. Eng. Sci., 21 (1966) 755.

37. P.M. Heertjes, L.H. De-Nie and H.J. De-Vries, Chem. Eng. Sci., 26 (1971) 441.

38. R.S. Barhate, G. Patil, N.D. Srinivas and K.S.M.S. Raghavarao, J. Chromatography A, 1023 (2004) 197.

39. S.O. Unverdi and G. Tryggvason, J. Comp. Phys., 100 (1992) 25.

40. J. Eggers and T.F. Dupont, J. Fluid Mech., 262 (1994) 205.

41. X. Zhang, Chem. Eng. Sci., 54 (1999) 1759.

42. X. Zhang, J. Colloid Interface Sci., 212 (1999) 107.

43. H.J. Subramani, H.K. Yeoh, R. Suryo, Q. Xu, B. Ambravaneswaran and O.A. Basaran, Phys. Fluids, 18 (2006) 032106.

44. A. Javadi, D. Bastani and M. Taeibi-Rahni, AIChE J., 52 (2006) 895.

45. M. Seifollahi, E. Shirani, N. Ashgriz, Eur. J. Mech. B, Fluids, 27 (2008) 1.

46. R. Miller, K.-H. Schano and A. Hofmann, Colloids Surfaces A, 92 (1994) 189.

47. R. Miller, M. Bree and V.B. Fainerman, Colloids Surfaces A, 141 (1998) 253.

48. P. Doshi, I. Cohen, W.W. Zhang, M. Siegel, P. Howell, O.A. Basaran, and S.R. Nagel, Science, 302 (2003) 1185.

49. C. Cramer, P. Fischer and E.J. Windhab, Chem. Eng. Sci., 59 (2004) 3045.

50. T. Kaully, A. Siegmann, D. Shacham and A. Marmur, J. App. Polymer Science, 106 (2007) 1842.

51. O.E. Yildirim and O.A. Basaran, J. Non-Newtonian Fluid Mech., 136 (2006) 17.

52. V.B. Fainerman and R. Miller, Colloids Surfaces A, 97 (1995) 255.

53. V.B. Fainerman, S.V. Lylyk, E.V. Aksenenko, J.T. Petkov, J. Yorke and R. Miller, Colloid Surfaces A, 354 (2010) 8.

54. A.F.H. Ward and L. Tordai, J. Phys. Chem., 14 (1946) 453.

55. R. Miller and K. Lunkenheimer, Z. Phys. Chem., 259 (1978) 863.

56. K.L. Sutherland, Austr. J. Sci. Res., A5 (1952) 683.

57. R. Miller, Colloids Surfaces, 46 (1990) 75.

58. F.W. Pierson and S. Whittaker, J. Colloid Interface Sci., 52 (1976) 203.

59. C.A. MacLeod and C.J. Radke, J. Colloid Interface Sci., 166 (1994) 73.

60. R. Miller, Colloid Polymer Sci. 258 (1980) 179.

61. D. Ilkovic, J. Chim. Phys. Physicochem. Biol., 35 (1938) 129.

62. J. Kloubek, J. Colloid Interface Sci., 41 (1972) 1.

63. P. Delahay and I. Trachtenberg, J. Amer. Chem. Soc., 79 (1957) 2355.

64. V.G. Levich, Physicochemical Hydrodynamics, Prentice-Hall, Englewood Cliffs, New York, 1962.

65. V.B. Fainerman, Koll. Zh., 41 (1979) 111.

66. L. Liggieri, F. Ravera and A. Passerone, J. Colloid Interface Sci., 169 (1995) 226.

67. R. Cini, G. Loglio and A. Ficalbi, Ann. Chim. (Rome) 62 (1972) 789.

68. P. Joos, Dynamic Surface Phenomena, VSP, 1999.

69. R. Miller, A. Hofmann, K.-H. Schano, A. Halbig and R. Hartmann, Seifen Öle Fette Wachse, 118 (1992) 435.

70. E.P. Kalogianni, E.-M. Varka, T.D. Karapantsios, M. Kostoglou, E. Santini, L. Liggieri and F. Ravera, Colloids and Surfaces A, 354 (2010) 353.

71. D. Tura, O. Failla, D. Bassi, S. Pedó and A. Serraiocco, Scientia Horticulturae, 118 (2008) 139.

72. F. Paiva-Martins and M.H. Gordon, J. Amer. Oil Chem. Soc., 79 (2002) 571.

73. K. Dopierala, A. Javadi, J. Krägel, K.-H. Schano, E.P. Kalogianni, M.E. Leser and R. Miller, Interfacial tensions of vegetable oils, Colloids Surfaces A, doi:10.1016/j.colsurfa.2010.11.027.

Studies in Capillary Pressure Tensiometry and Interfacial Dilational Rheology

V.I. Kovalchuk [a], F. Ravera [b], L. Liggieri [b], G. Loglio [c], A. Javadi [d],
N.M. Kovalchuk [a,d] and J. Krägel [d]

[a] Institute of Bio-Colloid Chemistry, Vernadsky str. 42, 03142 Kiev, Ukraine
[b] CNR — Istituto per la Energetica e le Interfasi, 16149 Genoa, Italy
[c] University of Florence, 50019 Sesto Fiorentino (Firenze), Italy
[d] MPI of Colloids and Interfaces, 14424 Potsdam/Golm, Germany

Contents

Bubble and Drop Interfaces
© Koninklijke Brill NV, Leiden, 2011

A. Introduction

Measurements of equilibrium and dynamic interfacial tension of liquid interfaces are of great importance for many practical applications. Many technological processes, e.g. involving liquid–liquid extraction, foams and emulsions, liquid metals and alloys, multiphase flows and many others, strongly depend on interfacial tension [1]. Interfacial tension is very important for the proper work of many biomedical systems, such as lung alveoli, wetting films on eyes [2, 3]. Moreover, the dynamic interfacial tension can be used for diagnostics of diseases [3, 4]. Also many environmental systems are strongly affected by detergents, which change the tension of water in nature. Thus, monitoring the interfacial tension of natural waters is an important issue also in environmental studies [5, 6].

The overwhelming majority of methods to measure equilibrium and dynamic interfacial tensions are based on the mechanical response of so-called capillary systems, i.e. systems where the capillary pressure is acting under a curved liquid interface [7]. Examples of such systems are drops and bubbles, menisci of liquids near solid walls, liquid jets, free liquid surfaces, deformed due to propagating capillary waves, thin liquid films, foams and emulsions. To a great extent the mechanical properties of all capillary systems are determined by the surface or interfacial tension acting at liquid–gas or liquid–liquid boundaries. The equilibrium state and dynamics behaviour of such systems are determined by the joint action of interfacial tension, which is of molecular nature, and external forces (hydrodynamic, gravitation, electric and others). Accordingly, the most important feature of capillary systems is interdependence of mechanical and physicochemical processes in them [8–10]. The change of mechanical conditions in the system (velocities, pressure) leads to deformations of interfaces, perturbation of the adsorption equilibrium, change of interfacial tension, which in turn influences the mechanical conditions in the system.

In studies of capillary systems two general tasks can be considered: the direct task — prediction of the mechanical behaviour and mechanical properties of the system, if the respective physical and chemical characteristics are known, and the reverse task — obtaining physical and chemical characteristics, if the mechanical response of the system is known. By studying the mechanical behaviour of capillary systems, it is possible to obtain important information about the dynamics of adsorption and dynamic interfacial tension, interaction of molecules at the interface, change of conformation and aggregation of molecules, kinetics of chemical reactions, kinetics of formation and disintegration of micelles, and other processes which take place at a molecular level [11, 12]. For these purposes often small harmonic perturbations of the interfaces are generated, which allows us to speak about methods of mechanical relaxation spectrometry [13–15].

In a large group of methods for measuring the interfacial tension the capillary force is balanced by an external force, most often this is the weight of liquid or hydrodynamic force. These are such methods as the drop shape and the capillary rise methods, the Du Noüy ring, Wilhelmy plate, drop weight/volume and capillary waves [16]. Most of these methods cannot be used in absence of gravity.

In another group of methods the capillary force is directly measured by a pressure transducer. Such methods are related to capillary pressure tensiometry (CPT) [17]. Examples are the pressure derivative method, oscillating, growing and expanded drop/bubble methods, and the maximum bubble pressure method. Over the last two decades, a significant progress in the development of pressure transducers for liquids was achieved. New transducers are able to measure very low pressures with high accuracy, giving a better possibility to evaluate accurately surface tensions by a direct measurement of the capillary pressure. Simultaneous the development of computer techniques allowed fast data acquisition, transmission and processing of the experimental data in automatic regimes. Due to this all methods related to capillary pressure tensiometry became nowadays one of the most promising and flexible tensiometric tools for dynamic adsorption studies. An important advantage is also that the capillary pressure methods can work under microgravity conditions and can be used for experiments in space [17, 18]. CPT is also particularly helpful for studying liquid/liquid interfaces. Indeed, only a few of the traditional surface tension measurement techniques are suitable for such interfaces.

A wide variety of CP tensiometers has been developed. For measuring the interfacial tension of pure liquid–liquid systems the Pressure Derivative method is the most suitable [19]. The dynamics of adsorption can be investigated by the Growing Drop/Bubble [20–25] and the Expanded Drop [26–28] methods. The Oscillating Bubble/Drop methods are widely used in surface rheological studies [29–31]. The same instruments can be used also in the low amplitude Stress/Relaxation experiments [32]. The Static Maximum Bubble Pressure method has been the first CP method, initially developed to study equilibrium surface tension of pure liquids [33]. However its dynamic modification allows also the measurements of dynamic surface tensions in a wide interval of characteristic times, ranging from 0.001 to 100 s [34, 35] (see also Chapter 5).

B. Experimental Techniques for Capillary Pressure Tensiometry and Dilational Rheology

The coupling between the interfacial mechanics and the interfacial tension is expressed by the normal and tangential stresses balances at the interface [7–9]. For uniform interfacial tension under static conditions the normal pressure balance includes the capillary pressure, acting across the curved interface

$$P_\mathrm{i} - P_\mathrm{e} = \Delta P_\mathrm{cap} = \gamma \left(\frac{1}{R_1} + \frac{1}{R_2} \right), \tag{1}$$

where R_1 and R_2 are the principal radii of curvature of the interface, γ is the interfacial tension, and P_i and P_e are the pressures in the two contacting phases close to the interface. For a spherical interface the two principal radii of curvature are equal $(R_1 = R_2 = R)$.

Under dynamic conditions the pressure difference across a stretched spherical interface can be derived by the normal stress balance at the interface as

$$\Delta P = \frac{2\gamma}{R} + \frac{4\kappa}{R^2}\frac{dR}{dt} - \frac{4(\mu_i - \mu_e)}{R}\frac{dR}{dt}, \tag{2}$$

where R is the drop radius, κ is the surface dilational viscosity and μ_i and μ_e are the viscosities of the two fluid phases, internal and external, respectively. For an interface at mechanical equilibrium ($dR/dt = 0$) the equation reduces to the Young–Laplace equation and ΔP is equal to the capillary pressure. The second and third terms arise from the dynamic viscosity of the interface and the bulk. For sufficiently slow variations of the drop/bubble radius their contributions can be neglected. The effect of the bulk viscosity is negligible for $dR/dt \ll \gamma/(\mu_i - \mu_e)$. For aqueous solutions this means $dR/dt \ll 10^3$ cm/s, and for usual experimental conditions the last term in Eq. (2) is not significant. It can become significant only for highly viscous liquids (by two orders of magnitude higher than water) [36]. The contribution of the second term in Eq. (2) is also negligible as for most common systems $(dR/dt)/R \ll \gamma/\kappa \cong 10^4$ s^{-1} holds. Thus, we come to a quasi-equilibrium version of the Young–Laplace equation, which can include time dependent variables:

$$\Delta P_{cap}(t) = \frac{2\gamma(t)}{R(t)}. \tag{3}$$

This equation opens the opportunity for γ to be evaluated as a function of time by the simultaneous measurement of the varying capillary pressure $\Delta P_{cap}(t)$ and drop/bubble radius $R(t)$. This is the basic principle of CP tensiometry. For very fast dynamic regimes or highly viscous liquids special corrections have to be taken into account.

In practice, a small drop or bubble is formed at the tip of a capillary inside another fluid phase, by means of a liquid dosing system (Fig. 1). Using an excitation system the drop/bubble volume and interfacial area can be varied according to certain time law. The pressure difference between the two fluids is changed due to both the curvature radius change and the interfacial tension change (in the presence of surfactants). This pressure change is monitored using a pressure sensor. At the same

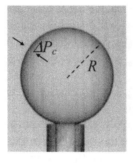

Fig. 1. Hexane-in-water droplet in a Capillary Pressure Tensiometer.

time the drop/bubble radius is determined by using either direct imaging or calculations from the injected liquid volume. Finally, the interfacial tension evolution is reconstructed according to Eq. (3).

Gravitational environment complicates the capillary pressure measurements because under gravity conditions the drops and bubbles are generally not spherical. The relative influence of these effects on the shape of an interface can be characterized by the Bond number

$$Bo = \frac{|\Delta\rho|gd^2}{\gamma}, \tag{4}$$

where g is the gravity acceleration, d is the characteristic length scale, and $\Delta\rho$ is the density difference between two fluid phases.

For $Bo = 0$, drops (or bubbles) are spherical. This can be achieved either for isodense fluids ($\Delta\rho = 0$) or for $g = 0$. The latter condition can be realized in a free falling system, where the weight is compensated by the inertia force. Thanks to the development of such facilities as drop-towers, aircrafts flying parabolic trajectories, sounding rockets and space missions, conditions of very low gravity can be achieved (with residual acceleration up to 10^{-2}–10^{-5} m/s^2). This stimulates the development of special CP tensiometers for investigations in microgravity conditions [18]. Moreover, experiments involving mass transport phenomena can largely benefit from microgravity conditions due to the strong attenuation of gravity driven convection. This opens opportunities to study purely diffusive processes where transport is driven only by gradients of the chemical potential. In particular, microgravity represents an ideal tool for studying the dynamic aspects of adsorption of soluble surfactants and their mixtures with other surface active compounds (polymers, proteins, micro- and nano-particles), both for liquid–liquid and liquid–gas interfaces.

Under normal gravity conditions the Bond number can be minimized by reducing the drop/bubble size. Drops and bubbles with a curvature radius of the order of 100 μm are almost spherical and can be used in ground experiments. For manipulations with such small volumes of liquids very precise dosing systems are required.

Depending on the particular aims of the experiment, properties of the studied systems and external conditions several configurations of CPTs utilising different methodologies have been developed.

A typical CP tensiometer is however composed by two fluid volumes connected by a capillary tube. A drop/bubble is formed at the tip of the capillary, creating the interface subjected to investigation. One of the volumes is closed and contains the pressure sensor, while the other is open to the atmospheric pressure. The closed volume of the tensiometer is usually in connection to the volume control system of the droplet/bubble. Such device can be either a precision syringe or a piezoelectric-driven piston. The latter is most frequently utilised in modern CPT, since it provides high flexibility and a maximum in accuracy. Two configurations are possible for the tensiometer, as sketched in Fig. 2. In the configuration A the liquid forming the

Capillary Pressure Methods

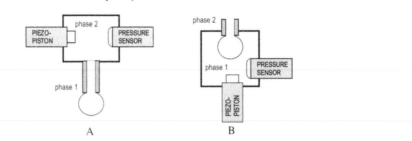

Fig. 2. General configurations of Capillary Pressure tensiometer set-ups.

droplet is contained in the closed chamber. Drops are therefore formed by injecting the fluid through the capillary. *Vice versa* in configuration B, the open chamber contains the fluid forming the drop/bubble which is now formed by "sucking" the fluid through the capillary.

Usually configuration A allows for smaller volume of the closed part of the cell and easier operations, but have longer capillary as compared to configuration B, which can be a drawback for dynamic measurements.

Droplets/bubbles of sub-millimetre diameter are typically used to minimise their gravitational deformation and consider them as spherical caps. In addition the small radius of curvature enhances the values of the measured capillary pressures, which are thus of the order of some hundreds Pascal. Often the tensiometer is coupled with a video camera to monitor the drop/bubble and, where required, for the drop/bubble radius acquisition and analysis. More details are given elsewhere [37, 38].

Owing to the possibility to apply different area stimuli, the CPT is a very flexible technique which can be applied with different methodologies. Further below in this section some of these methodologies are described in details and their critical aspects are discussed.

1. The Pressure Derivative Method

This methodology has been one of the first application of CPT, developed to measure the interfacial tension, γ_0, of pure liquids [19]. It relies on the observation that, if the interfacial tension is constant, the Laplace equation predicts a linear dependence between the capillary pressure and the curvature. Acquiring a set of data $(P, 1/R)$ during the continuous growth of a droplet, γ_0 is thus calculated as the proportionality constant. R data can be directly acquired by image processing or calculated from the injected volumes. To this aim, starting from a droplet smaller than a hemisphere its volume is increased at a constant flow rate, w.

An example of the measured pressure, P, is given in Fig. 3a. Provided the flow rate is sufficiently small to neglect dynamic effects on the pressure, P is given by

$$P(t) = \Delta P_{\text{cap}}(t) + P_0 = \frac{2\gamma_0}{R(t)} + P_0, \tag{5}$$

where P_0 is a constant offset arising from the hydrostatic contribution due to the immersion depth of the capillary. Accordingly, the pressure maximum in Fig. 3a

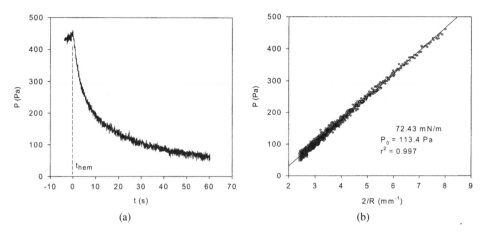

Fig. 3. Measured pressure by the Pressure Derivative method (a). The interfacial tension is obtained from the slope of the pressure *vs.* curvature data (b). The reported example refers to the measurement of the surface tension of pure water.

corresponds to the hemispherical size of the drop, i.e. the minimum R. Beyond the hemisphere R increases again causing the decrease of P.

When the droplet is larger than the hemisphere, its radius can be calculated at each time by equating the supplied volume with the volume of a spherical cap having a radius R and a constant base radius a

$$w(t - t_{hem}) = \frac{2}{3}\pi(R^3 - a^3) + \frac{\pi}{3}\sqrt{R^2 - a^2}(2R^2 + a^2). \qquad (6)$$

Thus after determining t_{hem} from the maximum in the $P(t)$ data, it is possible by solving Eq. (6) to associate a value of R to each P, as shown in Fig. 3. γ_0 can be then obtained from the slope of the linear best fit.

It is worth noting that the pressure term P_0 is also evaluated by the best fit procedure together with γ, thus the method does not require any calibration with respect to the pressure offset.

In principle, the above described methodology does not require the visualisation of the drop. The method could be therefore utilised also for measurements in opaque liquids. In addition the method is intrinsically suitable for interfacial tension measurements of isodense liquids.

In spite of the simple principle, evaluating the radius from Eq. (6) requires to know with great accuracy t_{hem} and a. While t_{hem} is relatively easy to estimate from the pressure signal, the evaluation of a poses more problems. In fact the base radius of the drop may differ from the actual radius of the capillary due to slightly uncontrolled wetting. That may occur even with the right choice of materials because of geometrical imperfections of the capillary edges. An error on the value of a reflects in errors in the estimation of R from Eq. (6) which are larger for smaller radii (i.e. close to the hemisphere), resulting in an apparent deviation from linearity of the $P(2/R)$-plot.

Another factor complicating the estimation of R from Eq. (6) is the presence of a residual compressibility of the liquids and/or of the cell. In this case, an additional flow term, depending on the pressure and on the compressibility, must be considered in the right hand side of Eq. (6). Errors in the estimation of the compressibility also results in an apparent deviation from the linear $P(2/R)$ relationship at radii closer to the hemisphere, i.e. at larger pressures.

These apparent deviations from linearity can be effectively utilised for the estimation of the above parameters, as it will be discussed in more detail in the section concerned with critical aspects of CPT.

In order to overcome some of these critical points, modifications of the Pressure Derivative method exploiting the direct measurement of the radius by image analysis have also been proposed. In this case the method has been also shown effective to measure the equilibrium interfacial tensions in the presence of surfactants [17, 18, 37]. With the increase of computer speed and calculation capacity and the progress in image analysis technologies, such approach will become more and more effective for CPT.

The Pressure Derivative method has been efficiently applied to measure values of the interfacial tension for different couples of immiscible liquids [17]. More important is the fact that this method is utilised for derived measurements in support of other CP methods, such as the Oscillating Drop [37, 38].

2. The Expanded Drop Method

The method allows for the investigation of the dynamic interfacial tension during the adsorption of surface active species at nearly fresh interfaces (that is, $\Gamma \approx 0$). To this aim, the volume of a nearly hemispherical droplet is abruptly increased, causing a large dilation of the interfacial area. The growth is then stopped in order to follow the evolution of dynamic interfacial tension on a steady interface. Usually the surfactant solution is the phase external to the droplet. Different arrangements are however possible in order to investigate different phenomena, such as surfactant transfer and depletion. Different authors contributed to conceive and develop this method [21, 26, 39].

A typical pressure signal recorded during the experiment is shown in Fig. 4. At first the pressure increases as the droplet radius decreases approaching the hemisphere. Right after achieving the hemisphere, the drop volume is abruptly expanded, which is revealed by the sharp drop in the capillary pressure consequent to the large radius increase. After that the radius is constant and the pressure decreases only because of the interfacial tension decrease due to the adsorption process.

In a first version of this method [26], the expansion was obtained by a sudden expansion of a compressed gas bubble in the experimental cell. In this case, in fact, a specific instability develops in the system causing an abrupt increase of the droplet volume immediately after the hemispherical shape is achieved. From that point in fact any even small increase in the drop volume (i.e. in the drop radius) causes a decrease in the pressure, resulting in a further increase in the drop volume. Thus an

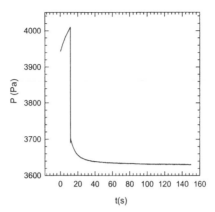

Fig. 4. Typical pressure signal recorded during an Expanded Drop experiment. The steep drop of the pressure corresponds to the abrupt expansion of the droplet volume.

explosive growth of the drop volume can be triggered which stops only when a new equilibrium state is established.

It has been shown [39] that the instability appears when the Bubble Stability Number (BSN)

$$\text{BSN} = \frac{27\pi P_{\text{atm}} a^4}{8 V^* \gamma_{\text{hem}}} < 1, \tag{7}$$

where V^* is the gas volume at the maximum pressure (hemispherical drop), γ_{hem} is the corresponding interfacial tension and P_{atm} is the absolute pressure. For BSN $\ll 1$, the relative variation of the surface area caused by the instability is well approximated by

$$\frac{\Delta A}{A} = \frac{2}{q^2} - \frac{1}{q} - 1, \tag{8}$$

where q is the solution of $81q^3 - 16\zeta q^2 - 16\zeta q - 16\zeta = 0$, with ζ being the value of BSN.

Due to the strong dependence on the radius of the capillary, instabilities with large $\Delta A/A$ are easily obtained for V^* of the order of 1 cm^3 in the normal experimental conditions ($a \approx 0.2$ mm).

The dynamics of drop expansion is controlled by the Poiseuille flow through the capillary. For liquids with kinematic viscosity of the order of 1 cSt (i.e., water, hexane, decane) however $\Delta A/A$ of the order of 50 are easily achieved in less than 0.3 s.

In the new versions of CPT [37], a similar area variation is produced by means of the expansion of a piezoelectric actuator directly embedded in the closed part of the tensiometer, as shown in Fig. 2B.

This solution allows for a wide dynamics of the surface area variation that can be easily controlled by a voltage signal, offering a large flexibility in the use of the tensiometer. In addition, the absence of mechanical parts warrants a dynamic response of the interfacial area change reflecting at best the patterns of the applied voltage signals.

As the measurement deals with drops of some mm^3, only a few drops can be formed *via* the piezo-electric actuator. This can be extended by a syringe pump connected to the cell, which provides the coarse management of the liquid flow to form the drop during the initialisation of the experiment and cut off during the real experiment by a valve. This reduces the residual elasticity of the system and any volume changes due to temperature fluctuations.

Examples of measurements performed according to the ED method are given in Figs 5 and 6. As shown, the method allows to measure with great accuracy the interfacial tension in the sub-second timescale, which represents a remarkable achievement for liquid–liquid interfaces.

An interesting alternative technique to investigate adsorption kinetics at a fresh interface by a CP tensiometer was proposed by Horozov *et al.* [40] In this case, the

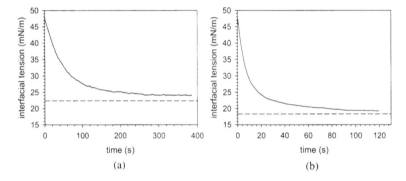

Fig. 5. Dynamic interfacial tension according to the Expanded Drop method, during adsorption of Triton X100 at the water–hexane interface; (a) $7.7 \cdot 10^{-6}$ M; (b) $2.0 \cdot 10^{-5}$ M.

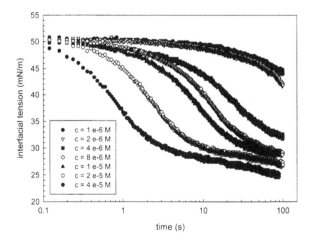

Fig. 6. Dynamic interfacial tension measured by the Expanded Drop method during adsorption at the water–hexane interface of $C_{10}E_8$ at different concentrations.

fresh interface was created by the breaking of a liquid jet obtained by stopping the flow through the capillary.

3. The Growing Drop Method

In contrast to the above described two methods, where either the surface tension (PD method) or the radius of curvature (ED method) remain constant, in the growing drop (GD) method all three parameters in Eq. (3) change continuously with time [20–24]. When the capillary pressure ΔP_{cap} and drop radius R are known as functions of time, then the dynamic interfacial tension γ can also be determined for each time moment during the drop growth process.

The variation of interfacial tension with time in GD experiments results from the competition between two simultaneously running processes: adsorption of surfactant molecules from the solution bulk and expansion of the interfacial area. The dilation process puts the system in a state out of equilibrium, whereas the adsorption process tends to restore the equilibrium state. During the initial stage, while the drop area is small, the dilation process is much more significant and the interfacial tension increases. However, when the drop becomes sufficiently large, the relative interfacial area increase becomes slower and cannot compensate the interfacial tension decrease caused by adsorption of new molecules. This results in a typical maximum in the $\gamma(t)$ dependencies, as shown in Fig. 7 [24, 25]. For high flow rates the initial expansion of the drop interface can be so strong that the interfacial tension at the maximum can increase almost up to the value corresponding to pure interface (as for curve $\gamma(t)$ corresponding to the highest flow rate 8.5 mm³/s in Fig. 7). The similar fast initial expansion of the drop interface takes place also in ED experiments. However, in GD experiments the subsequent adsorption of molecules occurs at the interface that continues to expand.

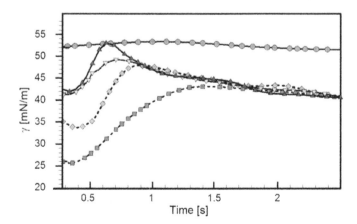

Fig. 7. Dynamic interfacial tension (filled symbols) *vs.* time for growing water drops in hexane (CGD protocol) measured for different flow rates: $Q = 0.85$ mm³/s (●) for pure water/hexane system and $Q = 0.85$ mm³/s (■), 2 mm³/s (◆), 4.8 mm³/s (○), and 8.5 mm³/s (▲) for water/solution system (10^{-4} mol/l Span80 in hexane); according to [25].

In the GD method the capillary pressure changes due to the variation of drop radius and interfacial tension. Because of the varying interfacial tension the maximum in the capillary pressure *vs.* time dependency does not coincides with the minimum in the drop radius [24, 25]. The shift between the maximum capillary pressure and the minimum drop radius increases with flow rate and is negligible only at sufficiently low flow rates [25].

Similar to PD and ED instruments the typical GD instrument includes a closed chamber, connected to a capillary, pressure sensor and liquid dosing system. In a typical experiment the dynamic interfacial tension measurements are made in a periodic way in which drops are formed, grow and detach from the capillary under the conditions of a continuous supply of drop forming liquid by the dosing system. These techniques are suitable for measuring dynamic interfacial tension in liquid–liquid or liquid–gas systems over a time scale from about 10 ms to several hours. In some of these measuring systems the drop radius is evaluated from the known injected volume at a given flow rate [20, 22, 23]. In [24] it was proposed to find the drop radius from a video image acquired by a CCD camera which makes the determination of the drop size completely independent of the variation of pressure and the volumetric flow rate.

An improved version of the capillary pressure tensiometer for GD experiments was presented recently in [25]. Its schematic view is shown in Fig. 8. This equipment allows obtaining faster dynamic regimes (flow rates about one order of magnitude higher than in previous studies) which is very important for the analysis of short time properties of liquid–liquid interfacial layers. Three particular experimental protocols can be realized with this instrument:

– continuously growing drop (CGD);
– pre-aged growing drop (PGD);

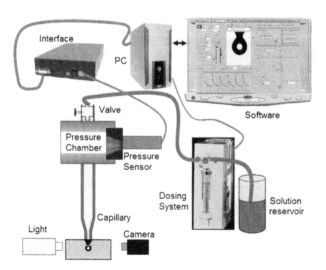

Fig. 8. Capillary pressure tensiometer for GD and oscillating drop experiments; according to [25].

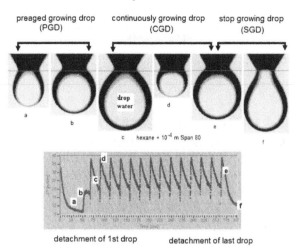

preaged growing drop   continuously growing drop   stop growing drop
(PGD)   (CGD)   (SGD)

detachment of 1st drop   detachment of last drop

Fig. 9. Experimental protocols for the capillary pressure tensiometer shown in Fig. 8; stages a and b: pre-aged growing drop (PGD); stages c and d: continuously growing drop (CGD); stages e and f: stop growing drop (SGD) protocol; according to [25].

– stopped growing drop (SGD).

These experimental protocols are illustrated by Fig. 9. The advantage of the CGD protocol is that it gives access to almost fresh interfaces. Therefore, it is the best procedure for studies of higher concentrated solutions. However a quantitative modelling of experimental results obtained with this procedure is difficult because of the unknown initial load of the drop surface. The PGD protocol is therefore a good alternative as it can provide the initial amount of adsorbed molecules. At the same time it has the advantage of a large surface expansion during the experiment. The SGD procedure is similar to the ED method as here also a drop with a fresh interface is formed by a sudden stop of the dosing system in a certain time moment. The most important advantage of the SGD protocol is the negligible hydrodynamics and convection effects. Thus, all three protocols are complementary.

4. The Oscillating Bubble/Drop Method

In contrast to other CPTs, in the oscillating bubble and drop methods only small-amplitude variations of the interfacial area are generated:

$$A(t) = A_0 + \delta A(t), \tag{9}$$

where $\delta A \ll A_0$. In this case a linear approximation can be used which allows to assume that all system characteristics have only small deviations from their equilibrium values. In particular, interfacial tension only slightly deviates from the equilibrium interfacial tension γ_0:

$$\gamma(t) = \gamma_0 + \delta\gamma(t). \tag{10}$$

Such approach allows applying the methods of mechanical spectrometry [13] to study the dynamic properties of interfaces. In particular, the dilational interfacial modulus can be introduced as [41, 42]:

$$E = \frac{d\gamma}{d \ln A}. \tag{11}$$

where $d\gamma$ and $d \ln A$ are the frequency dependent complex amplitudes for the interfacial tension and relative interfacial area variations, defined from the Fourier transforms of the respective time functions $\delta\gamma(t)$ and $\delta \ln A(t)$ [43, 44]. In the general case, the modulus E represents a complex quantity $(E = E_r + iE_i)$ the real and imaginary parts of which are functions of frequency. For harmonic oscillations of the interfacial area the modulus E can be obtained as

$$E = \frac{\Delta\gamma}{\Delta \ln A} e^{i\varphi}, \tag{12}$$

where $\Delta\gamma$ and $\Delta \ln A$ are the real amplitudes of interfacial tension and relative interfacial area oscillations, and φ is the phase shift between them.

Studies of the dilational interfacial modulus allow obtaining important information about the relaxation processes at the interface [15, 45]. The first attempt to use small spherical drops and bubbles for studies of the dilational rheology was made already in 1970 [29]. At this time, however, electric pressure transducers were not available yet and the capillary pressure variations had been obtained in an indirect way through the volume variation produced by an acousto-mechanical excitation system and the meniscus height variation measured by a photoelectric detection system [30]. Based on these pioneering ideas, the presently used capillary pressure method for oscillating drops and bubbles was developed in a modern and effective instrumentation by the contributions of different authors [30, 37, 38]. After the development of suitable modelling of the cell fluid-dynamics [46–48] measurements of the dilational viscoelasticities are possible with these instruments up to frequencies of the order of hundreds Hz.

The capillary pressure variations can be produced by generation of volume variations in a closed cell filled with a liquid and connected to a capillary, at the tip of which a drop or bubble is formed. The meniscus (drop or bubble) can be formed either inside or outside the closed cell (Fig. 10).

During drop or bubble small-amplitude oscillations the capillary pressure, $\Delta P_{\text{Cap}} = \frac{2\gamma}{R}$, can vary because of both the variation of the curvature radius, $R = a_0 + \delta a$, and the variation of the interfacial tension, $\gamma = \gamma_0 + \delta\gamma$:

$$\delta P_{\text{Cap}} = -\frac{2\gamma_0}{a_0^2}\delta a + \frac{2}{a_0}\delta\gamma. \tag{13}$$

Using the dilational interfacial modulus, Eq. (11), the capillary pressure variations can be related to interfacial area variations

$$d P_{\text{Cap}} = \left(-\frac{2\gamma_0}{a_0^2}\frac{da_0}{d \ln A} + \frac{2E}{a_0} \right) d \ln A. \tag{14}$$

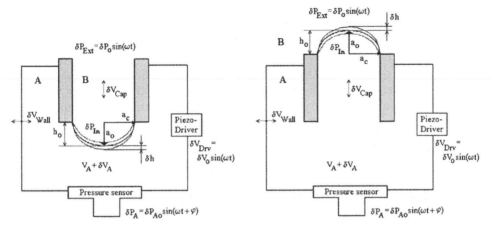

Fig. 10. Two experimental configurations for the measurement of the interfacial tension response by deformations of drops or bubbles: the meniscus is formed at the capillary tip either inside or outside the closed cell; according to [45].

Thus, the dilational interfacial modulus can be obtained from the measured capillary pressure response as

$$E = \frac{a_0}{2} \frac{d P_{Cap}}{d \ln A} + \frac{\gamma_0}{a_0} \frac{d a_0}{d \ln A}. \qquad (15)$$

For spherically shaped drops or bubbles, from geometrical consideration one obtains

$$\frac{d a_0}{d \ln A} = \frac{h_0^4 - a_C^4}{4 h_0^3} \quad \text{and} \quad a_0 = \frac{a_C}{2} \left(\frac{a_C}{h_0} + \frac{h_0}{a_C} \right), \qquad (16)$$

where h_0 is the equilibrium height of the drop or bubble, measured from the capillary tip, and a_c is the base radius (Fig. 10).

In practice, the small-amplitude oscillations of drops or bubbles can be obtained with the same equipment as for GD experiments. In particular, the capillary pressure tensiometer shown in Fig. 8 can be used also for oscillating drop experiments. Similar equipment (but working according to the first configuration in Fig. 10) was used by Wantke *et al.* [31] for oscillating bubble experiments. These set-ups have been coupled with innovative optical techniques, such as Second Harmonic Generation (SHG) and ellipsometry, to perform interesting studies on the kinetics of adsorbed layers [49].

All set-ups so far mentioned in this section were explicitly developed to perform studies of the dilational rheology at relatively large frequencies. For this reason they do not utilise the direct measurement of the drop radius during the oscillation. Such measurement which simplify considerably the calculation of the dilational viscoelasticity has instead been adopted for some low frequency studies. In these cases, following the original idea of Zhang *et al.* [24], capillary pressure tensiometry was integrated in Drop Shape set-ups [50], in order to exploit image analysis capabilities. Even not yet described in literature, potential CP tensiometry developments in this direction are today made possible by the easy accessibility of high speed image acquisition systems. It is however worth to mention that recently an interesting application of the concept of Spectrum Compression has been suggested [51] for the integration in CP Oscillating Drop tensiometer. Such application allows for direct measurement of the drop geometrical characteristics during high frequency oscillations using standard image acquisition systems.

5. *The Maximum Bubble Pressure Method*

In contrast to other CP methods, in the Maximum Bubble Pressure method the capillary pressure and the curvature radius are not monitored continuously. Instead it is accepted that when a growing bubble approaches a hemispherical shape its curvature radius is equal to the inner capillary radius and the capillary pressure is equal to the pressure in a closed reservoir connected to the opposite capillary end [34, 35]. For this specific time moment the dynamic surface tension is calculated according to Eq. (3).

The time during which the bubble grows from initial to hemispherical size corresponds to the aging time of the interface. A sequence of bubbles with different lifetimes (i.e. formed at different flow rates) allows obtaining the full dynamic surface tension *vs.* time curve for a given solution. An approach using the concept of effective surface age allows accounting for expansion of the interface during the bubble growth [11, 52].

In fast bubbles formation regimes (with lifetimes below 0.1 s) some additional dynamic effects become significant. Therefore, to obtain accurate dynamic surface tension data and lifetimes of the interface for such regimes special corrections accounting for these dynamic effects are necessary [35, 53]. The Maximum Bubble Pressure method is discussed in detail in Chapter 5.

6. *Studies of Film Tension*

The capillary pressure studies can be realised not only for single liquid–fluid interfaces but also for thin liquid films comprised of two interfaces. For a very thin, spherically shaped films the difference between the outer and inner film radii is negligible compared to the film radius R_f. Therefore, under static conditions the Young–Laplace equation can be written as [21, 54]:

$$\Delta P_{cap} = \frac{2\gamma_f}{R_f},\tag{17}$$

where the film tension γ_f is equal to the sum of the interfacial tensions for the inner and outer interfaces

$$\gamma_f = \gamma_i + \gamma_e.\tag{18}$$

The last equation holds for relatively thick films (above 30 nm thickness) when the effect of the disjoining pressure on the film tension can be neglected.

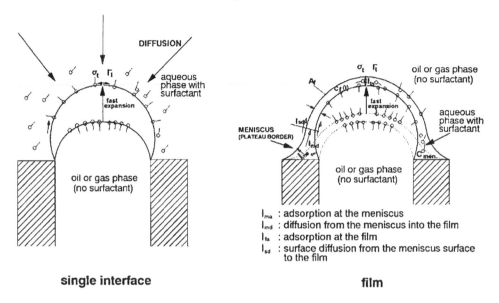

single interface **film**

I_{ma} : adsorption at the meniscus
I_{md} : diffusion from the meniscus into the film
I_{ta} : adsorption at the film
I_{sd} : surface diffusion from the meniscus surface to the film

Fig. 11. Difference in the exchange of matter for adsorption layers and thin liquid films; according to [54].

Similarly to the case of single interfaces, the expansion or contraction of a spherical film should be accompanied by a relaxation of the film tension, which can be monitored through the variation of the capillary pressure [54]. Such studies can provide important information about the dynamic properties of foam and emulsion films. A computer controlled apparatus for such studies was developed by Soos *et al.* [21]. Also the dilational rheological properties of thin liquid films can be investigated by using the same approach [55–57]. For more precise control and manipulation of the foam films an additional dosing system can be used, as proposed in [57] and applied in [58]. Such additional dosing system is necessary for accurate changes of the volume of liquid in the Plateau borders surrounding the film which is simultaneously a tool to control the film thickness and the film drainage rate.

However, the mechanical behaviour of a liquid film cannot be assumed to be just twice of the respective adsorption layer. The principle difference is shown in Fig. 11 and consists in the reservoir of surfactant. When the area of an adsorption layer is changed the perturbation relaxes due to intrinsic interfacial processes (such as molecular orientations, change of conformation or aggregation) and mainly due to the exchange of matter with the bulk phase. In contrast, for a thin liquid film the situation at the interfaces is comparable but the reservoir of surfactant in the interior of the film is limited and depends strongly on the thickness of the film. This leads to a substantial surfactant depletion inside the film during the fast expansion of the film [54]. As a consequence the dynamic film tension is no more equal to twice the interfacial tension as it was under static conditions. The dilational film elasticity is also strongly affected by the surfactant depletion effect inside the film [59].

C. Theoretical Basis of Capillary Pressure Tensiometry

1. The Pressure Measurements

From the previous considerations it is clear that accurate knowledge of the capillary pressure is one of the most important preconditions for the accurate determination of the interfacial tension. However, the pressure measured by the pressure transducer does not give immediately the capillary pressure because it includes also some additional contributions, such as static components, hydrodynamic contributions, contribution from the bulk compressibility effects, external noise etc. These additional terms have to be removed from the measured pressure signal.

What concerns the hydrostatic contributions or any other constant pressure terms internal to the measurement device (related to the pressure offset), there are different procedures developed to determine or evaluate them [17] which depend on the particular equipment.

Additional care should be taken to avoid external noise, the primary source of which are vibrations of the drop [22]. Any electronic noise, present during signal processing, should also be removed [21, 25].

Hydrodynamic contributions are negligible in slow dynamic regimes. However, increasingly dynamic regimes, utilised in recent times to study very fast relaxation processes, require detailed accounting for hydrodynamic effects. In usual configuration of CPTs the hydrodynamic contributions originate from hydrodynamic pressure losses in the capillary and from the drag force of the external phase onto the growing drop.

Typical experimental conditions are characterized by a non-stationary (transient or oscillatory) flow regime in the capillary. Under such conditions the particular velocity profile in the capillary cross-section depends on the ratio of the characteristic time of the pressure variation and the characteristic hydrodynamic time. The hydrodynamic relaxation time is

$$t_h = a_C^2/\nu, \tag{19}$$

where $\nu = \eta/\rho$ is the kinematic viscosity of the liquid or gas passing through the capillary. When the pressure changes proceed in times much longer than the hydrodynamic relaxation time then a quasi-stationary (parabolic) velocity profile establishes in the capillary cross-section, and the hydrodynamic pressure losses in the capillary can be estimated from the Poiseuille law. In particular, for oscillatory flow with a frequency $f = \omega/2\pi \ll t_h^{-1}$, the hydrodynamic contribution (which includes viscous and inertia effects) can be written as [46, 53]:

$$\delta P_h = (i\omega G_1 - \omega^2 G_2)\delta V_{Cap}, \tag{20}$$

where i is the imaginary unit, ω is the angular oscillation frequency, δV_{cap} is the volume of liquid (gas) passing though the capillary into the cell (which is equal to the meniscus volume variation, $\pm \delta V_m$), and G_1 and G_2 are the viscous and inertial coefficients, respectively, which can be calculated approximately as:

$$G_1 \approx \frac{8\eta_{B,A}l}{\pi a_C^4}, \tag{21}$$

$$G_2 = \frac{4\rho_{B,A}l}{3\pi a_C^2} + \frac{\rho_{A,B}}{\pi\sqrt{2(h_0^2 + a_C^2)}}. \tag{22}$$

Here $\eta_{A,B}$ and $\rho_{A,B}$ are the dynamic viscosities and densities of the phase A and B, respectively, and l is the capillary length. The expressions (21) and (22) depend on what phase is inside the capillary, A or B, i.e. what configuration we are using (see Fig. 10). The last term in Eq. (22) includes also the contribution of the added mass of the liquid surrounding the drop or bubble. The contribution of the viscous stresses acting at the interface (the last term in Eq. (2)) is usually small, as it has been discussed above.

For oscillation frequencies $f \geqslant t_h^{-1}$ a quasi-stationary velocity profile has not sufficient time to establish. In these cases non-stationary velocity profiles in the capillary have to be considered [46, 60]. This leads to a more general equation for the hydrodynamic contribution

$$\delta P_h = \left(\frac{B_{B,A}\beta \tanh \beta l}{\pi a_C^2} - \frac{\rho_{A,B}\omega^2}{\pi\sqrt{2(h_0^2 + a_C^2)}} \right) \delta V_{Cap}, \tag{23}$$

where

$$\beta^2 = -\frac{\rho_{B,A}\omega^2}{B_{B,A}} \frac{I_0(a_C\sqrt{i\omega/\nu})}{I_2(a_C\sqrt{i\omega/\nu})}. \tag{24}$$

Here $I_0(x)$ and $I_2(x)$ are modified Bessel functions of zero and second order, $B_{B,A}$ and $\rho_{B,A}$ are the bulk elasticity modulus and the density of phase B or A, depending on the system configuration (Fig. 10).

The equations similar to Eqs (20) and (23) can be also used for transient relaxation experiments if written for the Laplace–Fourier images of the respective time functions. The equations for the time domain can be then obtained by applying the inverse Laplace–Fourier transformations to these equations, which requires however the respective initial conditions to be specified.

The hydrodynamic relaxation time t_h depends on the capillary radius: the larger is the capillary radius, the larger is t_h, accordingly, the smaller is the frequency limit for the applicability of Eq. (20). For water ($\nu = 10^{-6}$ m^2/s) flowing through the capillary of radius $a_c = 0.2$ mm this frequency limit is $f < 25$ Hz, while for air ($\nu = 1.5 \cdot 10^{-5}$ m^2/s) we get $f < 400$ Hz.

2. The Bulk Elasticity Effects

Though the bulk elasticity for liquids is very high (and compressibility is very low), the liquid in the cell cannot be considered as incompressible because in the opposite case the pressure variation δP_A would be infinitely large even for very small volume variations. Due to pressure changes the liquid in the cell is periodically compressed and expanded during the formation and detachment of drops or during oscillations of drops (or bubbles). The inner volume in the cell can also periodically change because of wall deformations. Though the volume changes due to the liquid's compressibility and walls deformations are very small, they can be comparable to the drops or bubbles volume changes which are also very small. This can lead to a certain time lag between the externally applied volume variations and the meniscus volume variations. Neglecting these effects can lead to erroneous determination of the meniscus size, and, accordingly, to a decrease in accuracy of the interfacial tension measurements.

The small pressure variations inside the liquid in the cell due to its compression/expansion are characterized by the equation

$$\delta P_A = \frac{1}{V_A} \frac{dP}{d\ln V} \delta V_A = -\frac{B}{V_A} \delta V_A, \tag{25}$$

where V_A is the volume of the liquid in the cell, and $B = -dP/d\ln V$ is the bulk elasticity of the liquid ($K = B^{-1} = -d\ln V/dP$ being the bulk compressibility coefficient). For water in the isothermic regime the bulk elasticity is $B = 2.04 \cdot 10^9$ Pa, while in the adiabatic regime we get $B = 2.22 \cdot 10^9$ Pa, which are much higher than for a gas (10^5 Pa and $1.4 \cdot 10^5$ Pa, respectively). For oscillations at high frequencies or fast transient relaxations the adiabatic bulk elasticity should be used because the heat transfer has practically no time to proceed.

Considering the volume balance in the cell we should take into account: (i) the externally applied volume variation produced by a piezo-driver (or syringe pump), δV_{Drv}, (ii) the volume of liquid (gas) passing though the capillary, δV_{Cap}, and (iii) the inner cell space variations due to wall deformations, δV_{Wall}. The signs will be chosen here such that δV_{Cap} is positive, when the flow though the capillary is directed into the cell, and δV_{Drv} is positive, when the piezo-driver recedes back from the cell (see Fig. 10). When unwanted small gas bubbles occasionally remain in the cell they can also contribute to the volume change:

$$\delta V_{Gas} = -\frac{V_{Gas}}{B_{Gas}} \delta P_A, \tag{26}$$

where V_{Gas} and B_{Gas} are the volume and the bulk elasticity modulus of the gas. With all these contributions the full volume balance can be written as

$$\delta V_A = \delta V_{Drv} - \delta V_{Cap} + \delta V_{Wall} - \delta V_{Gas}. \tag{27}$$

For small-amplitude oscillations the wall deformation is proportional to the pressure variation

$$\delta V_{Wall} = \frac{dV_{Cell}}{dP} \delta P_A \tag{28}$$

and with account for Eqs (25)–(28) the difference between the volume supplied to the cell and that passing through the capillary can be expressed as

$$\delta V_{Drv} - \delta V_{Cap} = -\left(\frac{V_A}{B} + \frac{dV_{Cell}}{dP} + \frac{V_{Gas}}{B_{Gas}}\right)\delta P_A = -\frac{V_A}{B_{ef}}\delta P_A, \tag{29}$$

where the effective cell elasticity, B_{ef}, is introduced as

$$B_{ef} = \frac{B}{1 + \frac{B}{V_A}\frac{dV_{Cell}}{dP} + \frac{B}{V_A}\frac{V_{Gas}}{B_{Gas}}}. \tag{30}$$

This coefficient is an intrinsic characteristic of the cell and includes contributions from all compressibility effects. For an incompressible liquid in the cell and non-deformable cell walls we get the expression

$$B_{ef} = V_A\frac{B_{Gas}}{V_{Gas}}. \tag{31}$$

Using this equation it is possible to express B_{ef} in terms of an equivalent volume of gas, $V_{eq} = V_A\frac{B_{Gas}}{B_{ef}}$, apparently present in the cell with incompressible liquid and walls, which gives the same volume variation as the real cell with intrinsic elasticity B_{ef}. Such approach is also widely used in literature (see, e.g. [61] and Subsection D.4).

It should be noted, however, that if even small entrapped gas bubbles (including those condensed from the liquid phase) occur in the cell, they can strongly influence the effective cell elasticity because of the high gas compressibility. Moreover, relaxation at the interface of such bubbles can contribute to the total pressure relaxation in the cell. Therefore, for better reproducibility of results it is very important to avoid the occurrence of any additional bubbles inside the cell. For this aim the liquid in the cell should be carefully degassed.

The volume flux through the capillary is determined by the complex resistance (impedance) of the capillary with the attached meniscus

$$R_C = \frac{\delta P_{Ext} - \delta P_A}{\delta V_{Cap}}. \tag{32}$$

Combining Eqs (29) and (32), one obtains the expression for the volume flowing though the capillary:

$$\delta V_{Cap} = \frac{\delta P_{Ext} + (B_{ef}/V_A)\delta V_{Drv}}{B_{ef}/V_A + R_C} \tag{33}$$

which is equal to the meniscus volume variation, $\pm\delta V_m$, (as the fluid compressibility inside the capillary can usually be neglected). Thus, it follows from Eq. (33) that even at constant external pressure ($\delta P_{Ext} = 0$) the meniscus volume variation, δV_m, and the applied volume variation, δV_{Drv}, should be different, which is the consequence of compressibility effects. They can be close to each other only by a very small resistances R_C as compared to B_{ef}/V_A. However, the capillary resistance increases with increasing frequency. Therefore, for fast hydrodynamic regimes the difference between δV_m and δV_{Drv} should be accounted for. This is important also for GD experiments, as due to compressibility effects the liquid can be partially accumulated in the cell and then released depending on the pressure variations on different stages of the drop formation.

From Eqs (29) and (32), one obtains also the expression for the pressure variation

$$\delta P_A = \frac{B_{ef}}{V_A} \cdot \frac{\delta P_{Ext} - R_C \delta V_{Drv}}{B_{ef}/V_A + R_C}. \tag{34}$$

For constant external pressure ($\delta P_{Ext} = 0$) the pressure variation in the cell is proportional to the volume variation δV_{Drv}. By measuring the pressure variation δP_A produced by the applied volume variation δV_{Drv} one obtains the resistance R_c that contains information about the relaxation processes in the capillary and at the meniscus interface. The resistance R_c is given by [18]:

$$R_C = i\omega G_1 - \omega^2 G_2 - \frac{2\gamma_0}{a_0^2} \frac{da}{dV_m} + \frac{2E}{a_0} \frac{d\ln A}{dV_m}. \tag{35}$$

Substituting Eq. (35) into Eq. (34) one obtains an expression for the dilational elasticity modulus:

$$E = \frac{a_0}{2} \frac{dV_m}{d\ln A} \left[-\left(\frac{\delta V_{Drv}}{\delta P_A} + \frac{V_A}{B_{ef}} \right)^{-1} - i\omega G_1 + \omega^2 G_2 + \frac{2\gamma_0}{a_0^2} \frac{da}{dV_m} \right]. \tag{36}$$

This equation allows finding the dilational elasticity modulus when the pressure variation δP_A and the applied volume variation δV_{Drv} are known. It includes also the corrections for the compressibility effects in the cell and hydrodynamic effects.

D. Critical Aspects in Capillary Pressure Tensiometry

1. Characteristics of Pressure Transducers

Among other measuring principles the use of piezo-resistive strain gauges is the most commonly employed sensing technology for general purpose of pressure measurements. Pressure transducer working according to this principle take use of the piezo-resistive effect of bonded or formed strain gauges to detect strain due to the applied pressure and can be used to measure absolute and differential pressures. The piezo-resistive effect only causes a change in electrical resistance; it does not produce an electric potential. It is known that the piezo-resistive effect of semiconductor materials are several orders of magnitudes larger than the geometrical effect in metals and is present in materials like germanium, polycrystalline silicon, amorphous silicon, silicon carbide, and single crystal silicon.

The origin of the strong change in resistivity of semiconductors are changes in inter-atomic spacing resulting from strain which affects the band gaps and making it easier (or harder depending on the material and strain) for electrons to be raised into the conduction band. The resistance of silicon changes not only due to the stress dependent change of geometry, but also due to the stress dependent resistivity of the material. This results in larger gauge factors. Therefore piezo-resistive silicon devices can be used to create much more sensitive strain gauges. Due to the basic mechanism they are in general also more sensitive to environmental conditions, especially against temperature. The variations in temperature will cause a multitude of effects. The transducer will change in size by thermal expansion, which will be detected as a strain by the gauge. In addition the resistance of gauge will change and the resistance of the connecting wires will change too. Therefore a careful calibration of the sensor at a given constant temperature is needed.

High performance millivolt output pressure transducers have been developed, e.g. the Druck PDCR 4000 Series from GE Sensing and Inspection Technology, which offer enhanced levels of measuring accuracy, long time stability and reliability [62]. For example the most sensitive sensor of this series has an operating pressure range up to 70 mbar. Such sensitive sensors are very susceptible to overpressures. Only very small overpressure will be tolerated with a negligible effect on the calibration. Higher overpressures implicate an irreversible destruction of the pressure transducer. The hearts of such sensors are micro-machined silicon assemblies. The pressure sensitive silicon element is mounted within a high integrity glass-to-metal seal and is fully isolated from the pressure media. The chip size is in the order of 5×5 mm^2 and is sealed in a tube and positioned such that both sides are independent. The top-side of the sensor, which is directed to the pressure to be measured, the sensor surface is covered by a thin layer of special coupling oil. This side of sensor is closed by a thin stainless steel membrane. The back-side of the sensor where the strain gauges are situated is connected to the surrounding atmosphere. The strain gauges are bonded and connected to a specific electronic. The advanced stress sensor derived from piezo-resistors is working like a Wheatstone bridge. The ready fabricated transducer is packaged with conditioning electronics into a Hastelloy® and stainless steel enclosure and completed by special pressure and electrical connections. The features of these transducers are very precise, highly linear and long-term stable.

Due to the applied strain on one side of the sensor by the pressure, the sensor chip will be deformed which causes changes in the electrical resistance in the strain gauges on the sensor's back-side. The changes in resistance are measured by a Wheatstone bridge with an out-put signal as small voltage (between 0 and 20 mV). Therefore for the transfer into pressure data a calibration of the whole transducer is needed. The aim of such calibration is to obtain the values of parameters for the transfer of the acquired pressure expressed in voltage (mV) into the pressure expressed in pressure units (Pa). These parameters typically give a linear relationship with the transducer's slope and offset. Both parameters are specific of each transducer and depend on temperature and gain. The accurate determination of these parameters is fundamental for the interpretation of experimental results and for their quantitative analysis. The calibration procedure trusts on a reference pressure provided by a pressure calibrator. During the calibration routine the reference pressure source is connected with the pressure measurement chain (transducer and amplifier). The reference pressure is increased stepwise and the associated electrical output signal is registered. The voltage output signal is plotted over the reference pressure. The linear regression of this plot gives the two important quantities of the pressure transducer, the slope and the offset. The slope is a kind of conversion constant.

2. Drop/Bubble Deformations

Equation (3) is applicable only for spherical interfaces, which are formed for Bond numbers close to zero. With increasing Bond number, due to the increasing drop deformation, the mean curvature of the interface becomes non-uniform and the capillary pressure, described in this case by Eq. (1), becomes different in each point of the interface. Under the conditions of deformed interfaces any accurate measurement of the capillary pressure becomes impossible.

Usually for drops and bubbles used in CPTs the Bond numbers is in the range between 0 and 10^{-1}. In [22] MacLeod *et al.* calculated the ratio between the mean curvature and the radius of the equivalent sphere for a pendant drop depending on the Bond number. Such calculations were performed for the local curvatures at the apex, at the equator and at the capillary tip. For $Bo < 10^{-2}$ the drop deformations are practically negligible. However, above this value, the ratios at the apex and at the tip deviate rapidly from unity, and, at $Bo \approx 10^{-1}$, the deviations become of the order of 10%. Though the deviation is less significant for the equatorial ratio still it is of the order of few percents. Such calculations represent an important guideline for the preparation of experiments.

3. Drop Instability

If measurements with capillary pressure instruments are performed at a liquid/liquid interface and the surfactant is soluble in both phases, then convective Marangoni instabilities due to surfactant transfer through the interface can come into play and affect the results. According to [63], for example, the growth of a droplet of pure water in toluene is accompanied by a slow recirculation of the interface toward the rear of the droplet. If, however, a solute (diacetone alcohol, propionic acid or acetone) is added to the toluene phase, then chaotic spontaneous motion of the interface is observed. Marangoni instabilities were reported for both stationary drops of constant volume [64] and growing drops [65].

For a stationary interface, the criteria enabling the prediction of a Marangoni instability have been derived based on the results of linear stability analysis, supposing that in the initial stable state the surfactant is transferred solely by diffusion. A review of available criteria is given in [59]. Most of those criteria deal with planar interfaces and only in [67] a spherical interface, more suitable to the geometry of capillary pressure methods was considered. The necessary conditions for an instability are based on the relations between the kinematic viscosities of contacting liquid phases and the surfactant diffusion coefficients in them. The system remains stable if the surfactant is transferred from the phase with larger the surfactant diffusion coefficient. For spherical interfaces the criteria differ for the transfer out of the drop and into the drop. Instabilities caused by transfer out of the drop develop only at sufficiently high values of the interfacial tension whereas it can develop at any interfacial tension for a transfer into the drop [67].

The situation becomes much more complicated when the interface moves, which is the case for growing or oscillating drop techniques. In this case the surfactant transfer to and from the interface occurs not only by diffusion but also by convection which can dominate the mass transfer.

A comprehensive experimental study on Marangoni instabilities appearing in measurements with growing drops is presented in [68]. The study was performed with the tensiometer PAT-1 equipped with the ODBA module (SINTERFACE Technologies, Berlin) [69] for water–hexane system (see also Chapter 2): aqueous drops growing into a continuous hexane phase. Capillaries with tip diameters of 0.3, 0.5 and 0.8 mm were used in the experiments with four experimental protocols:

  (i) pure water — pure hexane;
  (ii) pure water — 10 mM solution of hexanol in hexane;
  (iii) 10 mM aqueous solution of hexanol — pure hexane;
  (iv) 10 mM aqueous solution of hexanol — 20 mM solution of hexanol in hexane.

Marangoni instabilities, which appear as an uncontrolled shaking of the drop (Fig. 12), were observed only for the experimental protocol (iii).

The instability appeared to depend on the drop growth rate and was more pronounced for smaller drop growth rates (Fig. 13) and larger capillary diameters at the same drop growth rate (Fig. 14). For the thinnest capillary used, the only weak instability was observed at the smallest flow rate of 0.16 mm^3/s^2, whereas there was no instability at higher flow rates.

The strength of the instability decreased with increasing flow rate (Fig. 13) and decreasing capillary diameter (Fig. 14). The instability starts at a certain drop size, which increases with increasing flow rate, and terminates after a certain time, but before drop detachment. It should be stressed that the onset of an instability results

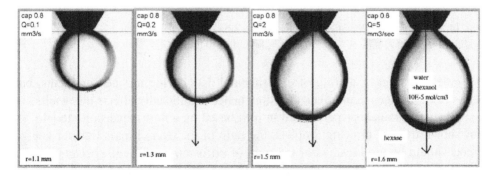

Fig. 12. Shaking of a droplet due to a Marangoni instability caused by transfer of hexanol from water to the hexane phase according to the experimental protocol (iii); according to [68].

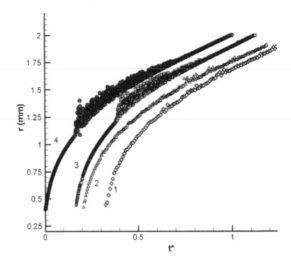

Fig. 13. Development of Marangoni instabilities depending on the liquid flow rate: (1) 2 mm^3/s, (2) 1 mm^3/s, (3) 0.2 mm^3/s, (4) 0.16 mm^3/s; diameter of capillary tip 0.8 mm; according to [68].

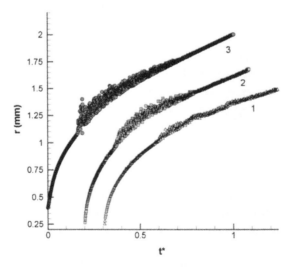

Fig. 14. Development of Marangoni instability depending on the capillary tip diameter: (1) 0.3 mm, (2) 0.5 mm, (3) 0.8 mm; flow rate 0.16 mm^3/s; according to [68].

not only in a large scattering of experimental data during the measurements, but also causes changes in interfacial tension, hence significantly affects the results.

A theoretical analysis performed in [68] based on a flow expansion model [70] has shown that for growing droplets, the ratio of the overall mass transfer coefficients should be used instead of the ratio of diffusion coefficients, because of the significant contribution of convection in the surfactant mass transfer. In contrast to the diffusion coefficient, the mass transfer coefficient changes with time and varies in space depending on local values of convection and mixing strength.

According to [70] the overall mass transfer coefficient k for a growing drop can be estimated as

$$k = k'CDSc^{1/3}Re_N^n R_N^{-1}(t^*)^{(-n-1)/3},$$ (37)

where k' is a calibration constant, C is the concentration, D is the diffusion coefficient, $Sc = v/D$ is the Schmidt number, Re_N is the Reynolds number for the liquid flow through the capillary, R_N is the capillary radius, $t^* = t/T_N$ is the dimensionless time with $T_N = R_N/U_N$, U_N is the liquid velocity at the nozzle.

If an instability occurs during capillary pressure measurements, then Eq. (37) and the experimental results presented in [70] provide a guideline for changes of the experimental conditions to prevent the onset of instabilities.

4. Measurement of Pressure Offsets and Cell Compressibility

As discussed previously, the volume compressibility represents a critical aspect for CPT. Such compressibility originates both from the residual amount of gas entrapped or dissolved in the liquids, and from the mechanical elasticity of the cell itself. Reducing as much as possible this compressibility is an important target in the design of the cell.

As described for the Expanded Drop method, when using small capillaries, with diameters below 0.4 mm, the droplet volume may even exhibit an instable behavior, which may have a seriously impact on the measurements. It is therefore suitable to reduce as much as possible the utilization of materials such as elastomers, as for example o-rings for the mounting. In addition it is worth to utilize for the experiments degassed liquids, which can be easily obtained by different standard procedures. However, due to very small volumes of the droplets in the experiments, the compressibility can never be reduced to negligible values. It is therefore quite important to estimate and account for it in the experiment modelling, as, for example, discussed for the Oscillating Drop method.

In order to perform such estimation it is useful to assume that the compressibility originates from an equivalent gas volume. Assuming a perfect gas behavior and taking as reference the gas volume, V_B, at the pressure corresponding to the hemispherical drop, at a given capillary pressure, ΔP_{cap}, the compressibility contribution, ΔV_B, to the droplet volume is given by

$$\Delta V_B = V_B \left(\frac{\Delta P_{hem} - \Delta P_{cap}}{\Delta P_{cap} + P_{atm}} \right).$$ (38)

where ΔP_{hem} is the capillary pressure at hemisphere.

Different procedures have been proposed to estimate V_B. The most effective is probably to compare during a Growing Drop experiment the droplet radii measured from the image with those expected from the liquid flow. This, however, requires to equip the tensiometer with an effective analysis technique.

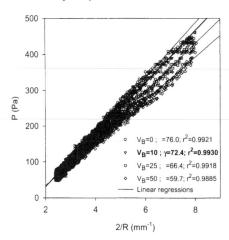

Fig. 15. Sets of pressure *vs.* curvature data obtained for different values of V_B (see text) from the P data of Fig. 3a; the correct value of V_B maximizes the correlation coefficient of the linear best fit.

An equivalent method not requiring radius measurement is instead based on the application of the Pressure Derivative method. In the presence of compressibility the Eq. (6) can be rearranged to

$$w(t - t_{hem}) + \Delta V_B = \frac{2}{3}\pi(R^3 - a^3) + \frac{\pi}{3}\sqrt{R^2 - a^2}(2R^2 + a^2). \tag{39}$$

When solving this equation to obtain R, it is obvious that only the correct value of ΔV_B yields a linear relationship between the measured P and $2/R$. The value of ΔV_B can be then evaluated as the one maximizing the correlation coefficient, r^2, of the linear best fit of the P *vs.* $2/R$ data.

The utilization of this method to obtain simultaneously the interfacial tension, the volume of the equivalent bubble, V_B, and P_0 is illustrated in Fig. 15. The figure shows the different sets of $P(2/R)$ data, where R is obtained from the data reported in Fig. 3a by solving Eq. (40) with different values of V_B. The maximum in the correlation coefficients corresponds to the set plotted in Fig. 3b, which provides in addition the expected value of the interfacial tension.

The procedure allows also for the simultaneous evaluation of P_0, which is the constant term obtained from the linear best fit and corresponds to the hydrostatic contribution to the measured pressure. It is however worth to note that P_0 depends only weakly on V_B.

In spite of being developed for constant interfacial tension, similar concepts can be utilized to evaluate the compressibility and P_0 for surfactant solutions.

In particular P_0 can be effectively evaluated by a long and slow growing drop from the best fit of the $P(2/R)$ data corresponding to the large radii. In fact, for a sufficiently slow flow rate the drop growth can be considered at a constant interfacial tension.

Another similar procedure is commonly utilized [37, 38] to evaluate the system's compressibility and equilibrium interfacial tension of the reference state in Oscillating Bubble/Drop experiments to measure the dilational visco-elasticity. Starting from a hemispherical droplet, the volume is increased in steps by few percents each. After each step the droplet is left to equilibrate for interfacial tension and the values of P and $2/R$ are then evaluated. The duration of each step must be sufficiently long to achieve equilibration. For usual surfactants that occurs within few minutes due to the small relative variation of the interfacial area at each step. In this way a set of 10 to 20 data points of $P(2/R)$ is obtained which can be treated in a similar way as illustrated in Fig. 15.

5. Capillary Size and Wetting

One of the most important elements in CPT is the capillary at the tip of which the drop or the bubble is formed. The geometric characteristics and wetting properties of the capillary can strongly influence the experimental results. The capillary inner and outer radii should be sufficiently small to provide the formation of small spherical drops or bubbles. The capillary inner radius and length determine the hydrodynamic resistance of the capillary, which should not be too high. Usually one uses capillaries (or needles) with an inner radius in the range of 0.1–0.5 mm and a length of about 3 mm.

It should be noted that the entrance regions of the capillary contribute to the total hydrodynamic resistance, which is not accounted for in the above presented equations. To correct this situation, in practice one uses an effective capillary length which is slightly larger than the actual one.

As already underlined, the capillary size has a major impact also in determination of the conditions for the sudden growth of the drop/bubble volume after reaching the hemispherical shape [17]. For the application of all methods illustrated so far, except for the Expanded Drop method, the size of the capillary should be then large enough to avoid this effect. As general indication, the radius of the capillary must provide values of the Bubble Stability Number (Eq. (7)) of the order of 0.1.

Finally, the capillary radius (as the drop base radius) is included in the geometric characteristics of the meniscus, such as $\frac{da_0}{d\ln A}$, $\frac{da}{dV_m}$ and $\frac{d\ln A}{dV_m}$, which are necessary for calculations of the interfacial tension and dilational elasticity.

It should be noted, however, that in spite of their importance detailed studies of the effects of the geometric characteristics of the capillaries on experimental results have not been performed or not published yet.

The material of which the capillaries are made is also very important. The wetting properties of the capillary surface determine where the three-phase contact line will be fixed. It can be fixed either at the inner edge of the capillary tip [20–22], or at its outer edge [22–25]. Two difficulties can arise during the measurements. When the inner (i.e. drop forming) liquid perfectly wets the capillary then it can climb the outside walls, and small spherical drop can not be formed at the capillary tip [22–24]. When the inner liquid only partially wets the capillary then the contact line can slide across the capillary tip between its inner and outer edges during the growth of the drop [22, 24]. In [24] it is stated that in principle measurements are possible even in the case of a sliding contact line provided that the imaging system is sufficiently fast to give the instantaneous location of the contact line (note, however, that any drop or bubble needs to be axisymmetric). Nevertheless the authors find it preferable to keep the contact line fixed at the capillary edge. To avoid the contact line sliding one usually looks for a suitable material for the capillaries, modify the wetting properties of the capillary surface, makes sharper the edges of the capillary tip or makes the tip face rough [22, 24]. In [23] a composite capillary has been used to avoid the climbing and contact line sliding.

6. Evaporation/Dissolution of Drop Liquid

One of the critical aspects in capillary pressure tensiometry is evaporation of the liquid forming the drop. This leads to loss of the drop material and therefore to uncontrolled changes in the drop size during the experiment. In liquid–liquid system dissolution of the drop material can lead to the same results. Moreover, these processes can lead to changes of the local temperature and concentrations at the interface between two contacting phases. Therefore it is very important to use a gas phase saturated by the drop forming liquid and to use mutually saturated liquids when studying liquid–liquid interfaces.

E. Capillary Pressure Instruments for Space Experiments

As stated above microgravity represents much more favourable conditions for CPT measurements because of both minimization of the Bond number and preventing buoyancy driven convection. At the same time many other methods for studying dynamic properties of liquid interfaces are not applicable in absence of gravity. This was the stimulus to develop special equipments for CPT under microgravity conditions.

The first interfacial tension measurement with the PD method were performed by Passerone *et al.*, in 1990 in a sounding rocket of the European Space Agency in the framework of MITE (Measurement of Interfacial Tension Experiment) [19]. A new experiment was designed in 1994, flown again on a sounding rocket, with a new module and experimental design (MITE-2). These studies formed the basis for wider experiments on interfacial dynamics during the two FAST (Facility for Adsorption and Surface Tension) missions in 1998 and 2003 on space shuttles of NASA. These included ED, oscillating drop and stress–relaxation experiments with a non-ionic surfactant, $C_{10}EO_8$, at the water–hexane interface and oscillating bubble experiments with another non-ionic surfactant, $C_{12}DMPO$, at the water–air interface [32]. The interfacial relaxations and rheological interfacial properties were investigated for a wide range of surfactant concentrations at three different temperatures. The results confirmed that the CPT is a reliable technique for microgravity investigations.

The present accommodation of the FASTER (Facility for Adsorption and Surface Tension Research) facility on the Columbus Laboratory of the International Space

Station involves a new detailed design, manufacturing and assembly. It is capable of dosing increasing amounts of two different surfactants and of observing surface tension phenomena as responses to different kinds of stimuli, like oscillations or trapezoidal pulses at liquid/liquid or liquid/gas interfaces. More details about this facility were given in a recent review [18].

F. Experimental Results

A huge number of experimental results obtained with CPTs has been published in literature. Here we discuss only few selected experimental results to demonstrate the capabilities of these techniques.

The values of the interfacial tension of some organic liquids *vs.* water at 20°C measured by the PD method are presented in [17]. Reference values from literature are in a good agreement with the measured values.

An example of ED experimental results is presented above in Fig. 5 showing the plot of the measured dynamic interfacial tension for the adsorption process of Triton X-100 at a fresh water/hexane interface for two surfactant concentrations. These results have been confirmed by microgravity experiments in a sounding rocket [71].

Many experimental data obtained with the GD tensiometer are presented in [22, 23]. They concern the dynamic interfacial tension of liquid–liquid and liquid–air systems for several surfactant solutions. In [23] a theoretical model has been developed predicting the dynamic behaviour of the interfacial tension on a growing drop. The model has been specified for diffusion-limited, kinetic-limited and combined diffusion–kinetic-limited adsorption. The effect of initial adsorption has been demonstrated. The model allows describing the experimentally observed maxima in dynamic surface tension *vs.* time data. The comparison between theory and experiments provides information about the mechanism governing the adsorption of the surfactants and their diffusion coefficients.

Particular interesting results have been obtained in [22] with the GD method for SDS/dodecanol mixtures, which are reproduced in Fig. 16. The non-monotonous behaviour of the lower curve is explained by the presence of dilute dodecanol impurity in the aged SDS solution.

Figure 17 demonstrates the applicability of small-amplitude transient perturbations of the interfacial area to study the 2D rheology at liquid interfaces. In this particular example the response to a trapezoidal perturbation of the interfacial area for a $C_{12}DMPO$ solution is shown. The results are obtained by CPT under microgravity conditions during the FAST mission [72]. The theoretical curve describes the expected response by the assumption of a diffusion controlled adsorption.

Figure 18 represents an example of rheological studies on a model particle–surfactant system performed by the oscillating drop method [73]. In this study a drop of aqueous dispersion of silica nanopartices plus CTAB was formed in hexane. It has been shown that a reorganisation process at the interface has to

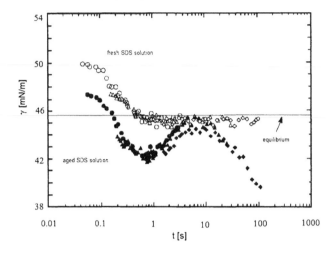

Fig. 16. Dynamic surface tension of fresh and aged 6 mM aqueous SDS solutions obtained with growing drops in pre-equilibrated air at 23°C; $Q = 100$ (○), 15 (△) and 3 (◇) mm³/min; according to [22].

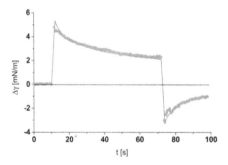

Fig. 17. Surface tension response to a small-amplitude trapezoidal perturbation of the interfacial area. C_{12}DMPO at water/air interface ($c = 1.7 \cdot 10^{-8}$ mol/cm³, $T = 15°$C). The theoretical curve is calculated according to the model of diffusion controlled adsorption; according to [72].

be accounted for in order to explain the viscoelastic characteristics of the mixed particle-surfactant interfacial layer.

G. Conclusions

The above considerations show that the capillary pressure tensiometry is represented by a large group of experimental methods where the capillary force under a curved interface is directly measured by a pressure transducer. This principle provides a number of advantages compared to other tensiometric techniques. First of all the modern transducers allow fast and accurate acquisition of pressure values inside very small liquid volumes. At the same time the modern piezo actuators allow precise manipulations with such small volumes, thus giving the opportunity for precise variations of the interfacial area according to practically any chosen time law. Com-

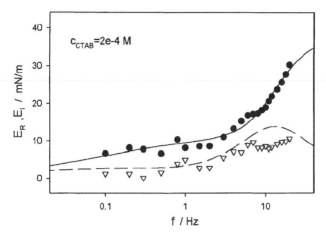

Fig. 18. Real (•) and imaginary (∇) part of the dilational viscoelasticity for the interface between aqueous dispersion (1% of silica nanopartices in presence of 2×10^{-4} M CTAB) and hexane; according to [73].

bining such stimuli and detection elements with novel computer techniques allows one to produce fully automatic systems working according to a chosen protocol.

The CPT technique is suitable to investigate different dynamic aspects of non-equilibrium adsorption layers at liquid interfaces. It allows monitoring dynamic interfacial tension changes which are affected by various relaxation processes, such as diffusion exchange, surface re-arrangement of the adsorbed molecules, etc. The time-scale of such relaxation processes can range from milliseconds up to thousands of seconds. The applicability of the CPT technique was demonstrated for various liquid–liquid and liquid–gas systems including thin liquid films. An important advantage of CPT is also its applicability under reduced gravity conditions. All these factors explain the increasing interest into this methodology.

There are however some common critical aspects inherent to CPT methods, such as drop/bubble shape deformations, high dependency on the capillary properties, complicated calibration procedures, high sensitivity to external disturbances, drop/bubble explosive growth, interfacial instability at high flow rates, etc. The effect of such critical aspects can be minimized by a careful preparation of the experiments.

H. Acknowledgements

This work was performed within the framework of MAP AO-99-052, FASES of the European Space Agency. The authors acknowledge also financial support by projects of COST (D43 and P21), the DFG (Mi418/16-2 and Mi418/18-1), the Italian Space Agency (ASI I/002/10/0), and the German Space Agency (DLR 50WM0941).

I. References

1. T.F. Tadros, Colloid Stability, in "Colloid and Interface Science Series", Vol. 1, Wiley-VCH, Weinheim, 2007.

2. R. Wüstneck, J. Perez-Gil, N. Wüstneck, A. Cruz, V.B. Fainerman and U. Pison, Adv. Colloid Interface Sci., 117 (2005) 33.

3. V.N. Kazakov, O.V. Sinyachenko, V.B. Fainerman, U. Pison and R. Miller, Dynamic surface tension of biological liquids in medicine, in "Studies in Interface Science", D. Möbius and R. Miller (Eds), Vol. 8, Amsterdam, Elsevier, 1999.

4. V.N. Kazakov, A.F. Vozianov, O.V. Sinyachenko, D.V. Trukhin, V.I. Kovalchuk and U. Pison, Adv. Colloid Interface Sci., 86 (2000) 1.

5. G. Loglio, N. Degli-Innocenti, U. Tesei, R. Cini and W. Qi-Shan, Nuovo Cimento, 12 (1989) 289.

6. J.C. DaSilva, S.A. Ermakov, I.S. Robinson, D.R.G. Jeans and S.V. Kijashko, J. Geophys. Research, 103 (1998) 8009.

7. V.V. Krotov and A.I. Rusanov, Physicochemical hydrodynamics of capillary systems, Imperial College Press, London, 1999.

8. V.G. Levich, Physicochemical Hydrodynamics, Prentice-Hall, Englewood Cliffs, NJ, 1962.

9. D.A. Edwards, H. Brenner and D.T. Wasan, Interfacial transport processes and rheology, Butterworth-Heinemann, Boston, 1991.

10. L.G. Leal, Laminar flow and convective transport processes. Scaling principles and asymptotic analysis, Butterworth-Heinemann, Boston, 1992.

11. P. Joos, Dynamic surface phenomena, VSP, Dordrecht, 1999.

12. S.S. Dukhin, G. Kretzschmar and R. Miller, Dynamics of adsorption at liquid interfaces. Theory, experiment, application, in "Studies in Interface Science", Vol. 1, Elsevier, Amsterdam, 1995.

13. B.A. Noskov and G. Loglio, Colloids Surfaces A, 143 (1998) 167.

14. B.A. Noskov, Physical chemistry of capillary waves, D. Sc. thesis, Saint Petersburg University, 1999.

15. Interfacial rheology, in "Progress in Colloid and Interface Science", R. Miller and L. Liggieri (Eds), Vol. 1, Brill, Leiden-Boston, 2009.

16. A.I. Rusanov and V.A. Prokhorov, Interfacial Tensiometry, in "Studies of Interface Science", D. Möbius and R. Miller (Eds), Vol. 3, Elsevier, Amsterdam, 1996.

17. L. Liggieri and F. Ravera, Capillary pressure tensiometry and applications in microgravity, in "Drops and Bubbles in Interfacial Research", D. Möbius and R. Miller (Eds), Vol. 6, pp. 239–278, Elsevier, Amsterdam, 1998.

18. V.I. Kovalchuk, F. Ravera, L. Liggieri, G. Loglio, P. Pandolfini, A.V. Makievski, S. Vincent-Bonnieu, J. Krägel, A. Javadi and R. Miller, Adv. Colloid Interface Sci., 161 (2010) 102.

19. A. Passerone, L. Liggieri, N. Rando, F. Ravera and E. Ricci, J. Colloid Interface Sci., 146 (1991) 152.

20. R. Nagarajan and D.T. Wasan, J. Colloid Interface Sci., 159 (1993) 164.

21. J.M. Soos, K. Koczo, E. Erdos and D.T. Wasan, Rev. Sci. Instrum., 65 (1994) 3555.

22. C.A. MacLeod and C.J. Radke, J. Colloid Interface Sci., 160 (1993) 435.

23. C.A. MacLeod and C.J. Radke, J. Colloid Interface Sci., 166 (1994) 73.

24. X. Zhang, T. Harris and O.A. Basaran, J. Colloid Interface Sci., 168 (1994) 47.

25. A. Javadi, J. Krägel, P. Pandolfini, G. Loglio, V.I. Kovalchuk, E.V. Aksenenko, F. Ravera, L. Liggieri and R. Miller, Colloids Surfaces A, 365 (2010) 62.

26. L. Liggieri, F. Ravera and A. Passerone, J. Colloid Interface Sci., 169, (1995) 226.

27. L. Liggieri, F. Ravera, A. Passerone, A. Sanfeld and A. Steinchen, Lecture Notes in Physics, 467 (1996) 175.

28. P.J. Breen, Langmuir, 11 (1995) 885.

29. G. Kretzschmar and K. Lunkenheimer, Ber. Bunsenges Phys. Chem., 74 (1970) 1064.

30. K.-D. Wantke, K. Lunkenheimer and C. Hempt, J. Colloid Interface Sci., 159 (1993) 28.

31. K.-D. Wantke and H. Fruhner, in "Studies in Interface Science", D. Möbius and R. Miller (Eds), Vol. 6, Elsevier Science, Amsterdam, 1998, p. 327.

32. L. Liggieri, F. Ravera, M. Ferrari, A. Passerone, G. Loglio, R. Miller, J. Krägel and A. Makievski, Microgravity Sci. Technol. XVI-I (2005) 196.

33. M. Simon, Ann. Chim. Phys., 32 (1851) 5.

34. K.J. Mysels, Colloids and Surfaces, 43 (1990) 241.

35. V.B. Fainerman and R. Miller, Adv. Colloid Interface Sci., 108–109 (2004) 287.

36. N. Alexandrov, K.G. Marinova, K.D. Danov and I.B. Ivanov, J. Colloid Interface Sci., 339 (2009) 545.

37. L. Liggieri, V. Attolini, M. Ferrari and F. Ravera, J. Colloid Interface Sci., 252 (2002) 225.

38. F. Ravera, M. Ferrari, E. Santini and L. Liggieri, Adv. Colloid Interface Sci., 117 (2005) 75.

39. L. Liggieri, F. Ravera and A. Passerone, J. Colloid Interface Sci., 140 (1990) 436.

40. T. Horozov and L. Arnaudov, J. Colloid Interface Sci., 219 (1999) 99.

41. J. Lucassen and M. van den Tempel, Chem. Eng. Sci., 27 (1972) 1283.

42. J. Lucassen and M. van den Tempel, J. Colloid Interface Sci., 41 (1972) 491.

43. G. Loglio, U. Tesei and R. Cini, J. Colloid Interface Sci., 71 (1979) 316.

44. G. Loglio, U. Tesei and R. Cini, Colloid Polymer Sci., 264 (1986) 712.

45. R. Miller, J.K. Ferri, A. Javadi, J. Krägel, N. Mucic and R. Wüstneck, Colloid Polymer Sci., 288 (2010) 937–950.

46. V.I. Kovalchuk, J. Krägel, R. Miller, V.B. Fainerman, N.M. Kovalchuk, E.K. Zholkovskij, R. Wüstneck and S.S. Dukhin, J. Colloid Interface Sci., 232 (2000) 25.

47. F. Ravera, G. Loglio, P. Pandolfini, E. Santini and L. Liggieri, Colloids and Surfaces A: Physicochem. Eng. Aspects, 365 (2010) 2–13.

48. F. Ravera, G. Loglio and V.I. Kovalchuk, Current Opinion in Colloid & Interface Science, 15 (2010) 217–228.

49. J. Örtegren, K.-D. Wantke, H. Motschmann and H. Möhwald, J. Colloid Interface Sci., 279 (2004) 266.

50. S.C. Russev, N. Alexandrov, K.G. Marinova, K.D. Danov, N.D. Denkov, L. Lyutov, V. Vulchev and K. Bilke-Krause, Rev. Sci. Instr., 79 (2008) 104102.

51. G. Loglio, P. Pandolfini, R. Miller and F. Ravera, Langmuir, 25 (2009) 12780.

52. N.A. Mishchuk, S.S. Dukhin, V.B. Fainerman, V.I. Kovalchuk and R. Miller, Colloids and Surfaces A, 192 (2001) 157.

53. V.I. Kovalchuk and S.S. Dukhin, Colloids Surfaces A, 192 (2001) 131.

54. Y.-H. Kim, K. Koczo and D.T. Wasan, J. Colloid Interface Sci., 187 (1997) 29–44.

55. H. Bianco and A. Marmur, J. Colloid Interface Sci., 158 (1993) 295.

56. V.I. Kovalchuk, A.V. Makievski, J. Krägel, P. Pandolfini, G. Loglio, L. Liggieri, F. Ravera and R. Miller, Colloids Surfaces A, 261 (2005) 115.

57. A.V. Makievski, V.I. Kovalchuk, J. Krägel, M. Simoncini, L. Liggieri, M. Ferrari, P. Pandolfini, G. Loglio and R. Miller, Microgravity — Science and Technology Journal, 16 (2005) 215.

58. D. Georgieva, A. Cagna and D. Langevin, Soft Matter, 5 (2009) 2063.

59. V.I. Kovalchuk, J. Krägel, P. Pandolfini, G. Loglio, L. Liggieri, F. Ravera, A.V. Makievski and R. Miller, Dilatational rheology of thin liquid films, in "Interfacial rheology", Progress in Colloid and Interface Science, R. Miller and L. Liggieri (Eds), Vol. 1, Brill, Leiden-Boston, 2009, p. 476.

60. V.I. Kovalchuk, J. Krägel, E.V. Aksenenko, G. Loglio and L. Liggieri, Oscillating bubble and drop techniques, in "Novel Methods to Study Interfacial Layers", Studies in Interface Science, D. Möbius and R. Miller (Eds), Vol. 11, Elsevier, Amsterdam, 2001, p. 485.

61. F. Ravera, L. Liggieri and G. Loglio, Dilational Rheology of Adsorbed Layers by Oscillating Drops and Bubbles, in "Interfacial Rheology", R. Miller, L. Liggieri (Eds), Progress in Colloid and Interface Science series, Vol. 1, Brill, 2009, p. 137.

62. PDCR 4000 Data-sheet, 920-214B_E; 2005, GE Sensing and Inspection Technology.

63. J.D. Thornton, T.J. Anderson, K.H. Javed and S.K. Achwal, AIChE J., 31 (1985) 1069.

64. B. Arendt and R. Eggers, Int. J. Heat Mass Transfer, 50 (2007) 2805.

65. M. Wegener, A.R. Paschedag and M. Kraume, Int. J. Heat Mass Transfer, 52 (2009) 2673.

66. N.M. Kovalchuk and D. Vollhardt, Adv. Colloid Interface Sci., 120 (2006) 1.

67. T.S. Sørensen, J. Chem. Soc. Faraday II, 76 (1980) 1170.

68. A. Javadi, D. Bastani, J. Krägel and R. Miller, Colloids Surfaces A, 347 (2009) 167.

69. G. Loglio, P. Pandolfini, R. Miller, A.V. Makievski, F. Ravera, M. Ferrari and L. Liggieri, Drop and bubble shape analysis as tool for dilational rheology studies of interfacial layers, in "Novel Methods to Study Interfacial Layers", D. Möbius, R. Miller (Eds), Studies in Interface Science, Vol. 11, Elsevier, Amsterdam, 2001, p. 439.

70. A. Javadi, D. Bastani and M. Taeibi-Rahni, AIChE J., 52 (2006) 895.

71. L. Liggieri, F. Ravera and A. Passerone, in "Proceedings of The Second European Symposium on Fluids in Space — Naples — April 1996", A. Viviani (Ed.), p. 135.

72. L. Liggieri, F. Ravera, M. Ferrari, A. Passerone, G. Loglio, R. Miller, A. Makievski and J. Krägel, Microgravity Sci. Technol. Journal, 16 (2006) 112.

73. F. Ravera, M. Ferrari, L. Liggieri, G. Loglio, E. Santini and A. Zanobini, Colloids Surfaces A, 323 (2008) 99.

Direct Determination of Protein and Surfactant Adsorption by Drop and Bubble Profile Tensiometry

V.B. Fainerman [a] **and R. Miller** [b]

[a] Donetsk Medical University, 16 Ilych Avenue, 83003 Donetsk, Ukraine
[b] Max-Planck-Institut für Kolloid- und Grenzflächenforschung, Am Mühlenberg 1,
14424 Potsdam, Germany

Contents

A. Introduction

The direct measurement of protein adsorption at liquid interfaces is a quite compli-cated technical problem. The most common methods here are radiotracer technique, neutron reflection and interfacial ellipsometry. Advantages and drawbacks of these methods were extensively discussed in [1–7]. For example, to apply the ellipsom-etry, one has to assume a model for the surface layer structure, in particular the refractive index profile of the protein solution adjacent to the interface, which restricts the applicability of the method. Extensive experimental studies of sur-factant adsorption layers at liquid interfaces were performed by neutron reflection and radiotracer techniques [8–11]. Additional optical methods, such second har-monic generation, sum frequency spectroscopy, infrared and Raman spectroscopy, X-ray reflection are thoroughly analysed and widely referenced in [10, 11]. De-spite some deficiencies and model assumptions, these methods have considerably improved our understanding of the adsorption mechanism at liquid-fluid inter-faces.

The interrelation between the chemical structure of a surfactant and its ability to be adsorbed at an interface and modify its properties is one of the main questions of

Bubble and Drop Interfaces
© Koninklijke Brill NV, Leiden, 2011

surfactant science [11–24]. The comparison of experimental data with various theoretical models allows to decide which of the theoretical models is most appropriate and can be suggested as valid adsorption mechanism for the particular surfactant. However, more reliable are conclusions about the suitability of a theoretical model when in addition to equilibrium and dynamic surface (interfacial) tensions also data from other experimental techniques (surface rheology, ellipsometry, neutron and X-ray reflection, Brewster angle microscopy and other optical methods) are available.

The experimental techniques based on drop or bubble profile analysis can be employed to obtain extensive information about the adsorbed amount of highly surface active surfactants (e.g., oxyethylated alcohols) and proteins. From comparison of surface tension isotherms obtained by drop shape and bubble shape experiments and employing the mass balance for the solution bulk and the surface area of the drop or bubble, respectively, the amount of adsorbed surfactants or proteins can be calculated [25–28].

In this chapter, the adsorption of some non-ionic surfactants [26, 28] and proteins [25, 27] at the solution/air interface studied by bubble and drop profile analysis tensiometry are presented, and the adsorption values obtained are compared with those predicted by respective theoretical models.

B. Experiments and Determination of Adsorption

The experiments were performed with bubble/drop profile analysis tensiometers (PAT-1 and PAT-2P, SINTERFACE Technologies, Germany) as described elsewhere [29, 30]. The determination of the adsorbed amount requires studies with the emerging bubble (with volume of the measuring glass cell $V = 20$ ml) and pendant drop methods for the same solutions. The combination of the two tensiometry methods, each having a significantly different ratio of total surface area to solution bulk volume (drop and bubble shape analysis) represents a new methodology for the determination of the adsorbed amount.

Let us consider a drop of a protein or surfactant solution with the initial concentration c_D. During the adsorption process, some amount of the substance becomes accumulated in the surface layer, therefore the concentration inside the drop is reduced. Finally, when the adsorption equilibrium is attained, the concentration inside the drop becomes equal to c_0. The balance of protein or surfactant in the drop bulk and its surface at equilibrium reads:

$$\frac{\Gamma S_D}{V_D} = c_D - c_0, \tag{1}$$

where Γ is the equilibrium adsorption value, S_D and V_D are the area and the volume of the drop, respectively. In bubble shape experiments (and also in adsorption experiments from a large solution volume to a plane surface) the mass of the adsorbed protein or surfactant can be neglected, i.e., the initial concentration in the surrounding solution c_B is practically equal to the equilibrium concentration c_0.

When in these two experiments the equilibrium surface or interfacial tensions γ are equal to each other, the corresponding equilibrium adsorption values Γ are also equal to each other. With this assumption ($\Gamma = $ const if $\gamma = $ const) one can calculate the adsorbed amount Γ from the difference between the initial concentration in the drop and that in the solution surrounding the bubble, c_D and c_B, respectively:

$$\Gamma\left(\frac{A_D}{V_D} - \frac{A_B}{V_B}\right) = (c_D - c_B)_{\gamma=const}, \tag{2}$$

where A_B is the area of the bubble, and V_B is the volume of the solution surrounding the bubble. Typically, $V_D/A_D = 0.5$–0.7 mm, while for the bubble shape method the ratio of the volume of the solution surrounding the bubble to the bubble surface area is very large, $V_B/A_B > 100$ mm. Therefore, the variation in the concentration of the ambient solution caused by adsorption at the bubble surface, i.e. the second term in the parentheses in the left hand side of Eq. (2), can be neglected and we obtain [26]:

$$\Gamma = \frac{V_D}{A_D}(c_D - c_B)_{\gamma=const}. \tag{3}$$

Instead of the bubble shape method, other methods can be used, for which the ratio A/V is small enough, e.g., ring or plate tensiometry. If the solution depth in the measuring cell is high, say 50 mm (for this case the ratio $V/A = 50$ mm), we can use c_0 instead of c_B in Eq. (3) with a relative error less than 2%.

C. Results and Discussion

1. Surfactant Solutions

In Figs 1 and 2 the dynamic surface tensions of solutions of the ethoxylated surfactant $C_{14}EO_8$ at several concentrations are shown, as measured by the drop and bubble profile methods (at temperature 25°C) [28]. Note, the results obtained by the two methods differ from each other significantly.

For example, at a $C_{14}EO_8$ concentration of 0.02 μmol/l (see Fig. 1) the surface tension measured by the pendant drop method is almost equal to that of pure water, while at the same time the emerging bubble method shows an essential decrease in surface tension. With increasing $C_{14}EO_8$ concentration the difference between the results obtained by the two methods persists: the surface tension values for the bubble method remains significantly lower than those measured by the pendent drop method. This phenomenon was firstly discussed in [25] and it was shown that the surfactant concentration in the drop interior can become essentially lower than the initial concentration. In contrast, the bubble is surrounded by a large volume of solution so that the respective decrease in surfactant concentration due to adsorption is negligibly small.

The equilibrium surface tensions of $C_{14}EO_8$ solutions estimated from the extrapolation of the dynamic surface tension curves for all concentrations studied by the two methods to infinite time are shown in Fig. 3. One can see that the two

Comparison of Drops and Bubbles

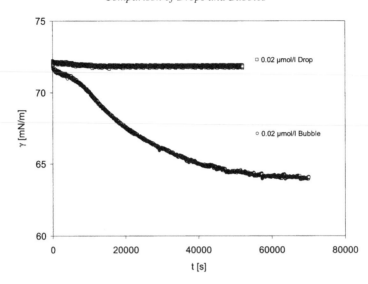

Figure 1. The dependence of dynamic surface tension of $C_{14}EO_8$ solution at the initial concentration 0.02 μmol/l measured by drop shape method (□) and bubble shape method (○).

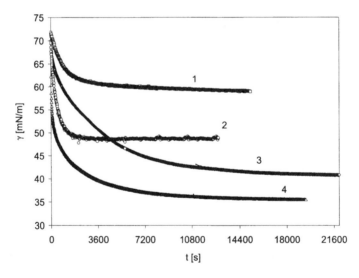

Figure 2. The dependence of dynamic surface tension of $C_{14}EO_8$ solutions at the initial concentrations of 2 μmol/l (1, □; 2, ◇) and 5 μmol/l (3, ○; 4, △) measured by drop shape method (1, □; 3, ○) and bubble shape method (2, ◇; 4, △).

methods yield quite different results, and one obtains identical surface tensions if the initial concentration in the drop is 1 or even 2 orders of magnitude higher than in the corresponding solution which surrounds the bubble. It should be noted that our equilibrium surface tensions measured for $C_{14}EO_8$ using the emerging bubble method are quite consistent with those reported in literature [31–33].

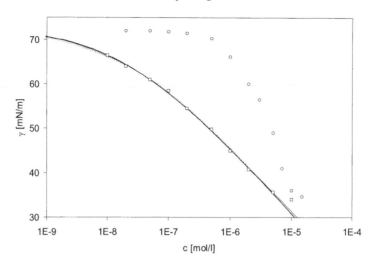

Figure 3. The dependence of equilibrium surface tension of $C_{14}EO_8$ solutions on initial concentration measured by drop shape method (○) and bubble shape method (□). The theoretical curves were calculated from models A and B (thick solid line), C (thin solid line) and D (thin dotted line).

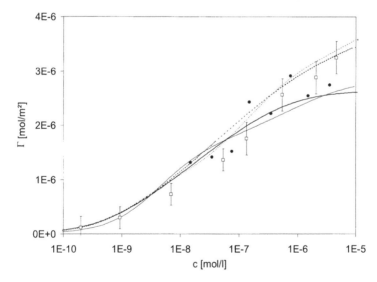

Figure 4. The $C_{14}EO_8$ adsorption isotherm calculated from two surface pressure dependences shown in Fig. 3 using Eq. (3) (□) and from the bubble shape method data using the Gibbs equation (●). Theoretical curves were calculated using model A (thick solid line), model B (thick dotted line), model C (thin solid line) and model D (thin dotted line).

Figure 4 illustrates the adsorption values of $C_{14}EO_8$ obtained from the two surface tension isotherms shown in Fig. 3 using Eq. (3). Note that the Γ values were calculated from the finite (equilibrium) values of the drop volume V_D and drop surface area A_D. During the pendant drop experiments, the drop volume was controlled, and the evaporation losses were compensated by injecting the evaporated solvent only (water) into the drop through the so-called coaxial capillary [28]. Therefore, the total surfactant amount in the drop was kept constant during the experiment, while the surface area of the drop somewhat increased due to changes in shape caused by the surface tension decrease.

Besides the adsorption Γ calculated *via* Eq. (3) from the experimental points given in Fig. 3 those obtained directly from the Gibbs adsorption equation as the slope of the data measured by the emerging bubble method

$$\Gamma = -\frac{1}{RT}\frac{d\gamma}{d\ln c} \tag{4}$$

are shown in Fig. 4 as well. Both sets of data are in a satisfactory agreement.

A theoretical analysis of the experimental dependencies shown in Figs 3 and 4 was performed in [28] using four different models which were described in detail elsewhere [21, 34–36].

1.1. The Frumkin Model without (Model A) and with Two-Dimensional Compressibility (Model B)

It was shown elsewhere, that the molar area of the surfactant molecules ω_S depends linearly on the surface pressure [35, 36]:

$$\omega_S = \omega_0(1 - \varepsilon\Pi). \tag{5}$$

The proportionality factor ε in Eq. (5) is the relative two-dimensional compressibility coefficient of the surfactant surface layer, $\Pi = \gamma_0 - \gamma$ is the surface pressure, γ_0 and γ are the surface tension of solvent and solution, respectively, ω_0 is the molar area of the surfactant at $\Pi = 0$ or the molar area of the solvent. Assuming the ideality of entropy, the respective equation of state and adsorption isotherm read [35]:

$$\Pi = -\frac{RT}{\omega_0}[\ln(1-\theta) + a\theta^2], \tag{6}$$

and

$$bc = \frac{\theta}{(1-\theta)}\exp[-2a\theta]. \tag{7}$$

Here R is the gas law constant, T is the temperature, $\theta = \omega_S\Gamma$ is the surface coverage, Γ is the adsorption, c is the concentration of the surfactant in the solution bulk, a is the intermolecular interaction constant, and b is the adsorption equilibrium constant. With Eq. (5) we can express θ in Eqs (6) and (7) as:

$$\theta = \Gamma\omega_S = \Gamma\omega_0[1 - \varepsilon\Pi]. \tag{8}$$

For $\varepsilon = 0$ we obtain $\omega_S = \omega_0$ and of course the Eqs (6) and (7) become identical with the classical equation of state and adsorption isotherm of the Frumkin model [37].

1.2. Reorientation Model (Model C)

For molecules like oxyethylated surfactants we can assume that they have two orientations in the adsorption layer with different molar areas ω_1 and ω_2 (for definiteness we assume $\omega_1 > \omega_2$). This reorientation model, derived for the case of ideal enthalpy and entropy of mixing of the surface layer, yields the equation of state [21, 35]:

$$\Pi = -\frac{RT}{\omega_S}\ln(1-\theta) \tag{9}$$

and adsorption isotherm

$$b_2c = \frac{\Gamma_2\omega_S}{(1-\theta)^{\omega_2/\omega_S}}. \tag{10}$$

Here Γ $\Gamma_1 + \Gamma_2$ is the total adsorption and ω_S is the mean molar area defined by $\omega_S\Gamma = \theta = \omega_1\Gamma_1 + \omega_2\Gamma_2$.

1.3. Combined Reorientation Model with Two-Dimensional Compressibility in State 2 (Model D)

The mechanism responsible for the variation of the molar area in the model called Frumkin + Compression and given by Eqs (5)–(8) is different from that characteristic to the reorientation model. Hence, it is possible to arrange a combined model, which assumes both the reorientation of molecules and the variation of the molar area in the state with lower area 2 caused by a two-dimensional compressibility according to Eq. (5), i.e., $\omega_2 = \omega_{20}(1 - \varepsilon\Pi)$.

In the application of the theoretical models to the experimental dependencies given in Figs 3 and 4 the following model parameters were used [28].

As one can see from Fig. 3 all models describe the equilibrium surface tension data (obtained by the emerging bubble method) adequately. The experimental adsorption values as a function of the $C_{14}EO_8$ concentration are also in satisfactorily agreement with all models as demonstrated in Fig. 4. The best agreement, however, is obtained by the models B and D which assume a two-dimensional compressibility. While the models A and C yield adsorption values of about $2.6 \cdot 10^{-5}$ mol/m²

Table 1.

Summary of model parameters

Parameter	Model A	Model B	Model C	Model D
$\omega_0 \cdot [10^5 \text{ m}^2/\text{mol}]$	3.8	3.8		
$\omega_1 \cdot [10^5 \text{ m}^2/\text{mol}]$			6.5	10
$\omega_2 \cdot [10^5 \text{ m}^2/\text{mol}]$			3.5	4.4
a	−1.7	−1.7		
α			2.2	0.9
b $[10^5 \text{ m}^3/\text{mol}]$	3.0	3.0	0.32	1.0
ε [m/mN]		0.006		0.008

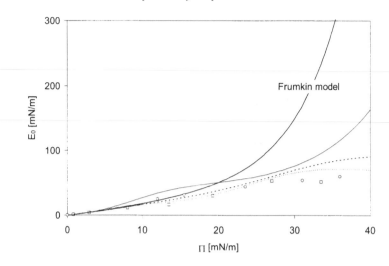

Figure 5. Dependence of limiting (high frequency) elasticity E_0 on the surface pressure of $C_{14}EO_8$ solutions: ($\bigcirc$) — experimental data [28] from oscillating bubble experiment; ($\square$) — experimental data from dependence γ *vs* Γ; theoretical curves — calculated using model A (thick solid line), B (thick dotted line), C (thin solid line) and D (thin dotted line).

close to the CMC at 10^{-5} mol/l, the corresponding values from the models B and D are significantly higher: $3.4 \cdot 10^{-5}$ mol/m^2, and $3.5 \cdot 10^{-5}$ mol/m^2, respectively, and close to the value of $3.7 \cdot 10^{-5}$ mol/m^2 obtained by neutron reflectivity [11] for $C_{14}EO_6$, a very similar surfactant.

For a verification of the adsorption values obtained by the comparative drop and bubble method the dilational elasticity method is suitable. The limiting (high frequency) elasticity $E_0 = -d\gamma/d \ln \Gamma$ of $C_{14}EO_8$ solutions was obtained in [28] using harmonic oscillations of the drop area at various frequencies, and plotted in Fig. 5 *vs* the surface pressure Π. Independent experimental values of E_0 can be calculated from the dependences of γ and Γ on the surfactant concentration from Figs 3 and 4. These dependences can be transformed into a dependence $\gamma(\Gamma)$. These limiting elasticity values are also shown in Fig. 5 and compared with those calculated for the four adsorption models discussed above.

It is clearly seen that only the models B and D which assume an intrinsic compressibility of the adsorbed layer yield a good correspondence with both experiments in the entire range of surface pressure. In contrast, the Frumkin model predicts very high limiting elasticities for $\Pi > 20$ mN/m.

The equilibrium surface tension values of tridecyl dimethyl phosphine oxide ($C_{13}DMPO$) solutions as a function of the initial bulk concentration, as obtained in [26] by two methods (the drop profile analysis tensiometer PAT1 and the ring method TE2, Lauda, Germany), are shown in Fig. 6. As one can see, similar to Fig. 3 the results obtained by the drop shape method are essentially higher than those obtained by ring tensiometry. The difference between the two sets of results

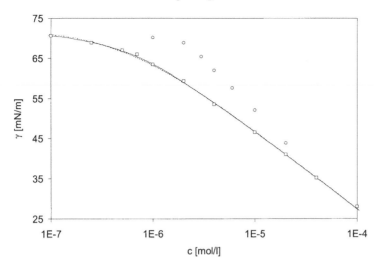

Figure 6. Equilibrium surface tension of the C_{13}DMPO solutions, measured by the ring method ($\square$) and drop shape method ($\diamond$). The theoretical curves were calculated from the Reorientation (C) model (solid line) and from the Frumkin (A) model (dotted line).

is especially pronounced at low initial bulk concentrations. The data obtained in [23] with the ring method agree well with those given in [38].

The data obtained from ring tensiometry were used to calculate the parameters of two theoretical models: the Frumkin model A and the reorientation model C. Note, as mentioned above the equilibrium concentration of the surfactant solution in this method is equal to the initial concentration. Best fits between the theoretical and experimental dependencies in Figs 6 and 7 were obtained with model parameters listed below [26]:

Model A: $\omega_0 = 2.82 \cdot 10^5$ m^2/mol, $a = -0.15$, $b = 1.96 \cdot 10^3$ m^3/mol;

Model C: $\omega_2 = 2.88 \cdot 10^5$ m^2/mol, $\omega_1 = 1.17 \cdot 10^6$ m^2/mol, $\alpha = 0.5$, $b_2 = 2 \cdot 10^3$ m^3/mol.

For the two models analysed, a perfect agreement with the experimental γ vs c data is observed, and the theoretical isotherm calculated from model A is almost identical to that calculated for the model C, cf. Fig. 6.

The adsorption of C_{13}DMPO can be calculated from the two isotherms in Fig. 6 *via* Eq. (3). As mentioned above, the area or the volume of the drop are continuously controlled by the instrument PAT1. In the given experiments we had $V_D/A_D = 0.6 \pm 0.01$ mm [26]. The theoretical models and experimental dependencies of adsorption on the equilibrium concentration are presented in Fig. 7. Both theoretical models show good agreement with the experimental data, but for the reorientation models the agreement is slightly better.

In Fig. 7 the relative adsorption Γ calculated from the experimental points in Fig. 6 (the ring method) with using of the Gibbs adsorption equation (4) is also shown. These Γ values are in satisfactory agreement with those obtained by the proposed drop and bubble method and with theoretical calculations. It should be

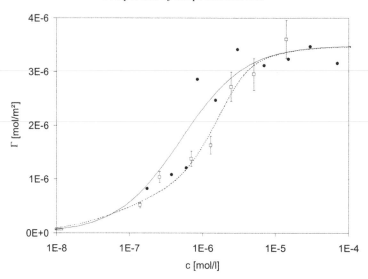

Figure 7. Dependence of $C_{13}DMPO$ equilibrium adsorption on bulk concentration as obtained from experiments ($\square$) and calculated from the Frumkin (A) model (dotted line), the Reorientation (C) model (solid line), and the Gibbs equation ($\bullet$).

noted, that only for weak solutions of non-ionic surfactants the Gibbs adsorption equation in the presented simple form is valid. However, in many other cases (solutions of protein, mixtures of surfactants, aggregation in the monolayer, as well as the ionisation and non-ideality of surfactant in the bulk) the problem, as it is known, is the form of the Gibbs adsorption equation. This open point is, in particular, the reason for the development of new methods for the direct determination of adsorbed amounts.

2. *Protein Solutions*

The results obtained from simultaneous experiments with the two methods (the drop and bubble profile methods) for β-casein [25] are in many aspects similar to those found for $C_{14}EO_8$ and $C_{13}DMPO$. The measured dynamic surface tensions for a $5 \cdot 10^{-8}$ mol/l β-casein solution, is shown as example in Fig. 8 [25]. As one can see the results from the two methods, similar to Figs 1 and 2 for $C_{14}EO_8$, differ significantly from each other.

Let us compare the experimental results for β-casein and HSA (BSA) [1–4, 11] with those obtained by the radiotracer and ellipsometric methods. Figure 9 illustrates surface pressure measured for β-casein by different methods: Wilhelmy plate [1, 3], drop [27] and bubble shape tensiometry [25]. Quite expectedly, the Wilhelmy plate and bubble shape data are almost identical, while those obtained by drop shape tensiometry are shifted by 1–2 orders of magnitude towards larger β-casein concentrations. The theoretical curves shown here and in subsequent figures correspond to the model described in [39]. This theory assumes that protein molecules can adsorb

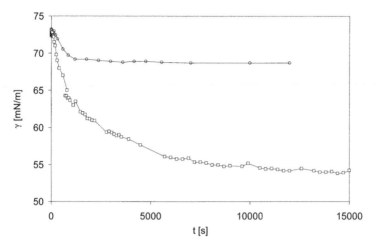

Figure 8. Dynamic surface tension γ of β-casein solutions for the concentration $5 \cdot 10^{-8}$ mol/l as a function of time; using drop ($\bigcirc$) and bubble ($\square$) profile tensiometry.

in a number of states with different molar area, varying from a maximum (ω_{max}) to a minimum value (ω_{min}).

For the isotherm obtained by the drop shape method (see Fig. 9), the β-casein equilibrium concentration values $c_0 = c_B$ were obtained *via* the Wilhelmy plate and bubble shape data taken at the same surface pressure. Then, Eq. (3) was applied to calculate the adsorption of the β-casein. In drop shape experiments for all β-casein concentrations a value of $V/S = 0.59 \pm 0.03$ mm was kept. The calculated adsorption values are shown in Fig. 10. For comparison, the adsorption of β-casein measured elsewhere [1, 3] by ellipsometry and radiotracer technique is also shown. We can see that the different methods used to determine the adsorption value lead to quite consistent results.

In Figs 11 and 12 similar results for the HSA (BSA) solutions using experimental data of the Wilhelmy plate method [3] and drop shape method are given [40]. The maximal adsorption of HSA (BSA) is lower than that of β-casein; therefore, the concentration shift in Fig. 11 is lower than that in Fig. 9. The calculated adsorption values are shown in Fig. 12 together with data measured in [1, 3] by ellipsometry and radiotracer technique. It can again be concluded that different methods yield the same adsorption values of the protein.

Another modification of the proposed method is to use two drops of different size [27], with the assumption that the smaller drop is 1 and the larger drop 2. To attain the same equilibrium adsorption values, the initial concentration of protein in the smaller drop should exceed that in the larger drop. The mass balance of the protein in the two drops yields:

$$\Gamma\left(\frac{S_1}{V_1} - \frac{S_2}{V_2}\right) = c_{D1} - c_{D2}. \tag{11}$$

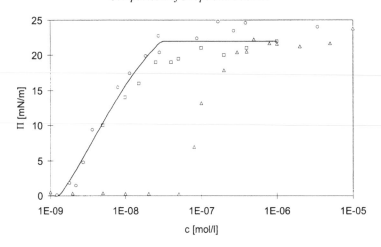

Figure 9. Dependence of surface pressure on the concentration of β-casein. Experimental data from the Wilhelmy plate method [3] (○); data from the bubble shape method [25] (□); data obtained in [27] from the drop shape method (△).

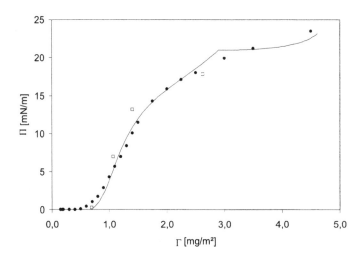

Figure 10. Dependence of surface pressure on adsorption for β-casein from the data obtained in [1, 3] (●) and in [27] (□).

If the larger drop has more than 2 times larger a volume, the accuracy in the determined adsorption values *via* this two-drop approach is as good as that corresponding to the procedure based on the comparison of a drop and a bubble. As in both drops the initial protein concentration exceeds the equilibrium one, Eq. (1) can be used to determine c_0 from the adsorption calculated from Eq. (14).

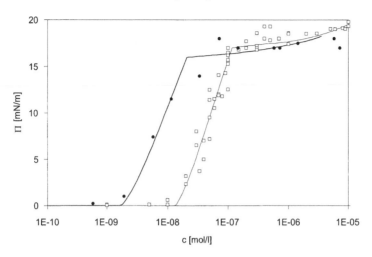

Figure 11. Dependence of surface pressure on concentration of BSA (HSA). Experimental data from the Wilhelmy plate method for BSA [3] (●); data from the drop shape method for HSA [39] (□).

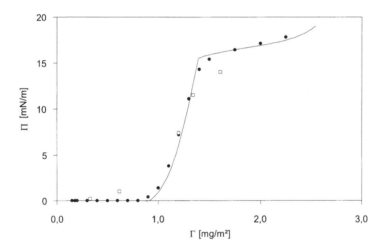

Figure 12. Dependence of surface pressure on adsorption for BSA from the data obtained in [1, 3] (●) and in [27] (□).

D. Conclusions

At low surfactant or protein concentration the amount of molecules depleted from the bulk phase due to adsorption can be significant if the surface area to solution bulk volume ratio is large. This loss of molecules, however, can be useful for the determination of the adsorbed amount of surfactant or protein if the experiments are performed under well controlled conditions. The method to directly determine the adsorption of surfactant or protein at the liquid/fluid interface is discussed, based on the comparison between the equilibrium surface tension values obtained from experiments with the drop and bubble shape tensiometry.

The adsorption values for β-casein and BSA (HSA), obtained using the proposed method, are in a satisfactory agreement with those measured by the ellipsometry and radiotracer technique.

For the aqueous solutions of $C_{13}DMPO$ the adsorption is determined *via* surface tension measurements with drop shape and ring (or bubble shape) tensiometry. The experimental adsorption and surface tension values are in good agreement with those predicted from the reorientation model. The surface tension of $C_{14}EO_8$ solutions was also studied by drop and bubble profile tensiometry at the solution/air interface in order to determine of adsorbed amount. The complete set of experimental data for $C_{14}EO_8$ solutions (including adsorption, surface tension and rheological parameters) can be well described by a combined reorientation plus intrinsic compressibility model, i.e. assumes the reorientation of the EO groups and hydrocarbon chains.

E. Acknowledgements

The work was financially supported by DLR 50WM0941 and the DFG SPP 1273 (Mi418/16-2).

F. References

1. J. Benjamins, J.A. de Feijter, M.T.A. Evans. D.E. Graham and M.C. Phillips, Disc. Faraday Soc., 59 (1975) 218.
2. J.A. Feijter, J. Benjamins and F.A. Veer, Biopolymers, 17 (1978) 1760.
3. D.E. Graham and M.C. Phillips, J. Colloid Interface Sci., 70 (1979) 415.
4. D.E. Graham and M.C. Phillips, J. Colloid Interface Sci., 70 (1979) 427.
5. J.R. Hanter, P.K. Kilpatrick and R.G. Carbonell, J. Colloid Interface Sci., 142 (1991) 429.
6. R. Douillard, M. Daoud, J. Lefebvre, C. Minier, G. Lecannu and J. Coutret, J. Colloid Interface Sci., 163 (1994) 277.
7. D.O. Grigoriev, V.B. Fainerman, A.V. Makievski, J. Krägel, R. Wüstneck and R. Miller, J. Colloid Interface Sci., 253 (2002) 257.
8. K. Tajima, M. Muramatsu and T. Sasaki, Bull. Chem. Soc. Japan, 43 (1970) 1991.
9. K. Tajima, Bull. Chem. Soc. Japan, 43 (1970) 3063; 49 (1976) 3403.
10. D. Möbius and R. Miller (Eds), Novel Methods to Study Interfacial Layers, in "Studies in Interface Science", Vol. 11, Elsevier, Amsterdam, 2001.
11. J.R. Lu, R.K. Thomas and J. Penfold, Adv. Colloid Interface Sci., 84 (2000) 143.
12. M.J. Rosen, Surfactants and Interfacial Phenomena, John Wiley & Sons, New York, 1978.
13. M.J. Schick, Nonionic Surfactants: Physical Chemistry, Surfactant Science Series, Vol. 23, Marcel Dekker, New York, 1986.
14. J.T. Davies and E.K. Rideal, Interfacial Phenomena, Acad. Press, New York, 1963.
15. E.H. Lucassen-Reynders, Anionic Surfactants: Physical Chemistry of Surfactant Action, Surfactant Science Series, Vol. 11, Marcel Dekker Inc., New York–Basel, 1981.
16. C.-H. Chang and E.I. Franses, Colloids Surfaces A, 100 (1995) 1.
17. P.M. Holland, in "Relation between Structure and Performance of Surfactants," M.J. Rosen (ed.), ACS Symposium Series. Vol. 253, 1984. p. 141.

18. D.N. Rubingh and P.M. Holland, Cationic Surfactants: Physical Chemistry, Surfactant Science Series, Vol. 37, Marcel Dekker, New York, 1991.

19. N.M. van Os, J.P. Haak and L.A.M. Rupert, Physico-Chemical Properties of Selected Anionic, Cationic and Nonionic Surfactants, Elsevier, 1993.

20. M. Aratono and N. Ikeda, in "Structure-Performance Relationships in Surfactants," Surfactant Science Series, Vol. 70, Marcel Dekker Inc., New York, Basel, Hong Kong, 1997, p. 83.

21. V.B. Fainerman, R. Miller, E.V. Aksenenko and A.V. Makievski, in "Surfactants — Chemistry, Interfacial Properties and Application", V.B. Fainerman, D. Möbius and R. Miller (Eds), Studies in Interface Science, Vol. 13, Elsevier, 2001, pp. 189–286.

22. V.B. Fainerman, R. Miller and H. Möhwald, J. Phys. Chem., 106 (2002) 809–819.

23. M. Mulqueen and D. Blankschtein, Langmuir, 15 (1999) 8832.

24. Y.J. Nikas, S. Puvvada and D. Blankschtein, Langmuir, 8 (1992) 2680.

25. A.V. Makievski, G. Loglio, J. Krägel, R. Miller, V.B. Fainerman and A.W. Neumann, J. Phys. Chem. B, 103 (1999) 9557.

26. V.B. Fainerman, S.V. Lylyk, A.V. Makievski and R. Miller, J. Colloid Interface Sci., 275 (2004) 305–308.

27. R. Miller, V.B. Fainerman, A.V. Makievski, M. Leser, M. Michel and E.V. Aksenenko, Colloids Surfaces B, 36 (2004) 123–126.

28. V.B. Fainerman, S.A. Zholob, J.T. Petkov and R. Miller, Colloids Surfaces A, 323 (2008) 56–62.

29. G. Loglio, P. Pandolfini, R. Miller, A.V. Makievski, F. Ravera, M. Ferrari and L. Liggieri, Drop and Bubble Shape Analysis as Tool for Dilational Rheology Studies of Interfacial Layers, in "Novel Methods to Study Interfacial Layers", D. Möbius and R. Miller (Eds), Studies in Interface Science, Vol. 11, Elsevier, Amsterdam, 2001, pp. 439–484, ISBN: 0-444-50948-8.

30. R. Miller, C. Olak and A.V. Makievski, SÖFW (English version), (2004) 2–10.

31. Y.C. Lee, S.Y. Lin and H.S. Liu, Langmuir, 17 (2001) 6196.

32. S.Y. Lin, Y.C. Lee, M.J. Shao and C.T. Hsu, J. Colloid Interface Sci., 244 (2003) 372.

33. Y.C. Lee, K.J. Stebe, H.S. Liu and S.Y. Lin, Colloids Surfaces A, 220 (2003) 139–150.

34. V.B. Fainerman, E.H. Lucassen-Reynders and R. Miller, Colloids & Surfaces A, 143 (1998) 141.

35. V.B. Fainerman, V.I. Kovalchuk, E.V. Aksenenko, M. Michel, M.E. Leser and R. Miller, J. Phys. Chem., 108 (2004) 13700–13705.

36. V.I. Kovalchuk, R. Miller, V.B. Fainerman and G. Loglio, Adv. Colloid Interface Sci., 114-115 (2005) 303–313.

37. A. Frumkin, Z. Phys. Chem., Leipzig, 116 (1925) 466–485.

38. A.V. Makievski and D. Grigoriev, Colloid Surfaces A, 143 (1998) 233.

39. V.B. Fainerman, E.H. Lucassen-Reynders and R. Miller, Adv. Colloid Interface Sci., 106 (2003) 237.

40. A.V. Makievski, R. Wüstneck, D.O. Grigoriev, J. Krägel and D.V. Trukhin, Colloids and Surfaces A, 143 (1998) 461.

Coaxial Capillary Pendant Drop Experiments with Subphase Exchange

James K. Ferri [a], **Ashley D. Cramer** [a], **Csaba Kotsmar** [b], **and Reinhard Miller** [c]

[a] Department of Chemical and Biomolecular Engineering Lafayette College, Easton, Pennsylvania 18042, USA

[b] Department of Chemical and Biomolecular Engineering, University of California at Berkeley, Berkeley, California 94720, USA

[c] Max Planck Institut für Kolloid und Grenzflächenforschung D-14424 Potsdam, Germany

Contents

A. Introduction

The study of the rheology, equilibrium and exchange kinetics, and assembly of macromolecules and surfactants at soft and immiscible fluid interfaces are of significant interest for both scientific and technological applications. Foams and emulsions are of particular concern with regards to the non-equilibrium properties that arise, as discussed in [1–3]. Most practical systems are based on multi-component adsorption layers. Reported investigations include adsorption kinetics from mixed solutions [4], penetration of one species into an established adsorbed layer [5], and viscoelastic phenomena in well-defined interfacial nanomembranes.

Bubble and Drop Interfaces

Very few experimental methods exist that are capable of investigating equilibrium and exchange kinetics or rheology and constitutive behavior of liquid-supported surface material assemblies. The classic method for the preparation of interfacial composites requires a Gibbs or Langmuir monolayer at the interface of a Langmuir trough, followed by transfer of the monolayer from one reservoir to another via a translating barrier to exchange the subphase [6, 7]. Highly stable surface layers are requisite because of the interfacial hydrodynamic shear, which arises from the motion of the adjacent bulk during transfer. This limits the experiments that can be undertaken. Also, the large subphase volume required for this method considerably restricts the amount of studies that can be performed in this manner.

Techniques that are more suitable for such experiments are the drop and bubble profile methods. They allow for more stringent control of the environmental conditions, and therefore more uniform pressure, temperature, and concentration at the interface, a much higher interface/volume ratio than in conventional Langmuir troughs, and smaller amounts of material required. A modified pendant drop method has been described by Wege *et al.* [8–10]. This method consists of a coaxial double capillary which allows for *in-situ* internal subphase exchange in single pendant drops.

Numerous applications of the coaxial capillary pendant drop (CCPD) will be subsequently presented. A summary of the method, dynamics of subphase exchange, and underlying equations is given in Section B. Section C describes adsorption, desorption [11], reversibility of adsorbed species [12], and sequential adsorption [13] for surfactants and small amphiphiles. Experiments concerning macromolecules and biomacromolecules, as well as interfacial mixtures of small and large molecules, are described in Section D. Protein desorption kinetics [14, 15], molecular displacement, specifically of proteins by surface active molecules [16], and the penetration of surfactants into an existing surface layer [17] are discussed. The fabrication of soft elastic surfaces and nanomembranes using non-specific adsorption of polymers [18] and polyelectrolytes using electrostatic layer-by-layer assembly [19] are detailed in Section E. Section F introduces additional applications of the CCPD method that allow for more sophisticated studies, for example investigations of the effect of solvent conditions such as pH and/or ionic strength [20], the composition of mixed solvents [21] on interfacial rheology, and interfacial and interphase transport phenomenon.

B. A New Drop Methodology: Coaxial Capillary Pendant Drop (CCPD) Method

A technique was recently introduced in which coaxial capillaries are used to exchange the volume of the drop interior, while maintaining a constant volume or area [8–10]. This method employs axisymmetric drop shape analysis (ADSA), which has emerged as a powerful tool for the study of equilibrium and dynamic adsorption at liquid/fluid interfaces [22]. A commercially available pendant drop tensiometer

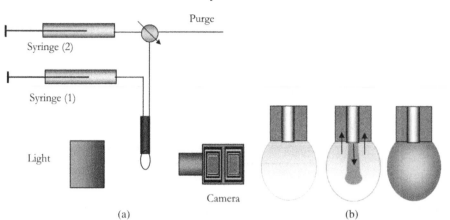

(a)   (b)

Figure 1. Schematic of coaxial capillary pendant drop (CCPD) method, in [15]; (a) experimental set-up; (b) subphase exchange for an experiment.

PAT-1D (Sinterface Technologies, Berlin, Germany) was used for the majority of experiments detailed [23] in this chapter.

A drop ($5 < V_D < 15$ µl) of surfactant solution is formed at the tip of a coaxial capillary for hydrostatic pendant drop experiments. An image of the drop is cast on a CCD camera and digitized. These silhouettes are recorded over time and fit to the Young–Laplace equation to resolve the surface tension (± 0.1 mN/m). The concentric capillaries described by Cabrerizo-Vilchez *et al.* [10] are used for experiments involving subphase exchange. A schematic of the pendant drop method is shown in Fig. 1.

A typical drop subphase exchange experiment is performed as follows: a drop of species concentration $C_{1,\infty}$ is formed using Syringe 1 and allowed to quiesce. Exchange then occurs by the injection of a second liquid of species concentration $C_{2,\infty}$ using Syringe 2. The concentration of each species in the drop, i.e. $C_1(t)$ and $C_2(t)$, evolve continuously from the initial distribution, $C_1(t = 0) = C_{1,\infty}$ and $C_2(t = 0) = 0$, to the final distribution, $C_1(t = \infty) = 0$ and $C_2(t = \infty) = C_{2,\infty}$ during the exchange. Drop profile analysis tensiometry is used to determine the surface tension before, during, and after the exchange. Constant feedback control allows for the maintenance of a constant drop volume V_D or area A_D by drop shape and the withdrawal of liquid from the droplet interior at the same volumetric flow rate (R_E) using Syringe 1. Simple liquids such as DMSO and chlorobenzene [24] and mixtures of water and ethanol [11] can be used for experiments to measure the rate of exchange, see Fig. 2.

The evolution of the subphase concentration can be derived under the assumption that perfect mixing occurs. The concentration of species 1 is described by [11]

$$C_1(t) = C_{1,\infty} \exp(-t/\tau), \tag{1}$$

where τ is the residence time of the liquid in the drop $\tau = V_D/R_E$.

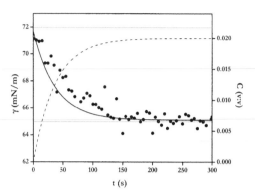

Figure 2. Surface tension (γ) *versus* time for water exchanged with 2% (v:v) ethanol. $C_{1,\infty} = 0$ and $C_{2,\infty} = 2\%$ (v:v) ethanol for exchange flow rate (R_E) = 0.2 µl/s. Description of the macroscopic subphase exchange rate γ(●), the best fit of Eq. (1) (-), and the calculated development of the bulk concentration of ethanol from Eq. (2) (- -).

The concentration of species $C_{2,\infty}$ (i.e. $C_{1,\infty} = 0$) is

$$C_2(t) = C_{2,\infty}(1 - \exp(-t/\tau)). \qquad (2)$$

The experimentally measured characteristic exchange time constant $\tau > V_D/R_E$. It has been shown that the assumption of perfect mixing does not hold for data showing the deviation of the actual time constant from the theoretical time constant [10, 25].

Details of the dynamics of the subphase exchange are important to kinetic and equilibrium studies. The dynamics of the subphase exchange must be differentiated from the interfacial exchange kinetics. This distinction can be made by describing the subphase concentration as a function of the extrinsic conditions, such as average and local fluid velocity, subphase concentration, and capillary tip geometry. Quantifying the volume of liquid necessary for complete exchange is important for equilibrium.

The spatiotemporal species distribution can be calculated to determine the rate at which compositional uniformity is reached. The solution of the Navier–Stokes equation describes the velocity distribution

$$\rho\left(\frac{\partial \underline{v}}{\partial t} + \underline{v} \cdot \nabla \underline{v}\right) = -\underline{\nabla} P + \mu \nabla^2 \underline{v} + \rho \underline{g} \qquad (3)$$

which describes the velocity vector field $\underline{v}(t, x, y, z)$ in terms of pressure P and the intrinsic fluid properties, density ρ and viscosity μ. The species continuity equation accounts for both diffusive and convective transport of each species $C_i(t, x, y, z)$ in the drop subphase by

$$\frac{\partial C_i}{\partial t} + \underline{v} \cdot \underline{\nabla} C_i = D_i \nabla^2 C_i, \qquad (4)$$

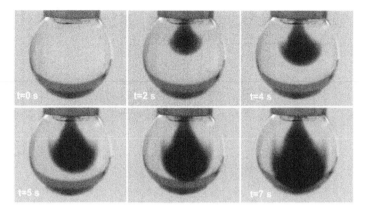

Figure 3. Depiction of the microscopic subphase exchange dynamics. Exchange with Brilliant Green ($C_{2,\infty} = 3.3$ mg/ml) at $R_E = 0.44$ µl/s.

where D_i is the diffusion coefficient of the species in the liquid phase. Solution of Eqs (3) and (4) subject to appropriate boundary and initial conditions yields $\underline{v}$ and C_i.

Experiments using an aqueous solution of a low molecular weight dye (Brilliant Green, $C_{2,\infty} = 3.3$ mg/ml, $D_2 \approx 6 \times 10^{-6}$ cm²/s) were run in order to visualize the spatiotemporal distribution of the subphase concentration; see [25] for details. The evolution of the dye distribution for the exchange flow rate $R_E = 0.44$ µl/s is shown in Fig. 3.

C. Surfactant and Small Amphiphile Exchange

Exchange of surfactants and small amphiphiles at fluid interfaces is an area of great interest for scientists and engineers, especially with respect to the equilibrium and kinetics of these systems; see [26–30]. The Langmuir adsorption isotherm relates the equilibrium surface excess concentration $\Gamma_{1,eq}$ of small molecules to their equilibrium bulk concentration $C_{1,\infty}$

$$x_1 = \frac{\Gamma_{1,eq}}{\Gamma_{1,\infty}} = \frac{C_{1,\infty}}{(a_1 + C_{1,\infty})}, \tag{5}$$

where $a_1 = \frac{\alpha_1}{\beta_1}$ is the equilibrium adsorption constant, $\Gamma_{1,\infty}$ is the maximum packing of the surfactant in a monolayer, and x_1 is the fractional coverage of the interface.

A mass balance at the interface [31] characterizes exchange kinetics

$$\frac{\partial \Gamma_1(t)}{\partial t} = \beta_1 C_{1,s}(t)(\Gamma_{1,\infty} - \Gamma_1(t)) - \alpha_1 \Gamma_1(t). \tag{6}$$

The rate of change of the surface excess concentration $\Gamma(t)$ is the difference between the adsorptive flux, i.e. the difference between the maximum surface concentration $\Gamma_{1,\infty}$ and the surface concentration $\Gamma_1(t)$, which is shown by Eq. (6).

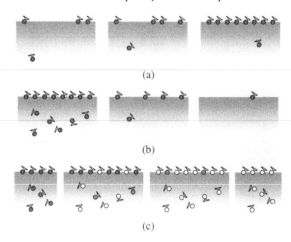

Figure 4. Adsorption and desorption equilibrium and kinetic studies; (a) adsorption of species A, (b) desorption of species A from the surface, (c) competitive adsorption displacement of species A by B.

The adsorptive flux is the product of the adsorption kinetic rate constant β_1, the instantaneous subphase concentration $C_{1,s}(t)$, and the space available at the interface. The desorptive flux is the product of the desorption kinetic constant α_1 and $\Gamma_1(t)$. Solving (6) when the time dependence of $C_{1,s}(t)$ leads to the surface concentration of surfactant; see Section B for details.

The surface equation of state can be used to describe the equilibrium and dynamic surface tensions γ_{eq} and $\gamma(t)$, which corresponds to Langmuir adsorption as shown by Eqs (7) and (8)

$$\gamma_{eq} = \gamma_0 + RT\Gamma_{1,\infty} \ln\left[1 - \frac{\Gamma_{1,eq}}{\Gamma_{1,\infty}}\right] \tag{7}$$

and

$$\gamma(t) = \gamma_0 + RT\Gamma_{1,\infty} \ln\left[1 - \frac{\Gamma_1(t)}{\Gamma_{1,\infty}}\right], \tag{8}$$

where γ_0 is the surface tension of the surfactant-free interface and RT is the product of the gas constant and the temperature.

This section explores the adsorption dynamics of surfactants and small amphiphiles. Figure 4 summarizes the three different experiments investigated within the subsequent sections using the CCPD method. These experiments used phosphine oxide surfactants, which can be relatively well described by the Langmuir model, to study the equilibrium and kinetics of adsorption, desorption, and sequential adsorption and displacement. Details of equilibrium data and model constants are presented in [32].

1. Convection-Enhanced Adsorption

Convection works to increase the rate at which equilibrium is achieved, and can sometimes even affect the extent to which it is reached. In performing convection-enhanced adsorption experiments a drop of surfactant solution is first formed, and then the drop subphase is continuously exchanged with the same surfactant solution. (The withdrawal and injection flow rates are the same during this process to ensure the volume remains constant). A description of these two effects and representative experiments are presented below.

Consider a semi-infinite surfactant solution immediately after the formation of an interface. A concentration gradient from the interface to the bulk appears when surfactant adsorbs at the interface. Bulk diffusion is most often the mechanism controlling the rate of adsorption. The characteristic diffusion time scale τ_D for surfactant adsorption depends on the diffusion coefficient of surfactant and the ability of the interface to deplete the bulk surfactant

$$\tau_D = \frac{h_1^2}{D_1}, \tag{9}$$

where the adsorption depth, h_1, can be defined by the ratio of the equilibrium surface concentration of a surfactant to the bulk concentration, i.e.

$$h_1 = \frac{\Gamma_{1,eq}}{C_{1,\infty}} \tag{10}$$

which can be calculated from an equilibrium model such as Eq. (5) [30]. From Eqs (9) and (10) it can be seen that larger adsorption depths and diffusion time scales occur with lower concentrations. Thus, a long period of time is required for equilibration at smaller concentrations. At low concentrations the surfactant adsorption can also shift the apparent equilibrium due to the depletion of the bulk phase because there is not an infinite reservoir of surfactant. The depletion number, S_b, can be used to describe this effect

$$S_{b,1} = \frac{A_D \Gamma_{1,eq}}{V_D C_{1,\infty}}, \tag{11}$$

where A_D is the total interfacial area and V_D is the total volume of the pendant drop. The depletion number relates the equilibrium mass of surfactant adsorbed at the interface to the total amount of surfactant in the bulk. When $S_b \ll 1$, the drop can be treated as an infinite reservoir. There is a shift in the apparent adsorption equilibrium when S_b is large; the shift is greater as S_b increases. This effect can be exploited to directly calculate adsorbed mass as described in Chapter 8. Here it is shown convective exchange of the drop subphase during adsorption experiments mitigate both of these effects.

The surfactant $C_{10}DMPO$ demonstrates convection-enhanced adsorption. At a bulk concentration of $C = 1 \times 10^{-4}$ mol/l the adsorption depth, h, is 2.5×10^{-5} m. The corresponding diffusion timescale for surface tension equilibration is around 6 seconds. The depletion number is 5×10^{-2} under normal experimental conditions ($V_D = 15$ µl, $A_D = 25$ mm²), which shows the drop will act as an infinite surfactant reservoir. These expectations are confirmed by experiments; the surface tension

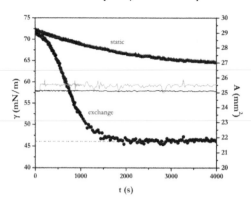

Figure 5. Surface tension of C_{14}DMPO (bulk concentration $C = 6 \times 10^{-6}$ mol/l) for static adsorption and convection exchange experiments and area *versus* time; static adsorption (▲), convection-enhanced adsorption (●), drop area (-), and the Langmuir predicted equilibrium surface tension (- -).

does not change over the observed times and the values agree with those forecasted by the Langmuir isotherm, see [33].

The surfactant C_{14}DMPO at a bulk concentration of $C = 6 \times 10^{-6}$ mol/l has an adsorption depth of around 6×10^{-4} m as predicted by the constants for the Langmuir model detailed in [32]. This leads to a diffusion timescale for adsorption equilibration to be on the order of several hours. The depletion number S_b is nearly unity for these experiments. Figure 5 shows the surface tension relaxation for static and convection enhanced transport. For static experiments it is shown that the bulk depletion affects the apparent equilibration tension. The continuous exchange helps to decrease the time necessary for equilibrium to be reached and aids in the approach of the predicted equilibrium tension. The long time asymptote for the exchanged subphase is represented by the Langmuir model.

Acceleration of adsorption kinetics can be accomplished in situations where there is no bulk depletion effect. These situations usually appear for macromolecules with long timescales at low (to moderate) concentration, cf. Section D for examples.

2. Desorption

Both the rate and extent of desorption can be measured from properly designed experiments. This can be accomplished by preparing a drop of surfactant solution and then replacing the subphase with a surfactant free aqueous solution. A discussion of theory and experiments follows.

Consider an interface of a pendant drop having an equilibrium distribution of surfactant adsorbed from the bulk phase. By replacing the bulk phase with water a gradient is formed, which acts as a driving force for desorption. The dynamic surface concentration can be described by integrating Eqs (6) and (1) using the Runge–Kutta method, subject to the initial condition, $\Gamma'_1(t' = 0) = 1$. The desorption process is a function of the product of $\alpha_1\tau$, the ratio of the convection and desorption timescales, the adsorption number, k, and the fractional interfacial coverage, x_1. The surface concentration is desorption-controlled [34] when the convection timescale is considerably smaller than the desorption timescale

$$\Gamma_1(t) = \Gamma_{1,\text{eq}} \exp(-\alpha_1 t). \tag{12}$$

The surface tension increases as desorption proceeds. The rate of this increase can be cast in dimensionless form by scaling the surface tension by its equilibrium lowering

$$\theta(t) = \frac{\gamma(t) - \gamma_{\text{eq}}}{\gamma_0 - \gamma_{\text{eq}}}. \tag{13}$$

Consider the variation of surfactant physical chemistry at a fixed exchange rate. As the desorption coefficient increases, the increase in θ becomes more rapid, although there does exist an upper limit. There is a local equilibrium between the interface and the bulk when desorption is practically instantaneous. This observation suggests that the study of a homologous series of surfactants, viz. n-alkanes conjugates of the same hydrophilic moiety, at a fixed exchange rate should be able to identify the presence of a desorption barrier and demonstrate its dependence on hydrocarbon chain length. The exchange rate should also affect the process; as the rate is increased the bulk concentration approaches zero faster. Although, desorption controlled processes are independent of the rate of exchange.

A plot of surface tension and drop area *versus* time for C_8, C_{10}, C_{12}, and C_{14}DMPO at bulk concentrations having an equilibrium surface tension of 45 mN/m are shown in Fig. 6. The convection timescale is the same in each experiment, which is accomplished by holding the drop volume and exchange rate constant. Thus, the only adjustable parameter is the desorption coefficient α_1. The reported desorption coefficients for each surfactant [11] are 4.1×10^{-3} s^{-1} (C_8), 2.8×10^{-3} s^{-1} (C_{10}) 2.1×10^{-3} s^{-1} (C_{12}) and 5.5×10^{-4} s^{-1} (C_{14}).

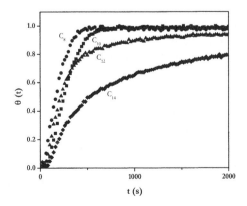

Figure 6. Scaled surface tension (θ) *versus* time; kinetics of desorption for surfactants C_nDMPO: C_8 (●), C_{10}(■), C_{12}(▲) and C_{14}(◆).

3. Sequential Adsorption and Replacement

Non-equilibrium surface concentrations and non-equilibrium surface tensions can stem from finite surfactant sorption kinetics during sequential subphase exchange. Consider an interface having an equilibrium distribution of the first surfactant (component 1) adsorbed from the adjacent bulk phase. A mass balance can be written around each species at the interface, describing the rate of change of the surface concentration, Γ_i, of each species as a difference between adsorptive and desorptive kinetic fluxes. The adsorptive flux of each species is proportional to the bulk concentration, C_i, of that species and the space available at the interface; i.e. the difference between the maximum surface concentration, Γ_∞, and the total instantaneous surface concentration, $\Gamma_1 + \Gamma_2$. The desorptive flux of each component is linearly proportional to its surface concentration if it is assumed negligible interactions occur between the different adsorbed species. For a binary system, the sorption kinetic equations are:

$$\frac{\partial \Gamma_1(t)}{\partial t} = \beta_1 C_{s,1}(t)(\Gamma_\infty - \Gamma_1(t) - \Gamma_2(t)) - \alpha_1 \Gamma_1(t), \tag{14}$$

$$\frac{\partial \Gamma_2(t)}{\partial t} = \beta_2 C_{2,s}(t)(\Gamma_\infty - \Gamma_1(t) - \Gamma_2(t)) - \alpha_2 \Gamma_2(t), \tag{15}$$

where β_i and α_i are the adsorption and desorption kinetic rate constants of each component respectively. The maximum surface concentration is assumed to be the same for each component in Eqs (14) and (15), which is reasonable for homologous surfactants possessing the same polar moiety.

The surface concentration dictates the development of the interfacial tension *via* the equation of state, which relates the dynamic surface tension $\gamma(t)$ to the surface concentrations $\Gamma_1(t)$ and $\Gamma_2(t)$

$$\gamma(t') = \gamma_0 + RT\Gamma_\infty \ln\left[1 - \frac{\Gamma_1(t')}{\Gamma_\infty} - \frac{\Gamma_2(t')}{\Gamma_\infty}\right], \tag{16}$$

where γ_0 is the surface tension of the surfactant-free interface.

The equilibrium interfacial coverage and the surface tension are equal prior and subsequent to the exchange when the adsorption number for each component is the same. Similar behavior arises when the sorption kinetic constants of both species are the same. The interfacial tension during an exchange of surfactants with different desorption coefficients evidence a temporal dependence, relying on both magnitude of the difference and the sequence in which the exchange occurs. Figure 7 illustrates representative experiments in which minima and maxima in the dynamic surface tension exist during subphase exchange. This observation was shown to be consistent with interfacial over- and under-population during the exchange process using theory previously described within this section, assuming species concentrations of the form given in Eqs (1) and (2). Further details were reported earlier by Gorevski *et al.* [13].

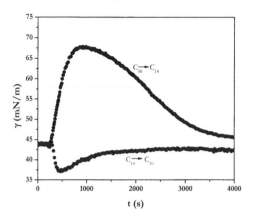

Figure 7. Surface tension *versus* time; sequential adsorption of surfactants: C_{10}DMPO and C_{14}DMPO.

D. Polymers, Biomacromolecules and Mixed Systems

The adsorption of proteins and their interaction with synthetic macromolecules or low-molecular-weight surfactants are of extreme importance in various industries, including pharmaceutics, foods, and cosmetics. There are difficulties associated with polymers and proteins at fluid interfaces, such as dynamic surface tension, adsorption and desorption kinetics, adsorption and thermodynamic equilibrium, and interactions in interfacial mixtures of surfactants and macromolecules [35–43].

Macromolecules desorb from interfaces very slowly [44–47]. This behavior is most likely associated with the high Gibbs free energy of adsorption. It can be assumed that the adsorption kinetics of larger species at surfaces is an irreversible process, which is similar to a kinetically unidirectional reaction [48, 49].

The adsorption of macromolecules at solid or liquid interfaces is often kinetically irreversible. Along with the concentration of macromolecule solution, the condition under which an adsorption layer is created affect both the calculated adsorption and surface tension. The adsorption can be promoted by convection and can even cause more molecules of the same bulk concentration to adsorb at equilibrium because of the shorter time needed for unfolding of the adsorbed macromolecules [49]. Due to this behavior, the derived adsorption and interfacial tension isotherms for macromolecules fall short of characterizing these convection-enhanced systems.

The adsorption kinetics of β-casein (BCS), β-lactoglobulin (BLG), and human serum albumin (HSA) proteins were investigated at the air/water interface both with and without forced convection by Fainerman *et al.* [49]. A bubble profile analysis tensiometer and CFT similar to Svitova *et al.* [50] were used to perform these experiments. Different solution concentrations for the proteins both with and without forced convection were used for the described adsorption experiments. A significant difference between the adsorption rates in the case of all proteins at all concentrations in the different types of experiments is shown. The rate of interfacial tension decrease and the adsorption rate are approximately one order of magnitude larger when forced convection was employed. The equilibrium surface tensions, i.e. the values extrapolated to infinite time, are independent of adsorption rate. The limiting surface tension values are the same within experimental error for experiments with BCS at a concentration of 5×10^{-9} mol/l and similar findings were obtained for the other two solutions of HSA and BLG.

The equilibrium surface tension of a solution is not affected by the protein adsorption rate. Although the protein adsorption kinetics are practically irreversible, the process is thermodynamically reversible [51]. This behavior may be attributed to the characteristic time for conformational changes, which is essentially shorter than the time to reach the equilibrium state in experiments with forced convection.

The adsorption equilibrium of poly(ethylene oxide)-poly(propylene oxide)-poly(ethylene oxide), hereafter PEO/PPO/PEO, was investigated with and without forced convection [18]. Equilibrium is reached on comparable timescales at higher concentrations for both experiments. At low concentrations, convection greatly reduces the timescale at which equilibrium is attained, as compared to the diffusion limited case; see Fig. 8a. The long time asymptotes for both cases are equivalent, which suggests a thermodynamic equilibrium is reached. As shown in Fig. 8a, accelerated adsorption kinetics was observed for convection, even when adsorption equilibrium is accessible by diffusion as discussed in Section C. These studies suggest that polymers are highly mobile in the adsorbed layer and readily rearrange at the interface, despite the fact that polymers are thought to have frozen, unresponsive structures at interfaces. Similar results have been reported; for additional discussion see [52] and Section E.

The energy of the adsorbed layer can be determined from the desorption kinetics of macromolecules. For example, the desorption of PEO/PPO/PEO was investigated [53]; these macromolecules are kinetically irreversible on the timescale of available experiments. Figure 8b presents data from dynamic surface tension experiments during subphase washout with the CCPD method. Desorption experiments with BCS and BLG were performed to determine the role of the initial surface concentration. The kinetics and extent of adsorption were studied for protein solutions at different concentrations [34]. Desorption of proteins from liquid interfaces was found to depend on the conditions of adsorption. It was previously discussed that at low concentrations, the adsorption process is longer.

During this time molecules have a chance to unfold at the interface. At higher concentrations the adsorption is faster and there is competition for adsorption to the interface. The data shown in Fig. 9 for BCS support the case that the extent and rate of desorption are functions of the surface coverage.

Only a small amount of molecules were shown to desorb for both proteins at all concentrations. The observed change in the surface tension was not significant, signifying irreversible adsorption, which is in contradiction with previously published data [54]. Theoretical analysis shows that the relative desorption is 10^4–10^8 times slower than that for surfactants. Thus, experiments run for 10^4 s cannot discern between the reversibility and irreversibility of adsorbed proteins.

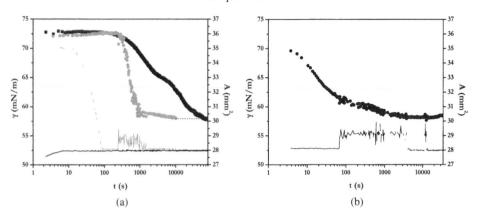

Figure 8. Surface tension *versus* time for PEO/PPO/PEO; (a) $C_{1,\infty} = 1.6 \times 10^{-7}$ mol/l; subphase exchange $R_E = 0.5$ μl/s (●), diffusion only (■), and drop area (-); (b) desorption; $C_{1,\infty} = 9.5 \times 10^{-7}$ mol/l. γ (●) and drop area (-); experimental timescales indicate a kinetically irreversible process.

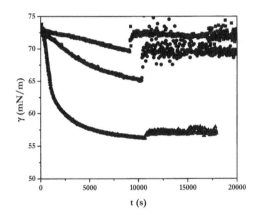

Figure 9. Surface tension *versus* time for the desorption of BCS with different monolayer concentrations; comparison of desorption kinetics from adsorption layers formed from $C_{1,\infty} = 5 \times 10^{-8}$ (■), 1×10^{-7} (●), and 5×10^{-7} (▲) mol/l; at $R_E = 0.2$ μl/s.

E. Ultrathin Surface Composites and Membranes

Soft surface nanocomposites have both practical and scientific applications ranging from electro-optical [55–58] and mechano-sensitive materials [59–61] to biofunctional surfaces for the stimulation of cell proliferation, differentiation, and gene expression [62–68].

The nature and strength of bonding forces, such as physical forces and interfacial chemistry, can be used to categorize ultrathin surface materials [69, 70]. The requirements for nanomembrane templating using CCPD method are discussed within this section. Representative experiments such as electrostatic templating and layer-by-layer assembly of polyelectrolytes, adsorption to insoluble monolayers by non-ionic macromolecules, and representative interfacial covalent cross-linking chemistry of polysaccharides, peptides, and proteins are further detailed. Results for polyelectrolyte multilayer assemblies including poly-(allylamine hydrochloride) (PAH) with poly-(styrene sulfonate) (PSS) or poly-(acrylic acid) (PAA), hyaluronic acid (HA)-poly(L-lysine) (PLL), and fibrin-based nanomembranes are presented. The formation of supramolecular networks which confer mechanical rigidity that is outside the description of equilibrium surface thermodynamics; i.e. Gibbs elasticity can arise from electrostatic complexation, hydrophobic association, or covalent crosslinking at the air/water interface.

1. Interfacial Nanocomposite Synthesis

The CCPD method allows for subphase exchange with a relatively low convective disturbance to the interface. Fabrication of nanocomposites with well-defined composition, architecture, and processing conditions is possible. Two essential requirements for fabrication exist: (1) a driving force for interfacial assembly and (2) a kinetic hindrance to disassembly; i.e. slow desorption kinetics. An additional prerequisite is (3) the constituents of the composite should partition relatively weakly to the solid interfaces contacted by the exchange liquid in the experimental set-up to prevent large scale fouling. These requirements are addressed for representative systems of strong and weak polyelectrolytes and protein-based nanomembranes.

Strong polyelectrolytes are soluble in aqueous systems because they have a high degree of counter-ion dissociation, thus satisfying (3). Therefore, a charged template at the air/water interface is required. In order to create an electrostatic monolayer, insoluble molecules with a charged head group are deposited on the surface of the drop. Surface charge density can be adjusted by changing the drop size using Syringe 1. Various phospholipids [8, 18, 71] are able to endure the compression process under a wide range of experimental conditions, viz. drop volume, exchange flow rate, and monolayer film pressure.

A microdepositer is used to create a monolayer of lipid on the pendant drop surface and brought to the desired compression state. The subphase can then be replaced with a polyelectrolyte solution with the opposite charge of lipid; polycation (PC) and polyanion (PA) are alternately exchanged after intermittent monovalent electrolyte washing. This process leads to a freestanding polymer film defined by the amount of adsorbed layers, shown in Fig. 10.

Nanocomposites of well-defined composition and transverse direction are produced from the assembly of strong polyelectrolytes. Drop profile analysis is used to measure the free energy changes during the adsorption cycle, which offers data on the dynamics of structure development, see Fig. 11. Further details are reported in [19].

Weak polyelectrolytes have a solution chemistry-dependent degree of dissociation. Therefore, the aqueous solubility depends on the conditions of synthesis, presenting an experimental obstacle associated with (3). A change in solution conditions between layers can lead to desorption and simplex formation as shown below

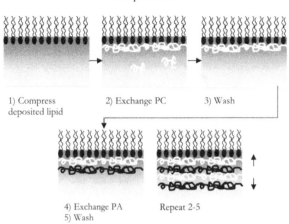

1) Compress
deposited lipid

2) Exchange PC

3) Wash

4) Exchange PA
5) Wash

Repeat 2-5

Figure 10. Schematic of the assembly of freestanding asymmetric polyelectrolyte nanocomposites at the air/water interface with lipid templating.

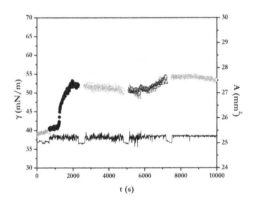

Figure 11. Surface tension and area *versus* time for strong polyelectrolyte assembly: DMPG-(PAH/PSS)$_1$. $C_{PAH,\infty} = 1$ mg/ml, $C_{PSS,\infty} = 1$ mg/ml; $R_E = 0.2$ μl/s; pH 6.0; $C_{NaCl} = 0.5$ mol/l; γ for DMPG-PAH adsorption (●), NaCl wash (○), and DMPG-PAH/PSS-adsorption (△); drop area (-).

in Fig. 12. The surface tension increases concomitant with PAH (pH 6.0) adsorption as in Fig. 11. There is a reduction in the interfacial tension after exchange with PAA (pH 9.0), consistent with PAH/PAA simplex formation and desorption. The surface tension of the interface during compression and expansion may serve as an indication of polymer adsorption because of the surface equation of state of the lipid monolayer, which is sensitive to impurities.

The expansion and compression of the lipid monolayer before and after contact with the polyelectrolyte solutions are shown in Fig. 13. The reversible nature of the adsorbed polymer layer is highlighted by the reversal of the adsorbed PAH. The importance of maintaining the solution chemistry, which is well known in any layer-by-layer assembly of weak polyions [72, 73], is also emphasized.

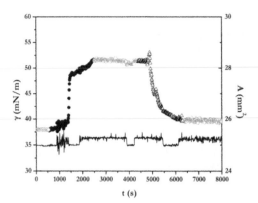

Figure 12. Surface tension and area *versus* time for the desorption of a weak polyelectrolyte bilayer DMPG-(PAH/PAA)$_1$ and simplex formation; $C_{PAH,\infty} = 1$ mg/ml, $C_{PAA,\infty} = 1$ mg/ml; $R_E = 0.2$ µl/s; pH 6.0 for PAH; pH 9.0 for PAA $C_{NaCl} = 0.5$ mol/l; γ during DMPG-PAH adsorption (●), NaCl wash (○), and DMPG-PAH/PAA-adsorption (△); drop area (-).

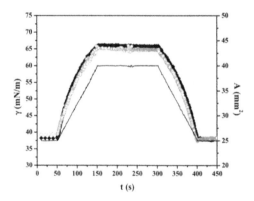

Figure 13. Surface tension and area *versus* time for the lipid template and the desorption of a weak polyelectrolyte bilayer DMPG-(PAH/PAA)$_1$; γ for DMPG (◆), γ for the desorbed (PAH/PAA)$_1$ (○), and area (-).

Viruses, proteins, or reactive nanoparticle-based nanomembranes [74, 75] can be synthesized by covalent network formation through non-specific adsorption, viz. using divalent complexing agents such as Ca^{2+}, polycondensation reactions, or biochemically specific reactions, such as the cleavage of fibrinogen by thrombin to form fibrin fibers [76]. The fluid interface can be used to template the transverse confinement necessary to prepare a nanomembrane because most proteins have low rates of desorption and high surface affinity. The kinetics of desorption are *a priori* unknown, but can be measured using methods described in Section C. For biomacromolecules, adsorption is driven by the hydrophobicity of the reactants, satisfying (1). In some instances, condition (3) is problematic, mainly when the subphase is not fully exchanged.

2. Penetration of Lipid Monolayers by Non-ionic Polymers

Immunoassay specificity [77] and reverse osmosis membrane filtration fouling [78, 79] represent some of the technologically important obstacles that arise due to non-specific adsorption of macromolecules at soft and solid interfaces. Interfacial scientists are still challenged by the kinetics of adsorption and equilibration of soft matter. The kinetics and structure of macromolecular adsorption at soft surfaces can be investigated by using representative insoluble lipid templates at the air/water interface. The ambiguity in the initial condition of the mixed monolayers provides complexities when investigating lipid/polymer interactions by adsorption onto a monolayer at the air/water interface. Mixed monolayers were studied in [80] by depositing lipid onto a Langmuir trough with a subphase free of polymer, and subsequently injecting a concentrated polymer solution into the subphase below the lipid monolayer.

Adsorption equilibrium is attained when the polymer diffuses to uniformity with the bulk. Spreading the lipid on a subphase of polymer solution is also precluded because the preexisting adsorbed phase lowers the surface tension and attenuates the Marangoni effect, by which homogeneous spreading of the lipid is accomplished.

Experiments investigating the equilibrium penetration of a wide variety of polymers and copolymers into phospholipid monolayers as a function of polymer bulk concentration can be performed with the CCPD method. The lipid is first spread to form a monolayer on the pendant drop and the subphase is replaced by a polymer solution, which allows for uniform adsorption. Compression isotherms can be determined by studying the response in the surface tension to periodic area perturbations. Information can be gleaned about the rate of exchange of matter between the surface and the subphase and the relaxation kinetics within the monolayer.

Representative data for the adsorption of PEO/PPO/PEO copolymer onto a DPPC monolayer is depicted in Fig. 14. The surface pressure of the penetrated monolayer is approximately the same as for adsorption of polymer only because of the relatively low surface pressure of the preexisting monolayer.

The lipid is in the liquid expanded state. As the lipid molecular area decreases due to compression, the surface pressure increases, which is associated with the squeeze out of the polymer from the lipid monolayer. These observations suggest reversible polymer adsorption occurs and the same desorption timescale for the same copolymer at the air/water interface as previously shown in Fig. 8a is greater than that of experimental observation. The viscoelasticity of the interfacial nanocomposites prepared using this method for PEO/PPO/PEO–DPPC system was measured and suggests an interaction between the lipid and copolymer in the adsorbed layer [18].

3. Layer-by-Layer Assembly

Polyelectrolyte nanomembranes have received much attention for their prospective uses as actuators, micromechanical sensors, and barrier materials [81–83]. Thin films exhibit significantly different behavior as compared to bulk materials, i.e.

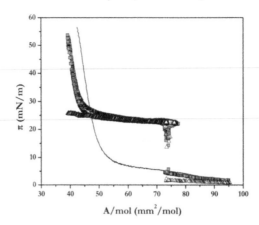

Figure 14. Surface pressure (π) *versus* lipid area per molecule for the adsorption of PEO/PPO/PEO ($C_{1,\infty} = 2.4 \times 10^{-5}$ mol/l) onto 1-DPPC; PEO/PPO/PEO ($\triangle$), PEO/PPO/PEO adsorption onto 1-DPPC (■); 1-DPPC alone (-), $R_E = 0.2$ µl/s.

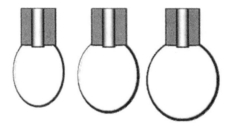

Figure 15. Schematic of the inflation of a pendant drop in order to probe the mechanics of the synthesized PEM.

transport, glass transition, and stress to failure, although the mechanisms responsible for these phenomena are not fully understood.

The fabrication and characterization of the mechanics of strong polyelectrolyte nanocomposites was performed using the CCPD method and was previously described. A polyelectrolyte multilayer (PEM) is first synthesized on the surface of a pendant drop. The mechanical response of the multilayers can be obtained by increasing the membrane area by inflation and measuring the surface tension response, as shown in Fig. 15.

The mechanical response of strong polyelectrolyte multilayers DMPG-(PAH/PSS)$_n$ as a function of thickness is shown in Fig. 16.

Strong polyelectrolyte nanomembranes are likely to be elastomeric structures that stretch semi-reversibly upon large deformation, with an increasing dependence of the film surface elastic modulus on film thickness and template charge density (see [84] for details). The effect of solvent ionic strength and molecular weight of the strong polyanion were also studied by Cramer and Ferri [85]. Figure 17 shows the dependence of surface and bulk elastic moduli on solvent concentration. There

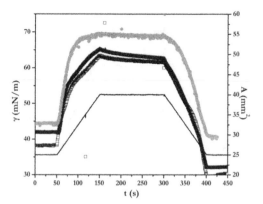

Figure 16. Mechanics of strong polyelectrolyte membranes at air/water interface under the condtions: $C_{1,\infty} = 1$ mg/ml; $C_{NaCl} = 0.5$ mol/l; $R_E = 0.2$ μl/s; surface tension and drop areas *vs* time for DMPG-(PAH/PSS)$_n$; $n = 1$ (□), 2(▲), and 3(●); drop area (-).

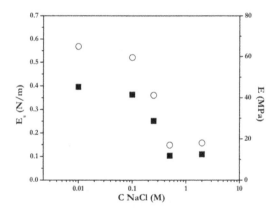

Figure 17. Surface modulus (E_s) and bulk modulus (E) *versus* solvent concentration; effect of solvent strength on elasticity of strong polyelectrolyte multilayers DMPG-(PAH/PSS)$_3$; surface modulus E_s (○) and bulk modulus E (■).

is a transition from a relatively high to low modulus exhibiting a saloplastic effect [86].

As the ionic strength increases both the yield stress and surface elastic modulus are shown to decrease. The bulk elastic modulus E is related to the surface elastic modulus E_s via $E = E_s/h$, where h is the nanomembrane thickness. It was reported that for strong polyelectrolyte multilayers, out of plane coupling is limited to about two bilayers. The magnitude of the elastic moduli for strong polyelectrolytes is consistent with that of elastomeric rubbers.

A viscoelastic response and some amount of reversibility would be expected for these materials, and both were experimentally observed [84]. Other polyelectrolyte systems were investigated using CFT at the air/water interface and presented in [87, 88].

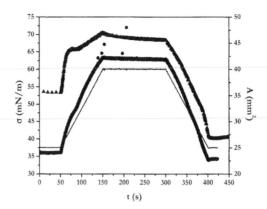

Figure 18. Surface tension and area *versus* time. DMPG-(PAH/PSS)$_2$ (▲), DMPG-(PLL/HA)$_2$ (●), and drop area (-).

Strong polyelectrolytes have a high degree of dissociation and therefore interlayer interactions are stronger; film growth is linear in the number of layers in opposition to those of weak polyelectrolyte interactions. Weak polyelectrolytes have a lower degree of dissociation and weaker interlayer interactions than those of strong polyelectrolyte systems. Film growth of weak polyelectrolytes is exponential in the number of layers [89]. Various techniques have been used to probe the mechanical properties of weak polyelectrolyte nanocoposites [90–93].

Strong and weak polyelectrolyte films show significant differences in their mechanical properties as studied by the CCPD method. Consider the data shown below, Fig. 18 compares the surface tension response to dilation for two bilayers ($n = 2$) for strong (PAH/PSS) nanocomposites and weak polyelectrolytes hyaluronic acid (HA) and poly(L-lysine) (PLL). A linear (elastic) relationship between stress and area for up to 10% area dilation is shown for the strong PE pair, whereas plastic flow over a relatively low deformation is shown for the weak PE pair.

For ($2 < n < 6$) the surface modulus of DMPG-(HA/PLL)$_n$ as a function of film thickness was shown to be approximately constant [94]. This demonstrates that for an increase in membrane thickness there is not an increase in the surface elastic modulus, which corresponds to a decrease in the bulk modulus. Experiments performed independently using the colloidal probe AFM technique showed similar trends for the (HA/PLL)$_n$ system [64]. Polycondensation of HA and PLL to form an interlayer amide can be used to adjust the mechanical strength of the membrane [64]. The surface tension as a function of deformation for DMPG-(HA/PLL)$_2$ as assembled and after covalent crosslinking are shown in Fig. 19.

The surface tension (i.e. stress response) of the interface during dilation exceeds the limit as defined for a Gibbs surface layer as illustrated in Fig. 19. When the surface tension is a function of the surface density of adsorbed species only, the surface stress has an upper bound of the surface tension of the pure solvent subphase; $\gamma_0 = 72$ mN/m for air/water. The crosslinked composite exhibits elastic behavior

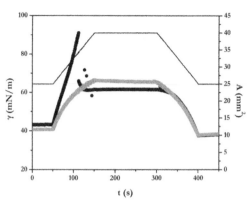

t (s)

Figure 19. Surface tension and drop area *versus* time. DMPG-(PLL/HA)$_2$ (■), and DMPG-(PLL/HA)$_2$ after crosslinking (●) for 12 hours under constant drop area (25 mm^2); drop area (-).

over a significantly broader range of strain and also fractures, as shown in the surface stress response for the crosslinked and non-crosslinked (HA/PLL)$_2$.

4. Interfacial Crosslinking with Proteins

It has been shown that cells are affected and can be extremely sensitive to changes in the mechanical properties of their substrates under chemostatic conditions [95]. Proteins are critical for cell function and are responsible for many tasks, which include mechanical support, signal transduction, and metabolic and catalytic functions [96]. Mechanical forces are capable of causing protein unfolding and protein domain deformation; which is important in molecular biomechanics. The viscoelasticity of fibrin is unique and noteworthy among biopolymers [76, 97, 98]. The surface stress response to inflation of fibrin nanomembranes was examined. The interfacial tension during deformation of a fibrinogen monolayer and a fibrin nanomembrane at the air/water interface are compared in Fig. 20.

The surface tension once again exceeds that of the pure solvent γ_0 showing the impact of covalent structure on surface stress response. The non-linear nature of the surface stress as a function of strain and the absence of fracture also illustrates the role of internal structure on material response.

F. Other Applications

Further studies of surfactant and protein adsorption reversibility and kinetics are relatively straightforward. However, the CCPD method allows for a wide variety of systematic investigations on the mechanics of well-defined thin films and structure-property relationships for materials beyond those discussed in Section E. Studies of diffusion and molecular transport in these materials are also possible with the CCPD method. A description of both types of experiments follows. Fabrication of surface materials of well-defined composition can be achieved by sequential and co-

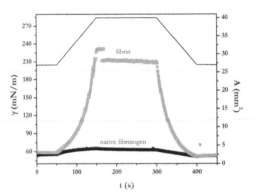

Figure 20. Surface tension response of native fibrinogen before and after crosslinking with thrombin to form a fibrin network, synthesized under the conditions: $C_{\text{fibrinogen},\infty} = 1$ mg/ml; $C_{\text{thrombin},\infty} = 20$ units/ml; $C_{\text{CaCl2}} = 0.010$ mol/l. Native fibrinogen (●), native fibrinogen crosslinked with thrombin (■), and drop area (-).

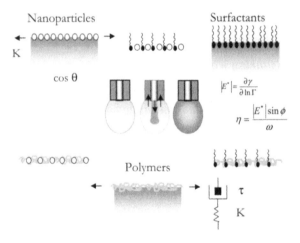

Figure 21. Interfacial mechanical properties of nanomaterials: surfactant, polymer, and nanoparticles. Possible combinations of different materials at fluid interfaces.

adsorption processes. Illustrated in Fig. 21 are some of the permutations between the basic building blocks available for surface modification.

The interface can be designed to be a kinetically irreversible structure that can provide engineered functionality for a variety of applications, which are previously referenced. The ability of the CCPD method to allow for subphase exchange makes it possible to measure dilational moduli of interfacial nanocomposite materials. This capability also allows the intrinsic surface mechanics to be separated from the kinetic effects, which arise from the dynamics of desorption between the interface and elevated bulk concentrations necessary for synthesis. A wide variety of constitutive behaviors described in Section E would also be directly accessible. Interfacial transport phenomena, also remains a relatively unexplored area [99–101]. Fick's law is the simplest constitutive law that describes molecular transport, relating the flux of a material through a continuous medium to both the diffusion coefficient and the concentration gradient of the material. The continuum approximation may be reasonable for the transport of small solutes through nanocomposite materials.

This assumption necessitates that the characteristic length scale of the solute $l(A)$ be much smaller than the transverse dimension of the interface h.

The rate of transport of A per interfacial area (or simply flux of A) through the interface (i) in the limit dilute solutions is given by

$$N_A = -D_{i,A} \frac{\partial C_A}{\partial r}, \tag{17}$$

where N_A is the flux with respect to a fixed frame of reference, i.e. the drop interfacial area S, $D_{i,A}$ is the diffusion coefficient of species A through the interfacial composite, and $\frac{\partial C_A}{\partial r}$ is the concentration gradient normal to the interface. The gradient can be written in terms of the concentration in the (α) phase C_A^α, the concentration in the (β) phase C_A^β, and the interfacial thickness, h, using $\frac{\partial C_A}{\partial r} \approx \frac{C_B^\alpha - C_A^\beta}{h}$. In this, resistance to mass transfer is associated with the interface, although boundary layer effects can be readily incorporated.

The driving force must be adjusted to reflect the difference in solute partitioning between phases; $C_A^{*,\alpha} = f(C_A^\beta)C_A^\beta$ or interphase transport, where $C_A^{*,\alpha}$ is the (α) phase concentration in equilibrium with the (β) phase concentration and $f(C_A^\beta)$ the thermodynamic function that describes equilibrium partitioning of species A between the phases (α) and (β). Upon substitution into Eq. (17) an expression for the permeability of the interface p is

$$p = \frac{N_A}{C_A^\alpha - C_A^{*,\alpha}}, \tag{18}$$

where p is the product of $D_{A,i}$ and h. For a given interfacial composite, p is an intrinsic material property and a technologically relevant parameter. It provides additional information about the internal structure of the material.

Subphase exchange in pendant drops provides a means to tailor driving forces for adsorption and desorption in liquid/fluid systems and in some cases, fabricate surface materials of well-defined composition at these interfaces. The CCPD method allows for a wide variety of experimental possibilities since it is capable of complete subphase exchange, much unlike traditional techniques. The CCPD method is not limited and provides a new framework for the assessment of mechanics and transport of soft surface materials.

G. Acknowledgements

This work was financially supported by projects of the DLR (50WM 0640 and 0941), DFG SPP 1273 (Mi418/16-2), NSF (CMMI) Award 0729403, and NASA support in cooperation with ESA through AO-2009-0813.

H. References

1. E. Dickinson and R. Miller (Eds), Food Colloids — Fundamentals of Formulation, Special Publication No 258, Royal Society of Chemistry, 2001.
2. M.A. Bos and T. van Vliet, Adv. Colloid Interface Sci., 91 (2001) 437.
3. V.B. Fainerman, M.E. Leser, M. Michel, E.H. Lucassen-Reynders and R. Miller, J. Phys. Chem. B, 109 (2005) 9672.
4. C. Kotsmar, V. Pradines, V.S. Alahverdjieva, E.V. Aksenenko, V.B. Fainerman, V.I. Kovalchuk, J. Krägel, M.E. Leser, B.A. Noskov and R. Miller, Adv. Colloid Interface Sci., 150 (2009) 41.
5. J. Zhao, D. Vollhardt, J. Wu, R. Miller, S. Siegel and J.B. Li, Colloids Surfaces A, 166 (2000) 235.
6. D. Vollhardt and M. Wittig, Colloids and Surfaces, 47 (1990) 233.
7. S. Sundaram, J.K. Ferri, D. Vollhardt and K.J. Stebe, Langmuir, 14 (1998) 1208.
8. H.A. Wege, J.A. Holgado-Terriza and M.A. Cabrerizo-Vilchez, J. Colloid Interface Sci., 249 (2002) 263.
9. H.A. Wege, J.A. Holgado-Terriza, A.W. Neumann and M.A. Cabrerizo-Vilchez, Colloids Surfaces A, 156 (1999) 509.
10. M.A. Cabrerizo-Vilchez, H.A. Wege, J.A. Holgado-Terriza and A.W. Neumann, Rev. Scientific Instruments, 70 (1999) 2438.
11. J.K. Ferri, N. Gorevski, C. Kotsmar, M.E. Leser and R. Miller, Colloids Surfaces A, 13 (2008) 319.
12. V.B. Fainerman, R. Miller, J.K. Ferri, H. Watzke, M.E. Leser and M. Michel, Adv. Colloid Interface Sci., 123 (2006) 163.
13. N. Gorevski, R. Miller and J.K. Ferri, Colloids Surfaces A, 12 (2008) 323.
14. J. Maldonado-Valderrama, H.A. Wege, M.A. Rodriguez-Valverde, M.J. Galvez-Ruiz and M.A. Cabrerizo-Vilchez, Langmuir, 19 (2003) 8436.
15. J. Maldonado-Valderrama, M.J. Galvez-Ruiz, A. Martin-Rodriguez and M.A. Cabrerizo-Vilchez, Langmuir, 20 (2004) 6093.
16. C. Kotsmar, J. Krägel, V.I. Kovalchuk, E.V. Aksenenko, V.B. Fainerman and R. Miller, Journal of Physical Chemistry B, 113 (2009) 103.
17. C. Kotsmar, D.O. Grigoriev, F. Xu, E.V. Aksenenko, V.B. Fainerman, M.E. Leser and R. Miller, Langmuir, 24 (2008) 13977.
18. J.K. Ferri, R. Miller and A.V. Makievski, Colloids Surfaces A, 39 (2005) 261.
19. J.K. Ferri, W.F. Dong and R. Miller, J. Phys. Chem. B, 109 (2005) 14764.
20. A.L. Caro, M.R.R. Nino and J.M.R. Patino, Colloids Surfaces A, 332 (2009) 180.
21. P. Li, G.H. Xiu and A.E. Rodrigues, Aiche Journal, 53 (2007) 2419.
22. Y. Rotenberg, L. Boruvka and A.W. Neumann, J. Colloid Interface Sci., 93 (1983) 169.
23. G. Loglio, P. Pandolfini, R. Miller, A.V. Makievski, F. Ravera, M. Ferrari and L. Liggieri, Drop and Bubble Shape Analysis as Tool for Dilational Rheology Studies of Interfacial Layers, in "Novel Methods to Study Interfacial Layers", D. Möbius and R. Miller (Eds), Studies in Interface Science, Vol. 11, 2001, p. 439.
24. H.A. Wege, J.A. Holgado-Terriza, M.J. Galvez-Ruiz and M.A. Cabrerizo-Vilchez, Colloids Surfaces B, 12 (1999) 339.
25. A. Javadi, J.K. Ferri and R. Miller, Colloids Surfaces A, 365 (2010) 145.
26. B.A. Noskov, Adv. Colloid Interface Sci., 69 (1996) 63.
27. C.H. Chang and E.I. Franses, Colloids and Surfaces A, 1 (1995) 100.
28. D. Langevin, Current Opinion in Colloid & Interface Sci., 3 (1998) 600.
29. J. Eastoe and J.S. Dalton, Adv. Colloid Interface Sci., 85 (2000) 103.

30. J.K. Ferri and K.J. Stebe, Adv. Colloid Interface Sci., 85 (2000) 61.

31. J.F. Baret, J. Colloid Interface Sci., 1 (1969) 30.

32. E.V. Aksenenko, A.V. Makievski, R. Miller and V.B. Fainerman, Colloids and Surfaces A, 143 (1998) 311.

33. J.K. Ferri, C. Kotsmar and R. Miller, Adv. Colloid Interface Sci., 161 (2010) 29.

34. R. Miller, D.O. Grigoriev, J. Krägel, A.V. Makievski, J. Maldonado-Valderrama, M.E. Leser, A. Michel and V.B. Fainerman, Food Hydrocolloids, 19 (2005) 479.

35. V.B. Fainerman, E.H. Lucassen-Reynders and R. Miller, Adv. Colloid Interface Sci., 106 (2003) 237.

36. E.D. Goddard, J. Colloid Interface Sci., 256 (2002) 228.

37. R. Miller, V.B. Fainerman, A.V. Makievski, J. Kragel, D.O. Grigoriev, V.N. Kazakov and O.V. Sinyachenko, Adv. Colloid Interface Sci., 86 (2000) 39.

38. E. Dickinson, Colloids and Surfaces B, 15 (1999) 161.

39. E. Dickinson, J. Chemical Society-Faraday Trans., 94 (1998) 1657.

40. A.V. Makievski, V.B. Fainerman, M. Bree, R. Wüstneck, J. Krägel and R. Miller, J. Phys. Chem. B, 102 (1998) 417.

41. B.C. Tripp, J.J. Magda and J.D. Andrade, J. Colloid Interface Sci., 173 (1995) 16.

42. R. Miller, P. Joos and V.B. Fainerman, Adv. Colloid Interface Sci., 49 (1994) 249.

43. P. Walstra and A.L. Deroos, Food Reviews International, 9 (1993) 503.

44. J.J. Ramsden, D.J. Roush, D.S. Gill, R.G. Kurrat and R.C. Willson, J. Am. Chem. Soc., 117 (1995) 8511.

45. M.A. Cohen Stuart, Biopolymers at Interfaces, M. Dekker, 1999.

46. W. Norde and C.A. Haynes, Interfacial Phenomena and Bioproducts, M. Dekker, 1999.

47. J.J. Ramsden, Biopolymers at Interfaces, M. Dekker, 1999.

48. Z. Adamczyk, J. Colloid Interface Sci., 229 (2000) 477.

49. V.B. Fainerman, S.V. Lylyk, J.K. Ferri, R. Miller, H. Watzke, M.E. Leser and M. Michel, Colloids Surfaces A, 282–283 (2006) 217.

50. T.F. Svitova, M.J. Wetherbee and C.J. Radke, J. Colloid Interface Sci., 261 (2003) 170.

51. V.B. Fainerman, M.E. Leser, M. Michel, E.H. Lucassen-Reynders and R. Miller, J. Phys. Chem. B, 109 (2005) 9672.

52. A.F. Xir and S. Granick, Nat. Mater., 1 (2002) 129.

53. T.F. Svitova and C.J. Radke, Industrial & Engineering Chemistry Research, 44 (2005) 1129.

54. F. Mac Ritchie, Protein at Liquid Interface, in "Studeis in Interface Science", D. Möbius and R. Miller (Eds), Vol. 7, Elsevier, 1998, p. 149.

55. S.J. Kang, C. Kocabas, H.S. Kim, Q. Cao, M.A. Meitl, D.Y. Khang and J.A. Rogers, Nano Letters, 7 (2007) 3343.

56. P. Ravirajan, S.A. Haque, J.R. Durrant, D.D.C. Bradley and J. Nelson, Adv. Functional Materials, 15 (2005) 609.

57. V. Pardo-Yissar, E. Katz, O. Lioubashevski and I. Willner, Langmuir, 17 (2001) 1110.

58. J.A. He, L. Samuelson, L. Li, J. Kumar and S.K. Tripathy, Langmuir, 14 (1998) 1674.

59. A. Rogach, A. Susha, F. Caruso, G. Sukhorukov, A. Kornowski, S. Kershaw, H. Möhwald, A. Eychmuller and H. Weller, Adv. Materials, 12 (2000) 333.

60. F. Caruso, Adv. Materials, 13 (2001) 11.

61. D.J. Schmidt, F.C. Cebeci, Z.I. Kalcioglu, S.G. Wyman, C. Ortiz, K.J. Van Vliet and P.T. Hammond, Acs Nano, 3 (2009) 2207.

62. W.J. Li, R.L. Mauck, J.A. Cooper, X.N. Yuan and R.S. Tuan, Journal of Biomechanics, 40 (2007) 1686.

63. C.R. Wittmer, J.A. Phelps, C.M. Lepus, W.M. Saltzman, M.J. Harding and P.R. Van Tassel, Biomaterials, 29 (2008) 4082.

64. C. Picart, A. Schneider, O. Etienne, J. Mutterer, P. Schaaf, C. Egles, N. Jessel and J.C. Voegel, Adv. Functional Materials, 15 (2005) 1771.

65. A.S. Hoffman, P.S. Stayton, O. Press, N. Murthy, C.A. Lackey, C. Cheung, F. Black, J. Campbell, N. Fausto, T.R. Kyriakides and P. Bornstein, Polymers Adv. Techn., 13 (2002) 992.

66. D.E. Discher, P. Janmey and Y.L. Wang, Science, 310 (2005) 1139.

67. P.C. Georges and P.A. Janmey, J. App. Phys., 98 (2005) 1547.

68. D. Mertz, C. Vogt, J. Hemmerle, J. Mutterer, V. Ball, J.C. Voegel, P. Schaaf and P. Lavalle, Nature Materials, 8 (2009) 731.

69. H. Rehage, M. Husmann and A. Walter, Rheologica Acta, 41 (2002) 292.

70. A. Burger and H. Rehage, Angew. Makromolekulare Chemie, 202 (1992) 31.

71. J. Ruths, F. Essler, G. Decher and H. Riegler, Langmuir, 16 (2000) 8871.

72. V. Ball, E. Hubsch, R. Schweiss, J.C. Voegel, P. Schaaf and W. Knoll, Langmuir, 21 (2005) 8526.

73. P.T. Hammond, Current Opinion in Colloid & Interface Sci., 4 (1999) 430.

74. J.K. Ferri, P. Carl, N. Gorevski, T.P. Russell, Q. Wang, A. Boker and A. Fery, Soft Matter, 4 (2008) 2259.

75. J.T. Russell, Y. Lin, A. Boker, L. Su, P. Carl, H. Zettl, J.B. He, K. Sill, R. Tangirala, T. Emrick, K. Littrell, P. Thiyagarajan, D. Cookson, A. Fery, Q. Wang and T.P. Russell, Angew. Chemie-Intern. Edition, 44 (2005) 2420.

76. J.P. Collet, H. Shuman, R.E. Ledger, S.T. Lee and J.W. Weisel, Proceedings of the National Academy of Sciences of the United States of America, 102 (2005) 9133.

77. H.Y. Shen, J. Watanabe and M. Akashi, Analytical Chemistry, 81 (2009) 6923.

78. M. Ulbricht, Polymer, 47 (2006) 2217.

79. W.Q. Jin, A. Toutianoush and B. Tieke, Langmuir, 19 (2003) 2550.

80. S.A. Maskarinec, J. Hannig, R.C. Lee and K.Y.C. Lee, Biophysical J., 82 (2002) 1453.

81. K.C. Krogman, J.L. Lowery, N.S. Zacharia, G.C. Rutledge and P.T. Hammond, Nature Materials, 8 (2009) 512.

82. C.Y. Jiang and V.V. Tsukruk, Adv. Materials, 18 (2006) 829.

83. M. Nolte, I. Donch and A. Fery, Chemphyschem, 7 (2006) 1985.

84. J.K. Ferri, W.F. Dong, R. Miller and H. Möhwald, Macromolecules, 39 (2006) 1532.

85. A.D. Cramer and J.K. Ferri, in preparation.

86. C.H. Porcel and J.B. Schlenoff, Biomacromolecules, 10 (2009) 2968.

87. E. Guzman, H. Ritacco, F. Ortega, T. Svitova, C.J. Radke and R.G. Rubio, J. Phys. Chem. B, 113 (2009) 7128.

88. E. Guzman, H. Ritacco, J.E.F. Rubio, R.G. Rubio and F. Ortega, Soft Matter, 5 (2009) 2130.

89. P. Lavalle, C. Gergely, F.J.G. Cuisinier, G. Decher, P. Schaaf, J.C. Voegel and C. Picart, Macromolecules, 35 (2002) 4458.

90. L. Richert, F. Boulmedais, P. Lavalle, J. Mutterer, E. Ferreux, G. Decher, P. Schaaf, J.C. Voegel and C. Picart, Biomacromolecules, 5 (2004) 284.

91. A.J. Nolte, M.F. Rubner and R.E. Cohen, Macromolecules, 38 (2005) 5367.

92. L. Richert, A.J. Engler, D.E. Discher and C. Picart, Biomacromolecules, 5 (2004) 1908.

93. S. Markutsya, C.Y. Jiang, Y. Pikus and V.V. Tsukruk, Adv. Functional Materials, 15 (2005) 771.

94. J.T. McRuiz, N. Adje and J.K. Ferri, in preparation.

95. M.E. Chicurel, C.S. Chen and D.E. Ingber, Current Opinion in Cell Biology, 10 (1998) 232.

96. G. Bao and S. Suresh, Nature Materials, 2 (2003) 715.

97. M. Guthold, W. Liu, E.A. Sparks, L.M. Jawerth, L. Peng, M. Falvo, R. Superfine, R.R. Hantgan and S.T. Lord, Cell Biochemistry and Biophysics, 49 (2007) 165.

98. W. Liu, L.M. Jawerth, E.A. Sparks, M.R. Falvo, R.R. Hantgan, R. Superfine, S.T. Lord and M. Guthold, Science, 313 (2006) 634.

99. F. Ravera, M. Ferrari, L. Liggieri, R. Miller and A. Passerone, Langmuir, 13 (1997) 4817.

100. M. Ferrari, L. Liggieri, F. Ravera, C. Amodio and R. Miller, J. Colloid Interface Sci., 186 (1997) 40.

101. L. Liggieri, F. Ravera, M. Ferrari, A. Passerone and R. Miller, J. Colloid Interface Sci., 186 (1997) 46.

Wetting Dynamics of Aqueous Solutions on Solid Surfaces

Victoria Dutschk

Engineering of Fibrous Smart Materials (EFSM), Faculty of Engineering Technology (CTW),
University of Twente, P.O. Box 217, 7500 AE Enschede, The Netherlands

Contents

A. Essentials of Wetting

In 1876 Gibbs [1] elaborated the fundamentals of the thermodynamic theory of capillarity; his paper 'On the Equilibrium of Heterogeneous Substances' was taken as a basis for all subsequent theoretical and experimental wetting studies. Since then, diligent work has been performed to describe the wetting behaviour of heterogeneous systems, thereby determining the surface energies of liquid and solid bodies, and in this manner predicting their adhesion behaviour important in technology applications. Over a period of time, plenty of literature have been accumulated, proposing various measurement techniques and different evaluation possibilities including even criticism of one or the other computational algorithm or fundamental idea.

Bubble and Drop Interfaces
© Koninklijke Brill NV, Leiden, 2011

1. Surface Tension and Surface Energy

It is common knowledge that in the interior of a liquid, a molecule undergoes a different equilibrium position than that at the surface, due to its neighbours, and that work has to be done to direct this molecule toward the surface. As this takes place, the surface will increase by this molecule which now has a potential energy elevated by the amount of this work. The corresponding increase in energy, being related to the unit of area, is referred to as specific surface energy. The force needed to do this work and related to the unit of length is referred to as surface tension. Most textbooks assume the surface tension to be identical to the specific surface energy.

Strictly speaking these terms are not identical. While the term surface tension (a unit of measurement: force per length, the force being determined by specifying amount and direction) originates from classical mechanics, the term surface energy (a unit of measurement: energy per area, the energy being fully considered by specifying a number) results from the energy approach to solving physical problems [2]. The thermodynamic description of the surface of a real, i.e. deformable, solid object has to take into account the tensor character of the surface tension, which in the general case is a function of not only the surface energy [3].

2. Wetting and Wettability

Both wetting and de-wetting of liquids on different surfaces play an important role in many natural and technological processes. In many applications, surface wettability is macroscopically described by the equilibrium contact angle Θ. The final state of a wetting process is characterized by the Young equation (1), as shown in Fig. 1:

$$\gamma_{SV} = \gamma_{SL} + \gamma_{LV} \cdot \cos \Theta, \qquad (1)$$

where γ_{SV}, γ_{SL} and γ_{LV} are interfacial solid–vapour, solid–liquid and liquid–vapour tensions, respectively. Since the quantities γ_{SV} and γ_{SL} are generally inaccessible to experiments, in contrast to γ_{LV}, the Young equation is often used for solving the inverse problem, namely to determine the difference ($\gamma_{SV} - \gamma_{SL}$), which is referred to as wetting tension or adhesive tension, by means of experimental values for static or quasi-static contact angle θ and interfacial tension γ_{LV}. The magnitude of the contact angle depends on the strength of molecular interactions between liquid molecules inside the drop as well as between a liquid and a solid surface.

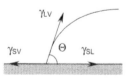

Figure 1. Liquid drop on a solid surface: Θ is the Young contact angle; γ_{LV} is the interfacial tension liquid–vapour; γ_{SV} and γ_{SL} are interfacial tensions solid–vapour and solid–liquid, respectively.

In general, when a liquid drop is placed on a solid surface, either it spreads over the surface, i.e. it completely wets it, or it builds a finite contact angle with the surface. If the contact angle is between 0 and 90° the situation is referred to as partial wetting. However, if the contact angle is larger than 90°, the liquid does not wet the surface and the situation is referred to as non-wetting. A more detailed description of advancing, receding, Young's contact angles as well as problems of experimental and theoretical verification of equilibrium contact angle is recently provided by Chibowski [4].

B. Wetting Dynamics

If the phase boundary liquid-solid — three-phase contact (TPC) line — moves relative to an adjacent solid surface, a dynamic contact angle will be observed. The dynamic contact angle is a contact angle as a function of time, which can significantly differ from the static contact angle. A unified approach to a theoretical description and measurement of dynamic contact angles does not exist in the literature.

The Young equation (1) is valid for smooth, chemically homogeneous surfaces and pure liquids, excluding absorption, evaporation and other effects. Furthermore, the thermodynamic consideration of wetting processes is necessary, but it is not sufficient to describe a lot of technological processes as kinetic aspects are not considered. In such applications, dynamic wetting and de-wetting processes are of crucial importance.

The dynamic behaviour of a pure liquid on an ideal solid surface can often be successfully described by the equilibrium contact angle, dynamic time-dependent contact angle as well as spreading velocity. In the hydrodynamic consideration [5, 6] disturbed equilibrium leads to the spreading force $\gamma_{LV}(\cos\theta_0 - \cos\theta(t))$, where θ_0 is initial contact angle. Work is necessary to expanding the solid–liquid interface, and energy will dissipate due to viscose shear in the liquid. Therefore, this theory is applicable to the description of a slow spreading near equilibrium. The molecular-kinetic theory [7] assumes, however, particular displacements of the TPC line at the molecular level as a possible reason for the spreading force; it is suitable for describing high spreading velocities far from equilibrium. Although this theory, in contrast to the hydrodynamic theory, includes surface effects, its application to predict the spreading velocity is rather problematic since molecular parameters such as the density of adsorption centres and the distance between them on real surfaces is unknown und generally inaccessible to experiments.

1. Spreading and Spreading Velocity

Spreading of evaporating droplets is determined by the spreading rate law dr/dt and evaporation rate law $V(t)$, where r is the base radius of a spreading drop and V is the drop volume. In many cases, the spreading rate law was found to be bi-exponential [8, 9], while the evaporation has a well documented proportionality to the TPC line length [10–12]. If these two laws are in force, the spreading kinetics of a liquid drop, i.e. the dependence of the base radius r and contact angle θ on time, can be predicted. A more general step-mechanism of the TPC line motion was recently proposed [13]. Here, surface energy fluctuations serve as energy barrier of the spreading process, described as a nucleation process driven by capillary waves at liquid surfaces. It was shown, that such a mechanism leads to a spreading rate law similar to one obtained by molecular-kinetic treatment, without being based on adsorption/desorption mechanism of spreading [7–9]. The well-documented exponential law of spreading [8, 14–16] can also be interpreted by this mechanism.

2. *Spreading as a Rate Process*

Spreading rate law dr/dt can be physically determined by bulk friction or by friction in the vicinity of the TPC zone [8, 17]. In other words, spreading of droplets is a hydrodynamic problem with slip boundary conditions at the TPC line [18] such as bulk properties, existing in the hydrodynamic equations, or TPC region properties, occurring in their boundary condition, that are the decisive factors for spreading. In the following, the case, where spreading rate is determined mainly by the TPC line region properties, is assumed. A general form of the spreading rate law in such case is given by Eyring's bi-exponential form [19]

$$\frac{dr}{dt} = A_A e^{-b_A \cos\theta} - A_R e^{b_R \cos\theta}, \tag{2}$$

where θ is dynamic contact angle, i.e. the contact angle as a function of time $\theta = \theta(t)$. The indices A and R describe the parameters of advancing and receding movement of the TPC line. For $\theta > \Theta$, the TPC-line is advancing, e.g. in the case of spreading of aqueous surfactant solutions; for $\theta < \Theta$, the TPC-line is receding, e.g. during evaporation of a water drop. The spreading law (2) reflects the barrier character of the TPC line movement [20], which was confirmed for the adsorption–desorption mechanism of the TPC movement [21, 22]. Most of wetting characteristics such as mobility and immobility of the TPC line — pinning effect [23]), quasi-static advancing and receding contact angles, characteristic spreading velocity — can be interpreted with the help of equation (2). The parameters $A_{A,R}$ and $b_{A,R}$ depend on a particular mechanism, but in the case of equilibrium $dr/dt = 0$, the following relationship is valid

$$\cos\Theta = \frac{1}{b_A + b_R} \ln\frac{A_A}{A_R}. \tag{3}$$

The cosine of Θ is defined by the Young equation (1) and does not depend on any mechanisms.

From a simple geometrical consideration [24, 25], the base radius r for a small drop can be expressed as

$$r^3 = \frac{3}{\pi} \frac{(1 + \cos\theta)^{3/2}}{(1 - \cos\theta)^{1/2}(2 + \cos\theta)} V, \tag{4}$$

where V is the drop volume, which is supposed to remain constant, and θ is the contact angle changing with time. In summary, the cosine of dynamic contact angle $\theta(t)$ is determined by the spreading law (2), drop geometry (4) and evaporation rate law $V(t)$. The problem of the determination of $\cos\theta(t)$ and $r(t)$, assuming that drop volume remains constant, was solved numerically [26], where the exponential factors b_A und b_R were assumed to be identical. In this case, the spreading law of Eq. (2) reduces to

$$\frac{dr}{dt} = A \sinh(b(\cos\Theta - \cos\theta)). \tag{5}$$

3. Spreading of Aqueous Surfactant Solutions

If a surface-active substance and a real (i.e. rough, inhomogeneous) solid surface is of interest, some attempts to describe the wetting behaviour theoretically are rather of a speculative nature.

Aqueous surfactant solutions differ from pure liquids by the fact that their surface tension γ_{LV} and solid–liquid interfacial tension γ_{SL} are functions of time, and the molecular orientation influences surfactant-solid surface interactions. In the study of aqueous surfactant solutions, along with the solid surface state (chemical and morphologic nature), additional factors such as the solution concentration, the chemical nature of a surfactant (non-ionic, anionic, cationic, amphoteric) have to be taken into account. In the last decade, various authors, using dynamic contact angle measurements, found that the spreading velocity of aqueous surfactant solutions is strongly affected by the solid surface energy.

An adequate interpretation of the results of contact angle measurements is additionally complicated because a solid surface is able to adsorb water vapour from humid air. So, the surface energy value γ_{SV} depends on the thickness of the adsorption film. Disjoining pressure isotherms in the presence of surfactants are well investigated in the case of free liquid films [27], much less is known in the case of liquid films on solid substrates [28]. At the present, an answer to the question how surfactant molecules are transferred in the TPC line vicinity is not given. In the case of aqueous surfactant solutions, the knowledge of the transition zone behaviour from meniscus to thin films in front is very limited as referred in [29].

Hitherto, either only the solid–vapour interfacial tension γ_{SV} or only the surface tension γ_{LV} was assumed in the literature, though variations of the solid–liquid interfacial tension γ_{SL} was not excluded in this case. Depending on these assumptions, two fundamentally different spreading mechanisms of aqueous surfactant solutions were proposed for hydrophobic surfaces [29]. Following von Bahr *et al.* ideas [30] the surfactant molecules adsorb at the freshly formed solid–liquid interface behind the advancing wetting front (cf. Fig. 2a).

The transfer of molecules from the liquid–vapour interface occurs very quickly. The replenishment of this interface happens by diffusion of surfactant molecules from the volume phase and depends linearly on the root of time (diffusion-controlled).

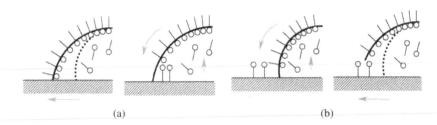

(a)   (b)

Figure 2. Spreading mechanisms of aqueous surfactant solutions over hydrophobic surfaces according to (a) von Bahr, Tiberg and Yaminsky [30] and (b) Starov, Kosvintsev and Velarde [24], according to [29].

In contrast with it, Starov *et al.* [24] assumed that the surfactant molecules adsorb onto the solid–liquid interface before the wetting front as illustrated in Fig. 2b. As this takes place, interfacial tension γ_{SV} increases. The transfer of molecules from the liquid–vapour interface occurs very slowly. The reason for the spreading force is the difference $(\Gamma_S(t) - \Gamma_e)$, where $\Gamma_S(t)$ is the current value of the adsorption quantity on the solid surface and Γ_e is the corresponding equilibrium value.

If a surface is rough, the topographic inhomogeneities have to be considered as well. For that, the existing theories for pure liquids according to Wenzel (homogeneous surface) [31], Cassie and Baxter (heterogeneous surface) [32], Johnson and Dettre (composite surface) [33], Extrand (contact line approach) [34] can be taken into account.

4. *Dynamic Surface Tension*

To explain and to predict the behaviour of aqueous surfactant solutions on the solid–liquid interface, information of their dynamic behaviour at the liquid–vapour interface is absolutely necessary. The dynamic surface tension of aqueous solutions can be measured according to different time windows by suitable methods: (i) bubble pressure tensiometry; (ii) drop volume tensiometry; (iii) drop/bubble profile analysis, described in more details in Chapters 5, 6 and in Chapters 2 and 3, respectively.

C. Dynamic Contact Angle Measurements

Dynamic contact angle measurements are possible either force-driven, if the drop volume will be increased/decreased or as time-dependent contact angle measurements with a constant volume. In the former case, advancing or receding angles[1] are formed to analyse the contact angle hysteresis, i.e. analysis of chemical and mechanical heterogeneities. In the latter case, the temporal contact angle change because of spontaneous spreading of the liquid is measured. If the spreading velocity is limited by the resistance of the TPC line, this phenomenon is referred to as wetting dynamics or wetting kinetics. The contact angle depending on the con-

---

[1] Strictly speaking, these are quasi-static measurements in the physical sense.

tact time of solid surface with measuring liquid is called dynamic contact angle (cf. Section 2).

1. Measuring Possibilities

The wetting dynamics is generally examined by the sessile drop method. Such measurements can be done with any suitable measuring units. For industrial purposes (especially for the paper industry) the Swedish Fibro System Company developed a tailor-made absorption and contact angle tester FibroDAT. The device is completed with software to automate a lot of measuring steps thus avoiding accidental measuring errors. The operation of this device is covered by patents and the measuring specification standardized (ASTM[2] D 5725-99 and TAPPI[3] 558 om-97). The advantages of this measuring device are also:

- Fully automated measurement of the surface tension of liquids by the pendant drop method with the indication of a time delay (in particular, helpful in studies on aqueous surfactant solutions, because they show time-dependent behaviour); it is also possible to carry out separate dynamic surface tensions measurements independently of dynamic wetting measurements;

- Setting of a desired drop volume by means of a micro-dosing system;

- Defined deposition of the drop on the surface with a short stroke upon the syringe tip from an electromagnet (the stroke strength as well as its time delay can be optimized);

- Setting of a desired drop-surface distance;

- Application of PTFE syringes with PTFE coated needles (in particular, helpful in studies of highly concentrated aqueous surfactant solutions).

If the dynamic contact angle measurements are carried out in open air, surfactant solutions spread on solid surfaces to a lenticular drop with a large surface (Fig. 3), governed by competition between evaporation and spreading (Fig. 4).

While spreading, the drop volume remains constant with its radius becoming larger.

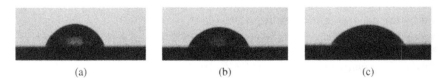

(a)　　　　　　　(b)　　　　　　　(c)

Figure 3. Liquid drop on a solid surface: (a) at the beginning $t = t_0$; (b) during spreading process; (c) at the end (quasi-equilibrium).

---

[2] American Society for Testing and Materials (USA).
[3] Physical Properties Committee of the Process and Product Quality Division (USA).

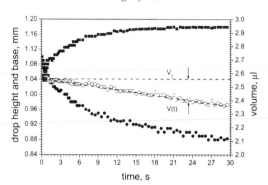

Figure 4. Drop base, height and volume of a water drop, containing a pure non-ionic surfactant ($C_{12}EO_5$) on a hydrophobic surface (Teflon AF): (●) drop height; (■) drop diameter; (○) drop volume; V_0 initial drop volume; $V(t)$ drop volume as a function of time; [25] with permission of Carl Hanser Verlag, München.

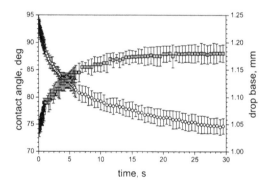

Figure 5. Contact angle and drop base of a water drop, containing a pure non-ionic surfactant ($C_{12}EO_5$) on a hydrophobic surface (Teflon AF): (○) contact angle; (□) drop diameter; [25] with permission of Carl Hanser Verlag, München.

If necessary, dynamic wetting processes can be recorded by methods different from the sessile drop method. The dynamic wetting and penetration behaviour of liquids into porous structures such as powder, fibres and textile materials can be estimated using a tensiometer by studying the increase in the weight of a capillary caused by the progression of the liquid inside it.

Figure 5 shows an example of measurement of dynamic wetting parameters [25].

2. Wetting Dynamics in Technology Applications

For many applications, surfactants are introduced into the aqueous phase to increase the rate and uniformity of wetting. Despite their enormous technical importance, there is a lack of data in the literature about the spreading dynamics of aqueous surfactant solutions [29]. The knowledge of how surfactant adsorption at the surfaces involved affects the spreading mechanism and dynamics is also limited.

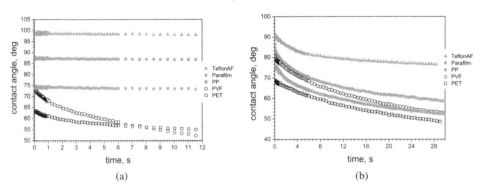

Figure 6. Dynamic contact angles of aqueous solutions of (a) an anionic sodium dodecyl sulfate SDS and (b) a non-ionic pentaethylene glycol monododecyl ether $C_{12}EO_5$ on different polymer surfaces of technological relevance: Teflon AF, Parafilm, polypropylene (PP), polyvinyl fluoride (PVF) und polyethylene terephthalate (PET).

Dynamic wetting measurements allow studying surfactants, polyelectrolytes or other surface-active substances as well as their mixtures and engineered surfaces.

2.1. *Screening Technical Surfaces Using Model Surfactants*

Based on the dynamic behaviour of pure surfactants on technologically relevant polymer surfaces (Fig. 6), the conclusion can be drawn: polymer surfaces may be divided into different classes with reference to the spreading behaviour of aqueous surfactant solutions (see Table 1).

Hydrophobic surfaces are surfaces such as Teflon AF, Parafilm and PP (polyethylene) with the surface free energy of 11.8, 18.4 and 23.2 mJ/m², respectively. Moderately hydrophobic surface are surfaces such as PVF (polyvinyl fluoride) and PET (polyethylene terephthalate) with the surface free energy of 36.2 and 36.7 mJ/m², respectively.

With concentrations $c < \frac{1}{2}cmc$ (critical micelle concentration), aqueous solutions of non-ionic C_nEO_m surfactants do not spread on hydrophobic polymer surfaces such as PTFE, PP and PE. However, they do spread on moderately hydrophobic polymer surfaces in the whole range of concentrations. They spread well over highly hydrophobic surfaces as well, but in this case, the effect is less pronounced. Non-ionic ethoxylated alcohols C_mEO_n enhance spreading in aqueous solutions

Table 1.

Feature	Classes	
Ionogeneity	ionic	non-ionic
Concentrations range	$<\frac{1}{2}cmc$	$\geqslant\frac{1}{2}cmc$
Polymer surfaces of technological relevance	hydrophobic surfaces such as PTFE, PP, PE	moderately hydrophobic surfaces such as PVF, PVC, PET

on both highly hydrophobic and moderately hydrophobic surfaces, demonstrating rather universal behaviour which seems to be independent of the surface nature.

Ionic surfactants, however, spread on hydrophobic polymer surfaces at any concentration investigated. From the concentration $c = \frac{1}{2}cmc$, they spread on moderately hydrophobic surfaces with a pronounced dependency of the dynamic contact angle on the air humidity: the higher the air humidity, the slower the spreading process [35].

Considering the spreading rate, the results indicate that spreading, if it occurs, may be divided into two regimes, the short time regime (fast spreading) and the long time regime (slow spreading). In the first regime, until approximately 1 s, the base radius depends linearly on time. The analysis of time dependencies of the drop base radius reveals that the slow wetting dynamics observed for both ionic and non-ionic surfactants on hydrophobic surfaces can be explained neither by surfactant diffusion from the bulk of the drop to the expanding liquid–vapour interface nor in terms of viscous spreading (Fig. 7).

Indeed, in the first case the drop base radius has to be linearly dependent on the square root of time, whereas in the second case Tanner's law [36] has to be valid according to which $r \propto t^{0.1}$. Obviously, another process associated with the surfactant adsorption near the expanding TPC line has to be taken into account for the unusually slow drop spreading that was observed in the experiments.

As was shown earlier [25], that spreading occurs only at high surfactant concentration, of the order of the *cmc* and higher; otherwise the (TPC) line stayed pinned. Spreading typically took place in two marked consecutive regimes (similar behaviour was found in [37]). Usually, after a first, more intensive spreading regime, a second slower stage took place with a contact angle tending to a stationary value, corresponding to evaporation-driven receding contact angle at a constant velocity. This behaviour could not be explained with the familiar power law regimes [37, 38],

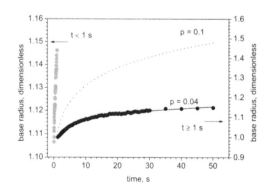

Figure 7. Reduced base radius r/r_0 of a $C_{12}EO_5$ drop at the *cmc* on Parafilm (fast spreading in grey and slow spreading in black) *versus* time, where r is the current value of the base radius and r_0 is its initial value. Dependency of experimental base radius on time is fitted by a power function ($p = 0.04$) being shown by a solid line. The case of $p = 0.1$ corresponding to Tanner's law [36] is shown by a dotted line; [35] with permission of Carl Hanser Verlag, München.

as power law was either not fitting the experimental data, or the obtained power values were in considerable dissonance with the theoretically expected one [35]. The second slow regime was found close to exponential, but the characteristic time of the exponential spreading was of values far too high to be explained with slow adsorption on the liquid–vapour interface [25]. Similarly, high characteristic time values of the exponential spreading were found also by other authors, and were reasonably interpreted as characteristic times for surfactant molecule surface diffusion to the dry solid–vapour interface in front of the TPC line [24]. However, this interpretation could not explain the fast initial spreading regime observed. In addition, at surfactant concentration above the *cmc*, the characteristic time of spreading was found continued to decrease (surfactant solution at the 2 *cmc* was found to spread faster than the one at the *cmc*), while the surface diffusion mechanism should not be strongly affected by the presence of micelles. To analyze these observations, another viewpoint will be adopted and developed [13], based on Eyring's approach [19] to rate processes of barrier nature, to which TPC line propagation belongs (cf. Section 2.2).

Ellipsometric investigations of adsorption kinetics on selected solid surfaces confirm the common spreading trends on these surfaces established by the dynamic contact angle measurements. With increasing surfactant concentration in an aqueous solution to the critical micelle concentration, the solid–liquid interfacial tension γ_{SL} calculated from the contact angle decreases, whereas the amount of adsorption will rise (Figs 8 and 9).

Investigations of wetting dynamics with aqueous solutions of pure surfactants in a wide range of concentrations on technologically relevant surfaces with different hydrophobicity and roughness allow various analyses on the reproducibility of their manufacture, on the influence of manufacturing methods (e.g., solution, injection moulding process or compressed air procedure, foil) and pre-treatment of various material surfaces (polymer, metal, wood and glass/ceramic surfaces) and working out of cleaning instructions for them.

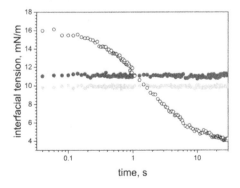

Figure 8. Interfacial tension calculated from water contact angles and dynamic contact angles for ( ) anionic SDS; (●) cationic dodecyl trimethyl ammonium bromide DTAB; (○) non-ionic $C_{12}EO_5$ on Teflon AF.

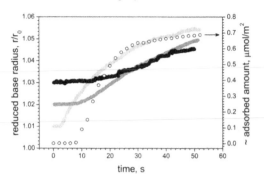

Figure 9. Reduced base radius for non-ionic $C_{12}EO_5$ solution on Teflon AF surface at different relative humidity (◯) 40% RH; (◑) 60% RH; (●) 80% RH and adsorbed amount (○) which was estimated in a aqueous phase (100% RH) using null ellipsometry (Optrel, Germany).

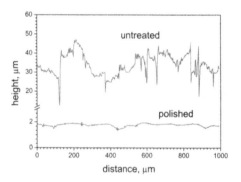

Figure 10. Roughness profiles of an untreated and polished titanium surface, measured with an optical sensor MicroGlider (FRT, Germany).

The influence of surface roughness on the wetting behaviour of aqueous surfactant solutions is illustrated on examples of metallic surfaces (moderately hydrophobic to moderately hydrophilic) (Figs 10–12). In order to do this, the metallographic preparation of metal samples (aluminium, aluminium–magnesium, titanium and high-grade steel) with the succeeding estimation of topography was carried out; an example is shown in Fig. 10 for the titanium surface.

The metal surfaces were characterised before and after their treatment by quasistatic wetting measurements with water in respect to their changes in hydrophilicity/hydrophobicity (Fig. 11). The quasi-equilibrium contact angles of diluted aqueous surfactant solutions on polished metallic surfaces show higher values than those of water. The polished metal surfaces were possibly hydrophobized because of surfactant molecules adsorption. No such phenomenon was observed on hydrophobic and moderately hydrophobic polymer surfaces (PET, PVF). For the common trend of the quasi-equilibrium contact angle for the aqueous solutions of pure surfactants with the equal alkyl residue length but with a different ionogeneity on untreated and polished metallic surfaces refer to Fig. 12.

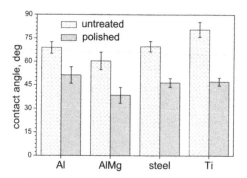

Figure 11. Quasi-static water contact angle for different metal surfaces (untreated, ground and polished).

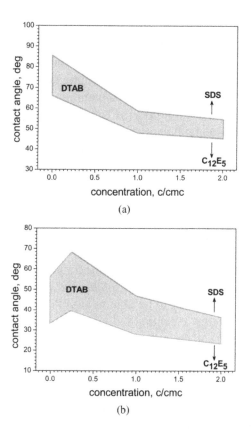

Figure 12. A general tendency of quasi-equilibrium contact angle for aqueous solutions of pure surfactants on polished metal surfaces; DTAB: dodecyl trimethyl ammonium bromide; SDS: sodium dodecyl sulfate; $C_{12}EO_5$: pentaethylene glycol monododecyl ether. To make surfactants comparable with each other, the reduced concentration (c/cmc) was used for the illustration of results.

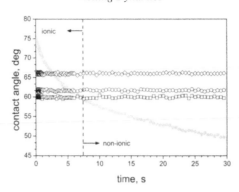

Figure 13. Dynamic contact angle of aqueous surfactant solutions with the same alkyl chain length and different ionogeneity ($c = cmc$) on PP surface: ($\Diamond$) cationic DTAS; ($\bigcirc$) anionic SDS; ($\square$) cationic DTAB; ($\bullet$) non-ionic $C_{12}EO_5$.

2.2. Screening Technical Surfactants Using Model Surfaces

Surfactants are responsible for the control of wetting and de-wetting processes in various applications. For surfactant solutions, the contact angle as a function of time may be significant as illustrated in Fig. 13. In the time interval of 0 to 7 seconds, the most effective (the lowest contact angle) surfactant for PP surface is cationic DTAB (dodecyl trimethyl ammonium bromide). Beginning at approximately 7 s, $C_{12}EO_5$ becomes more effective than DTAB, DTAS (cationic dodecyl trimethyl mmonium sulfate) and anionic SDS (sodium dodecyl sulphate). Therefore, the special attention should be paid to wetting dynamics. Selecting the correct product governs the quality of processes, as illustrated in Fig. 13. Remaining on the polypropylene surface up to about 7 seconds, the drop of an ionic surfactant shows a lower contact angle than that of the non-ionic one. After that the non-ionic ethoxylated alcohol proves to be a better wetting agent for highly hydrophobic polypropylene surfaces.

The dynamic contact angle and spreading velocity can be a criterion, on which basis the efficiency of technical surfactants can be estimated. Based on a host of suitable solid surfaces and dynamic wetting studies, technical surfactants can be systematized with regard to the interaction of their aqueous solutions with these surfaces. On the one hand, thus a surfactant suitable for a special application, selected as precise as possible is allowed to be able to significantly increase the efficiency of a large number of dyeing, cleaning and printing processes. On the other hand, new fields of application can be opened for existing technical surfactants, based on catalogued physicochemical interaction parameters.

Investigations concerning the wetting dynamics on very hydrophilic surfaces such as glass (heterogeneous), silicon wafers with a native SiO_x or a thermally oxidized SiO_2 oxide layer (homogeneous) as well as titanium nitride layers revealed that the quasi-equilibrium contact angles of diluted aqueous surfactant solutions on hydrophilic surfaces show higher values than those of water. Interestingly, the cationic surfactant DTAB spreads hardly on very hydrophilic surfaces (Fig. 14). On the contrary, the drop moves back a little (autophobic wetting behaviour). However,

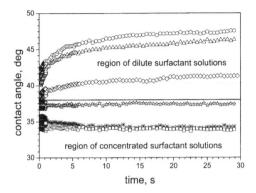

Figure 14. Contact angle of aqueous cationic DTAB on glass surface.

this is valid for diluted solutions up to $\frac{1}{2}cmc$ on the surfaces of glass and Si wafers only. Obviously, there is no dependency of the contact angle of aqueous surfactant solutions with glass surfaces on the concentration.

Hydrophilic surfaces of glass and Si wafers are wetted by technical surfactants better than by pure surfactants with comparable properties, i.e. alkyl rest length and ionogeneity (Fig. 15).

Systematic studies of the dynamic surface tension and contact angles of aqueous solutions of anionic surfactants in the surfactant classes AS (alkyl or fatty alcohol sulphates) and LES (lauryl ether sulphates), a cationic polyelectrolyte HEC (hydroxyethylcellulose) as well as their mixtures led to an indirect estimation of their interactions both on the liquid–vapour and solid–liquid interfaces — Parafilm (highly hydrophobic) and PET (moderately hydrophobic). Increasing concentration of aqueous LAS and AS solutions caused a better wettability of unmodified polymer surfaces. Here, the differences in the wetting kinetics depend on adsorption mechanisms.

Modifying these polymer surfaces with solutions and mixtures allowed the conclusion, that hydrophilisation of the both polymer surfaces occurs, if modifying them with positively charged surfactant-polyelectrolyte mixtures (depending on the concentration of both components). If the surfaces were modified with negatively charged mixtures, no change in the wettability of Parafilm surface and a low hydrophobisation of PET surface were observed.

An essential conclusion is that the degree of wettability of both highly and moderately hydrophobic polymer surfaces with aqueous surfactant solutions and their mixtures can be assessed from the measurements of dynamic surface tensions (Fig. 16). The corresponding isotherms of the surface tension and the contact angle show almost the same shape if the measuring points are recorded at the same time (non-equilibrium states).

Evidently, hydrophobic surfaces show the same or similar adsorption kinetics on the solid–liquid interface as to be expected in the case of the liquid–vapour interface. Besides, the current state of the air interface (dynamic surface tension)

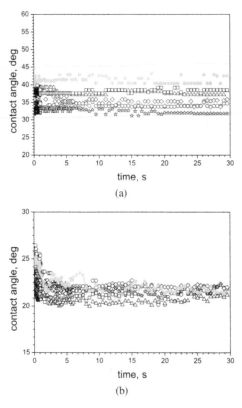

(a)

(b)

Figure 15. Contact angle of aqueous SDS solutions on the glass surface (a); contact angle of aqueous solution of technical surfactant Marlon ARL (Sasol Germany, Marl) on the glass surface (b): ( ) $c = cmc$.

at the instance of the drop-surface contact determines the quasi-equilibrium contact angles on hydrophobic surfaces, as clearly visible in Fig. 16.

2.3. Some Practical Examples of Technology Applications

Additionally, a brief look at two practical examples including technical surfactants and surfaces of technical relevance seems to be worthwhile.

Dynamic contact angle of aqueous surfactant and polymer solutions were measured on textile surfaces in order to examine their soil-repellent properties. From the results for untreated textile surfaces and those impregnated with soil release polymers (SRP), the impact of the classical textile parameters and the topographic textile structure on the surface wettability before and after impregnation was quantified [39, 40]. By modifying the chromatic aberration method, a transparent SRP film was successfully visualized on the surfaces impregnated and its relative surface area was quantified. A comparison of the results from wetting measurements, topographic studies and the literature allowed describing the mechanism of soil release by SRP. Furthermore, common interrelations between the spreading rate of water

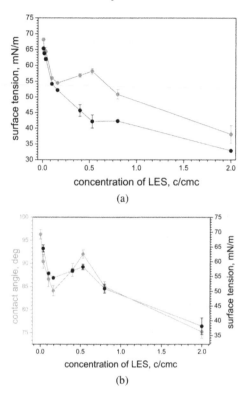

(a)

(b)

Figure 16. Surface tension isotherms of LES/HEC mixtures, measured at $t = 30$ s (black) as a single value and within 60 s, then extrapolated against infinity (a); contact angle and surface tension isotherms ($t = 30$ s) of LES/HEC mixtures on Parafilm (b).

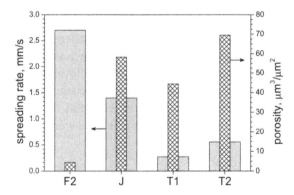

Figure 17. Relationship between spreading rate of aqueous solutions of a technical surfactant and porosity of different textile surfaces: (F2): woven fabric; (J) weft-knitted fabric; (T1 and T2) warp-knitted fabric; [40] with permission of Carl Hanser Verlag, München.

and aqueous surfactant solutions and the porosity (depending on the pore size and shape) were demonstrated using textile polyester materials (Fig. 17).

Investigations of the wetting dynamics on polyester textile materials serves to improve the soil release either by hydrophilic finishing of these textile materials *via* wet-chemical modification, e.g., impregnation with SRP [39, 40] or by hydrophobic finishing via the variation of structure and topography with a constant chemistry [41, 42].

Dynamic penetration studies on differently modified human hair makes it possible to estimate the possibilities of water-soluble polymers to act as care components both in conventional shampoo formulations (free of the oil component) and in micro-emulsions. To estimate the spreading behaviour of water in terms of penetration rate, the capillary penetration method was used and improved. Methodological studies were accompanied by a study to determine water-absorbing capacity, water release by evaporation, topographic properties such as roughness and lustre of hair surfaces differently treated, compared to the reference hair available as damaged (bleached) and undamaged. A comparison of the parameters investigated reveals that some polymers under investigation are well suitable as care components in a shampoo formulation. Moreover, it is possible to evaluate the influence of the oil component (paraffin, jojoba and silicone oil) in regard to the care effect. Jojoba and paraffin oil, in connection with the 'care polymers' bring a positive effect for damaged hair, i.e., a considerable increase in the surface hydrophobicity, an improvement in lustre accompanied by surface smoothing and a time reduction in drying the hair. In the case of damaged hair, the conventional formulation with a care component gave better results than a micro-emulsion.

D. Summary

Basics of wetting including the concept of surface tension and surface energy were briefly mentioned. Spreading as a rate process was considered and described. Wetting dynamics of aqueous surfactant solutions over different surfaces was outlined. Both model surfaces and surfaces of technical relevance as well as pure and technical surfactants were taken into account. Two most relevant models from literature explaining the spreading behaviour of aqueous surfactant solutions over hydrophobic surfaces were displayed. Some practical examples of industrial applications based on the wetting dynamics of surfactant solutions were illustrated.

E. Acknowledgements

The author is indebted to Boryan Radoev (Sofia University, Bulgaria) for the careful attention, fruitful discussions about a possible description of spreading as a rate process according to Eyring's approach. The author is grateful to Radomir Slavchov (Sofia University, Bulgaria) for performing several numerical studies. This research was supported by Sasol Germany (Marl). V.D. is obliged to Burkhard Breitzke, Martin Stolz, Wulf Ruback and Herbert Koch for proving technical surfactants, solid materials of technological relevance, giving many useful advises in the field of surfactants. Financial support is also acknowledged from ESA (FASES, MAP AO-99-052, Fundamental and Applied Studies of Emulsion Stability).

F. References

1. J.W. Gibbs, On the Equilibrium of Heterogeneous Substances, Transactions of the Connecticut Academy, 3, 1873.
2. H. Poincaré, in "La Science et l'Hypothèse, Flammarion", Paris, 1902.
3. Y.S. Podstrigach and Y.Z. Povstenko, Introduction in Mechanics of Surface Phenomena in Deformable Solids, Naukova Dumka, Kiev, 1985 (in Russian).
4. E. Chibowski, Adv. Colloid Interface Sci., 133 (2007) 51.
5. R.G. Cox, J. Fluid Mech., 168 (1986) 169.
6. O.V. Voinov, Fluid Dyn., 11 (1976) 714.
7. T.D. Blake and J.M. Haynes, J. Colloid Interface Sci., 30 (1969) 421.
8. T.D. Blake, in "Wettability", C.B. John (Ed.), Surfactant Science Series, Vol. 49, 1993, pp. 291–309.
9. E. Ruckenstein, Langmuir, 8 (1992) 3038.
10. R.D. Deegan, O. Bakajin, T.F. Dupont, G. Huber, S.R. Nagel and T.A. Witten, Phys. Rev. E, 62 (2000) 756.
11. H. Hu and R.G. Larson, J. Phys. Chem. B, 106 (2002) 1334.
12. G. Guéna, P. Allancon and A.M. Cazabat, Colloids Surfaces A, 300 (2007) 307.
13. R. Slavchov, V. Dutschk, G. Heinrich and B. Radoev, Colloids Surfaces A, 354 (2010) 252.
14. M. Scheemilch, R.A. Hayes, J.G. Petrov and J. Ralston, Langmuir, 14 (1998) 7047.
15. J.G. Petrov, J. Ralston, M. Scheemilch and R.A. Hayes, J. Phys. Chem. B, 107 (2003) 1634.
16. J.G. Petrov and P.G. Petrov, Colloid Surfactants A, 64 (1992) 143.
17. P.G. De Gennes, Rev. Mod. Phys., 57 (1985) 827.
18. E.B. Dussan, V, Ann. Rev. Fluid Mech., 11 (1979) 371.
19. S. Glasstone, K.J. Laidler and H.J. Eyring, The Theory of Rate Processes, McGraw-Hill, New York, 1941.
20. J. Petrov and R. Radoev, Colloid Polymer Sci., 259 (1981) 753.
21. R.A. Hayes and J. Ralston, Colloids Surfaces A, 93 (1994) 15.
22. E. Ruckenstein and C.S. Dunn, J. Colloid Interface Sci., 59 (1977) 135.
23. J.R. Moffat, K. Sefiane and M.E.R. Shanahan, J. Nano Res., 7 (2009) 75.
24. V.M. Starov, S.R. Kosvintsev and M.G. Velarde, J. Colloid Interface Sci., 227 (2000) 185.
25. V. Dutschk, K.G. Sabbatovskiy, M. Stolz, K. Grundke and V.M. Rudoy, J. Colloid Interface Sci., 267 (2003) 456.
26. S. Semal, T.D. Blake, V. Geskin, M.J. de Ruijter, G. Gastelein and J. de Coninck, Langmuir, 15 (1999) 8765.
27. D. Exerowa and P. Krugliakov, in "Foam and foam films: theory, experiment, application", Studies in interface science, Vol. 5, Elsevier, New York, 1988.
28. N. Churaev and Z. Zorin, Adv. Colloid Interface Sci., 40 (1992) 109.
29. K.S. Lee, N. Ivanova, V.M. Starov, N. Hilal and V. Dutschk, Adv. Colloid Interface Sci., 144 (2008) 54.
30. M. von Bahr, F. Tiberg and V. Yaminsky, Colloid Surfaces A, 193 (2001) 85.
31. R.N. Wenzel, Ind. Eng. Chem., 28 (1936) 988.
32. A.B.D. Cassie and S. Baxter, Trans. Faraday Soc., 40 (1944) 546.
33. R.E. Johnson and R.H. Dettre, Adv. Chem. Ser., 43 (1964) 112.

34. C.W. Extrand, Langmuir, 19 (2003) 3793.
35. V. Dutschk and B. Breitzke, Tenside Surfactants Detergents, 42 (2005) 82.
36. L. Tanner, J. Phys. D: Appl. Phys., 12 (1979) 1473.
37. B. Lavi and A. Marmur, Colloids Surfaces A, 250 (2004) 409.
38. A.L. Biance, C. Clanet and D. Quere, Phys. Rev. E, 69 (2004) 016301.
39. A. Calvimontes, V. Dutschk, H. Koch and B. Voit, Tenside Surfactants Detergents, 42 (2005) 210.
40. A. Calvimontes, V. Dutschk, B. Breitzke, P. Offermann and B. Voit, Tenside Surfactants Detergents, 42 (2005) 17.
41. M.M.B. Hasan, A. Calvimontes and V. Dutschk, J. Surfactants Detergents, 12 (2009) 285.
42. M.M.B. Hasan, A. Calvimontes, A. Synytska and V. Dutschk, Textile Res. J., 78 (2008) 996.

Bubbles Rising in Solutions; Local and Terminal Velocities, Shape Variations and Collisions with Free Surface

K. Malysa [a], **J. Zawala** [a], **M. Krzan** [a], **M. Krasowska** [b]

[a] Institute of Catalysis and Surface Chemistry, Polish Academy of Sciences, ul. Niezapominajek 8, 30-239 Cracow, Poland. E-mail: ncmalysa@cyf-kr.edu.pl
[b] Ian Wark Research Institute, University of South Australia, Mawson Lakes Campus, South Australia 5095, Australia

Contents

A. Introduction

Bubble motion is an important problem for mass transfer applications and is encountered in many industrial applications. Recently, Kulkarni and Joshi underlined in their review [1] the importance of the gas bubbles and their rise, due to buoyancy, for various kinds of the gas–liquid reactors. They stated [1] that gas–liquid con-

Bubble and Drop Interfaces
© Koninklijke Brill NV, Leiden, 2011

tacting is one of the most important and very common operations in the chemical process industry, petrochemical industry, and mineral processing. In applications such as absorption, distillation, and froth flotation, the interaction of two phases occurs through dispersing the gas into bubbles and their subsequent rise in the liquid pool. The physicochemical properties of the continuous liquid phase (viscosity, surface tension, density, etc.) and the dispersed phase (bubble size, bubble rise velocity, adsorption coverage over the bubble surface, etc.) govern the hydrodynamics as well as flow pattern in the system. In froth flotation the bubble acts as a carrier of the attached grain(s) having density larger than the continuous liquid medium (pulp). Collectors and frothers are two essential types of the reagents added to the flotation pulp [2–4] for modification surface properties of the liquid/solid and liquid/gas interfaces, respectively. Collectors are expected to make grains of the useful components more hydrophobic, by selective adsorption at their surfaces, while frothers should modify properties of the liquid/gas interface to assure higher degree of the gas phase dispersing and formation of a foam layer of definite properties. As a result of the frother addition the bubbles formed are smaller and their rise velocities are lower. Magnitude of the bubble rise velocity is an important factor in the particle–bubble collision and collection probability, i.e. formation of the bubble grain aggregates. Mechanism and kinetics of the three phase contact formation and the bubble attachment was the subject of many studies [5–15], which were described and summarized in the monograph of Nguyen and Schulze [16]. However, there are still a lot of unanswered questions. For example a significant role of air presence at hydrophobic surfaces (nano-bubbles) [17–22] in facilitation of the three phase contact formation was demonstrated only recently [15, 23, 24] — similarly as some data on influence of the bubble kinetic energy (rising velocity) on its bouncing from various interfaces [25].

Bubble rising velocity in liquids is determined mainly by the bubble's size, viscosities and densities of the liquid and of the gas phases as well as by the properties of the gas/liquid interface [26–29]. The properties of the gas/liquid interface of the bubble are one of the main factors determining magnitude of the bubble rising velocity. A deep theoretical analysis of influence state of adsorption layer and adsorption kinetics on motion of the rising bubbles was presented in the monograph by Dukhin, Miller and Logio [27]. In the case of really pure liquids the bubble surface is fully mobile and therefore the bubble velocity is higher than that of a solid sphere of identical diameter and density. Adsorption layer at the rising bubble surface retards fluidity of the gas/liquid interface and the viscous drag is increased. Simultaneously, uneven adsorption coverage along the interface of the rising bubble is developed, as a result of the viscous drag exerted by continuous medium on the surface. Such adsorption layer is called the dynamic adsorption layer (DAL) [27]. In the DAL the adsorption coverage (surface concentration) is at minimum at the upstream pole of the moving bubble, while at the rear pole is higher than the equilibrium coverage [28]. The surface concentration difference between the poles of the rising bubble is due to its movement and disappears when the bubble is at rest. This gradient of the surface concentration reduces mobility of the bubble interface and consequently the bubble velocity is lowered.

Numerous theoretical approaches regarding various aspects of formation and properties of the dynamic adsorption layer and physicochemical hydrodynamics of the rising bubble have been described by Dukhin *et al.* [27, 29]. This work aims at describing current state of knowledge about bubbles rising in surfactant solutions with a focus on new experimental data and relevance of various theoretical and semi-theoretical models describing the bubble terminal velocities. Phenomena occurring at various stages of the bubble life, from the formation through acceleration and attainment the terminal velocity till the collision with solution surface are described. New experimental data, which were not available when the monographs [27, 29] were written, referring especially to: (i) initial acceleration of the bubbles, (ii) the bubble local velocity at various distances from a point of its formation, (iii) time-scale of inducement of the DAL, (iv) experimental evidences on the DAL formation, (v) minimum concentration and adsorption coverage of surfactants needed for the bubble surface immobilization, and (vi) dynamics of the bubble collisions with free surface are presented.

B. Bubble Formation

Gas bubbles in liquids can be formed either by dispersion methods or as a result of nucleation in oversaturated liquids. The dispersion methods are most commonly used and the gas phase can be dispersed either by injecting (pressing) the compressed gas through various kinds of porous diaphragms or due to turbulent liquid motion caused by various kind of rotors (for example in so-called mechanical flotation machines). Generally, the process of dispersing the gas phase into bubbles is highly dynamic and hard to control when there are various multi-body interactions between the bubble streams generated. However, in the case of the bubble formation at a single capillary orifice there is no multi-body interactions and the formation can be well controlled and described, as well as is widely used in surface science (maximum bubble pressure methods). According to Tate [30] the diameter, d_b, of the bubble detaching from the capillary orifice is determined by a balance between the buoyancy force (responsible for the bubble detachment) and surface tension (attachment force). Assuming the bubble sphericity its diameter (d_b) is given as:

$$d_b = \sqrt[3]{\frac{6d_c\sigma}{g\Delta\rho}}, \tag{1}$$

where: d_c — capillary diameter, σ — surface tension, g — gravity acceleration, and $\Delta\rho$ is the density difference between gas and liquid phases.

1. Adsorption Kinetics at Surface of the Growing Bubble

In surfactant solution the bubble formation is accompanied by adsorption of surfactant molecules at the expanding bubble surface. Equilibrium adsorption coverage at the surface of the detaching bubble is attained only in the cases when the adsorption kinetics is much faster than the bubble surface growth. Thus, degree of the adsorption coverage at the surface of the detaching bubble depends on rate of adsorption of surfactant molecules at the bubble/solution interface and on velocity of the bubble surface expansion. Warszynski *et al.* [31] elaborated model for the adsorption kinetics at the growing spherical surface and applied it in [32] for calculation the degree of adsorption coverage at the interface of the bubble detaching from the capillary orifice. It was assumed that at time $t = 0$ the bubble, having the initial size equal to the capillary radius, starts to expand rapidly with the constant rate, $v_s(t) = V_s = $ const, up to the moment of its detachment from the capillary orifice [32]. Therefore, the instantaneous radius of the bubble was given by the linear dependence:

$$R_b(t) = R_c + v_s t, \tag{2}$$

where R_c — radius of the capillary.

The transport of surfactant molecules to the interface of the growing bubble was described by the convective–diffusion equation, which was expressed in the spherical coordinates [26, 31, 33] as:

$$\frac{\partial c}{\partial t} = D \frac{1}{R_0^2} \frac{\partial}{\partial R_0} R_0^2 \frac{\partial c}{\partial R_0} - v \frac{\partial c}{\partial R_0}, \tag{3}$$

where c — surfactant concentration at a given point, R_0 — distance of this point from the center of the bubble, D — the surfactant diffusion coefficient, v — the fluid velocity. For a uniform expansion of the bubble surface, the fluid velocity can be expressed, using the continuity equation for motion of incompressible fluid [31, 33], in the form:

$$v = v_s(t) \frac{R_b(t)^2}{R_0^2}. \tag{4}$$

When the coordinates relative to the instantaneous position of the bubble surface: $\rho = R_0 - R_b(t)$, were introduced into Eq. (3) then the following relation was obtained [32]:

$$\frac{\partial c}{\partial t} = D \frac{1}{(\rho + R_b(t))^2} \frac{\partial}{\partial \rho} (\rho + R_b(t))^2 \frac{\partial c}{\partial \rho} + v_s(t) \left[1 - \frac{R_b(t)^2}{(\rho + R_b(t))^2} \right] \frac{\partial c}{\partial \rho}. \tag{5}$$

Far from the expanding bubble surface, i.e. for $R_0 \to \infty$ we have $c \to c_b$. The boundary conditions for equation (5) at the bubble surface i.e. for $\rho = 0$, was derived using the continuity equation for the surfactant flux:

$$\frac{1}{A} \frac{dN_s}{dt} = \frac{1}{A} \frac{d(\Gamma A)}{dt} = \frac{d\Gamma}{dt} + \frac{2v_s(t)}{R_b(t)} \Gamma = D \frac{\partial c}{\partial \rho}\bigg|_{\rho \to 0} \tag{6}$$

and simultaneously expressing the surfactant flux at the bubble interface as a balance between the adsorption/desorption fluxes:

$$\frac{1}{A}\frac{dN_s}{dt} = D\frac{\partial c}{\partial \rho}\bigg|_{\rho \to 0} = j_a - j_d = J(c_{\rho \to 0}, \Gamma), \tag{7}$$

where N_s is the number of moles of the surfactant adsorbed at the bubble surface A, j_a — adsorption, and j_d — desorption flux, respectively.

The balance between adsorption and desorption depends on the surfactant concentration in the "subsurface" and on the adsorption coverage at the bubble surface [34]. As at adsorption equilibrium $j_a = j_d$, so the equation (7) gives the adsorption isotherm. The Frumkin–Hinshelwood adsorption kinetic model, which at equilibrium is consistent with the Frumkin adsorption isotherm, was used to solve the equation (5), with the boundary conditions given by equations (6) and (7). Thus, the surface tension at the expanding bubble surface was expressed as:

$$\sigma = \sigma_0 + RT\Gamma_\infty\left[\ln(1 - \theta) + \frac{H_s}{RT}\theta^2\right], \tag{8}$$

where σ is the is the surface tension of surfactant solution, σ_0 — is the surface tension of water, $\theta = \Gamma/\Gamma_\infty$ (Γ_∞ — maximum surface concentration) and H_s is the Frumkin interaction parameter, which expresses the standard enthalpy of surface mixing. This model was used to calculate the degree of adsorption coverage at the surface of bubbles detaching from the capillary orifice in solutions of various surfactants.

2. Degree of Adsorption Coverage at Surface of the Detaching Bubble

Degree of adsorption coverage at surface of the detaching bubble depends on rates of the bubble growth and surfactant adsorption kinetics, and can be very different for various surfactants. Influence of time of the bubble formation on adsorption coverage at interface of the detaching bubble is showed in Fig. 1 for n-octanol solutions [32]. There are presented the adsorption coverage over surface of the bubble detaching after different formation times as a function of n-octanol concentration, as calculated from the model described above. Values of the adsorption coverage for time $t = t_{eq}$ are the equilibrium adsorption coverage. As seen there, even at adsorption time above 3 seconds the n-octanol surface coverages were significantly lower than the equilibrium ones and it is pronounced especially at low concentrations. A comparison of the adsorption coverages at the surface of bubbles detaching in solutions of homologous series of n-alkanols (C4, C5, C6, C8, C9) after the formation time of 1.6 s is presented in Fig. 2. There are presented the adsorption coverage over the detaching bubble surface as a function of the equilibrium adsorption coverage. In the case when the detaching bubble had attained the equilibrium adsorption coverages within the adsorption time equal 1.6 s the data points should follow the line 1:1. Lower are the points below the line 1:1, larger are the deviations from the equilibrium adsorption coverage. As seen in the Fig. 2 for whole range of

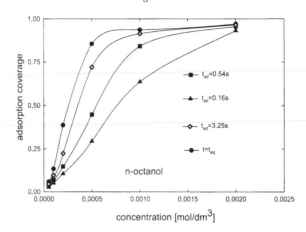

Figure 1. Dependence of the adsorption coverage at surface of the detaching bubble on concentration of n-octanol solution as calculated from the convective–diffusion model for various times of the bubble formation (redrawn from reference [32]).

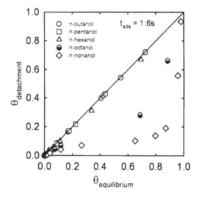

Figure 2. The adsorption coverage over surface of the detaching bubble ($t_{ads} = 1.6$ s) as a function of the equilibrium adsorption coverage for solutions of various n-alkanols (redrawn from references [35, 36]).

concentrations of n-butanol, n-pentanol and n-hexanol the data points are on the line 1:1. For n-octanol and n-nonanol the points are located below the line 1:1, i.e. the bubbles have detached with non-equilibrium adsorption coverage.

Data presented in Figs 1 and 2 illustrate that the adsorption coverage at surface of the detaching bubbles can be, in most of the cases, significantly lower than the equilibrium ones, especially in the cases when the times of the bubbles formation are shorter than 1 second. Deviations from the adsorption coverages are the largest at low concentrations of the surfactant solutions. Lack of the equilibrium adsorption coverage at interface of the detaching bubble affects the time of formation of DAL and in fact the bubble rising velocity.

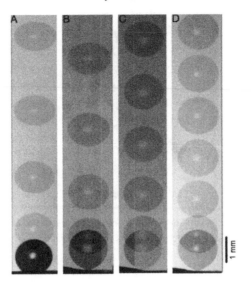

Figure 3. Images of the bubbles detaching from the capillary orifice in: A — distilled water, B — 1×10^{-3}, C — 3×10^{-3} and D — 5×10^{-3} M n-pentanol solutions. Frequency of stroboscopic illumination 100 Hz (redrawn from the reference [36]).

C. Initial Acceleration of the Bubble

When bubble grows at the capillary orifice in quiescent liquid and reaches a certain size, at which the buoyancy force prevails over the capillary force, then the detachment occurs. The moment of the detachment and the size of the detaching bubble depend mainly on the inner diameter of the capillary (d_c), solution surface tension (σ) and kinetics of the adsorption. Immediately after the detachment the bubble accelerates rapidly. Simultaneously, the bubble shape starts to be deformed due to increasing velocity of the rising bubble. Acceleration is one of the main stages of the bubble motion and is considered in the number of recent papers [35–40] which are dealing with the motion of the rising bubbles. However, some numerical data on the bubble acceleration immediately after its detachment were presented only in a few [35, 36]. The approximate values of the initial bubble acceleration evaluated in [35] were reported to be of an order 600–900 cm s^{-2}. It was estimated there that the highest initial acceleration (ca. 900 cm s^{-2}) was found to be in clean water, while in concentrated n-butanol solutions the acceleration was lowered to approximately 600 cm s^{-2}. Recently, Krzan *et al.* [36] presented a more detailed analysis and numerical data on acceleration of the detached bubble.

Figure 3 presents photos of the bubble at the moment of detachment and immediately after the detachment from the capillary orifice in distilled water and n-pentanol solutions of different concentrations [36]. Since the frequency of the stroboscopic illumination was identical, then increase in the distance between the subsequent images of the rising bubble shows that the bubble velocity was increased. It can be

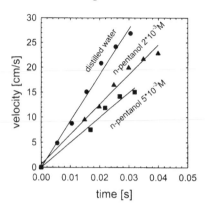

Figure 4. Local velocities of the detached bubble in distilled water and *n*-pentanol solutions. Points — experimental values, lines — linear regressions used for determination the acceleration values (redrawn from the reference [36]).

noted in Fig. 3 that in all cases the bubble velocity started to increase rapidly, and that the bubble acceleration was the utmost in distilled water, while was decreasing with *n*-pentanol concentration. It can also be observed that the bubble was spherical at the capillary orifice and its shape started to be deformed immediately after the detachment. Degree of this deformation (which can be expressed as the ratio of the bubble horizontal and vertical diameters — d_h/d_v) was the utmost in distilled water and diminished with the increase of *n*-pentanol concentration. The reasons of these variations in the bubble shape deformation will be discussed in details in paragraph 5. To determine the accelerations the bubble local velocities in the vicinity of the capillary were measured [36].

Figure 4 presents the local velocities as a function of time elapsed from the moment of the bubble detachment from the capillary orifice. As seen there for the time period of 30–50 ms these dependencies are almost linear and their slope is diminishing as *n*-pentanol concentration is increasing. Values of the bubble initial accelerations were calculated from the regression lines fitted to the experimental data and it was reported [36] that in distilled water the bubble acceleration was ca. 925 cm/s^2, while in *n*-pentanol solution of high concentration (0.005 M) was only 500 cm/s^2. The bubble detaching from the capillary accelerates and increases its velocity until the drag force exerted by the continuous phase on the bubble equals the acting buoyancy force. However, even immediately after the detachment the bubble motion is strongly influenced by presence of surface active substances (surfactants). The experimentally determined values of the bubble acceleration were significantly lowered in surfactant solutions. In surfactant solutions the fluidity of the liquid/gas interface was retarded as a result of the DAL formation at the bubble surface. Lower fluidity of the interface means larger drag force and therefore, in surfactant solutions the acceleration of the rising bubbles was lower.

D. Local Velocity Profiles

Local velocity profiles, i.e. variations of the bubble local velocity with distance from the point of the bubble formation, are different for pure liquids (e.g. clean water) and surfactant solutions. Generally, in clean water there can be distinguished two stages: (i) acceleration, and (ii) steady state motion, after terminal velocity is attained. In surfactant solutions there can be observed an additional stage i.e. a maximum followed by a deceleration until the terminal velocity is established. The bubble terminal velocity in surfactant solutions is lower than in clean water. The presence and the position of the maximum is a function of the surfactant concentration and both of them depend on type of the surfactant used.

Figure 5 shows the profiles of the local velocity of the bubbles of diameters 0.99, 1.22, 1.48 and 1.66 mm rising freely in distilled water. As seen there, the constant value of the terminal velocity was attained immediately after the acceleration stage and the values of the terminal velocity were dependent on the bubble size. Similar results were reported in recent papers (Duineveld [41–43], Zhang *et al.* [38, 44], and Krzan *et al.* [15, 35, 36, 45, 46], Alves *et al.* [47]) where "hyper clean" water was used. For the bubbles of diameters 1.4–1.5 mm the terminal velocity values of 32 cm/s (Duineveld [41–43]), 33.6 cm/s (Zhang *et al.* [44]), 34.8 cm/s (Krzan *et al.* [15, 35, 36, 45, 46]) and 35 cm/s (Alves *et al.* [47]) were reported. The bubbles rising in clean water moved along the straight path and large (ca. 50%) deformations of the bubble shape were observed when they moved with the terminal velocity. In earlier papers [37, 48–51] the terminal velocity values ranging between 14 and 38 cm/s for bubbles of diameter ca. 1.5 mm were reported. However, these experiments were not carried out in really clean water. For example Sam *et al.* [37] reported that the bubbles rising in tap water reached the velocity of 36.5 cm/s and later their velocity was diminishing with the distance down to a value ca. 20 cm/s. These results show that there were present surface active contaminations in tap water, which strongly affected the local velocity profiles and values of their terminal velocity.

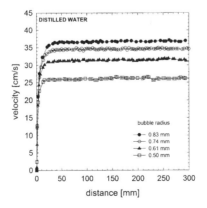

Figure 5. Profiles of the local velocities of the bubbles with diameter: ● — 1.66 mm, ○ — 1.47 mm, ▲ — 1.22 mm, □ — 0.99 mm, rising in distilled water (redrawn from the reference [25]).

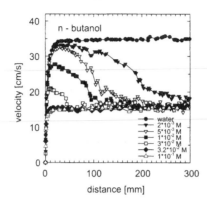

Figure 6. The bubble local velocity as a function of distance from the capillary orifice in *n*-butanol solutions of different concentrations (redrawn from the reference [35]).

Influence of various surfactants on motion of the rising bubbles was studied in many recent papers [15, 35, 36, 38, 39, 41–46, 52, 53]. Sam *et al.* [37] and Zhang *et al.* [38, 44] reported the local velocity profiles for solutions of popular flotation frothers; Dowfroth 250 (CH_3–$(O$–$C_3H_6)_4$–OH), MIBC ($C_6H_{13}OH$), pine oil ($C_{10}H_{17}OH$) and Triton X-100. Duineveld [41–43], and Fdhila and Duineveld [52] presented results of experiments carried out in hyper clean water and dilute solutions of poly-etoxy surfactant Triton X100 and sodium *n*-dodecylsulfate (SDS). Results of systematic studies on variations of the bubble local velocities in solutions of non-ionic and ionic surfactants were reported recently in a series of papers [15, 35, 36, 45, 46, 53]. The non-ionic surfactants, *n*-butanol, *n*-pentanol, *n*-hexanol, *n*-octanol and *n*-nonanol of homologous series of *n*-alkanols, α-terpineol, *n*-octyl-β-D-glucopyranoside, *n*-octyldimethylphosphine oxide and *n*-octanoic acid in 0.005 M HCl, and ionic surfactants *n*-octyltrimethylammonium (OTABr), *n*-dodecyltrimethylammonium (DDTABr) and *n*-cetyltrimethylammonium (CTABr) bromides were used in these experiments. Figures 6–9 present the local velocity profiles of the bubbles rising in *n*-butanol, *n*-octanol, *n*-octyl-β-D-glucopyranoside and *n*-cetyltrimethylammonium bromide solutions of different concentrations.

As seen there the local velocity profiles are strongly dependent on the concentration of the surfactant solutions. At low concentrations three stages can be distinguished: (i) rapid acceleration, similar to acceleration in pure water, (ii) a maximum value of the velocity followed by its monotonic decrease, and (iii) attainment the terminal velocity. With increasing concentration of the surfactants the height of the maximum at local velocity profiles was diminishing and its position was shifted towards shorter distances. At high concentrations of the solutions there was no maximum and after the acceleration period the bubbles attained levels of their terminal velocity. Maximum on the local velocity profile was interpreted [15, 35, 39, 45, 46, 53] as evidence that stationary non-uniform distribution of adsorption coverage over surface of the rising bubble was not established. The absence of the maximum shows that non-uniform distribution of the adsorption coverage

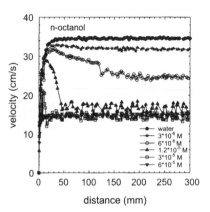

Figure 7. The bubble local velocity as a function of distance from the capillary orifice in *n*-octanol solutions of different concentrations (redrawn from the reference [36]).

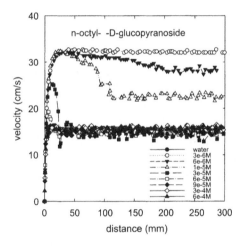

Figure 8. The bubble local velocity as a function of distance from the capillary orifice in *n*-octyl-β-D-glucopyranoside solutions of different concentrations (redrawn from the reference [53]).

(surface tension gradients) causing immobilization of the bubble interface was established immediately after the bubble detachment. As can be observed in Figs 6–9 the bubbles' terminal velocity was decreasing steadily with increasing concentration of different surfactants, from 34.8 ± 0.3 cm/s in distilled water down to a level of ca. 15 cm/s — similar for various surfactants at their highest concentrations.

Facts that: (i) starting form a definite concentration there was practically no further diminishing of the bubble terminal velocity with increasing surfactant concentration, and (ii) lowest values of the bubble terminal velocity were similar for various surfactants show that at these concentration the fluidity of the liquid/gas interface was fully retarded by the DAL formed. Similar findings that there existed a critical concentration above which the terminal velocity stopped to decrease, were reported by various authors Zhang *et al.* [39, 44, 54, 55]; Sam *et al.* [37]; Fdhila and

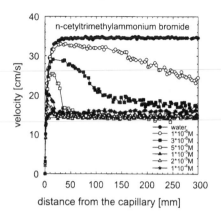

Figure 9. The bubble local velocity as a function of distance from the capillary orifice in *n*-cetyltrimethylammonium bromide solutions of different concentrations (redrawn from the reference [36]).

Duineveld [43, 52]; Krzan *et al.* [15, 35, 36, 45, 46, 53]. This problem is discussed in details below (Section F).

Dukhin *et al.* [27, 29] were the first to describe a transient period of the bubble interface relaxation between the bubble detachment from the capillary and the terminal velocity stage. For the bubble motion at low Reynolds numbers [27, 29] the drag and inertial forces have been taken into the account only, but an appearance of a characteristic transient period necessary to achieve the terminal velocity was predicted [27]. It was assumed [27, 56] that at the moment of detachment in surfactant solutions the bubble surface was almost free of the adsorbed molecules. Therefore, the maximum velocity was predicted to be close to the value given by the Hadamard–Rybczynski theory. Then, due to the surfactant adsorption and the DAL formation at the rising bubble surface the drag force was increased and its velocity was lowered down to a value predicted by the Stokes law. Two different types of the bubble behavior were considered, depending on the Marangoni number. The Marangoni number, *Ma*, characterize mutual importance of the surface and viscous effects, and is expressed as the ratio of the surface pressure that surfactant molecules exert under compression to the viscous forces tending to compress the surface layer:

$$Ma = RT\Gamma_0/\mu U. \tag{9}$$

where μ and U are the dynamic viscosity of the liquid and the bubble terminal velocity. It is considered [27] that at low Marangoni numbers only bottom part of the rising bubble is immobilized by the rear stagnant cap formed there [57–60], while at high Marangoni numbers the entire bubble interface is "solid".

In the case of moderate and high Reynolds numbers there does not exist, as far as we are aware, any analytical solution describing existence of maximum on the bubble local velocity profiles. For now, the problem is solved only by the numerical methods. Liao and McLaughlin [61, 62] attempted to describe theoretically the

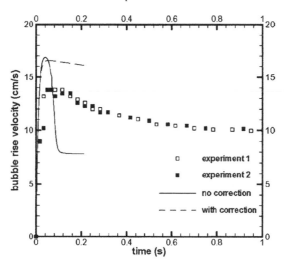

Figure 10. Bubble rise velocity as a function of time from the detachment in solutions of 2.3×10^{-7} mol/dm^3 Triton X-100 (redrawn from reference [63]).

existence of maximum on the local velocity profiles by computer simulations with a finite difference method. They found that in dilute surfactant solutions the value of the maximum velocity can be the bubble velocity in clean water even though the "...ultimate bubble speed may be smaller than this value by a factor roughly equal to two..." [61]. However, later in comments published in 2004 by Liao *et al.* [63] it was amended that the numerical algorithm used [61, 62] to solve the surfactant transport equation on the bubble surface was inaccurate and they presented results of the corrected calculations. Figure 10 presents a comparison of results of their previous and corrected numerical simulations. There are presented also some experimental data but, as one can see, the agreement is rather poor.

E. Inducement of Dynamic Structure of the Adsorption Layer (DAL) on the Rising Bubble

The adsorption coverage over the bubble surface at the moment of its detachment is uniform. The magnitude of the coverage is determined by the adsorption kinetics and rate of the bubble growth (bubble surface expansion). This uniform adsorption starts to be distorted by a viscous drag exerted by continuous medium on interface of the rising bubble and a dynamic structure of the adsorption layer is induced. The bubble motion with its terminal velocity means that a steady state non-uniform structure of the adsorption layer was established and this adsorption layer is called the dynamic adsorption layer (DAL) [27, 29, 64–67]. According to theoretical works by Dukhin [27, 29, 64–67], Frumkin and Levich [28] a steady state motion of the bubble induces adsorption-desorption exchange with the subsurface adjacent to the bubble interface. The amount of surface active substance adsorbed on one part of the bubble is equal to the amount desorbed from the other part. Thus, surface coverage varies along the rising bubble surface. The minimum adsorption coverage is at the leading pole and the maximum at the rear stagnation point, i.e. $\Gamma_{top} < \Gamma_{eq} < \Gamma_{rear}$, where Γ_{eq}, Γ_{top} and Γ_{rear} are the equilibrium surface concentration over a motionless bubble, the surfactant surface concentrations at the top pole and rear end of the rising bubble, respectively. Dukhin and Deryaguin showed [65–67] that when the ratio $\Gamma_{eq}/c > 10^{-4}$ cm (c — bulk concentration) then the top part of the bubble surface is practically devoid of any surfactant molecules, i.e. $\Gamma_{top} \approx 0$. Later, it was evaluated [60, 68] that even in the case when $\Gamma_{eq}/c < 10^{-4}$ cm the surface concentration of surfactant at the top pole of the rising bubble remains significantly lower than the equilibrium one. This disequilibration of surfactant concentration over the bubble surface depends on the bubble velocity and solute surface activity as was described in monograph by Dukhin at al. [29]. The inducement of non-uniform architecture of the adsorption layer leads to the Marangoni effect, which retards the bubble surface mobility.

The formation and properties of the dynamic adsorption layer (DAL) over the rising bubble surface was subject of numerous theoretical and experimental studies. For motion under creeping flow conditions the transient region between the fully mobile and stagnant stage of the bubble interface was theoretically studied in [27, 53, 56, 57, 59], where the increase of the drag coefficient due to the surface contamination (adsorption coverage) was theoretically evaluated and simulated numerically. Dukhin *et al.* [27] described the transient period of the bubble dynamic adsorption layer relaxation between the bubble acceleration, immediately after detachment, and terminal velocity stage. Zholkovskij *et al.* [56] presented a model predicting how the cap angle and bubble velocity would change with time of the motion and two types of the bubble behaviour were predicted. According to [27, 56] when the bubble detaches from the capillary in surfactant solutions the non-steady surfactant flux to the rising bubble surface overlaps with its non-steady movement caused by pure hydrodynamic reasons. For motion under higher Reynolds numbers ($Re \gg 1$) the problem is much more complicated and still not resolved, despite numerous papers [27, 35–39, 44, 54–56, 61, 63, 69, 70] on the bubble velocity variations in surfactant solutions.

1. Experimental Evidences on the DAL Formation

1.1. Correlation Between Variations of the Bubble Shape and Local Velocities
The existence of maximum on profiles of the local velocities of the detached bubble indicates, as described above (paragraph 4) that the dynamic adsorption layer was not established yet. The steady state bubble motion (terminal velocity) means that the DAL had been already established. State of the adsorption layer at surface of the rising bubble affects not only its shape velocity but also the bubble. Thus, a correlation between variations of the bubble shape and its local velocities was sought and found to exist [35, 36, 53]. Figure 11 shows that indeed there were variations in the bubble shape when the DAL was at the stage of formation. There are presented the photos of the bubble rising in *n*-pentanol solutions of concen-

trations 0.0015 and 0.005 M at different stages of the bubble motion. At higher *n*-pentanol concentration the bubble reached the terminal velocity immediately after the acceleration period, while at 0.0015 M *n*-pentanol solution there existed the maximum at the local velocity profile. Please note the different stages of the bubble motion and distances from the capillary, marked in Fig. 11. It can be noted there that immediately after the detachment the bubble shape started to be deformed and the deformation degree was larger at lower *n*-pentanol concentration. Moreover, variations of the bubble shape are easily noticeable on a quite long distance (ca. 100 mm) in 0.0015 M *n*-pentanol solution. At similar distance there was determined the existence of maximum on the local velocity profile. At the acceleration stage the bubble shape deformation was increasing and reached the maximum value at the point where the maximum velocity of the bubble was observed. Next, the deformation started to decrease (the deceleration stage) and a constant shape of the bubble was established when the terminal velocity was attained. As seen in Fig. 11 at higher *n*-pentanol concentration a a constant shape was attained much quicker (at distance below 15 mm) — similarly as much quicker the terminal velocity was attained.

Quantitative data on variations of the local velocities and bubble shape deformation published in [35, 36, 53] showed that the courses of their variations with distance from the capillary were very similar. As an example the comparison of variations of the bubble local velocity profiles and the bubble shape deformation (d_h/d_v ratio) are presented in Fig. 12 for *n*-pentanol solutions.

A good correlation between variations of the bubble shape and the local velocity values can be observed there. At initial stage of the bubble motion (acceleration) the local velocity values were increasing rapidly and similarly rapid was an increase of the bubble deformation degree. Both quantities show maxima at similar distances from the capillary and in a similar way the maxima are changing and shifting with the solution concentration. For example in the case of 0.0015 M *n*-pentanol solution (Fig. 12 — bottom part) the deformation degree (d_h/d_v ratio) increased from 1.0 (at the capillary orifice) to 1.29 at the point of the maximum velocity and then started to decrease till the constant value of the $d_h/d_v = 1.07$ was reached at the stage of the terminal velocity (at distance of ca. 150 mm from the capillary orifice). When there was no maximum on the local velocity profiles and the terminal velocity was attained after the acceleration period then a constant shape of the rising bubble was also established during the acceleration stage of the motion.

The existence of correlations between variations of the local velocity and shape deformation supplies strong experimental evidence that it is a period of formation of the dynamic structure of the adsorption layer in surfactant solutions. Simultaneous shape pulsations and local velocity variations indicate that a steady state distribution of the adsorbed molecules had not been established yet. Shape pulsation means variations of the interfacial area and adsorption–desorption processes counteracting expansion of the interfacial area. As a result of the surface area variations the surface tension gradients induced are changing and fluidity of the bubble interface is

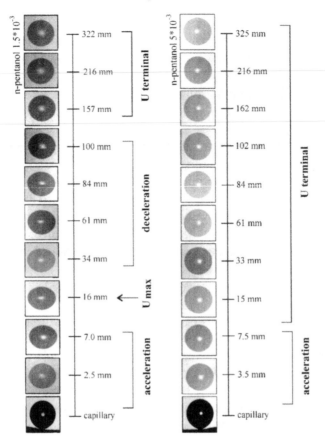

Figure 11. Sequences of photos showing the shapes of the bubbles at various distances from the capillary orifice in 0.0015 and 0.005 *n*-pentanol solutions (redrawn from the reference [36]).

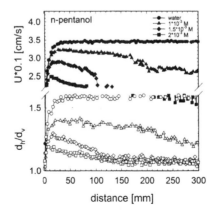

Figure 12. Variations of the bubble local velocity and shape deformation in *n*-pentanol solutions as a function of distance from the capillary (redrawn from the reference [36]).

varying. These variations in fluidity of the bubble interface lead to variations of the bubble local velocity.

1.2. Bubble Lifetimes at Solution Surface of Different Locations

Lifetime of the bubble arriving at free surface depends on the solution composition. In the case of single surfactant solutions the lifetime depends mainly on kind of the surfactant used and its concentration. More specifically the bubble lifetime is determined by properties of the liquid film (foam film) formed. Lifetime of a single foam film depends on many factors from which the most important are the following: (i) state of adsorption layer at both interfaces, (ii) type of the surfactant applied and electrolyte presence, (iii) surface activity of surfactant, (iv) solution concentration, (v) size of the film, (vi) adsorption kinetics, (vii) velocity of the film drainage, (viii) degree and frequency of external disturbances, (ix) environment conditions (humidity, atmospheric contaminations, surfactant vapour pressure, etc.), etc.

When the rising bubble arrives at solution surface then a liquid film starts to be formed. The top part of the rising bubble and a local area of the free surface compose the interfaces of the foam film formed. Equilibrium adsorption coverage is at solution surface forming the upper interface of the foam film. However, if the bubble rises with its terminal velocity then the DAL is formed and the adsorption coverage at the top part of the bubble is significantly lower than the equilibrium one. Thus, the adsorption coverage at the both interfaces of the foam film formed can be significantly different. The lifetime of such non-symmetrical foam film, having equilibrium adsorption coverage at the upper interface and much lower — resulting from the bubble motion — coverage at the bottom interface, should be different from that of the symmetrical film, having the equilibrium surfactant coverage on the both interfaces.

An original experimental method was elaborated in [32, 71, 72] to detect effect of the inducement of the DAL over surface of the rising bubble on the lifetimes of the single bubbles at solution surface. The principle of the method consisted in determining the lifetimes of single bubbles at the solution surface located "close" and "far" from the point of the bubble generation (capillary orifice). It was assumed [32, 71, 72] that when the distance traveled by the bubble was short (solution surface located "close") then a non-uniform distribution of the adsorbed surfactant over the rising bubble surface was not yet fully developed and a symmetric foam film was formed, i.e. the foam film having on its both interfaces equilibrium adsorption coverage. The non-symmetric foam film was assumed to be formed when the solution surface was located "far". Lower adsorption coverage at the "bottom" interface of the non-symmetric foam film should affect the lifetime of the bubble. It was shown [32, 71, 72] that indeed, the bubbles at the surface of aqueous n-alkanol solutions located "far" from the place of bubble generation had shorter lifetimes, i.e. the non-symmetrical foam film formed had lower stability. Figure 13 presents the dependences of the bubble average lifetimes at solution surface located "close" and "far" from the capillary on concentration of n-butanol solutions. As seen the average values of the bubble lifetime are systematically longer on n-butanol solution

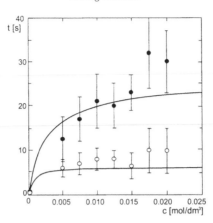

Figure 13. Lifetimes of single bubbles at *n*-butanol solution surface located "close" (upper line) and "far" (bottom line). Bars show standard deviation of the experimental data (redrawn from ref. [71]).

surface located close to the point of the bubble surface. The differences in bubbles lifetimes at these two locations of the solution surface show that stability of the symmetric foam films formed was higher than the non-symmetric ones. Shorter lifetimes of the bubbles at solution surface located "far" show that the bubble motion through the solution does result in a lowering of the adsorption coverage over the top part of the rising bubble. Thus, these data supply experimental evidences of the DAL formation over interface of the rising bubble.

1.3. Three Phase Contact Formation at Solid Surface of Different Locations

When the rising bubble collides with solution/solid interface then a thin liquid film (wetting film) is formed. Stability and time of drainage and rupture of the wetting film formed in surfactant solutions is affected by state of adsorption layer at the interacting solution/gas and solution/solid interfaces. Generally, stability of the wetting film is determined by the forces acting within the film, which in turn are affected by the state of the adsorption layer, solution composition and concentration, solid surface properties (hydrophobic/hydrophilic) and topography (chemical and/or physical heterogeneity, patterning, roughness, etc.). Repulsive forces stabilise the thin liquid film while the attractive forces and surface topography are factors facilitating the rupture. An unstable film will rupture spontaneously at a certain thickness, called the critical thickness of rupture [73]. The film stability and rupture can be related to the surface charge of the interacting interfaces and the solid hydrophobicity.

According to the classic DLVO theory the forces stabilizing/destabilizing colloidal system (wetting film) can be expressed as a sum of the electrostatic interactions between two overlapping double electric layers and van der Waals interactions:

$$V_{DLVO} = V_{el} + V_{vdW}, \tag{10}$$

where: V_{DLVO} is the total DLVO interaction potential energy, V_{el} is the electrostatic double layer contribution and V_{vdW} is the van der Waals component of the interactions. As the electrical potential at the gas/water interface (bubble surface) is negative, ca. -65 mV [74, 75] so the solid surface potential should be also negative to produce an electrostatic repulsion. Since most of the natural solids are negatively charged under neutral conditions, it is commonly accepted that the wetting film at such solids are stable. Opposite situation can be due to either: (i) a change of the bubble surface charge (for instance by cationic surfactant adsorption) or (ii) a change of the solid surface charge; for instance via Al^{3+} ions [76], polyelectrolyte adsorption [77] at the solid surface or simply by variation of the pH of the solution so it is below isoelectric point for the studied solid surface [78]. In such case, i.e. when the interfaces are oppositely charged, the attraction between the double electric layers should appear, leading to the film instability and rupture. As discussed above the bubble motion in surfactant solutions leads to formation of the DAL over the bubble surface and it can also be the factor affecting stability of the wetting film formed during the bubble collisions with the solids surfaces positioned at different location ("far" and "close") from the point of the bubble detachment.

The diagnostic technique to study factors affecting stability and of the wetting films formed during the bubble collisions with solid plates was described in [15, 23, 24, 79–81]. One of advantages of the method, beside easiness in controlling solution composition (which can affect surface charge of both the bubble and the solid plate), was a possibility to vary location of the solid surface in respect to the point of the bubble formation (capillary). In studies reported in [80] the solid plates were located either 10 mm (location "close") or 300 mm from the capillary (location "far"). The experiments were carried out in diluted solutions (10^{-5} M) of the cationic surfactant (DDTABr). Mica sheet was used in the experiments and its surface properties were modified by electrostatically driven sequential adsorption of polyelectrolytes (polycations and polyanions). When the polycation (PDADMAC) constituted the outer layer of the polyelectrolytes adsorbed at mica surface the electrical surface charge was positive. When the polyanions (PSS) constituted the outer layer the solid surface had a negative charge. Stability of the wetting films formed by the bubble rising in the DDTABr solution during collisions with the modified mica surface positioned at locations "close" and "far" were studied. The detection of the differences in stability of the wetting films at the mica sheets at locations "close" and "far" proves that the DAL is formed over surface of the bubbles rising in surfactant solutions.

In the case of cationic surfactants the electrical surface charge of the solution/gas interface is positive. Thus, when the solid surface was positioned at location "close" then the bubble surface was also positively charged because the DAL was not established yet. The location "far" (300 mm from capillary) meant that the bubble motion induced non-uniform bubble surface coverage and most of the surfactant molecules were "pulled down" from the top part of the rising bubble, i.e., the DAL was established. Thus, the top pole of the bubble was almost DDTABr free and

Table 1.
Schematic illustration of the electrical charge distribution over the rising bubble surface and at liquid/solid (modified mica) interface located at two different distances from the capillary orifice (redrawn from ref. [80])

Distance from the point of the bubble formation	PDADMAC	PDADMAC/PSS
	mica/PDADMAC	mica/PDADMAC/PSS
"close" 10 mm		
	mica/PDADMAC	mica/PDADMAC/PSS
"far" 300 mm		

therefore, negatively charged. These different situations are presented in sketches collected in Table 1.

There are also presented frames from the movies recorded in the experiments. It was found [80] that for location "close" only in the case of the bubble collision with negatively charged mica surface (mica/PDADMAC/PSS) the three phase contact (TPC) was formed (see Table 1). Taking into account that upper pole of the rising bubble at location "close" was positively charged due to the cationic surfactant adsorption the instability of the wetting film and the TPC formation was attributed to electrostatic attractions between overlapping double electric layers of oppositely charged solution/gas and solution/solid interfaces.

When at the location "close" the positively charged mica sheet (mica/PDADMAC) was mounted then the TPC was not formed, i.e. the wetting film formed between positively charged both bubble and solid surfaces was stabilised by the electrostatic repulsions. The opposite was observed when the modified mica was positioned at the location "far" (see Table 1), i.e. the wetting film rupture and the TPC formation occurred only in the case of the positively charged mica surface (mica/PDADMAC). As a result of the bubble motion over long enough distance the DAL was established prior to the collision, i.e. the upper pole of the bubble was depleted of the cationic surfactant and therefore the bubble upper pole was negatively charged. Having two oppositely charged interfaces the wetting film was destabilized due to the electrostatic attractions and therefore the TPC was formed. Whereas such bubble, having the DAL already established, collided with the negatively charged solid surface (mica/PDADMAC/PSS) then the wetting film was stable and the TPC was not formed (see Table 1). Thus, these data show directly that the DAL is formed over the bubble surface in surfactant solution when the distance covered by the rising bubble is long enough. Results of these experiments [80] can be treated as proof-of-principle for both motion induced non-uniformity of the DAL over the rising bubble surface and also contribution of electrostatic forces between two double electric layers for stability/instability of the wetting films.

F. Terminal Velocity of the Bubbles

Terminal velocity of the bubble means that the bubble moves with a constant velocity which is determined by the balance between the buoyancy force:

$$F_b = \Delta \rho V g \tag{11}$$

which acts against to the drag force:

$$F_D = 0.5 C_D \rho U^2 \pi R_b^2. \tag{12}$$

The dimensionless drag coefficient C_D is defined as:

$$C_D = \frac{4}{3} \frac{g d \Delta \rho}{\rho U^2} \tag{13}$$

where R_b, d, V and U are the bubble radius, diameter, volume and terminal velocity, ρ and $\Delta \rho$ are the liquid density and the density difference, g is the gravity acceleration. Numerical value of the drag coefficient depends on the conditions of motion in the liquids and is the main question discussed and addressed in various theoretical and semi-theoretical approaches since any universal theory describing the bubble motion in pure liquids and in surfactant solutions does not exist. To characterize the hydrodynamic conditions of the motion the dimensionless Reynolds number Re, the Weber number We, and the Morton number Mo are commonly used. They are defined as:

$$Re = \frac{d U \rho}{\mu}, \tag{14}$$

$$We = \frac{\rho U^2 d}{\sigma}, \tag{15}$$

$$Mo = \frac{g \mu^4}{\rho \sigma^3}. \tag{16}$$

1. Bubble Motion in Pure Liquids

The liquid motion around the bubble is described by the Navier–Stokes equation [26]. However, the analytical solution of Navier–Stokes equation — the Hadamard–Rybczynski equation [26, 82, 83] — was possible only for the creeping flow conditions, i.e. laminar flow at very low Reynolds numbers, $Re \ll 1$. Since the Reynolds number is a dimensionless quantity showing mutual importance of the inertial and the viscous forces therefore in the case of water the condition $Re \ll 1$ means that the bubble diameter is a small fraction of mm ($d_b < 0.01$ mm). The Hadamard–Rybczynski equation:

$$U_{H-R} = \frac{2(\rho - \rho_g)g R_b^2}{3\mu} \frac{\mu + \mu_g}{2\mu + 3\mu_g} \tag{17}$$

transforms into the Stokes equation when $\mu_g \gg \mu$, i.e. when we have the solid sphere instead of the gas bubble. Comparing the velocities predicted by the Stokes equation:

$$U_{St} = \frac{2}{9} \frac{(\rho - \rho_g)g R_b^2}{\mu} \tag{18}$$

to that ones of the Hadamard–Rybczynski (Eq. (17)) it is seen that the bubble velocity is by 50% higher. This higher velocity of the bubbles rising in pure liquids is due to the mobility of the liquid/gas interface, which reduces the drag force exerted on the rising bubble.

In the case of pure water the Hadamard–Rybczynski equation can be applied and it predicts properly the rising velocity of very small bubbles (of diameter of a fraction of millimeter and smaller). Recently, Parkinson et al. [84] measured the terminal rise velocity in ultra-clean water, of single bubbles with diameters ranging from 10 to 100 μm. They reported an excellent agreement between the experimental data and Hadamard–Rybczynski equation prediction. These experimental results and comparison with terminal velocity values predicted by Stokes' law and the Hadamard–Rybczynski equation are showed in Fig. 14. This excellent agreement of the experimental values with Hadamard–Rybczynski theory shows that in sufficiently clean water (liquids) the bubble surface is fully mobile, i.e. there is slip at the liquid–gas interface.

Shape of such small bubbles is spherical because of high internal gas pressure, which — according to the Laplace law — is inversely proportional to the bubble radius. Thus, when the bubble radius increases the internal gas pressure is decreasing and simultaneously the drag force (pressure) exerted on the bubble increases, due to the increasing bubble velocity. As a result the bubble shape starts to be deformed and degree of the shape deformation increases with the bubble size. Moreover, with the increasing bubble size and velocity its trajectory varies from linear via helical or spiral path to the zig-zag path. According to Haberman and Morton [85], Saffman [86] and Clift et al. [87] the rising bubble path is a straight line at the Reynolds

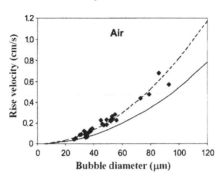

Figure 14. Terminal velocity of the air bubbles rising in clean water as a function of bubble diameter. Points – experimental values, dashed line — Hadamard–Rybczynski equation, solid line — Stokes law (redrawn from ref. [84])

numbers below 300. At Reynolds number of 300–3000 the bubble trajectory is helical and at $Re > 3000$ the bubble path in water has the zig-zag shape. There is no analytical solution of Navier–Stokes equation for the bubble motion at $Re > 1$. There exist only numerical solutions and/or semi-theoretical approximations, being mainly the mathematical correlations with the experimental data. Besides, majority of the available models is valid only for limited ranges of the Reynolds number values and often there are introduced additional limitations defined by values of the Weber and Morton numbers, i.e. related to strictly defined viscosity, density and surface tension regimes. The most often referred to, applied and attempted to be improved and extended are the models developed by Schiller and Naumann [88], Levich [26], Moore [89, 90], Taylor and Acrivos [91], Moore [92], Habeman and Morton [85], Clift *et al.* [87], Bhaga and Weber [93], Masliyah [94], Margaritis *et al.* [95], Rodrigue [96–98], Klaseboer *et al.* [99].

Levich [26] elaborated model for the potential flow of the spherical drops and bubbles in clean liquids. According to the Levich model [26] the velocity in the thin liquid layer near the bubble surface is constant at any cross-section and the inertia forces dominate over the viscous forces. In such case the drag coefficient is given as:

$$C_D = 48/Re. \tag{19}$$

The model is valid up to the Reynolds number ca. 50, i.e. for the bubble diameters below 0.5 mm for motion in clean water.

Moore [89, 90] extended the Levich model to higher Reynolds numbers. He assumed [89, 90] that a thin liquid layer at the bubble surface, of thickness of an order ca. $Re^{-1/2}$, was considerably thinner that one around the rigid sphere. This layer was assumed to be extended to larger thicknesses at the bubble bottom region. The assumed prolongation of the bubble interface, due to enlargement of the liquid layer thickness at the bottom part, could induce a thin wake. The wake caused a non-uniform velocity distribution around the bubble interface and in the consequence the zig-zag or helical path of the motion. From calculations of energy dissipation in all regions of the bubble interface the drag coefficient was evaluated to be:

$$C_D = \frac{48}{Re}\left(1 - \frac{2.21}{\sqrt{Re}} + O(Re^{-5/6})\right). \tag{20}$$

The model was claimed to be valid for Reynolds numbers from 15 till ca. 100, where the bubble shape deformations are still negligible.

The shape deformations depend on the bubble velocity, cross-sectional area, pressure difference, liquid density and surface tension. Taylor and Acrivos [91] took into account the influence of the deformation degree, which was related to the Weber number, on the bubble velocity. They stated that for Reynolds numbers within range from 1 to 10 the drag coefficient was equal to:

$$C_D = 2 + \frac{16}{Re}. \tag{21}$$

In 1965 Moore [92] extended his model by taking into account the effect of the bubble deformation and arrived to the following relation:

$$C_D = \frac{48}{Re}G(\gamma)\left\{1 + \frac{H(\gamma)}{\sqrt{Re}} + O\left(\frac{1}{\sqrt{Re}}\right)\right\}, \tag{22}$$

where $G(\gamma)$ and $H(\gamma)$ were functions of the bubble deformation degree, γ:

$$G(\gamma) = \frac{1}{3}\gamma^{4/3}(\gamma^2 - 1)^{3/2}\frac{\sqrt{\gamma^2 - 1} - (2 - \gamma^2)\sec\gamma^{-1}}{(\gamma^2\sec\gamma^{-1} - \sqrt{\gamma^2 - 1})^2}, \tag{23}$$

$$H(\gamma) = 0.0195\gamma^4 - 0.2134\gamma^3 + 1.7026\gamma - 1.5732. \tag{24}$$

The deformation degree γ was expressed in terms of the Weber number values as:

$$\gamma = 1 + \frac{9}{64}We + O(We^2). \tag{25}$$

According to Moore [92] the model is applicable within wide range of the Reynolds numbers, $100 < Re < 10000$. However, in reality the upper limit of the model applicability is much lower as a reasonable agreement of the model prediction with the experimental data was observed only for the Reynolds numbers below 1000 [46].

In the Clift *et al.* [87] monograph the comparison between experimental data and various theoretical models prediction for bubble motion in clean and contaminated liquids was presented. They reported that the Moore model [90] overestimated values of the drag coefficients. They underlined a strong impact of the bubble deformations on the drag coefficient values for bubbles rising in clean water. They proposed two relations for the bubble terminal velocity in pure liquids. According to them [87] the relation:

$$C_D = 14.9Re^{-0.78} \tag{26}$$

describes correctly the bubble motion for Reynolds numbers below 150, while the relation:

$$U = [(2.14\sigma/\rho d) + 0.505gd]^{1/2} \qquad (27)$$

was recommended for Re above 565 (U, ρ, σ, d and g are bubble terminal velocity, medium density, surface tension, bubble diameter and gravitational acceleration, respectively).

Ryskin and Leal [100–102] applied the finite difference numerical calculations to determine values of the local drag coefficient of the rising bubble. They used 3D coordinate configuration with its center in the bubble mass center. Their numerical technique allowed calculating the local temporary deformation of the bubble interface during each step of motion. The impact of surface and capillary forces was also taken into account considered. The motion of the liquid around the bubble was calculated by the finite difference method, from Laplace and Navier–Stokes equations, and the drag coefficient were continuously recalculated basing on the equation:

$$C_D = 2 \int_0^1 \left(-p_{dyn} + \frac{8}{Re} e_{rr} \right) \sigma \frac{\partial \sigma}{\partial \mu} \, d\mu, \qquad (28)$$

where p_{dyn} is the dynamic pressure. The term $(-p_{dyn} + 8e_{rr}/Re)$ describes the tangential stress on the bubble interface, caused by the static and dynamic pressure and viscous components. The computers limitation allowed them to obtain results only up to the Reynolds number 200. Their numerical calculations were later extended up to Re ca. 600 by McLaughlin and Ponoth [103, 104] who confirmed validity of the method.

Masliyah *et al.* [94] reported that their experimental data were well described by the relation:

$$C_D = \frac{16}{Re}(1 + 0.077Re^{0.65}) \qquad (29)$$

valid for $Re < 130$ and Weber number varying from 0.008 to 1.6.

Experimental results on bubble motion in clean liquids from 19 various research labs, published between 1900 and 1996, were reviewed by Rodrigue [96] and on this basis he proposed a generalized correlation for bubble motion:

$$C_D = \frac{16}{Re} \left[\left(\frac{1}{2} + 32\lambda + \frac{1}{2}\sqrt{1 + 128\lambda} \right)^{1/3} \right.$$
$$+ \left(\frac{1}{2} + 32\lambda - \frac{1}{2}\sqrt{1 + 128\lambda} \right)^{1/3}$$
$$\left. + \left(0.036 \left(\frac{128}{3} \right)^{1/9} Re^{8/9} Mo^{1/9} \right) \right]^{9/4}, \qquad (30)$$

where λ coefficient was equal:

$$\lambda = (0.018)^3 \left(\frac{2}{3} \right)^{1/3} Re^{8/3} Mo^{1/3}. \qquad (31)$$

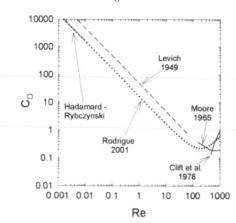

Figure 15. Bubble drag coefficient as a function of Reynolds number in pure liquids (redrawn from ref. [46]).

According to [96] this correlation can be used within a wide range of physical and hydrodynamic conditions: $722 < \rho < 1380 \text{ kg/m}^3$; $2.2 \times 10^{-4} < \mu < 18 \text{ Pa s}$; $15.9 < \sigma < 91.0 \text{ mN/m}$; $1.9 \times 10^{-7} < Re < 1.1 \times 10^4$; $1.0 \times 10^{-11} < Mo < 1.0 \times 10^7$.

Figure 15 presents a comparison of a few chosen theoretical semi-theoretical models proposed for description the bubble motion in pure liquids. There are given values of the drag coefficient as a function of the Reynolds number according to the Hadamard–Rybczynski theory [26, 82, 83] and for models elaborated by Levich [26], Moore [92], Clift *et al.* [87] and Rodrigue [96]. Limits of validity of these models (as stated by the authors) are marked as well.

As seen from the review presented above there is rather large number of different theoretical and semi-theoretical models and correlations (with experimental data) attempting to describe motion of the bubbles at higher Reynolds numbers. Such large number of models illustrate clearly that there does not exist any universal theory describing the bubble motion outside the creeping flow conditions.

2. Solutions of Surface Active Substances

In surfactant solutions the bubble terminal velocity is significantly lowered due to formation of the dynamic adsorption layer (DAL) over surface of the rising bubble. Steady state non-uniform distribution of the adsorbed surfactant molecules means that there are induced surface tension gradients over the rising bubble surface. These gradients of the surface tension retard mobility of the bubble interface (Marangoni effect) and consequently the bubble velocity is diminished in significant degree. Full retardation of the bubble surface mobility means that its behaviour is similar as a solid sphere of identical dimensions and density. Thus, under creeping flow conditions ($Re \ll 1$) the bubble rising velocity is described by the Stokes law [26, 29].

In the case of moderate or high Reynolds number there do not exist any exact analytical solutions of the Navier–Stokes equation. Similarly as the case of pure liquids, the different models recommended in literature are based either on numerical calculations or semi theoretical approximations and correlations with the experimental data. Additionally, the models assume, as a rule, that in surfactant solutions the bubble surface mobility is always completely retarded. This assumption is obviously correct for high surfactant concentrations, but at diluted solutions the degree of adsorption coverage can be too low for complete immobilization of the bubble surface. There is minimum solution concentration and/or adsorption coverage needed for full retardation of the bubble surface mobility. At low solution concentrations the values of the terminal velocity varied with surfactant concentration, as showed above in paragraph 4. In the case of bubbles of diameter ca. 1.5 mm rising in clean water with velocity of 34.8 cm/s the terminal velocity values were within the range between ca. 25 and 17 cm/s at lowest surfactant concentrations and were dropping down rapidly with the increase of the solution concentrations. At a definite solution concentrations of various surfactants the value of the bubble terminal velocity ca. 15 cm/s was reached (Zhang *et al.* [38, 44, 54, 55], Sam *et al.* [37], Fdhila and Duineveld [43, 52], Krzan *et al.* [15, 35, 36, 45, 46, 53]). Variations of the bubble terminal velocity with of *n*-hexanol and *n*-dodecyltrimethylammonium bromide concentration are presented in Fig. 16. As seen, at low concentrations the values of the terminal velocity decreased rapidly with concentration till the velocity of ca. 15 cm/s was attained. Further increase of the surfactants concentration had only a minute influence on the terminal velocity values. This influence was most probably caused by decreasing size of the detaching bubble due to the solution surface tension decrease. This is illustrated in Fig. 17, where variations of the bubble size as a function of *n*-hexanol and *n*-dodecyltrimethylammonium bromide concentration are presented. The bubbles equivalent diameter diminished almost linearly with concentration, but even at the highest concentrations the bubble diameter was smaller by 10–12% only (Fig. 17), while the terminal velocity values dropped down by over 50%, already at lowest surfactant concentrations (Fig. 16). It was reported by Zhang *et al.* [39, 44, 54, 55], Sam *et al.* [37], Fdhila and Duineveld [43, 52], Krzan *et al.* [15, 35, 36, 45, 46, 53] that there exist a critical concentration above which the terminal velocity almost stopped to decrease. Values of the critical concentrations and terminal velocities of bubbles with dimension 1.4–1.5 mm in solutions of various surfactants are collected in Table 2.

Dukhin *et al.* [27] described in details effect of adsorption layer on hydrodynamics of the bubble motion at low Reynolds numbers. Influence of the solute surface activity and adsorption kinetics on establishment of the dynamic adsorption layer (DAL) is analysed there in details.

Numerical solutions of axisymmetric flow around the bubble in surfactant solutions (contaminated liquids) for moderate Reynolds numbers were presented recently in [61–63]. Numerical calculations for bubble motion in surfactant solution, similar to that one used by Ryskin and Leal [100–102] and Ponoth and McLaughin

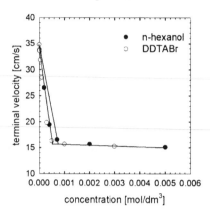

Figure 16. Terminal velocity of the rising bubble as a function of concentration of *n*-hexanol and *n*-dodecyltrimethylammonium bromide solutions (redrawn from ref. [35]).

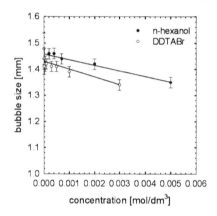

Figure 17. Variations of the bubble diameter with concentration of *n*-hexanol and *n*-dodecyltrimethylammonium bromide solutions (redrawn from ref. [35]).

[103, 104] for pure liquids, were applied in these studies. Influence of the surface tension gradient on the bubble hydrodynamic was additionally taken into account in numerical calculations reported in [61–63]. A good agreement with the experimental data was obtained for Reynolds number till ca. 300, i.e. for the bubbles of diameters up to ca. 1.5 mm.

Clift *et al.* [87] proposed the following relationship, based on correlations with 774 experimental data points:

$$U = \frac{\mu}{\rho d} Mo^{-0.149}(J_C - 0.857),$$ (32)

where:

$$J_C = 0.94 H^{0.757},$$ (33)

$$H = \frac{4}{3} Eo Mo^{-0.149}.$$ (34)

Table 2.
Critical concentrations and terminal velocity of bubbles with diameter 1.4–1.5 mm in solutions of various surfactants

Surfactant name	Terminal velocity cm/s	Critical concentration	Reference
Triton X-100	14.8	12×10^{-5} mol/m^3	Zhang *et al.* [39, 44, 54, 55]
Triton X-100	ca. 16.5	5×10^{-3} mol/m^3	Fdhila and Duineveld [42, 43, 52]
SDS	ca. 15	0.3 mol/m^3	Fdhila and Duineveld [42, 43, 52]
MIBC	16.5	ca. 0.06 ppm	Sam *et al.* [37]
Dowfroth 250	15.6	0.06 ppm	Sam *et al.* [37]
Pine oil	14.8	ca. 0.06 ppm	Sam *et al.* [37]
a-terpineol	15.4	5×10^{-5} mol/dm^3	Krzan and Malysa [45]
n-butanol	15.9	4×10^{-3} mol/dm^3	Krzan and Malysa [35]
n-pentanol	14.9	3×10^{-3} mol/dm^3	Krzan *et al.* [36]
n-hexanol	16.6	7×10^{-4} mol/dm^3	Krzan and Malysa [35]
n-octanol	14.4	3×10^{-5} mol/dm^3	Krzan *et al.* [36]
n-nonanol	15.1	1.2×10^{-5} mol/dm^3	Krzan and Malysa [35]
n-octyl-β-D-glucopyranoside	15.8	3×10^{-5} mol/dm^3	Krzan *et al.* [53]
n-octanoic acid (in 0.003 M HCl)	15.7	3×10^{-5} mol/dm^3	Krzan *et al.* [53]
n-octyldimethylphosphine oxide	15.6	3×10^{-5} mol/dm^3	Krzan *et al.* [53]
n-octyltrimethylammonium bromide	14.9	5×10^{-2} mol/dm^3	Krzan *et al.* [36, 53]
n-dodecyltrimethylammonium bromide	16.3	5×10^{-4} mol/dm^3	Krzan *et al.* [36]
n-cetyltrimethylammonium bromide	14.5	5×10^{-6} mol/dm^3	Krzan *et al.* [36]

The relationship (32) is valid in systems where surface-active contamination is inevitable and when the condition:

$$2 < H \leqslant 59.3 \qquad (35)$$

is fulfilled. Assuming for aqueous surfactant solutions that $g = 981$ cm/s^2, $\mu = 0.01$ g/(cm s), $\rho = 1$ g/cm^3, and σ is ca. 60 mN/m, the bubble terminal velocity U (Eq. (32)) can be expressed as a function of bubble diameter and density difference only:

$$U = 49.29 \Delta \rho^{0.495} d^{0.514} - 0.298 / \Delta \rho^{0.149} d, \qquad (36)$$

where the bubble diameter d is expressed in cm, and U in cm/s. Under these assumptions the relation 36 is valid within the bubble dimension range: $0.052 \leqslant d \leqslant 0.28$ cm.

The model described by Ng *et al.* [105] was based on the solution of the Oseen equation describing the object motion in viscous phases. For higher Reynolds numbers the following interpolation formula was applied [105] for the drag coefficient:

$$C_\mathrm{D} = \frac{12}{Re}\left(\frac{7}{6}Re^{0.15} + 0.02Re\right) \tag{37}$$

which provided correct match with the numerical solution of the Oseen equation. Under identical assumptions as above, for the liquid viscosity and density numerical values, the bubble terminal velocity was expressed as:

$$U = 5450\Delta\rho d^2\left(\frac{7}{6}Re^{0.15} + 0.02Re\right)^{-1} \tag{38}$$

and the model is claimed to be valid till Reynolds numbers of 1000. Figure 18 presents the experimental values of the bubble terminal velocity in *n*-alkanols solutions as a function of the bubble diameter. There are also presented the dependences of the terminal velocity on bubble radius as predicted from the models for clean water (Moore [92] — Eq. (22), Clift *et al.* [87] — Eq. (27)) and surfactant solutions (Ng *et al.* [105] — Eq. (38)). For clean water the models of Moore and Clift *et al.* are presented because their predictions were in a good agreement with the experimental data [15, 35, 36, 45, 46, 53]. Other models were in a poorer agreement. For example, according to the most recent model proposed by Rodrique [96] the terminal velocity in water of the bubble of diameter 1.5 mm should be equal ca. 26 cm/s, while the experimental value was ca. 35 cm/s. As seen in Fig. 18 for smaller bubble diameters, i.e. at higher *n*-alkanol solution concentrations, the experimental data are in a good agreement with the predictions of the Ng *et al.* model [105], so-called "contaminated" bubble model.

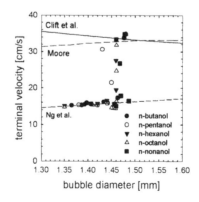

Figure 18. Terminal velocity as a function of the bubble diameter. Points — experimental data for various surfactants, lines — predictions of different models, as marked in the figure (redrawn from refs [35, 36]).

Thus, it can be concluded that here the bubble surface was fully immobilized. However, at low concentrations (bubble diameters > 1.4 mm) there are quite a lot of experimental points which are situated between the values predicted from the relations for the "clean" and "contaminated" bubble. It means that in these cases the adsorption coverage was not sufficient for complete immobilization of the bubble interface. There was still some fluidity at the bubble interface and therefore the bubble velocity was higher.

3. Minimum Adsorption Coverage for Complete Retardation the Bubble Surface Mobility

In surfactant solutions the bubble velocity is diminished due to formation of the dynamic adsorption layer, i.e. is determined by the system interfacial properties. Thus, an analysis of the bubble motion in terms of the interfacial properties is a sounder approach than in terms of the bulk concentration of various surfactants. A degree of adsorption coverage and time of formation of the DAL over surface of the rising bubble are of crucial importance for the local and terminal velocity values. As reported in [35, 36, 53] there were found minimum adsorption coverage values, different for various surfactants, which were needed for complete immobilization of the bubble interface. The method of calculation of adsorption degree over interface of the detaching bubble was described in details above (paragraph 2.2) and these values of the adsorption coverage were applied in the analysis [35, 36, 53]. It needs to be mentioned here about a simplification caused by the fact that adsorption coverage at the moment of the bubble detachment were used in the analysis. This means that an additional adsorption during the bubble motion is not taken into account. Importance of this simplification depends strongly on time of the bubble formation and kinetics of the surfactant adsorption as illustrated in paragraph 2.2 for homologous series of *n*-alkanols (see Fig. 2.). Figure 19 presents similar kind of the dependences for three members of the homologous series of *n*-alkyltrimethylammonium bromides. There are presented the adsorption coverage over the bubble surface, after the formation time of 1.6 s, as a function of the equilibrium coverage. Similarly as in the case of *n*-alkanols, the equilibrium adsorption coverage over the detached bubble surface were attained for lower members of the homologous series, while in the case of *n*-cetyltrimethylammonium bromide the coverage were significantly lower than the equilibrium ones.

Minimum adsorption coverage needed for complete immobilization of the bubble surface was evaluated from an analysis of the bubble terminal velocity dependence on the adsorption coverage [35, 36, 53]. Figure 20 presents the dependencies of the terminal velocity in *n*-alkanol solutions on adsorption coverage over surface of the bubbles at the moment of their detachment. A rapid decrease of the terminal velocity values occurred for the adsorption coverage smaller than 10%. Values of the minimum adsorption coverage needed for complete retardation of the fluidity of the bubble surface were evaluated from these dependences. The minimum adsorption coverage is the coverage value above which the terminal velocity is practically

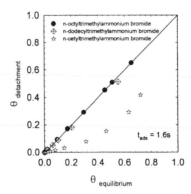

Figure 19. Adsorption coverage over surface of the detaching bubble ($t_{ads} = 1.6$ s) as a function of the equilibrium adsorption coverage for solutions of n-octyl-, n-dodecyl- and n-cetyltrimethylammonium bromides (redrawn from refs [36, 53]).

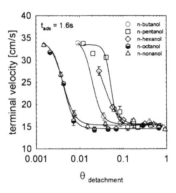

Figure 20. The terminal velocity as a function of adsorption coverage at the detaching bubble surface in n-alkanol solutions (redrawn from refs [35, 36]).

constant. As can be observed in Fig. 20 the minimum adsorption coverage (θ_{min}) needed to immobilize the bubble surface were ca. 6–9% for n-butanol, n-pentanol and n-hexanol, and ca. 2% for n-octanol and n-nonanol. The adsorption/desorption kinetics is much slower in the case of n-octanol and n-nonanol than in the case of n-butanol and n-hexanol. It seems to be the reason of these smaller values of the adsorption coverage needed to immobilize the bubble surface in n-octanol and n-nonanol solutions.

Figure 21 presents the terminal velocity values as a function of the adsorption coverage over surface of the bubbles in solutions of of n-octyltrimethylammonium bromide, n-octyldimethylphosphine oxide, n-octyl-β-D-glucopyranoside and n-octanoic acid (in HCl), i.e., the surface active compounds having in their molecule different polar groups and identical hydrocarbon chain length (C8) [36, 53]. As seen there the adsorption coverage of ca. 4% was sufficient for immobilization of the bubble surface in the case of n-octanoic acid, n-octyl-β-D-glucopyranoside and n-octyldimethylphosphine oxide, but in the case of n-octyltrimethylammonium

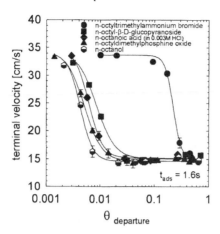

Figure 21. The bubble terminal velocity in *n*-octyltrimethylammonium bromide, *n*-octyldimethylphosphine oxide, *n*-octyl-β-D-glucopyranoside, *n*-octanol and *n*-octanoic acid (in HCl) solutions as a function of the adsorption coverage at surface of the detaching bubble (redrawn from refs [36, 53]).

bromide the adsorption coverage above 25% was needed. The surface activities of *n*-octyldimethylphosphine oxide, *n*-octyl-β-D-glucopyranoside and of *n*-octanoic acid are very similar, while *n*-octyltrimethylammonium bromide (cationic surfactant) shows significantly lower surface activity [36, 53]. Simultaneously, the adsorption kinetics of the ionic surfactant was much faster than the non-ionic ones, i.e. this situation is similar as in the case of homologous series of *n*-alkanols.

Values of the minimum adsorption coverage needed for complete immobilization are different in the case of various surfactants, ranging from ca. 2–30%. These variations of the θ_{min} values were attributed to differences in surfactant surface activity and adsorption kinetics [15, 36, 53]. Higher surface activity means simultaneously slower adsorption/desorption kinetics. For example the equilibrium adsorption coverage of 4% is obtained in 0.001 M *n*-pentanol and 6×10^{-5} M in *n*-octanol [36]. From the Ward and Tordai boundary solution of the diffusion equation, under assumption of zero surface concentration at the time t_0 and neglecting back diffusion, it was evaluated [36] that the time needed to attain this adsorption coverage was 280 times longer for *n*-octanol. Fast diffusion means that on every stage of the bubble motion there is a quick transport to and from the interface. Thus, any deviations from the equilibrium are much quicker compensated or even over-compensated. Larger bubble shape variations in solutions of surfactants of faster adsorption/desorption diffusion were used as an evidence supporting this hypothesis. Since the bubble shape variation means an enlargement or compression the surface area so the shape variations create an additional impulse for local adsorption exchanges. If the shape variations are too fast compensated and/or over-compensated then these processes prolong the time necessary for establishment the DAL. Thus, despite very fast adsorption kinetics the establishment of non-uniform steady state distribution of the adsorbed molecules can be in these cases prolonged due to some kind over-compensating ("overshooting") of the surface tension gradients induced. In contrary, in the case of surfactants of higher surface activity and slower adsorption kinetics (*n*-octanol or *n*-nonanol) there is a lack of the equilibrium adsorption coverage at the bubble surface. Bubble motion leads to inducement of the dynamic structure of the adsorption layer, but as the adsorption kinetics is much slower here so the convection over the bubble surface can prevail. Therefore, the adsorbed molecules are immediately transported and accumulated at the bottom part of bubble. In consequence, the strong surface tension gradient is created and bubble interface mobility is hindered at low adsorption coverage in the case of high surface activity solutes.

G. Bubble Collisions with Liquid/Gas Interface

When the rising bubble arrives at liquid/gas interface there is a thin liquid film formed prior to the bubble rupture. Properties and stability of the films formed can be dependent not only on composition of the solution but also on an actual (quite often non-equilibrium) state of the adsorption layers at both interfaces of the film formed and on the bubble impact velocity. It is rather commonly accepted that at surface of clean water the bubble bursts immediately and this is a simple and commonly accepted criterion to check if water doesn't contain any surface active contaminants. Bikerman stated in his textbook [106]: "... When the bubble reaches the upper surface of the liquid, and the liquid has no foaming tendency, the bubble bursts at once; that is the film separating it from the bulk gas phase immediately ruptures. When the liquid contains a foaming agent, the above film has a significant persistence, and the bubble lifts a "dome"...". As it was showed a few years ago [107] that in distilled water and/or in *n*-alkanol solutions the bubble colliding with liquid/air interface bounced backwards and its shape pulsated rapidly with frequency over 1000 Hz, so let's look closer into mechanism of the bubble collisions and the time-scale of this "immediate" burst at surface of clean water.

1. Influence of the Bubble Kinetic Energy

Bubble kinetic energy of the collision (impact velocity) is one of the important factors affecting the collision course and stability of the liquid film formed at liquid/gas interface. The bubble rising in clean (distilled) water is the simplest and most suitable system for investigation an influence of the bubble kinetic energy on its bouncing and time of the water film rupture. Kirkpatrick and Locket [108] were the first, as far as we are aware, who underlined the effect of bubbles kinetic energy on the time of its coalescence at water surface.

They measured a coalescence time of relatively large gas bubbles (5 mm diameter) colliding with the free surface of distilled water as a function of distance from the bubble formation point. They found that the time of the bubble coalescence increased form 5 to 180 ms when the separation was increased from 0.45 to 3.5 cm. The increase of the separation distance can be straightforwardly related to the increase in bubble impact velocity. According to [108] when the bubble had the highest velocity at the moment of collision then the rapid contact film area was increased and there was insufficient time for the film to drain to the critical thickness of rupture. Later, Chesters and Hofman [109] analyzed theoretically mechanism of the bubble coalescence and bouncing apart form the liquid/gas interface in pure liquids and proposed more detailed mechanism of the bubble bouncing. They considered the bubble bounce and/or coalescence as a competition of two processes: (i) the thinning of the liquid between the bubbles, and (ii) the increase of the free energy of the system resulting from the increase of the bubble's surface area. The free energy of the system increases, at the expense of the kinetic energy and they stated that: "...the bubbles therefore decelerate and eventually bounce apart..." if the thinning liquid layer did not reach earlier a critical thickness of rupture. Tsao and Koch [110] studied the collision and coalescence of two bubbles of different diameters (9 and 1 mm) in clean water and electrolyte solutions and reported that the bubbles coalescence was observed at low Weber number.

Klaseboer *et al.* [99] pointed out, considering the bouncing trajectory model of a drop and bubble impinging on a solid wall immersed in liquid, that in the acceleration period the transient and the added mass forces should be taken into account. The added mass force is defined as additional force acting on the drop or bubble due to the acceleration of the liquid around it. The added mass force acts as if the mass of the rising object has been increased and can be expressed by a coefficient related to geometry of the deformation [99, 111]. In the case of the rising bubble the added mass coefficient (C_m) is a function of the bubble deformation ratio [99] i.e. the ratio between the horizontal and vertical diameters, χ:

$$C_m = \frac{\alpha}{2 - \alpha}, \tag{39}$$

where

$$\alpha = \frac{2\chi^2}{\chi^2 - 1}\left(1 - \frac{1}{\sqrt{\chi^2 - 1}}\cos^{-1}\left(\frac{1}{\chi}\right)\right). \tag{40}$$

After linearization [99]:

$$C_m = 0.62\chi - 0.12. \tag{41}$$

Thus, kinetic energy, E_k, of the bubble at the moment of collision with interface can be expressed as:

$$E_k = 0.5 C_m \rho V_b U^2, \tag{42}$$

where ρ is the liquid density and V_b is the bubble volume. The added mass coefficient C_m calculated on the basis of Eq. (41) for a spherical bubble is equal to 0.5 [99, 111–113].

A systematic study on influence of the bubble kinetic energy on bouncing and time of rupture of the liquid film formed at water/air interface were carried out recently by Zawala *et al.* [25]. The bubble kinetic energy at the moment of collision

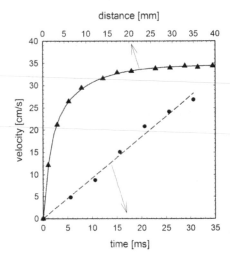

Figure 22. Variations of the bubble local velocity. immediately after detachment from the capillary in distilled water, as a function of time (full points) and distance (full triangles) from the capillary. Points show the experimental values, while lines are the regression lines (redrawn from the reference [25]).

(impact velocity) was varied in two ways: (i) by changing the diameter of the colliding bubble (see Fig. 5), and (ii) by changing the distance between a point of the bubble detachment and the interface (L), when the bubble was at the acceleration stage. Figure 22 presents the bubble local velocity in distilled water, immediately after detachment, as a function of the distance and time from the moment of the detachment. These dependences illustrate the approaches applied for precise determination the impact velocity of the bubble being at the acceleration stage. In the first approach, the bubble acceleration was calculated as the slope of the linear dependence of the bubble local velocity on time (Fig. 22 — dashed line). The impact velocity of the bubble U_i was calculated as the product of the acceleration and time interval (t) between the moment of detachment from the capillary and the moment of collision:

$$U_i = a \cdot t. \tag{43}$$

In the second approach, the following exponential regression equation:

$$U_i = A(1 - e^{-B \cdot s}) + C(1 - e^{-D \cdot s}) \tag{44}$$

was fitted to experimentally determined dependence of the bubble local velocity on distance, where s is the distance between the capillary orifice and the bubble bottom pole (Fig. 22). Values of the impact velocities obtained form the both approaches were in a good agreement, but the second approach (Eq. (44)) was considered to be sounder and was used for determination the bubble impact velocity.

Figure 23 presents sequences of photos illustrating phenomena occurring during the bubble's first and second collision with water surface located 300 and 2.2 mm from the capillary. Each subsequent picture shows the bubble position and shape af-

first collision

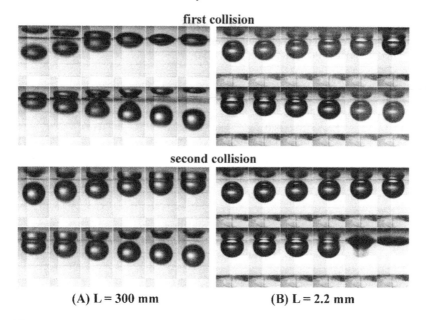

second collision

(A) L = 300 mm   (B) L = 2.2 mm

Figure 23. Comparison of the first and the second collision of the bubble ($d = 1.47$ mm) with water/air interface located at distance $L = 300$ and 2.2 mm from the capillary (redrawn from ref. [25]).

ter time interval of 0.845 ms. As seen the colliding bubble neither ruptured immediately nor was stopped at the water/gas interface located at the distance $L = 300$ and 2.2 mm. After the first collision the bubble bounced backward and simultaneously its shape started to pulsate rapidly, which is especially notable for $L = 300$ mm. The bubble shape changed completely between 2 subsequent photos (see the 2nd row of the sequence A), i.e. within time shorter than 0.845 ms. After bouncing, the bubble started its second approach and ruptured at the water/gas interface located at $L = 2.2$ mm, but bounced again from the interface located at $L = 300$ mm. In the case of the location $L = 300$ mm the bubble ruptured only during the fourth collision, due to its highest initial kinetic energy. It was stated [25] that higher shape deformations and tendency to bouncing are related to the kinetic energy (impact velocity) of the colliding bubble. This is illustrated by data presented in Table 3, where values of the horizontal and vertical diameter, impact velocity, kinetic energy, and number of the collisions prior to the rupture of the bubble colliding with water surface located at different distances are collected.

The increased tendency for bouncing with increasing the bubble kinetic energy was related [25] to enlargement of the radius (R_F) of the intervening liquid film being formed between the colliding bubble and water/air interface. Larger radius means longer time of the film syneresis [114–118] to reach its critical thickness of the rupture. Thus, these results confirm correctness of the mechanism of bouncing proposed by Chesters and Hofman [109]. Longer time needed for drainage to reach the critical thickness of the film rupture means a higher probability that the bubble kinetic energy is fully interchanged into surface energy. That is why the bubbles

Table 3.
Influence of kinetic energy of the rising bubble on its shape deformation and number of collisions with water/gas interface prior to the rupture (reference [25])

Bubble rupture at collision No:	Distance L [mm]	U_i of the first collision [cm/s]	d_h collision [mm]	d_v collision [mm]	E_k [J]
1	1.8	8.0	1.50	1.41	2.7×10^{-9}
2	2.2	11.0	1.53	1.35	5.0×10^{-9}
3	4.0	20.4	1.58	1.28	1.7×10^{-8}
4	300	34.6	1.69	1.11	1.0×10^{-7}

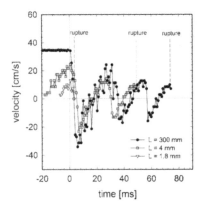

Figure 24. Variations of the local velocity and rupture times of the bubble of diameter 1.47 mm during collisions with water/air interface located at three different distances (1.8, 4 and 300 mm) from the capillary (redrawn from ref. [25]).

having higher kinetic energy performed more "approach-bounce" cycles prior to the rupture at water/gas interface.

Quantitative data on variations of the bubble local velocity and influence of the impact velocity on time of the bubble rupture at water/gas interface are presented in Fig. 24. Prior to the first collision the bubble approached the water/air interface with the constant terminal velocity (34.7 cm/s) only when the interface was at location at $L = 300$ mm (see Fig. 24). In the case of other locations the bubble was still at the acceleration stage and its velocities at the first collisions were significantly smaller. Differences in the impact velocity (kinetic energy) of the first collision had a profound influence on time of rupture of the bubble colliding with water/air interface. The rupture time of the colliding bubble was increased from 5 to 75 ms, i.e. by over order of magnitude, when the bubble impact velocity (kinetic energy) was increased from 8.0 to 34.6 cm/s (see Fig. 24 and Table 3).

Generally speaking the bouncing of the rising bubble from the interfaces is a consequence of competition between velocities of thinning of the intervening liquid film, and exchange of kinetic energy into surface energy of the deformed bubble.

It seems that the increased tendency for bouncing with increasing kinetic energy of the colliding bubble is due to enlargement of the radius of the thin liquid film being formed between the bubble and interface [119]. It needs to be underlined here that the collisions and bouncing of the rising bubbles are very rapid and in the case of clean water the time scale of these phenomena is shorter than 0.1 second. Thus, a popular criterion that at surface of really clean water the bubble bursts at fraction of seconds is valid and sound, but one should be also aware how many very rapid phenomena (bouncing and shape pulsations) occurs there within this short time.

2. Influence of Surface Active Substances on Bouncing and Time of the Bubble Rupture

Presence of adsorption layers, at interfaces of the thin liquid film formed by the colliding bubble, prolongs the lifetime of the bubble as a result of increased stability of the foam film. Under dynamic condition of the foam film formation the film properties and stability can depend not only on type of the surfactant and its concentration in solution, but also on an actual (quite often non-equilibrium) state of the adsorption layers at both interfaces of the films formed. As reported in [15, 107] the behavior of the bubble colliding with solution/air interface was qualitatively similar to that in distilled water, i.e. the bubble bouncing and rapid shape pulsations were observed. However, the bubble didn't rupture within a fraction of seconds, but lasted much longer time as a result of the increased stability of the foam film formed.

Moreover, the bubble shape fluctuations were diminishing with increasing surfactant concentration [107]. Doubliez [120] determined the thickness of the film formed by the colliding bubble in water and solutions of short-chain alcohols (methanol, ethanol) on the basis of examination of interference fringes shifts. He reported that the thinning rate of the foam film formed was higher for the second than for the first, and higher for the third than for the second collision. This thinning rate depended on the bubble approach velocity — the higher the velocity, the slower the initial thinning rate was. Moreover, the number of the bubble bouncing prior to the rupture depended on diameter of the colliding bubble and was increasing with the bubble size. Thus, these findings confirm the hypothesis [119] that the increased tendency for bouncing is due to enlargement of the radius of the thin liquid film formed between the bubble and interface, because larger radius of the thin liquid film means prolongation of the time of syneresis to a critical thickness of its rupture.

Figure 25 presents the local velocity variations for the bubbles rising in 0.001 and 0.005 M *n*-pentanol solutions during their collisions with the solution/air interface [15]. As seen, at higher *n*-pentanol concentration the bubble approach (terminal) velocity was lower and the amplitude of the bubble bouncing was smaller. Prior to the first collision the bubble approach velocities were ca. 30 and 15 cm/s in 0.001 and 0.005 M *n*-pentanol solutions, respectively. After first collision the maximum bouncing velocities reached were 26 cm/s in more diluted (0.001 M) solution,

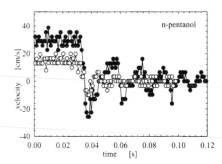

Figure 25. Velocity variations of the bubbles rising to and bouncing from n-pentanol solution surface of concentrations: • — 0.001, and ○ — 0.005 M (redrawn from ref. [15]).

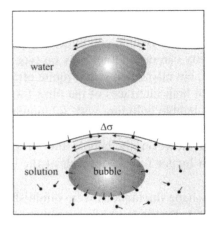

Figure 26. Schematic illustration of the surface tension gradients action during thinning of the foam film formed by the colliding bubble (redrawn from ref. [15]).

and 9.5 cm/s only at higher (0.005 M) n-pentanol concentration. In the case of 0.001 M n-pentanol solution at least 6 "approach-bouncing" cycles were detected. The amplitude of the bubble's velocity variations was decreasing with every cycle as a result of the energy dissipation.

Adsorption layer of surface active substance slows down the bubble rising velocity and diminishes its shape variations and bouncing from the solution/air interface. Simultaneously, the bubble lifetime at solution interface is prolonged. A mechanism of stabilization of the foam films formed by the colliding bubble was proposed in [15]. It was postulated there that surface tension gradients, induced as a result of the bubble motion and solution surface deformation (extension), were the force causing the damping of the bubble's shape pulsations and preventing the bubble rupture due to enhanced stability of the foam film being formed. Mechanism of the surface tension gradients action is showed schematically in Fig. 26.

In clean water there are no surface tension gradients at deformed liquid/gas interfaces (Fig. 26, top part). In surfactant solution (Fig. 26, bottom part) the surface tension gradients induced are directed in the opposite direction than the liquid outflow from the foam film being formed. These surface tension gradients tending to restore uniform adsorption coverage over deformed liquid/gas interface cause the motion of the surfactant molecules together with adjoining liquid layers. These liquid motions, induced by the surface tension gradients, counteract outflow of the synerising solution and lead to slowing down the film thinning and increase the film stability.

H. Bubble Velocity as a Tool to Monitor Organic Contaminations in Water

Velocity of the rising bubbles is very sensitive to a presence of surface active substances in water. In the case of surfactants of high surface activity the bubble velocity is lowered at adsorption coverage even below 1% (see Figs 20 and 21). Thus, monitoring the bubble motion in water can be used as sensitive tool for detection of surface active contaminations. Loglio *et al.* [121] were probably the first who applied measurements the time of bubbles rising at different heights of the column for detection surfactants presence in water. They pointed out that the velocity of the rising bubbles (2–3 mm diameter) decreased when an amount of surfactant increased. Ybert and di Meglio [38, 70] showed an importance of the surfactant adsorption on the bubble velocity variations with distance and postulated that the ascending velocity of the gas bubbles may be used for estimation of the adsorption rate constants. Recently, Zawala *et al.* [122] carried out systematic studies on influence of various contaminants on the bubble rising velocity and proposed a simple physicochemical method for detection of organic contaminations in water (SPMD).

The principle of the SPMD method consists in measurements of the time of the single bubble rise over the distance of 140 cm using a simple set-up consisting of a long glass tube with the capillary in bottom, air container and stop-watch. Precision of the detection of organic contaminations in water was found [122] to be satisfactory when the manual measurements of the time of the bubble rise over distance of 140 cm were repeated at least 20 times. Popular commercial detergent formulations "Ludwik" (cleaning liquid) and "Vizir" (washing powder), widely used in polish hoseholds, were applied in the tests. They were used as the reference formulations for calibration degree of water organic contaminations. Figure 27 presents the profiles of the bubble local velocity in distilled water "contaminated" by different amounts of the detergent "Ludwik", which were measured using the precise laboratory set-up [122]. As can be observed the local velocity profiles had a character similar to that observed in solutions of various surfactants and the bubble terminal velocity was decreasing with "Ludwik" concentrations.

A comparison of the bubble velocity variations with concentration of the detergents "Ludwik and "Vizir" is presented in Fig. 28. In the case of the "Ludwik" solutions there are showed besides the data obtained using the SPMD method also the data from the precise laboratory set-up. Some differences between values of these two sets of data were explained in [122] as due to the fact that values of the bubble velocity measured in the SPMD method were not the true terminal velocity

Rising Bubbles

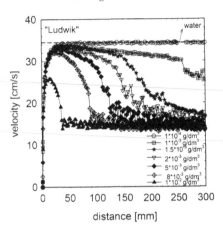

Figure 27. Profiles of the bubble local velocity in "Ludwik" solutions of different concentrations, determined using the laboratory set-up (redrawn from reference [122]).

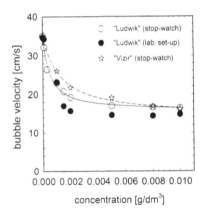

Figure 28. Variations of the bubble velocity with concentration of the commercial detergents "Ludwik" and "Vizir" obtained using SPMD method and precise laboratory set-up (redrawn from [122]).

values. In the SPMD method the time of the bubble rise was started to be measured when the bubble was passing the point situated 60 mm above the capillary orifice. It means that the maximum existing at the bubble velocity profiles (see Fig. 27) affected the determined velocity values. That it why the values of bubble velocity obtained from SPMD were called "average velocities". Nevertheless, the differences are not large and the character of the dependencies is practically identical. It is seen also in Fig. 28 that variations of the bubble average velocity with "Ludwik" and "Vizir" concentration showed similar features. There was a quick decrease of the bubble average velocity, at low concentrations, followed by almost a constant velocity value at higher concentrations. This rapid decrease of the bubble velocity at low solution concentrations indicates a suitability of the SPMD method for detection a presence of surface active contaminations in waters.

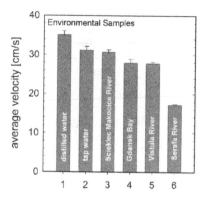

Figure 29. Bubble average velocity in distilled water, tap water and different water environmental samples as determined by the SPMD method (redrawn from [122]).

Sensitivity and reliability of the SPMD method was also tested on environmental samples, collected from different rivers in Poland and Baltic Sea. The results are presented in Fig. 29. High sensitivity of the method is clearly illustrated by the fact that even in the case of tap water the bubble velocity was lowered by ca. 10% in comparison to the distilled water — free of any organic contaminations. The most polluted water was in the sample collected from the Serafa River, where the bubble terminal velocity was lowered by over 50%. Thus, the bubble velocity is indeed a very sensitive tool for detection of organic contaminations in water. However, it does not supply a straightforward information about amounts of the organic contaminations in water. To estimate the amount of surface active contaminants in waters a simple method of calibration was proposed [122] and the detergents "Ludwik" and "Vizir" were arbitrary chosen as the references. Solid and broken lines in Fig. 28 present the exponential function fitted to the experimental data of the velocity as a function of concentration for "Ludwik" and "Vizir", respectively.

The function had a form:

$$\overline{U} = P \exp\left(\frac{Q}{c_{SAC} + S}\right),$$ (45)

where $\overline{U}$ was the bubble average velocity, c_{SAC} was a concentration of surface active contaminants, P, Q and S were the fitting parameters. For "Ludwik" $P = 15.6$ cm/s, $Q = 6 \times 10^{-4}$ and $S = 7.4 \times 10^{-4}$ g/dm^3, while for "Vizir" $P = 14.2$ cm/s, $Q = 1.8 \times 10^{-4}$ and $S = 2 \times 10^{-3}$ g/dm^3. Thus, knowing values of the bubble average velocity, $\overline{U}$, it is possible to calculate from Eq. (45) the so-called "equivalent concentrations of contaminations" (c_{SAC}) in respect to either "Ludwik" or "Vizir" used as reference "model contaminants".

The c_{SAC} values are collected in Table 4. There are also given values of the total concentrations of anionic surfactants (c_{Ref}) for water samples from rivers of Krakow region, which were determined by the Inspectorate of Environmental Protection in Krakow using the Methylene Blue Active Substances Method (MBASs) according to Polish standards [123]. Generally, the values of the c_{SAC} and the c_{Ref}

Table 4.
Equivalent concentrations of surface active contaminants (c_{SAC}) in environmental samples

Sample	c_{SAC} [mg/dm^3]		c_{Ref} [mg/dm^3]
	"Ludwik"	"Vizir"	
Tap water	0.12	0.29	–
Gdańsk Bay	0.29	0.67	–
Scieklec Makocice River	0.15	0.35	<0.10
Vistula River	0.30	0.70	0.70
Serafa River	5.3	7.2	0.98

were of similar order of magnitude, but in all cases the c_{SAC} values were larger than the c_{Ref}. This is quite reasonable because the c_{SAC} values reflect the presence of all kind of organic contaminations in water (bubble velocity is affected by presence of all kinds of surface active substances), while the c_{Ref} values show only concentration of anionic surfactants. This comparison shows that the bubble velocity can be applied for monitoring a quality of waters. According to [122], simplicity, swiftness, no need to use any additional (often toxic) reagents and a special sample and apparatus preparation are the main advantages of the SPMD method.

I. List of Symbols and Abbreviations

SYMBOLS

a — acceleration

A — bubble surface area

c — surfactant concentration

c_b — surfactant concentration in the bulk

C_D — dimensionless drag coefficient

C_m — added mass coefficient

c_{SAC} — equivalent concentrations of contaminations

c_{Ref} — total concentrations of anionic surfactants

D — surfactant diffusion coefficient

d — bubble diameter

d_c — capillary diameter

d_h — bubble horizontal diameter

d_v — bubble vertical diameter

E_k — kinetic energy

F_b — buoyancy force

F_D — drag force

g — gravitational acceleration

H_s — Frumkin interaction parameter which expresses the standard enthalpy of surface mixing

j_a — adsorption flux

j_d — desorption flux

J — summary flux, $j_a - j_d$

L — distance between capillary and interface

Ma — Marangoni number

Mo — Morton number

N_s — number of moles of the surfactant adsorbed at the bubble surface

P — fitting parameter

p_{dyn} — dynamic pressure

Q — fitting parameter

R — gas constant

R_0 — distance of this point from the center of the bubble

R_b — bubble radius

R_c — capillary radius

R_F — film radius

Re — Reynolds number

S — fitting parameter

s — distance between the capillary orifice and the bubble bottom pole

T — temperature

t — time

t_0 — start time, $t_0 = 0$ s

t_{ads} — time of bubble formation

t_{eq} — time necessary to obtain the equilibrium adsorption coverage

U — bubble terminal velocity

U_i — bubble impact velocity

$\overline{U}$ — bubble average velocity

U_{H-R} — velocity of the rising bubble from Hadamard–Rybczynski equation

U_{St} — velocity of the sphere (particle or bubble) in viscous liquid

V — volume

v — fluid velocity

v_s — rate of the bubble expansion

V_b — bubble volume

V_{DLVO} — total DLVO interaction potential energy

V_{el} — electrostatic double layer contribution in the DLVO interaction potential energy

V_{vdW} — van der Waals component of the DLVO interactions

We — Weber number

GREEK SYMBOLS

Γ_s — surface concentration

Γ_∞ — maximal surface concentration

Γ_{eq} — equilibrium surface concentration

Γ_{top} — surface concentration at the top pole of the bubble

Γ_{rear} — surface concentration at the rear pole of the bubble

γ — bubble deformation degree, i.e. the function of the We number

μ — dynamic viscosity of the liquid

μ_g — dynamic viscosity of the gas

θ — adsorption coverage

θ_{min} — minimum adsorption coverage

ρ — density of the fluid

ρ_g — density of the gas

$\Delta\rho$ — difference between liquid and gas density

σ — surface tension

σ_0 — surface tension of pure solvent

χ — bubble deformation degree, i.e. the ratio between the horizontal and vertical diameters

ABBREVIATIONS

CTABr — n-cetyltrimethylammonium bromide

DAL — dynamic adsorption layer

DDTABr — n-dodecyltrimethylammonium bromide

MBASs — Methylene Blue Active Substances Method

OTABr — n-octyltrimethylammonium bromide

PDADMAC — poly-diallyldimethylammonium chloride

PSS — poly-4-styrene sulfonate

SDS — sodium n-dodecylsulfate

SPMD — simple physicochemical method for detection of organic contaminations in water

TPC — Three Phase Contact

J. References

1. A.A. Kulkarni and J.B. Joshi, Ind. Eng. Chem. Res., 44 (2005) 5873.
2. J. Leja, "Surface Chemistry of Froth Flotation", Plenum Press, New York and London, 1982.
3. D.W. Fuerstenau and T.W. Healy, in "Adsorptive bubble separation techniques", R. Lemlich (Ed.), Academic Press, New York and London, 1972, Chapter 6, p. 91.
4. J.S. Laskowski, in "Frothing in Flotation — II", J. Laskowski and E.T. Woodburn (Eds), Gordon and Breach Publishers, 1998, Chapter 3, p. 81.
5. H.J. Schulze and G. Gotschalk, in "Developments in Mineral Processing", J. Laskowski (Ed.), Proc. 13th Intern. Miner. Process. Congr., Warsaw, Vol. 2, 1979, p. 63.
6. H.J. Schulze and G. Gotschalk, Aufbereitungstechnik, 22 (1981) 154.
7. H. Bergelt and H. Stechemesser, Intern. J. Miner. Process., 34 (1992) 321.
8. Z. Dai, D. Fornasiero and J. Ralston, J. Coll. Interface Sci., 217 (1999) 70.
9. A.V. Nguyen, H.J. Schulze, H. Stechemesser and G. Zobel, Intern. J. Miner. Process., 150 (1997) 113.
10. H. Stechemesser and A.V. Nguyen, Colloids & Surfaces A, 142 (1998) 257.
11. H.J. Schulze, K.W. Stockelhuber and A. Wenger, Colloids & Surfaces A, 192 (2001) 61.
12. G. Gu, Z. Xu, K. Nadakumar and J. Masliyah, Intern. J. Miner. Process., 69 (2003) 235.
13. R.-H. Yoon, Intern. J. Miner. Process., 58 (2000) 129.
14. W. Wang, Z. Zhou, K. Nandakumar, Z. Xu and J.H. Masliyah, Intern. J. Miner. Process., 68 (2003) 47.
15. K. Malysa, M. Krasowska and M. Krzan, Adv. Coll. Interface Sci., 114–115 (2005) 205.

16. A.V. Nguyen and H.J. Schulze, "Colloidal Science of Flotation", Marcel Dekker, Inc., New York, 2004.
17. N. Ishida, T. Inoue, M. Miyahara and K. Higashitani, Langmuir, 16 (2000) 6377.
18. P. Attard, Adv. Coll. Interface Sci., 104 (2003) 75.
19. A.V. Nguyen, J. Nalaskowski, J.D. Miller and H.-J. Butt, Intern. J. Miner. Proc., 72 (2003) 215.
20. R. Steitz, T. Gutberlet, T. Hauss, B. Klösgen, R. Krastev, S. Schemmel, A.C. Simonsen and G.H. Findenegg, Langmuir 19 (2003) 2409.
21. J. Yang, J. Duan, D. Fornasiero and J. Ralston, J. Phys. Chem. B., 107 (2003) 6139.
22. H. Zhang, N. Maeda and V.S.J. Craig, Langmuir, 22 (2006) 5025.
23. M. Krasowska, R. Krastev, M. Rogalski and K. Malysa, Langmuir, 23(2) (2007) 549.
24. M. Krasowska and K. Malysa, Intern. J. Mineral Process, 81 (2007) 205.
25. J. Zawala, M. Krasowska, T. Dabros and K. Malysa, Can. J. Chem. Engin., 85 (2007) 669.
26. V.G. Levich, "Physicochemical Hydrodymanics", Prentice-Hall, Englewood Cliffs, 1962.
27. S.S. Dukhin, R. Miller and G. Logio, "Physico-chemical hydrodynamics of rising bubble" — published in "Drop and Bubbles Interfacial Research", D. Mobius and R. Miller (Eds), Elsevier, New York, 1998.
28. A.N. Frumkin and V.G. Levich, Zh. Phys. Chim., 21 (1947) 1183.
29. S.S. Dukhin, G. Kretzschmar and R. Miller, "Dynamics of adsorption at liquid interfaces. Theory, Experiments, Application", Elsevier, 1995.
30. T. Tate, Philosophical Magazine, 27 (1864) 176.
31. P. Warszynski, K.-D. Wantke and H. Fruhner, Colloids & Surfaces A, 139 (1998) 137.
32. B. Jachimska, P. Warszynski and K. Malysa, Colloids & Surfaces A, 192 (2001) 177.
33. C.A. MacLeod and C.J. Radke, J. Coll. Interface Sci. 166 (1994) 73.
34. J.F. Baret, J. Coll. Interface Sci., 30 (1969) 1.
35. M. Krzan and K. Malysa, Colloids & Surfaces A, 207 (2002) 279.
36. M. Krzan, J. Zawala and K. Malysa, Colloids & Surfaces A, 298 (2007) 42.
37. A. Sam, C.O. Gomez and J.A. Finch, Intern. J. Miner. Process 47 (1996) 177.
38. C. Ybert and J.-M. di Meglio, Eur. Phys. J. B. 4 (1998) 313.
39. Y. Zhang, J.B. McLaughlin and J.A. Finch, Chem. Eng. Sci. 56 (2001) 6605.
40. J. Zhang and L.-S. Fan, Chemical Eng. J., 92 (2003) 169.
41. P.C. Duineveld, J. Fluid Mech., 292 (1995) 325.
42. P.C. Duineveld, Appl. Sci. Res., 58 (1998) 409.
43. P.C. Duineveld, "Bouncing and coalescence of two bubbles in water", PhD thesis, Univ. Twente, The Netherlands, 1994.
44. Y. Zhang and J.A. Finch, Column 96, (1996) 63.
45. M. Krzan and K. Malysa, Physicochem. Probl. Miner. Proces., 36 (2002) 65.
46. M. Krzan, "Local velocities, shape and size of bubbles rising in solutions of surface active substances", [in Polish], PhD thesis, Institute of Catalysis and Surface Chemistry PAS, Cracow, Poland, 2003.
47. S.S. Alves, S.P. Orvalho and J.M.T. Vasconcelos, Chem. Eng. Sci., 60 (2005) 1.
48. B. Stuke, Naturweissenchaften, 39 (1953) 325.
49. R. Hartunian and W.R. Sears, J. Fluid Mech. 3 (1957) 27.
50. N.M. Aybers and A. Tapucu, Wärme- Stoffübertrag., 2 (1969) 118.
51. H. Tsuge and S.-I. Hibino, J. Chem. Eng. Jpn., 10 (1977) 66.
52. R.B. Fdhila and P.C. Duineveld, Phys. Fluids A, 8 (1996) 310.
53. M. Krzan, K. Lunkenheimer and K. Malysa, Colloids & Surfaces A, 250 (2004) 431.
54. Y. Zhang and J.A. Finch, J. Fluid Mech., 429 (2001) 63.

55. Y. Zhang, J.A. Finch and A. Sam, Colloids & Surfaces A, 223 (2003) 45.
56. E.K. Zholkovskij, V.I. Kovalchuk, S.S. Dukhin and R. Miller, J. Coll. Interface Sci., 226 (2000) 51.
57. S.S. Sadhal and R.E. Johnson, J. Fluid Mech., 126 (1983) 237.
58. Z. He, Z. Dagan and C. Maldarelli, J. Fluid Mech., 222 (1991) 1.
59. Z. He, C. Maldarelli and Z. Dagan, J. Coll. Interface Sci. 146 (1991) 442.
60. J. Chen and K.J. Stebe, J. Coll. Interface Sci., 178 (1996) 144.
61. Y. Liao and J.B. McLaughlin, J. Coll. Interface Sci. 224 (2000) 297.
62. Y. Liao and J.B. McLaughlin, Chem. Eng. Sci., 55 (2000) 5831.
63. Y. Liao, J. Wang, R.J. Nunge and J.B. McLaughlin, J. Coll. Interface Sci., 272 (2004) 498.
64. S.S. Dukhin and B.V. Deryaguin, Zh. Fiz. Khim., 35 (1961) 1246.
65. S.S. Dukhin and B.V. Deryaguin, Zh. Fiz. Khim., 35 (1961) 1453.
66. S.S. Dukhin, B.V. Deryaguin and V.A. Lisichenko, Zh. Fiz. Khim., 33 (1959) 2280.
67. B.V. Deryaguin, S.S. Dukhin and V.A. Lisichenko, Zh. Fiz. Khim. 34 (1960) 524.
68. D.A. Saville, Chem. Eng. J., 5 (1973) 251.
69. G. Bozzano and M. Dente, Comp. Chem. Eng. 25 (2001) 571.
70. C. Ybert and J.-M. di Meglio, Eur. Phys. J. E, 3 (2000) 143.
71. P. Warszynski, B. Jachimska and K. Malysa, Colloids & Surfaces A, 108 (1996) 321.
72. B. Jachimska, P. Warszynski and K. Malysa, Colloids & Surfaces A, 143 (1998) 429.
73. H.J. Schulze, Physico-chemical elementary processes in flotation, in "Developments in Mineral Processing", Vol. 4, D.W. Fuerstenau (Ed.), Elsevier Science Publishing, Amsterdam, 1981, p. 348.
74. A. Graciaa, G. Morel, P. Saulner, J. Lachaise and R.S. Schechter, J. Coll. Interface Sci., 172 (1995) 131.
75. J.S.H. Lee and D. Li, Microfluid Nanofluid, 2 (2006) 361.
76. K.W. Stöckelheuber, H.J. Schulze and A. Wenger, Progr. Colloid Polym. Sci., 118 (2001) 11.
77. A. Michna, M. Zembala and Z. Adamczyk, Colloids & Surfaces A, 302 (2007) 467.
78. G. Heanly, University of South Australia, PhD Thesis, IWRI, 2008.
79. M. Krasowska, M. Krzan and K. Malysa, in "Proc. 5th UBC-McGill Bi-Annual International Symposium of Fundamentals of Mineral Processing", August 22–25, 2004, Hamilton, Ontario, Canada, J.S. Laskowski (Ed.), 2004, pp. 121–135.
80. M. Krasowska, M. Kolasinska, P. Warszynski and K. Malysa, J. Phys. Chem. C, 111 (2007) 5743.
81. M. Krasowska and K. Malysa, Adv. Coll. Interface Sci., 134–135 (2007) 138.
82. M.J. Hadamard, Comp. Rend., 152 (1911) 1735.
83. Rybczynski W., Bulletin International De L'Academie ds Sciences De Cracovie, Classe des sciences Mathematiques et Naturelles Serie A: Sciences Mathematiques, Cracovie, Imprimerie de L'Universite, Janvier 1A (1911) 40.
84. L. Parkinson, R. Sedev, D. Fornasiero and J. Ralston, J. Coll. Interface Sci., 322 (2008) 168.
85. W.L. Haberman and R.K. Morton, "An experimental investigation of the drag and shape of air bubbles rising in various liquids", David Taylor Model Basin, Rep. no. 807, 1953.
86. P.G. Saffman, J. Fluid Mech., 1 (1956) 249.
87. R. Clift, J.R. Grace and M.E. Weber, "Bubbles, drops and particles", Academic Press, New York, San Francisco, London, 1978.
88. N. Schiller and A. Naumann, Z. Ver. Deut. Ing. 77 (1933) 318.
89. D.W. Moore, J. Fluid Mech., 6 (1959) 113.
90. D.W. Moore, J. Fluid Mech., 16 (1963) 161.
91. T.D. Taylor and A. Acrivos, J. Fluid Mech., 18 (1964) 466.

92. D.W. Moore, J. Fluid Mech., 23 (1965) 749.

93. D. Bhaga and M.E. Weber, J. Fluid Mech., 105 (1981) 61.

94. J. Masliyah, R. Jauhari and G. Murray, Chem. Eng. Sci., 49 (1994) 1905.

95. A. Margaritis, D.W. te Bokkel and D.G. Karamanev, Biotech. Bioeng., 64 (1999) 257.

96. D. Rodrigue, A.I.Ch.E. J., 47 (2001) 39.

97. D. Rodrigue, Can. J. Chem. Eng., 80 (2002) 289.

98. D. Rodrigue, Can. J. Chem. Eng., 82 (2004) 382.

99. E. Klaseboer, J.-P. Chevaillier, A. Mate, O. Masbernat and C. Gourdon, Phys. Fluid, 13 (2001) 45.

100. G. Ryskin and L.G. Leal, J. Fluid Mech., 148 (1984) 1.

101. G. Ryskin and L.G. Leal, J. Fluid Mech., 148 (1984) 19.

102. G. Ryskin and L.G. Leal, J. Fluid Mech., 148 (1984) 37.

103. J.B. McLaughlin, J. Coll. Interface Sci., 184 (1996) 614.

104. S.S. Ponoth and J.B. McLaughlin, Chem. Eng. Sci., 55 (2000) 1237.

105. S. Ng, P. Warszynski, M. Zembala and K. Malysa, Min. Engng., 13 (2000) 1519.

106. J.J. Bikerman, "Foams", Springer-Verlag, Berlin, Heidelberg, New York, 1973, Chapter 2, p. 57.

107. M. Krzan, K. Lunkenheimer and K. Malysa, Langmuir, 19 (2003) 6586.

108. R.D. Kirkpatrick and M.J. Lockett, Chem. Eng. Sci., 29 (1974) 2363.

109. A.K. Chesters and G. Hofman, Appl. Sci. Research, 38 (1982) 353.

110. H.-K. Tsao and D.L. Koch, Phys. Fluids, 6(8) (1994) 2591.

111. H.-K. Tsao and D.L. Koch, Phys. Fluids A, 9 (1997) 44.

112. J. Magnaudet and D. Legendre, Appl. Sci. Res. 58 (1998) 441.

113. D. Legendre, R. Zenit, C. Daniel and P. Guiraud, Chem. Eng. Sci., 61 (2006) 3543.

114. E. Manev, S. Sazdanova and D.T. Wasan, J. Coll. Interface Sci., 97 (1984) 591.

115. A.K. Malhorta and D.T. Wasan, A.I.Ch.E. J., 33(9) (1987) 1533.

116. E. Manev, R. Tsekov and B. Radoev, J. Disp. Sci. Technol., 18 (1997) 769.

117. B.P. Radoev, A.D. Scheludko and E.D. Manev, J. Coll. Interface Sci., 95(1) (1983) 254.

118. D. Exerowa and P.M. Kruglyakov, "Foams and foam films — theory, experiment, application", Elsevier, Amsterdam, 1998.

119. K. Malysa and J. Zawala, in "Nanoscale Phenomena and Structures", D. Kaschiev (Ed.), Prof. Marin Dimov Academic Publ. House, Sofia, 2008, pp. 135.

120. L. Doubliez, Intern. J. Multiphase Flow, 17(6) (1991) 783.

121. G. Loglio, N. Degli Innocenti, U. Tesei, R. Cini and Q.-S. Wang, Il Nuovo Cimento Della Socieraie Italiana di Fisica C., 12 (1989) 289.

122. J. Zawala, K. Swiech and K. Malysa, Colloids & Surfaces A, 302 (2007) 293.

123. S. Pastewski and K. Medrzycka, Polish Journal of Environmental Studies, 12(5) (2003) 643.

Surface Tension Measurement of Polymer Melts in Supercritical Fluids

H. Wei [a], **R.B. Thompson** [a], **C.B. Park** [a], **and P. Chen** [b]

[a] Departments of Chemical Engineering, and Physics and Astronomy, University of Waterloo, 200 University Avenue, Waterloo, Ontario, Canada N2L 3GL
[b] Microcellular Plastics Manufacturing Laboratory, Department of Mechanical and Industrial Engineering, University of Toronto, 5 King's College Road, Toronto, Ontario, Canada M5S 3G8. E-mail: p4chen@uwaterloo.ca

Surface tension of a polymer melt in a supercritical fluid is a principal factor in determining cell nucleation and growth in microcellular foaming. This chapter introduces recent works on the surface tension measurement of two polymer melts in supercritical fluids under various temperatures and pressures. One is the amorphous polymer, polystyrene (PS), in supercritical CO_2 and the other is the crystalline polymer, high density polyethylene (HDPE), in supercritical N_2. The surface tension was determined by Axisymmetric Drop Shape Analysis-Profile (ADSA-P). The dependence of the surface tension on temperature, pressure and polymer molecular weight will be discussed. At temperatures above the polymer melting points, the surface tension of PS was found to be similar to that of HDPE; the surface tension decreased with increasing temperature and pressure. Self-consistent field theory (SCFT) calculations were applied to simulate the surface tension of corresponding systems and used to explain the results. Below the melting point, PS solidified and the surface tension did not change any further; but HDPE underwent the process of crystallization, where the surface tension dependence on temperature was different from that above the melting point, and the surface tension decreased with decreasing temperature. It was found that the amount of decrease in surface tension was related to the rate of temperature change and hence the extent of polymer crystallization.

Contents

Bubble and Drop Interfaces
© Koninklijke Brill NV, Leiden, 2011

A. Introduction

1. Polymer Melts in Supercritical Fluids

Surface tension of polymers is one of the most important parameters affecting the foaming and morphology of polymer products. This is because the surface tension between the polymer melt and the fluid is a principal factor in determining cell nucleation and growth [1]. Generally, low surface tension is desired in the polymer foaming process, to increase the nucleation rate and produce small and uniform cells [2]. A supercritical fluid (SCF) is a substance that is compressed beyond its critical pressure and heated above its critical temperature (see Fig. 1). Under these conditions, the vapor and liquid phases become indistinguishable and the fluid behaves as a single phase having advantageous properties of both a liquid and a gas. Supercritical fluids have been widely used as foaming agents in the production of microcellular polymer foams [3]. Specifically, carbon dioxide and nitrogen have advantages of being non-toxic and having relatively low critical points.

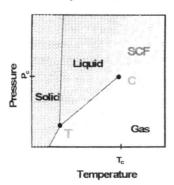

Figure 1. Schematic pressure-temperature phase diagram for a pure component showing the super-critical fluid (SCF) region.

Small amounts of supercritical fluids added to the polymer will result in dramatic changes in physicochemical properties, such as surface tension, viscosity, and solubility [4].

2. Methods of Surface Tension Measurement

Measurement of surface or interfacial tension of polymer melts can be divided into two groups: static and dynamic measurements. Static methods (e.g., pendant drop, sessile drop, and spinning drop) are based on the equilibrium shape of a droplet in a force field (e.g., gravitational or centripetal). These methods require an accurate measurement of the density difference between the polymer and its surrounding phase. Long waiting times are needed before the equilibrium is reached because of the high viscosity of polymers. This may lead to thermal degradation of the polymer. The dynamic methods follow the change in the shape of threads or elongated droplets to an equilibrium shape, including thread breakup, retraction of elongated droplets, and the dynamic shear rheometry on emulsions. The principles of static measurement techniques are briefly reviewed in the following sections.

2.1. Pendant Drop Method

Among the methods of surface tension measurement, the pendant drop method is most commonly used for polymers and liquids. Despite the theoretical simplicity of the pendant drop method, experimental determination of the surface tension of a high viscosity polymer has been difficult, due to the handling of highly viscous polymer melts at high temperatures and pressures [5–11].

The pendant drop method consists of immersing a drop of a molten polymer into the bulk of another fluid environment. The equilibrium drop profile is determined by the balance between gravity and interfacial tension. The interfacial tension evaluated by the Laplace equation of capillarity can be determined from the input of the drop profile and the density difference across the fluid interface. The pendant drop method has several advantages: it is useful for both Newtonian and viscoelastic fluids; no assumption about the rheological behavior of the component is made; the interfacial tension is not disturbed during measurements; it can be applied to liquid crystalline polymers; and the experiment setup is simple. On the other hand, the pendant drop method presents the following potential problems: It requires the knowledge of the density of the materials used, and such information is scarcely reported for polymeric materials. The density difference should be larger than 4–5% for polymer drops to reach the equilibrium shape in an acceptable time interval to avoid thermal degradation [12].

The classical Laplace equation is the basis for all static measurements of interfacial and surface tensions. It states that the pressure difference across a curved interface can be described as:

$$\frac{\Delta P}{\gamma} = \frac{1}{R_1} + \frac{1}{R_2}, \tag{1}$$

where R_1 and R_2 are the two principal radii of the drop, ΔP is the pressure difference across the curved interface, and γ is the interfacial tension.

The Axisymmetric Drop Shape Analysis (ADSA) approach is a pendant drop method; it relies on a numerical integration of the Laplace equation of capillarity to quantify surface tension. This numerical procedure applies to both sessile and pendant drops in shape analysis [13–15]. Recently, ADSA has been used for determining polymer melt surface tension in supercritical fluids at high temperature and high pressure [16]. In this chapter, the related theory and experimental approaches of ADSA are introduced based on the surface tension measurement of polystyrene (PS) in supercritical CO_2 and high density polyethylene (HDPE) in supercritical N_2. The details of this technique are discussed as below.

In the absence of external forces, other than gravity, the pressure difference is a linear function of the elevation.

$$\Delta P = \Delta P_0 + \Delta \rho g z. \tag{2}$$

In this expression, ΔP_0 is the pressure difference at a reference datum plane, $\Delta \rho$ is the difference in the density of the two bulk phases, g is the gravitational acceleration, and z is the vertical height measured from the reference plane.

When the axis x is tangent to the curved interface and normal to the axis of symmetry and the origin is placed at the apex as shown in Fig. 2, the Laplace equation can be rewritten as:

$$\frac{d\phi}{ds} = \frac{2}{a} - \frac{\Delta \rho g}{\gamma} z - \frac{\sin \phi}{x}. \tag{3}$$

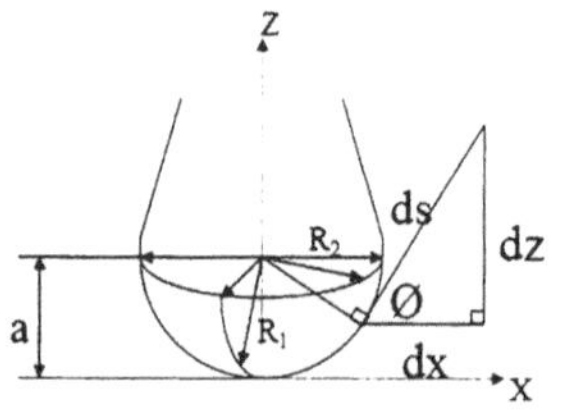

Figure 2. Definition of the coordinate system of a pendant drop.

Mathematically, the interface is described completely as $u = u(x, y, z)$. Due to the symmetry in the system, this may be reduced to the description of the meridian section alone. A suitable representation of the meridian curve is in a parametric form:

$$x = x(s) \quad \text{and} \quad z = z(s) \tag{4}$$

where s is the arc length measured from the origin. In this representation, both x and z are single-valued functions of s.

$$\cos\phi = \frac{dx}{ds}. \tag{5}$$

A geometrical consideration yields the differential identities. And the boundary conditions:

$$x(0) = y(0) = z(0) = 0 \tag{6}$$

from a set of first order differential equations for x, z, and ϕ as functions of the argument s. For given Ro and $\Delta\rho g/\gamma$, the theoretical drop given by the Laplace equation may be obtained by simultaneously integrating three equations (see also Chapter 3).

Once an experimental drop profile is obtained, the ADSA-P program digitizes the image with sub-pixel resolution and randomly selects 20 coordinates. The profile is compared with the theoretical drop profile, using a least square algorithm with interfacial tension as one of the adjustable parameters. The best fit between these two profiles identifies the correct, i.e. operative, interfacial tension. The procedure is repeated 10 times for each experimental drop profile and 95% confidence limits are reported. For this work, typically 95% confidence limits are around ± 0.01–0.02 mJ/m^2.

Besides the drop profile coordinates, the input information required are the acceleration due to gravity and the density difference across the liquid–fluid interface. The details of numerical methods and procedures can be found elsewhere [17] and Chapter 2.

2.2. Sessile Drop Method

The profile of the sessile drop is also that of a meniscus. The theoretical description of contact arises from the consideration of a thermodynamic equilibrium between the three phases: the liquid phase of the droplet (L), the solid phase of the substrate (S), and the gas/vapor phase of the ambient (V). The V phase could also be another liquid phase. At equilibrium, the chemical potential in the three phases should be equal. It is convenient to frame the discussion in terms of the interfacial energies. In Fig. 3, we define the solid–vapor interfacial energy as γ_{sv}, the solid–liquid interfacial energy as γ_{sl} and the liquid–vapor energy (i.e. the surface tension) as γ_{lv}, we can write an equation that must be satisfied in equilibrium (known as the Young Equation) [18]:

$$\gamma_{sv} - \gamma_{sl} - \gamma \cos\theta_e = 0. \tag{7}$$

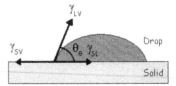

Figure 3. An illustration of the sessile drop method with a liquid droplet partially wetting a solid substrate. θ_e is the contact angle, and γ_{sv}. γ_{lv}. γ_{sl} represent the solid–gas, gas–liquid, and liquid–solid interfaces, respectively.

To calculate the surface tension, the method is similar to that of the pendant drop method mentioned above. The form of the Laplace Equation is the same as the one for the pendant drop method in Eq. (3), except for a change in the sign of the gravitational term [19]. An integration towards the three phase contact line (or point) provides a measure of the contact angle [12, 13, 15, 17]. This method is more accurate than most other approaches, such as those based on goniometry. For more details, see Chapter 10.

2.3. Drop Weight Method
The procedure, in its simplest form, is to form drops of the liquid at the end of a tube, allowing the drops to fall into a container until enough have been collected to accurately determine the weight of a drop [20]. The method was devised by Tate in 1864, and a simple expression for the weight W of a drop is given by what is known as Tate's law [19]:

$$W = 2\pi r \gamma, \tag{8}$$

where W is the weight of the drop and r is the inside radius of the tube. In employing this method, the tube should be ground smooth at the end and the drops should be formed slowly. γ, the surface tension can be obtained based on the equation (cf. Chapter 6).

2.4. Ring Method
This method is widely used and involves the determination of the force to detach a ring from the surface of a liquid. Like all detachment methods, one supposes that a first approximation to the detachment force is given by the surface tension multiplied by the periphery of the surface detached. Thus, the force balance, [20]

$$W_{tot} = W_{ring} + 4\pi r \gamma, \tag{9}$$

where W_{tot} is total weight, W_{ring} is the weight of the ring, r is the radius of ring, and γ is the surface tension. The surface tension is calculated from the diameter of the ring and the tear-off force.

2.5. Wilhelmy Plate Method
This method is a popular and straightforward technique for measuring surface tension and static contact angles. It consists of putting a thin plate, such as a microscope cover glass or platinum foil microscope slide in the test fluid and measure the

force acting on the plate when the system is at equilibrium. The basic observation is that the thin plate will support a meniscus. The force balance, [21]

$$W_{tot} = W_{plate} + p\gamma, \tag{10}$$

where W_{tot} is the total weight, W_{plate} is the weight of the plate, p is the perimeter and γ is the surface tension. The surface tension is calculated from the diameter of the plate and the force acting on the plate.

3. Factors Affecting the Surface Tension

Surface tension of polymers in supercritical fluids varies with many parameters, e.g., system temperature, pressure, the molecular weight of the polymer, the solubility of the supercritical fluids, and the degree of crystallinity of the polymer.

3.1. Surface Tension Dependence on Temperature and Pressure above Polymer Melting Point

Temperature and pressure are two important parameters during the polymer foaming process. Recent work [22, 23] showed several trends of the surface tension change with temperature and pressure for polymers above the melting points. In general, the surface tension decreases with increasing temperature and pressure. Self-consistent field theory (SCFT) calculations were used to explain these experimental trends [23].

3.2. Correlation Between Solubility and Surface Tension

Besides surface tension, the solubility of a supercritical fluid in a polymer is also an important parameter in determining the foaming quality. By examining the change of solubility, as well as surface tension, with the change of temperature and pressure, one can see that both surface tension and solubility depend on temperature and pressure. In this work, the solubility of CO_2 in PS and N_2 in HDPE at different temperatures and pressures will be compared and related with the change in the surface tension.

3.3. Surface Tension of Crystalline Polymers During Crystallization

The degree of crystallinity of a polymer can have a large impact on polymer properties. It is known that PS is a typical amorphous polymer, and HDPE is a typical crystalline polymer. Both types of polymers are often used in polymer microcellular foaming processes. In molten phase, crystalline polymers and amorphous polymers may behave similarly. When the temperature decreases below the melting point, amorphous polymers change into complete solid with disordered chain arrangements, but crystalline polymers would experience the process of crystallization: before a crystalline polymer turns completely solid, it enters a viscoelastic state, where micro-crystals form and grow into regions of ordered chain arrangements within a continuous polymer melt. Surface tension measurement of polymers undergoing such a transition can help understand the different behavior between amorphous and crystalline polymers [24, 25]. It has been found that the surface tension of amorphous polymers at temperatures below the melting point does not change significantly. The surface tension of a crystalline polymer may behave differently from that of an amorphous polymer, i.e., the crystalline polymer may respond to variations of temperature below the melting point. How temperature change, or the rate of temperature change, affects the surface tension, as well as polymer crystallization, will be discussed in this chapter [26–30].

3.4. Surface Tension Dependence on Polymer Molecular Weight

The effect of molecular weight on polymer properties and processing has been well documented in the literature. Two monodisperse polystyrenes of different molecular weight and one polydisperse polystyrene were used to find the effect of polymer molecular weight on surface tension. Monodisperse polystyrene of a higher molecular weight has a higher surface tension under all experimental conditions. The surface tension dependence on temperature and on pressure is more significant for the higher molecular weight polystyrene than that of a lower molecular weight. For the polydisperse polystyrene, high surface tension values seem to be related predominantly to its high molecular weight portion of polystyrene molecules [31].

It is generally the case that high molecular weight polymers will show a greater surface tension than low molecular weight polymers, all else being equal. The experimentally observed relationship is [32]

$$\gamma = \gamma(\infty) - c(1/M)^x, \tag{11}$$

where γ is the surface tension, M is the molecular weight, $\gamma(\infty)$ is the surface tension for infinite molecular weight and c is a constant. The power x has been observed to be $2/3$ for low molecular weight polymers, but switches to 1 for higher molecular weights. Self-consistent field calculations have been able to reproduce these low and high molecular weight regimes of polymer surface tension with molecular weight, and show that the transition between them is due to a depletion to effectively zero of the amount of polymer in the non-polymer side of the interface [33].

B. Experimental Section

Both amorphous and crystalline polymers were used in the experiments:

Polystyrene. Polystyrene (Dow Chemical Company $M = 312\,000$, $Mn = 120\,000$, polydispersity index $= 2.6$), High Molecular Weight Monodisperse PS (Polyscience Inc. $M = 400\,000$, $Mn = 377\,000$, polydispersity index $= 1.06$), Low Molecular Weight Monodisperse PS (Polyscience Inc. $M = 100\,000$, $Mn = 96\,000$, polydispersity index $= 1.04$).

High Density Polyethylene. High Density Polyethylene (Nova Chemicals, Calgary, Canada, melt flow index (MFI) $= 5.0$ g/10 min (ASTM D 1238)).

Supercritical fluids were made from carbon dioxide and nitrogen:

Carbon Dioxide. Carbon dioxide of 99.997% purity (SCF/SCE grade with a helium head) was used as received from Air Products and Chemicals.

Nitrogen. Nitrogen of 99.99% purity was purchased from PRAXAIR (Danbury, CT, USA).

1. Density Determination

The density of polymers not only is an input parameter for the determination of the surface tension of polymers in the Axisymmetric Drop Shape Analysis (ADSA) method, but is also important in the understanding of many polymer physics and engineering processes. However, density measurements are limited, time consuming and costly [34–36]. In previous research work, the sessile drop was employed to measure the surface tension and density simultaneously at high temperature [37].

The fact that the gas dissolution in the polymer melts caused volume swelling in the polymer melts must be considered for accurate solubility measurement and density calculation. The dilation of polymer samples caused by the plasticization effect of supercritical fluids was typically investigated at relatively low temperatures [38] because of the difficulty associated with the experiments at elevated temperatures and pressures. Therefore, it is a common practice to account for and predict the volume swelling at high temperatures of polymer melts using a thermodynamic approach, that is, the Equation of State (EOS). Among the equations of state, the Tait equation, the Sanchez and Lacombe (SL) EOS, the Simha and Somcynsky (SS) EOS are well known for predicting the density of polymers.

In a Tait equation method [39–41], the parameters for polymers in supercritical fluids are:

$$v_p = v_0 \left[1 - 0.0894 \ln \left(1 + \frac{P}{B(T)} \right) \right], \tag{12}$$

where v_p is the specific volume (cm^3/g), P is the pressure,

$$v_0 = 0.7884 \exp(5.79 \times 10^{-4} T),$$
$$B(T) = 887.2 \exp(-4.323 \times 10^{-3} T). \tag{13}$$

Sato et al. employed the Sanchez and Lacombe (SL) equation of state (EOS) to calculate the solubility of gases at high pressures and temperatures [42, 43]. The densities of polymers saturated with supercritical fluids were also determined by the SL-EOS as expressed below:

$$\tilde{\rho}^2 + \tilde{P} + \tilde{T} [\ln(1 - \tilde{\rho}) + (1 - 1/r)\tilde{\rho}] = 0, \tag{14}$$

where $\tilde{\rho}$ is the reduced density, $\tilde{P}$ is the reduced pressure, $\tilde{T}$ is the reduced temperature and r is the number of sites occupied by a molecule; they are defined as

$$\tilde{P} = \frac{P}{P^*}, \qquad \tilde{\rho} = \frac{\rho}{\rho^*}, \qquad \tilde{T} = \frac{T}{T^*}, \qquad r = \frac{M P^*}{R T^* \rho^*}, \tag{15}$$

where ρ is the density, P is the pressure, T is the temperature, M is the molecular weight and R is the gas constant. In the equation, the characteristic parameters, P^*, ρ^*, and T^*, of the SL-EOS for the mixture were evaluated using the following mixing rules:

$$P^* = \sum_i \sum_j \phi_i \phi_j P_{ij}^*, \qquad P_{ij}^* = (1 - k_{ij})(P_i^* P_j^*)^{0.5}, \qquad T^* = P^* \sum_i \frac{\phi_i^0 T_i^*}{P_i^*},$$

$$\frac{1}{r} = \sum_i \frac{\phi_i^0}{r_i^0}, \qquad \phi_i^0 = \frac{(\phi_i P_i^*/T_i^*)}{\sum_j (\phi_j P_j^*/T_j^*)}, \qquad \phi_i = \frac{w_i/\rho_i^*}{\sum_j w_i/\rho_i^*}, \tag{16}$$

where T_i^*, P_i^*, ρ_i^*, and r_i^0 represent the characteristic parameters of the component i in its pure state.

Recently, the Simha and Somcynsky (SS) EOS was employed by Li *et al.* to predict the PVT behavior of polymer/gas mixtures, and the results were compared with those obtained from the SL-EOS [44, 45]. The SS-EOS is expressed as below [46]:

$$\tilde{\rho}\tilde{V}/\tilde{T} = (1 - \eta)^{-1} + \frac{2yQ^2(1.011Q^2 - 1.2045)}{\tilde{T}}, \tag{17}$$

$$\left(\frac{s}{3c}\right)\left[\frac{s-1}{s} + \frac{\ln(1-y)}{y}\right] = \frac{\eta - 1/3}{1 - \eta} + \frac{y}{6\tilde{T}}Q^2(2.409 - 3.033Q^2), \tag{18}$$

where y is the occupied lattice site fraction; s is the number of segments dividing a molecule; c is the volume-dependent degrees of freedom; Q and η are defined as:

$$Q = (y\tilde{V})^{-1}, \qquad \eta = 2^{-1/6}yQ^{1/3}. \tag{19}$$

2. *Experimental Setup*

The apparatus developed for this research consists a high-temperature and high-pressure sample cell. This optical viewing cell in which a pendant drop is formed was connected with an electrical band heater and a pressure pump to simulate the polymer foaming conditions. An optical system was used to capture the image of the drop, and a data acquisition system with a PC was used to compute the interfacial tension from the drop profile [22].

Figure 4 shows the schematic diagram of the ADSA-P experimental setup. An isolated workstation (1) was used to isolate the instruments from floor motion or vibration. To illuminate the pendant drop, a light source (2) filtered by a diffuser (3) was used. The light diffuser provided a uniformly bright background, which resulted in images of high contrast. Pendant drops were obtained by extruding one polymer with a heated stainless steel syringe (4). An optical viewing cell (5) was fixed on a XYZ stage (6) that could finely adjust the position of drops in three directions. A microscope system, including a microscope (7) and a CCD camera (8), was used to take drop images and to export image signals to a monochromatic monitor (9) and the computer system (10), (11). The microscope and camera were mounted on another adjustable XYZ stage (12), which was used to change the angle between the microscope and the horizon. The whole optical system was mounted and aligned on an optical rail (13) [22].

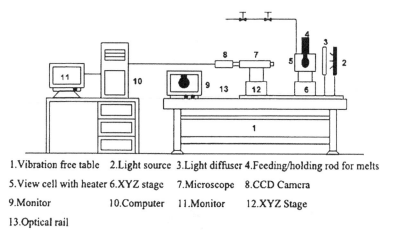

1.Vibration free table   2.Light source 3.Light diffuser 4.Feeding/holding rod for melts

5.View cell with heater 6.XYZ stage   7.Microscope   8.CCD Camera

9.Monitor   10.Computer   11.Monitor   12.XYZ Stage

13.Optical rail

Figure 4. Schematic diagram of the ADSA-P experimental setup.

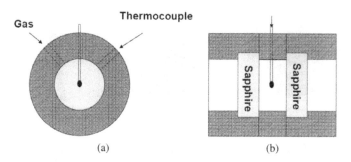

Figure 5. Schematic diagram of the optical viewing cell: (a) Front view; (b) Side view.

The optical viewing cell, where the melt sample was formed, has two sapphire windows, mounted perpendicular to the cell axis. It is believed that the distortion of the sapphire window under pressure of less than 30 Mpa is negligible. A schematic diagram of the cell is shown in Fig. 5. [22]

3. Stability of the Polymer Drops

A typical pendant drop image of the polymer melt in supercritical fluid taken by this system is shown in Fig. 6. This axisymmetric pendant drop is applicable to the ADSA surface tension measurement [22].

The polymer pellet size has to be determined before being attached to the rod. This is because the stability of the polymer drop is balanced by its gravity and surface tension. If the drop sizes are too large, the pendant drops are unstable and breakage may occur because of the dominant gravity (see Fig. 7(a)). However, when the drop was smaller than a certain critical volume, the drop liquid may climb up along the feeding rod (see Fig. 7(b)) [22].

Figure 6. A typical pendant drop image of the polymer melt in supercritical fluid.

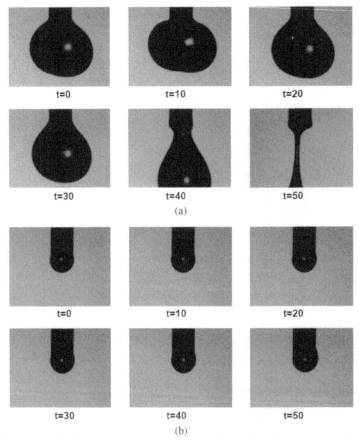

Figure 7. Evolution of polymer melts having (a) larger (b) smaller volume than a critical volume.

The stability of the polymer pendant drop may be related to the Bond number,

$$B_0 = \frac{\Delta \rho g R^2}{\gamma}, \tag{20}$$

where $\Delta \rho$ is the density difference between two phases; g is the gravitational acceleration; R is the drop radius; and γ is the interfacial or surface tension between the immiscible phases. The Bond number is the ratio of buoyancy force to surface force. The number is used to indicate whether breakage occurs or the stable pendant drop is maintained. Based on the experiment experience, one can calculate the stable range of the bond number to determine the size of the drop in order to avoid breakage and liquid climb-up [22].

C. Surface Tension of Polystyrene (PS) in Supercritical CO_2

1. Surface Tension Dependence on Temperature and Pressure above Polymer Melting Point

This experiment was performed at five different pressures: 500, 1000, 1500, 2000, and 2500 psi, and five different temperatures: 170, 180, 190, 200, and 210°C. The surface tension value under various conditions was taken at its steady-state, when the change was less than 0.0001 mJ m^{-2} s^{-1} for 1h. Thus the values obtained are regarded as equilibrium surf ace tensions. For each equilibrium surface tension reported, errors were on the order of 0.01 mJ m^{-2}. Figure 8 shows the equilibrium surface tension values as a function of temperature [23].

It is apparent that at a given pressure, the surface tension decreases with increasing temperature; at a given temperature, the surface tension decreases with increasing pressure. The trend observed of the surface tension change with temperature is consistent with Wu's work [47]. A linear relationship between surface tension and temperature was proposed for the polymer melts. However, in our experiments, pressure was another variable. To find how surface tension is related with both temperature and pressure, a second-order linear regression model was proposed and tested against the experimental results. Table 1 shows analysis of

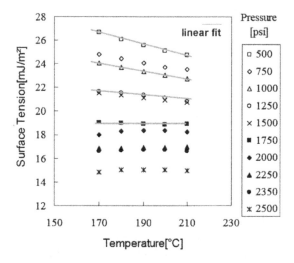

Figure 8. The equilibrium surface tension of polystyrene in carbon dioxide at various temperatures (170, 180, 190, 200, 210°C) and pressures (500, 1000, 1500, 2000, 2500 psi).

Table 1.

ANOVA (Analysis of Variance) table for a second-order linear regression model of PS in CO_2

	Sum of square (SS)	Degree of freedom	Mean square (MS)
Regression	560.83	3	180.95
Residual	7.62	46	0.165
Total	568.45	49	

$F_{obs} = 1129.9$, $F_{3,46,0.05} = 2.80$, R-Square $= 0.99$.

variance, or ANOVA, indicating the validity of the regression model: the observed F-value is larger than the tabulated F-value at the 95% confidence level. In Table 2, the validity of each parameter in the second order equation was also examined by using the t-test: all observed t values are greater than the tabulated t-value at the 95% confidence level. From these statistical investigations, we can propose the following equation for the surface tension γ [23]:

$$\gamma = 38.7032 - 0.0559T - 0.0100P + (2.596 \times 10^{-5})TP$$

$$(170°C < T < 210°C, 500 \text{ psi} < P < 2500 \text{ psi}),$$

(21)

where the surface tension of polystyrene in supercritical CO_2 γ is in mJ/m^2, the temperature T in °C, and the pressure P in psi. Note that the second-order term in T or P is absent; statistically, γ is linearly related to T and P. However, there is an interaction term in TP, indicating γ dependence on T or P is affected by P or T, respectively. This indicates that, for polymer melt processes, one has to adjust both T and P to control the value of γ.

From Eq. (21), the following equations can be derived:

$$\frac{\partial \gamma_{PS}}{\partial P} = -1 \times 10^{-2},$$

(22)

$$\frac{\partial \gamma_{PS}}{\partial T} = -5.6 \times 10^{-2}.$$

(23)

$$\frac{\partial^2 \gamma_{PS}}{\partial TP} = 2.6 \times 10^{-5}.$$

(24)

There are three main experimental trends presented in Eqs (22) to (24). These are the dropping of the surface tension as a function of temperature for the pressure being less than ~2153 psi, the dropping of the surface tension with increasing pressure for the temperature being less than ~385°C, and the flattening of the surface tension versus temperature curves with increased pressure (see the fit lines in Fig. 8).

Table 2.

t-test for evaluating each parameter of the proposed second-order linear regression model of PS in CO_2

Parameters	Coefficients	Standard error	$|t\text{-value}|$
Intercept	38.7032	2.0083	19.27
T	−0.0559	0.0105	5.30
P	−0.0200	0.0012	8.52
TP	2.5957E−5	6.13E−8	4.23

$T_{0.025,17} = 2.013$.

2. Self-Consistent Field Theory

To understand the surface tension and its dependence on temperature and pressure, experimentally determined surface tensions can be compared to surface tensions calculated using self-consistent field theory (SCFT). SCFT is an equilibrium statistical mechanical approach for determining structures in polymeric systems. It is based on a free energy functional, which is to be minimized in order to find the lowest free energy morphology. The procedure for deriving such functionals has been explained in depth in a number of reviews [47–51]. For the supercritical carbon dioxide–polystyrene system, an appropriate free energy functional has been derived in the canonical ensemble in Ref. [23].

In order to find the free energy of the system, a number of input parameters are needed. In the canonical ensemble, one needs the volume V as well as ϕ_s, ϕ_p, and ϕ_h representing the overall volume fractions of solvent molecules, polymer segments and "holes", respectively. One also needs the degree of polymerization N, which is based on a segment volume $1/\rho 0$. Since SCFT is a coarse-grained theory, a single segment may include many chemical monomers. In order to be consistent with the Sanchez–Lacombe equation of state [42, 43] being used experimentally to extract the surface tension in the supercritical carbon dioxide–polystyrene system, a Sanchez–Lacombe equation of state was used to model pressure in the SCFT. This approach was introduced by Hong and Noolandi [52] for SCFT and consists of treating a compressible system as an incompressible system together with vacancies, that is, holes. Higher pressure systems have fewer holes whereas lower pressure systems have more. The Sanchez–Lacombe equation of state thus relates the density to pressure for systems whose variable density is modeled in terms of holes. The volume fractions ϕ_s, ϕ_p, and ϕ_h, are therefore not all independent, rather $\phi_s + \phi_p + \phi_h = 1$. Other approaches for treating compressibility within SCFT are possible, in particular Binder *et al.* have studied solvent-polymer systems using a virial expansion to get an equation of state [53]. Flory–Huggins parameters are also required inputs; these are usually defined in terms of dissimilar constituents such as χ_{ps}, χ_{pk} and χ_{sk}. In Ref. [23] however, since the holes are fictitious, it was more meaningful to choose the three independent parameters as χ_{ps}, χ_{pp} and χ_{ss}. They were defined from first principles as

$$\chi_{ij} = \frac{\rho_0}{k_B T} \int d\mathbf{r} V_{ij}(|\mathbf{r}|), \tag{25}$$

where $V_{ij}(|r|)$ was two-body potential between species i and j with $i, j = p, s$ or h [52], and k_B is Boltzmann's constant. Since the potential between holes and anything else should be zero, all χ terms in the free energy involving h would vanish. The interpretation of these parameters was then no longer as the dimensionless change in energy upon exchange of segments between pure components, although the use of the term Flory–Huggins parameter would be maintained; they still arose as the first order in a gradient expansion of the potentials. Usually, the products χN are taken as the segregation parameters instead of just the parameters χ, and this was done in Ref. [23]. Lastly, one requires the ratio of the volume of a solvent molecule to a polymer molecule, which in [23] was given by α. The free energy functional was varied with respect to position dependent volume fraction functions and chemical potential functions to yield a set of equations that was solved numerically and self-consistently.

The free energy functional can be used to calculate the free energy for the whole system, as well as the free energies for two homogeneous systems equivalent to the bulk regions of the original system. Subtracting the homogenous free energies from the total free energy gives the excess free energy of the interface which, when divided by the surface area, is the surface tension. Details of the procedures can be found in Ref. [23].

To facilitate analysis of results, the surface tension can be broken up into thermodynamic components. The components chosen in Ref. [23] were the internal energy contributions to the free energy between polymer segments and solvent, solvent and solvent, and polymer and polymer, the translational entropy contribution to the free energy of the polymer, the configurational entropy of the polymer, the translational entropy of the solvent and the translational entropy of the holes. The configurational entropy accounts for the different conformations a polymer can take, whereas the translational entropy of the polymer accounts for the remaining positional degrees of freedom of the center of mass of a molecule. These components can be converted into excess free energy components by subtracting off the corresponding bulk free energy components of the homogenous phases on either side of the interface in exactly the same way as for the total free energy. Then by dividing by the interfacial area, these can be converted into components of the surface tension, just as the total excess free energy was expressed as a surface tension. These internal energy and entropic contributions to the surface tension were used in Ref. [23] to explain the trends observed experimentally in the supercritical carbon dioxide–polystyrene system, albeit only qualitatively.

The SCFT calculations were performed to find a dimensionless surface tension as a function of temperature at two different pressures. For the high pressure run, no holes were included and the overall volume fractions were taken as $\phi_p = 0.65$ and $\phi_S = 0.35$ for the polymer and solvent, respectively. This corresponds to an incompressible fluid, and thus is the highest pressure case possible. This was compared against a lower pressure run with $\phi_p = 0.60$ and $\phi_S = 0.30$, or in other words, with 10 percent holes by volume. In both cases and at all temperatures, the system size (considering one dimension) was $L = 12.0R_g$, where R_g is the unperturbed radius of gyration of a polymer. The ratio α of the volume of a solvent molecule to that of a polymer molecule was taken to be 0.1 for both pressure runs. This was not particularly realistic, as this ratio for the supercritical carbon dioxide–polystyrene system should be a much smaller number. Too great a size disparity between the different molecular species would however cause numerical difficulties. This resulted from the extremely high translational entropy of many, very small solvent molecules. This strongly favored mixing, and made it difficult to establish an interface unless the Flory–Huggins parameters were turned up extremely high. This in turn made it difficult to achieve numerical accuracy in the calculations. Rather, a qualitative approach was taken, making sure that trends observed experimentally were nonetheless still observed in the calculations despite a large value for α.

Given the qualitative philosophy, it sufficed to choose Flory–Huggins values that mapped the model system qualitatively onto the experimental structure. A relationship between χ (or in this case, χN) and temperature T that is commonly used is [54, 55]:

$$\chi N = \frac{A}{T} + B. \tag{26}$$

where A and B are constants. In Ref. [23] there were three different such parameters, namely $\chi_{ps}N$, $\chi_{ss}N$ and $\chi_{pp}N$, so there were three sets of constants, A_{ps}, B_{ps}, A_{ss}, B_{ss} and A_{pp}, B_{pp}. The constants B_{ps}, B_{ss} and B_{pp} were all set to zero for simplicity, as was A_{pp}. From Eq. (26), $A_{pp} = 0$ can only be satisfied for arbitrary T if $\chi_{pp}N = 0$, always. The remaining parameters were chosen as $A_{ps} = 100$ and $A_{ss} = 150$. The temperature T was ranged, in arbitrary units, from 2.0 to 2.5 giving $2.0 < T < 2.5$; $50 > \chi_{ps}N > 40$; $75 > \chi_{ss}N > 60$.

These values produced reasonable interfacial structures, as shown in Ref. [23]. To assign specific units to the temperature such as Kelvin or degrees centigrade, the parameters A should be specified in the desired units. The present values were chosen so as to reproduce an appropriate interface while at the same time allowing for numerically accurate calculations.

2.1. Temperature Dependence

Figure 9 shows the dimensionless surface tension results from Ref. [23] as a function of temperature at two different pressures. The results can explain the three main trends mentioned in section C.1. The temperature dependence of the model system can be seen to follow the trends of experiment and the empirical equation (21) at both pressures in that surface tension decreases with increasing temperature. In Fig. 10 the components of the surface tension are plotted. The two main com-

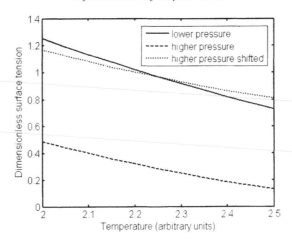

Figure 9. Dimensionless surface tension as a function of temperature for two different pressures. The lower pressure run is the solid curve while the higher pressure run is the dashed curve. The higher pressure run is also plotted a second time by a dotted curve where it is shifted upwards to more easily compare the slopes of the two runs.

ponents that can be seen to be contributing to the decrease of surface tension with temperature are the internal energy contribution to the surface tension (open circles on solid curve) and the polymer configurational entropy contribution to the surface tension (crossed dotted curve). The translational entropy of the holes contributes negligibly. The largest contribution is from the internal energy.

This contribution can in turn be split into the polymer–solvent, solvent–solvent, and polymer–polymer components of the internal energy contribution to the surface tension, as shown in Fig. 11. In that figure, the component that is clearly responsible for the overall drop of the total internal energy contribution is the polymer–solvent component; it is the only component with a slope in the correct direction. Translating this conclusion into polymer–solvent processes, one would concentrate on modifying the molecular interaction between the polymer and its solvent when making use of such temperature dependence of surface tension. Under this situation, modifications of polymer or solvent molecular properties alone could be less effective at reducing surface tension with an elevated temperature.

That the polymer–solvent internal energy contribution is responsible for the drop in surface tension makes perfect sense, in that the free energy of the system can be split into an internal energy part and an entropic part, the two parts having different signs, that is, they oppose each other. The entropic contributions promote mixing whereas the internal energy favors segregation. As the temperature is increased, the $\chi_{ps}N$ parameter decreases, reducing the segregation between polymer and solvent segments. This means the entropy becomes a larger relative portion of the free energy, more mixing takes place and the interface becomes more diffuse; this in turn means there would be a lower surface tension. This is a well known and understood effect which was correctly reproduced in the model system of Ref. [23].

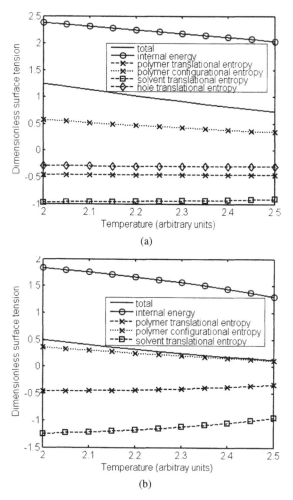

Figure 10. Components of the surface tension for (a) the lower pressure run and (b) the higher pressure run. Different contributions to the surface tension are shown in the legends.

2.2. Pressure Dependence

In Fig. 9 it can be seen that the surface tension versus temperature curve drops to lower surface tension for a higher pressure. This is again in agreement with the experimental findings and empirical equation (21). The components of the surface tension that drop are the internal energy, the configurational entropy of the polymer and the translational entropy of the solvent; this can be seen from Fig. 10 by comparing panels (a) and (b). Again, the largest single factor causing this drop is the internal energy contribution. In Fig. 11, however, it is seen that for the pressure induced surface tension drop, the responsible sub-component is not the polymer–solvent internal energy as for the temperature case, but rather the solvent–solvent sub-component. Translating this conclusion into industrial polymer–solvent processes, one could simply focus on modifying the molecular self-interaction among

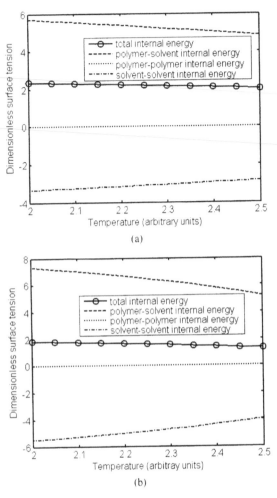

Figure 11. Sub-components of the internal energy contribution to the surface tension for the (a) lower pressure run and (b) the higher pressure run. Different contributions to the surface tension are shown in the legends.

solvent molecules when making use of such pressure dependence of surface tension.

The above conclusion can be understood in terms of a reduction of dilution by the holes. At higher pressure, there are fewer holes present. Since $\chi_{ss}N$ was chosen to be positive, solvent molecules prefer to be in an environment of holes rather than in an environment of other solvent molecules; in the former situation the unfavorable solvent–solvent contact energy is diluted by the holes. With the removal of holes at higher pressure, this dilution is reduced, the solvent–solvent contact energy goes up, and so does the free energy. This effect takes place predominantly in the bulk solvent side of the interface where the majority of solvent molecules can be found. This means the bulk free energy is increased. This increased quantity is subtracted off the total free energy to find the surface tension, therefore the surface tension would drop.

This last point may be understood in terms of density. The removal of holes is the same as an increase in density in the region where the holes are being removed. Thus the surface tension drops when the solvent phase increases in density to be more similar to the density on the polymer side of the interface. Thus one can say the drop in surface tension with increasing pressure is due to a reduction of the density difference between two sides of the interface.

The above analysis of pressure dependence required a χ_{ss} that was positive, and so it is appropriate here to discuss what might be the case if χ_{ss} were negative. This is important since from the first principles definition of χ_{ss} given in Eq. (25) one would expect that χ_{ss} would normally be less than zero, that is, the solvent molecules would have some slight attraction. For more realistic choices of α, the translational entropy of the solvent would not be negligible. Therefore instead of holes diluting the solvent phase for energetic reasons, the holes would dilute the phase for entropic reasons. The explanation would remain the same for the pressure dependence beyond this, and the density difference interpretation would still hold. As α is increased, the translational entropy of the solvent will become less important, and to maintain the interface structure, χ_{ss} must be made less negative. For a very large α, such as is being used here, χ_{ss} must become positive to draw the hole molecules into the solvent phase to reproduce the experimental configuration. At this point, χ_{ss} must be viewed entirely as a phenomenological parameter.

2.3. Change in Temperature Dependence with Pressure

In addition to an overall drop in surface tension upon increasing pressure, the temperature dependence of the surface tension is less pronounced at high pressures than at lower pressures. This is seen in Fig. 9 where the dotted curve is a repetition of the high pressure curve (dashed) shifted upwards to lie on top of the lower pressure curve (solid). One can clearly see the shallower slope with temperature of the high pressure results. This is again in agreement with the experimental findings and the empirical equation (21).

From Fig. 10, one can compute linear slopes for all the components of the surface tensions in order to find which components are responsible for this reduction in steepness. It was found that the translational entropy components of the polymer, solvent and holes all contributed to the overall reduction in steepness. The hole contribution was negligible compared to the other two and could safely be ignored. Thus it was the polymer and solvent translational entropy contributions to the surface tensions that caused the shallowness of the high pressure results.

This is explained in terms of the presence or absence of holes. The presence of holes can only affect the system in two ways: through energy dilution as discussed in the pressure dependence subsection, or through adding translational entropy. The latter was said to be insignificant, and so we are left with energy dilution alone. At low pressures, the solvent–solvent contacts are diluted by the holes, reducing the system free energy. At high pressures, solvent–solvent contacts cannot be reduced by holes anymore, so the only possibility for reducing these contacts is for the solvent to be near polymer segments. This can induce increased mixing, and thus increase translational entropy of both the solvent and the polymer. This increased mixing partially counteracts the internal energy segregation effect that is a function of temperature. Thus the surface tension profile with temperature is flatter at higher pressures than at lower pressures where this polymer–solvent mixing is unnecessary due to the presence of the holes. In other words, when the solvent is at higher density, there is a greater mixing effect that counteracts the formation of an interface due to a solvent–solvent internal energy reduction upon absorbing solvent into the polymer phase.

For small α values and negative χ_{ss} parameters, the same mechanism is expected to function, except that translational entropy would force the holes into the solvent phase rather than energetic considerations, along the lines explained in the pressure dependence section.

3. Surface Tension Measurement Below Glass Transition Temperature of PS

We already know that surface tension of a PS melt in CO_2 increases with decreasing temperature. But when the temperature is further decreased, sudden changes of surface tension are shown around 100°C in Fig. 12, which is the glass transition temperature of the sample [56, 57].

This is because once the temperature goes below 100°C, PS solidifies and hence surface tension detected by ADSA would not change any further. The glass transition temperature of amorphous polymers could be measured by differential thermal analysis, e.g., differential scanning calorimetry (DSC). Now, this surface tension measurement can also be applied to measure the glass transition temperature of amorphous polymers [58].

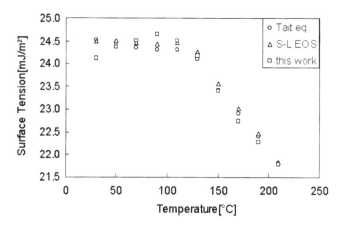

Figure 12. Surface tension of PS through its glass transition temperature.

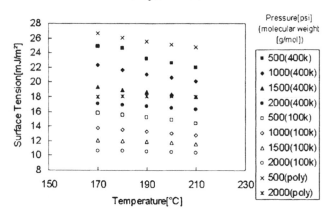

Figure 13. Surface tension as a function of temperature for two monodisperse polystyrenes under 4 different pressures and for one polydisperse polystyrene under 2 different pressures. Closed symbols refer to the monodisperse polystyrene of a weight average molecular weight of 400 000 g/mol; open symbols refer to that of a weight average molecular weight of 100 000 g/mol. Crossed and asterisk symbols refer to the polydisperse polystyrene. The lower molecular weight polystyrene shows lower surface tensions. The polydisperse polystyrene has a weight average molecular weight of 312 000 g/mol, and has slightly higher surface tensions than those of the high molecular weight monodisperse polystyrene.

4. Effect of Molecular Weight on the Surface Tension

The surface tension of polystyrene melts in supercritical nitrogen was measured at four different pressures, 500, 1000, 1500 and 2000 psi, and five different temperatures, 170, 180, 190, 200 and 210°C. The equilibrium surface tension values for two monodisperse polystyrenes of M 100 000 and 400 000, along with a polydisperse polystyrene are shown in Fig. 13. The results show that the higher molecular weight polystyrene has a higher surface tension under all pressure and temperature conditions tested [31].

To quantify the temperature and pressure influence on the surface tension, a second-order linear regression model was used. From statistical investigations, we can propose the following equations for the two monodisperse polystyrenes [31]:

$$\gamma(M \sim 100\,000) = 25.0362 - 0.0448T - 0.0068P + (1.97 \times 10^{-5})TP$$
$$(R^2 = 0.99), \tag{27}$$

$$\gamma(M \sim 400\,000) = 43.5497 - 0.0942T - 0.0120P + (3.91 \times 10^{-5})TP$$
$$(R^2 = 0.99). \tag{28}$$

Note that the second order terms in T and P are absent; statistically, γ is linearly related to T and P. There is an interaction term in TP, indicating γ dependence on T or P is affected by P or T, respectively. Comparison between the above two equations indicates that polystyrene of a higher molecular weight has a stronger temperature and pressure dependence of surface tension than polystyrene of a lower molecular weight. The cross interaction between temperature and pressure effects is also more significant for the higher molecular weight polystyrene.

5. *Effect of Polydispersity on the Surface Tension*

Similar to its monodisperse counterparts, the polydisperse polystyrene demonstrates three trends of surface tension variation: In Figure 13, it is noticed that the polydisperse polystyrene has a higher surface tension than the monodisperse polystyrene of M 400 000, even though its molecular weight is below 400 000. In a polydisperse polymer, a wide distribution of molecular weights exists; thus, it may not be surprising that a portion of polystyrene molecules possesses a molecular weight greater than 400 000. This large molecular weight portion of polystyrene molecules may contribute more influentially to a high surface tension. In other words, high surface tension values are mainly derived from polystyrene molecules of high molecular weights. This conclusion is also consistent with the fact that the surface tension of monodisperse polystyrene of M 400 000 is greater than that of M 100 000 [31].

D. Surface Tension of High Density Polyethylene (HDPE) in Supercritical N_2

1. *Surface Tension Dependence on Temperature and Pressure above Polymer Melting Point*

The equilibrium surface tension value of the HDPE melt in supercritical nitrogen was measured under various temperatures 125, 130, 140, 150, 160, 170, 180 and 190°C above the HDPE melting point, $\sim$125°C, and three different pressures 500, 1000 and 1500 psi. The surface tension value of HDPE in supercritical nitrogen under various conditions was taken at its steady-state.

Figure 14 shows the equilibrium surface tension at each temperature and pressure. The surface tension varies from 20.5 mJ/m^2 at 190°C, 1500 psi, to 25.5 mJ/m^2 at 125°C, 500 psi. It is apparent that similar with the work of PS in CO_2, at a given pressure, the surface tension decreases with increasing temperature; at a given temperature, the surface tension decreases with increasing pressure.

To find how surface tension is related with both temperature and pressure, a second-order linear regression model for the surface tension γ was proposed and tested against the experimental results.

$$\Gamma = 31.7534 - 0.04611T - 0.00165P$$
$$(125°C < T < 190°C, 500 \text{ psi} < P < 1500 \text{ psi}). \tag{29}$$

Table 3 shows the observed F-value is larger than the tabulated F-value at the 95% confidence level. In Table 4, the validity of each parameter in the second order equation was also examined by using the t-test: all observed t values are greater than the tabulated t-value at the 95% confidence level. This result shows the second-order term in T or P and the interaction term in TP are absent; statistically, γ is linearly related to T and P.

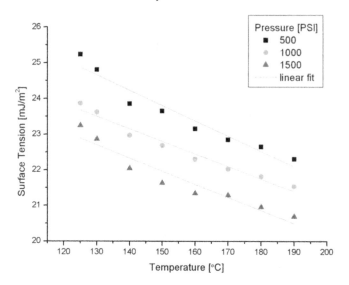

Figure 14. The equilibrium surface tension of HDPE in N_2 at various temperatures (125, 130, 140, 150, 160, 170, 180, 190°C) and pressures (500, 1000, 1500 psi) above HDPE melting point (125°C).

From Eq. (29), the following equations can be derived:

$$\frac{\partial \gamma_{HDPE}}{\partial P} = -1.65 \times 10^{-3}, \tag{30}$$

$$\frac{\partial \gamma_{HDPE}}{\partial T} = -4.61 \times 10^{-2}. \tag{31}$$

There is one different trend for HDPE in N_2 from PS in CO_2, Eq. (24): the interaction term in TP for PS in CO_2 shows the rate of the surface tension change of PS with temperature increases with increasing pressure, while this term is absent for HDPE in N_2 indicating the rate of the surface tension change of HDPE with temperature does not change much.

Table 3.
ANOVA (Analysis of Variance) table for a second-order linear regression model of HDPE in N_2

	Sum of square (SS)	Degree of freedom	Mean square (MS)
Regression	27.35163	3	13.67581
Residual	0.45054	17	0.02503
Total	27.80217	20	

$F_{obs} = 546.4$; $F_{3,17,0.05} = 3.2$, R-square $= 0.98$.

Table 4.
t-test for evaluating each parameter of the proposed second-order linear regression model of HDPE in N_2

Parameters	Coefficients	Standard error	|*t*-value|
Intercept	31.75343	0.29090	109.15406
T	−0.04611	0.00173	26.71155
P	−0.00165	8.46E−5	19.47416

$T_{0.025,17} = 2.11$.

2. The Change of Solubility with Temperature and Pressure

Besides surface tension, the solubility of a gas in a polymer is also an important parameter in determining the foaming quality. By relating the change of solubility, as well as surface tension, with the change of temperature and pressure, one can find that surface tension and solubility vary correspondingly to each other.

Let us compare the work of CO_2 in PS and N_2 in HDPE. First, if the temperature is maintained, as the pressure is increased, the solubility of CO_2 in PS and N_2 in HDPE increases and the surface tension decreases. This is reasonable when considering the fact that an increase in gas-phase pressure will likely induce more gas dissolution into the liquid phase. Comparing the surface tension dependence on pressure, from Eqs (22) and (30), it is found the surface tension drops more with the same amount of increase in pressure for PS in CO_2 than for HDPE in N_2. And correspondingly, the solubility dependence on pressure of CO_2 in PS is also stronger than that of N_2 in HDPE, which can be observed from Eqs (32) and (33) derived from the solubility data [59, 60]:

$$\frac{\partial C_{N_2}}{\partial P} = 7.5 \times 10^{-6}, \tag{32}$$

$$\frac{\partial C_{CO_2}}{\partial P} = 2.86 \times 10^{-5}. \tag{33}$$

From the experimental results, the surface tension at different temperatures begins to converge at higher pressures for PS in CO_2, while this phenomenon is not observed for HDPE in N_2. Figure 15 shows that the solubility of N_2 in HDPE increases slightly with increasing temperature, while to the contrary, the solubility for CO_2 in PS decreases with increasing temperature. For CO_2 in PS, there are two competing effects affecting the solubility: pressure tends to increase the solubility, while the temperature tends to decrease it. Thus, pressure and temperature compete and together determine the solubility. Correspondingly, the rate of the surface tension change of PS with temperature decreases at higher pressures, which can be observed through the fit lines in Fig. 8. But for the case of N_2 in HDPE, the increase in both pressure and temperature tends to increase the solubility. Correspondingly, the rate of the surface tension change of HDPE with temperature does not decrease at higher pressures, note the interaction term in Eq. (24).

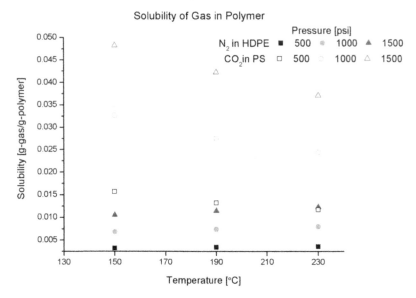

Figure 15. Solubility of gas in polymer at various temperatures (150, 190, 230°C) and pressures (500, 1000, 1500 psi): the solid symbols present the solubility data of N_2 in HDPE, and open symbols present the solubility data of CO_2 in PS.

3. Surface Tension Of HDPE in N_2 Accompanied by Crystallization

The reported melting point of HDPE, $\sim$125°C, is in the range of 120 to 130°C. Above the melting point, the polymer is liquid, and below it, the polymer starts to crystallize until it turns completely solid. To determine the melting point of the sample used in our experiments, differential scanning calorimetry (DSC) was used. The polymer starts melting at around 110°C. The peak point is found to be around 125°C, which is considered the melting point of the sample.

It has been found that the surface tension of amorphous polymers at temperatures below the melting point does not change significantly. It is plausible that the surface tension of a crystalline polymer may behave differently from that of an amorphous polymer, i.e., the crystalline polymer may respond to variations of temperature below the melting point. Thus, the surface tension of HDPE in N_2 during crystallization is measured. The system pressure was controlled at 500, 1000, or 1500 psi each time. To investigate the effect of HDPE crystallization, the system was cooled from 150 to 100°C in intervals of 10°C, during which the system was maintained at each condition for two hours, and the surface tension value was measured at its steady-state in each interval. The results of the experiment are shown in Fig. 16. With decreasing temperature, the surface tension first increases until temperature reaches the melting point of HDPE, $\sim$125°C, and then it drops sharply with further decreasing temperature. The surface tension eventually approaches a plateau, around 20 mJ/m^2 at 110°C for a pressure of 500 psi. The Differential Scanning Calorimetry results of melting HDPE show that the polymer starts to melt

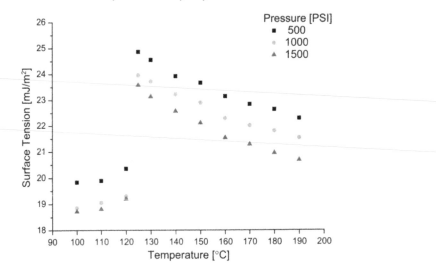

Figure 16. The equilibrium surface tension of HDPE in nitrogen at various temperatures and pressures (500, 1000, 1500 psi) through its crystallization region. The system was cooled from 150 to 100°C in intervals of 10°C, during which the system was maintained at each condition for two hours, and the surface tension value was measured at its steady-state.

at 110°C. This may explain why the surface tension of HDPE does not change any further when temperature goes under 110°C, since the polymer becomes completely solid at this point and below. Note, surface tension of a PS melt does not change any further when temperature reaches the glass transition temperature at 100°C. If comparing the surface tension results of these two polymers under their melting points, one may consider that the difference is due to the fact that polystyrene is an amorphous polymer, while HDPE is a crystalline polymer. Once the temperature goes below 100°C, PS solidifies and hence surface tension detected by ADSA would not change any further. This is similar to the case of HDPE under 110°C. However, different from PS, there is a decrease in surface tension between 110°C and 125°C observed for HDPE, which is the period for crystallization of HDPE. During the crystallization, HDPE assumes a state with micro-crystals immersed in the polymer melt.

Polymer crystallization can take time and occur with a range of temperatures, during which the polymer behaves viscoelastically with a high elasticity characteristic. When temperature is decreased to induce crystallization, small crystals form and grow. These crystals may act as, or be considered, nanoparticles, in the polymer melt. It is possible that nanoparticles in polymer melts decrease the surface tension. It is known that the presence of nanoparticles in polymer melts enhances the polymer interaction with foaming agents, which leads to an improved foaming quality [61, 62]. Thus, it may not be surprising that the surface tension decreases with decreasing temperature, when accompanied by the polymer crystallization.

A follow up question would be how temperature or the rate of temperature change, affects the surface tension, as well as polymer crystallization.

4. *Correlation of Surface Tension Decrease with Temperature Change Rate*

It is known that during crystallization, if the polymer is cooled down slowly, HDPE has enough time to crystallize, and thus more crystals can form and grow. This in turn should result in a greater decrease in surface tension. To the contrary, if the temperature change rate is high, HDPE does not have enough time to crystallize before becoming completely solid. Thus, less of a decrease in surface tension should be observed.

Two experiments with different cooling procedures when crystallization of HDPE occurred were performed to confirm this argument. For the slower cooling rate experiment, the temperature was decreased from 150°C to 100°C stepwise in 10°C intervals. The system was maintained at each interval until it reached steady-state and surface tension was measured. For the faster cooling rate experiment, the temperature was decreased from 150°C to 100°C steadily. The system was then maintained at 100°C until it reached steady-state. The results are shown in Fig. 17. The surface tension of HDPE in nitrogen at different temperature change rates was measured. The solid spots were the surface tension values obtained at the slower temperature change rate, while the open spots were obtained at the faster temperature change rate. It is seen that the faster a temperature change rate, the less of a decrease in surface tension. It is known that polymers with different degrees of crystallinity show different degrees of transparence [24]. We observed the HDPE sample that experienced a fast temperature change rate under 40 times microscope,

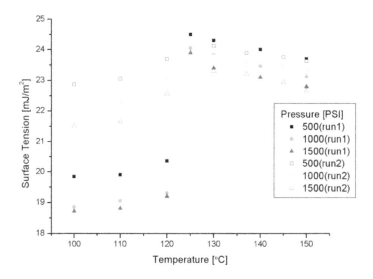

Figure 17. The surface tension of HDPE in Nitrogen at different temperature change rates: the solid symbols indicate experiments at slower crystallization cooling speeds. The open symbols indicate experiments at faster crystallization cooling speeds.

and found it was more transparent, indicating a lower extent of crystallinity. The sample that experienced a slow temperature change rate was more translucent, indicating a higher degree of crystallinity. These results support the above argument that the amount of decrease in surface tension is related to the rate of temperature change and the extent of polymer crystallization.

E. Conclusions

The surface tension of polymer melts in supercritical fluids under various temperatures and pressures were measured using the Axisymmetric Drop Shape Analysis-Profile (ADSA-P) method. The surface tension of the amorphous polymer, polystyrene, in supercritical CO_2 and the crystalline polymer, HDPE, in supercritical N_2 were discussed in this chapter. Surface tension dependence on temperature, pressure and polymer molecular weight was obtained experimentally and related with the solubility of gas in the polymers. At temperatures above the melting point, the trends of the surface tension change with temperature and pressure of PS are similar to those of HDPE, i.e., the surface tension decreases with increasing temperature and pressure. Self-consistent field theory (SCFT) calculations were applied to simulate the surface tension of a corresponding system and used to explain these trends. Surface tension dependence on polymer molecular weight was obtained based on the study of several different polystyrene. It has been found that the high molecular weight monodisperse polystyrene has a higher surface tension than the low molecular weight monodisperse one. The surface tension dependence on temperature, as well as on pressure, is stronger for the monodisperse polystyrene of higher molecular weight. For the polydisperse polystyrene, high surface tension values may be influenced more by high molecular weight portions of polystyrene molecules. For HDPE, when crystallization occurs, the surface tension decreases with decreasing temperature. During crystallization, polymer micro-crystals form and may act like nanoparticles in polymer melts, reducing the surface tension. It is found that the amount of decrease in surface tension is related with the temperature change rate, and hence the rate of crystallization; the surface tension decreases more with a slower temperature change rate, or a higher degree of crystallinity.

F. Acknowledgements

This work is supported by NSERC and CRC programs of Canada. The authors thank Research in Motion for the use of equipment and Dr. Bev Christian for advice and suggestions for many aspects of the research.

G. References

1. D. Myers, Surfaces, Interfaces, and Colloids: Principles and Applications, VCH Publishers, New York, 1991.
2. K. Nishioka and J. Kusaka, J. Chem. Phys., 96 (7) (1992) 5370–5376.

3. D.L. Tomasko, H. Li, D. Liu, X. Han, M.J. Wingert, L.J. Lee and K.W. Koelling, Ind. Eng. Chem. Res., 42 (2003) 6431–6456.
4. M. Lee, C.B. Park and C. Tzoganakis, Polym. Eng. Sci., 39 (1999) 99–109.
5. N.R. Demarquette and M.R. Kamal, Polym. Eng. Sci., 34 (24) (1994) 1823–1833.
6. R.-J. Roe, V.L. Bacchetta and P.M. Wong, J. Phys. Chem., 71 (1967) 4190.
7. S. Wu, J. Phys. Chem., 74 (1970) 632.
8. S. Wu, Polymer Interface and Adhesion, Marcel Dekker, New York, 1982.
9. N.R. Demarquette and M.R. Kamal, Polym. Eng. Sci., 34 (1994) 1823.
10. A.T. Morita, D.J. Carastan and N.R. Demarquette, Colloid Polym. Sci., 280 (2002) 857.
11. A. Xue, C. Tzoganakis and P. Chen, Polym. Eng. Sci., 44 (2004) 18.
12. S. Lahooti, O.I. Rio, P. Cheng and A.W. Neumann, in "Applied Surface Thermodynamics", A.W. Neumann, J.K. Spelt (Eds), Marcel Dekker, New York, 1996.
13. P. Cheng, D. Li, L. Boruvka, Y. Rotenberg and A.W. Neumann, Colloids Surfaces, 43 (1990) 151–167.
14. S.S. Susnar, H.A. Hamza and A.W. Neumann, Colloids Surfaces, 89 (1994) 169–180.
15. P. Cheng and A.W. Neumann, Colloids Surf., 62 (1992) 297–305.
16. H. Li, L.J. Lee and D.L. Tomasko, Ind. Eng. Chem. Res., 43 (2004) 409.
17. O.I. del Rio and A.W. Neumann, J. Colloid Interface Sci., 196 (1997) 136–147.
18. F. Bashforth and J.C. Adams, An Attempt to Test the Theory of Capillary Action, Cambridge University Press and Deighton Bell & Co., Cambridge, 1883.
19. A.W. Adamson and A.P. Gast, Physical Chemistry of Surfaces, John Wiley & Sons, Inc., New York, 1997.
20. J. Campbell, Journal of Physics D (Applied Physics), 3 (1970) 1499–1504.
21. A.W. Adamson and A.P. Gast, Physical Chemistry of Surfaces, Wiley, New York, 1997.
22. H. Park, C.B. Park C. Tzoganakis, K.H. Tan and P. Chen, Ind. Eng. Chem. Res., 45 (2006) 1650–1658.
23. H. Park, R.B. Thompson, N. Lanson, C. Tzoganakis, C.B. Park and P. Chen, J. Phys. Chem. B, 111 (2007) 3859–3868.
24. Zeus Industrial Products, Inc. Technical Whitepaper, 2007.
25. B. Wunderlich, J. Phys. Chem., 29 (1958) 6.
26. J.E.K. Schawe, Thermochimica Acta, 461 (2007) 145–152.
27. B. Wunderlich, Macromolecular Physics, Vol. 1, Academic Press, New York, 1973.
28. B. Wunderlich, Macromolecular Physics, Vol. 3, Academic Press, New York, 1980.
29. H.G. Zachmann, Fortschr. Hochpolym. Forsch., 3 (1964) 581.
30. T.F.J. Pijpers, V.B.F. Mathot, B. Goderis, R.L. Scherrenberger and E.W. vander Vegte, Macromolecules, 33 (2002) 3601.
31. H. Park, C.B. Park, C. Tzoganakis and P. Chen, Ind. Eng. Chem., 46 (2007) 3849–3851.
32. R.A.L. Jones and R.W. Richards, Polymers at Surfaces and Interfaces, Cambridge University Press, New York, 1999.
33. R.B. Thompson, J.R. MacDonald and P. Chen, Physical Review E 78, 030801(R) 2008.
34. W.W.Y. Lau and C.M. Burns, J. Colloid Interface Sci., 45 (1973) 295–302.
35. A.H. Alexopoulos, J.E. Puig and E.J. Franess, J. Colloid Interface Sci., 128 (1989) 26–34.
36. B. Song and J. Spinger, J. Colloid Interface Sci., 184 (1996) 77–91.
37. S.H. Anastasiadis, J.K. Chen, J.T. Koberstein, J.E. Sohn and J.A. Emerson, Polym. Eng. Sci., 26 (1986) 1410–1418.
38. S.L. Shenoy, T. Fujiwara and K. Wynne, J. Macromol., 36 (2003) 3380.
39. M. Hess, Macromolecules, 214 (2004) 361–379.

40. A. Quach and R. Simha, J. Applied Physics, 42 (1971) 4592–4606.

41. P. Zoller, P. Bolli, V. Pahud and H. Ackermann, Rev. Sci. Instrum., 47 (1976) 948–952.

42. I.C. Sanchez and R.H. Lacombe, J. Phys. Chem., 80 (1976) 2352–2362.

43. I.C. Sanchez and R.H. Lacombe, Macromolecules, 11 (1978) 1145–1156.

44. G. Li, H. Li, J. Wang and C.B. Park, Cell Polym., 25 (2006) 237.

45. G. Li, J. Wang, C.B. Park, P. Moulinie and R. Simha, Annu. Tech. Conf. Soc. Plast. Eng., 62 (2004) 2566.

46. R. Simha and T. Somcynsky, Macromolecules, 2 (1969) 342.

47. S. Wu, J. Colloid and Interface Sci., 31 (1969) 2.

48. M.W. Matsen, in "Soft Matter", Vol. 1, G. Gompper, M. Schick (Eds), Wiley-VCH, Weinheim 2005, pp. 87–178.

49. M.W. Matsen, J. Phys. Condens. Matter, 14 (2002) R21–R47.

50. G.H. Fredrickson, V. Ganesan and F. Drolet, Macromolecules, 35 (2002) 16–39.

51. F. Schmid, J. Phys. Condens. Matter, 10 (1998) 8105–8138.

52. K.M. Hong and J. Noolandi, Macromolecules, 14 (1981) 1229–1234.

53. K. Binder, M. Müller, P. Virnau and L.G. MacDowell, Adv. Polym. Sci., 173 (2005) 1–110.

54. S.-M. Mai, J.P.A. Fairclough, N.J. Terrill, S.C. Turner, I.W. Hamley, M.W. Matsen, A.J. Ryan and C. Booth, Macromolecules, 31 (1998) 8110–8116.

55. S.-M. Mai, W. Mingvanish, S.C. Turner, C. Chaibundit, F. Heatley, M.W. Matsen, A.J. Ryan and C. Booth, Macromolecules, 33 (2000) 5124–5130.

56. J.R. Royer, Y.J. Gay, J.M. Desimoneand and S.A. Khan, J. Polymer Sci.: Part B: Polymer Physics, 38 (2000) 3168–31380.

57. L.A. Utracki, J. Polymer Sci.: Part B: Polymer Physics, 45 (2007) 270–285.

58. H. Park, "Surface Tension Measurement of Polystyrenes in Supercritical Fluids", PhD thesis, University of Waterloo, 2007.

59. Y. Sato, M. Yurugi, K. Fujiwara, S. Takishima and H. Masuoka, Fluid Phase Equilibria, 125 (1996) 129–138.

60. Y. Sato, K. Fujiwara, T. Takikawa, Sumarno, S. Takishima and H. Masuoka, Fluid Phase Equilibria, 162 (1999) 261–276.

61. Y.H. Lee, C.B. Park and K.H. Wang, Journal of Cellular Plastics, 41 (2005).

62. Y.H. Lee, C.B. Park, M. Sain, M. Kontopoulou and W. Zheng, J. Appl. Polym. Sci., 105 (2007) 1993–1999.

Accumulation of Surfactant in the Top Foam Layer Caused by Ruptured Foam Films

S.S. Dukhin [a], **V.I. Kovalchuk** [b], **E.V. Aksenenko** [c] **and R. Miller** [d]

[a] New Jersey Institute of Technology, Newark, USA
[b] Institute of Bio-Colloid Chemistry, Vernadsky str. 42, 03142 Kiev, Ukraine
[c] Institute of Colloid Chemistry and Chemistry of Water, Vernadsky str. 42, 03142 Kiev, Ukraine
[d] Max Planck Institut für Kolloid- und Grenzflächenforschung, D-14424 Potsdam, Germany

Contents

A. Introduction

Foam science has been developed very systematically. Attention was paid to the structure, to the forces, to surface rheology, and to numerous sub-processes [1–5], however, stability remains the main target. The progress achieved in these analytical studies are the prerequisites for the elaboration of a model which would describe the stability of a foam as a whole.

About 35 years ago first attempts to develop a general theory of foams were made by Krotov and Kann [6–8]. Unfortunately, the set of conservation equations derived in these studies was solved only for some special and most simple cases.

Bubble and Drop Interfaces
© Koninklijke Brill NV, Leiden, 2011

The obtained results were later used in the theory developed by Ruckenstein and co-workers [9, 10]. Ruckenstein's model for foam stability is one of the most elaborated. Recently, also Neethling and co-workers have published some studies [11, 12] aiming at the modelling of the stability of a foam as a whole.

It can be stated that these models comprise the main processes which govern foam stability, however, under some conditions certain additional sub-processes can play a role. In particular, for some systems the existing foam stability model should be supplemented by the conservation equation for surfactant transport, as the surfactant concentration within a foam can be non-uniform, in contrast to the usual approach considering a uniform surfactant distribution within the foam (see Fig. 1a). In addition there can be an evolution of surfactant distribution in the foam with time [13].

This chapter is an attempt to reveal the main regularities governing the surfactant distribution within a steady pneumatic (according to Ruckenstein [9]) or dynamic foam (term introduced by Exerowa and Kruglyakov in [2] to denote this hydrodynamic mode of foam). The generalisation of the results obtained in this way towards a non-steady hydrodynamic modes is rather straightforward.

It turned out that the current foam model which assumes both time and spatial invariance of the surfactant concentration does not require any essential corrections if the initial surfactant concentration c_0 is sufficiently high, i.e. essentially when it exceeds the CMC. On the contrary, incorporation of the surfactant transport into the model of foam stability is mandatory if c_0 is rather low.

When a bubble is formed near the foam bottom, surfactant molecules adsorb and cause a decrease of surfactant concentration in the adjacent solution to a value $c_1 < c_0$ (cf. Fig. 1). The rising bubbles cause a convective transport of adsorbed surfactant to the foam top. This continuous surfactant transport to the foam top results in a surfactant accumulation within the top layer of bubbles, and leads to an increase in the local concentration c. The mechanism causing this increase is described below. The surfactant consumption by bubbles near the foam bottom and its release accompanied by bubble ruptures on the foam top causes a monotonous

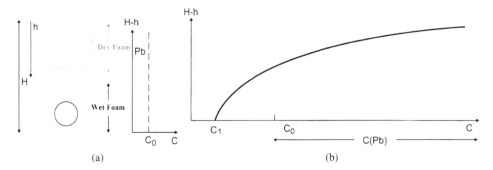

Figure 1. Models for surfactant distribution in a foam; according to [13].

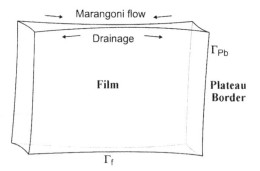

Figure 2. Marangoni flow caused by the surfactant accumulation retards film thinning by drainage; for explanation see text; according to [13].

increase of the surfactant concentration with decreasing distance h from the foam top (Fig. 1b).

As soon as the surfactant concentration within a dry foam increases in comparison with c_0, the bulk concentration and the surface concentration Γ_{Pb} within the Plateau borders (Pb) are higher than in the film Γ_f (Fig. 2)

$$\Gamma_{Pb} > \Gamma_f, \qquad \Gamma_{Pb} - \Gamma_f \ll \Gamma_f. \tag{1}$$

Therefore, the surface tension at the boundary between Pb and film becomes lower than the value in the film centre. This gives rise to a Marangoni flow from the Pb into the film, thus retarding drainage.

The conditions for coalescence in the top layer of bubbles are drastically different from those inside a dry foam [14, 15]. It is generally accepted that bubbles rupture mainly within the top layer. It means that the suppression of bubble rupture in this top layer is very important for the stabilisation of a foam as a whole and determines the importance of the effect of surfactant accumulation. As the surfactant concentration within the top layer becomes higher, this promotes the stabilisation of bubbles which in turn leads to the stabilisation of the foam as a whole.

There is a large qualitative difference between the phenomenon of surfactant accumulation, as considered here, and the well-known process of surfactant concentrating during foam fractionation [16–20].

At sufficiently high ratios of Γ/c the surfactant content in the film during drainage decreases only slightly, while the water content can become many times lower. As a result, the foam films become enriched in surfactant during foam drying (during film and Pb drainage). It is sufficient to separate dry foam from the liquid which drained out of the foam (foamate), to achieve large surfactant accumulation within a dry foam (Fig. 3a). The particular feature of this accumulation process is that only the films become enriched in surfactant, while the concentration in the Pbs remains low.

On the contrary, both the films and the liquid in Pbs accumulate surfactant within the top bubble layer of a foam. The concentration increases here are due to film rupturing, while the traditional surfactant accumulation process does not lead to such

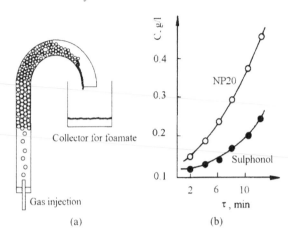

Figure 3. Modes of surfactant accumulation: (a) surfactant accumulation due to film drying and dry foam separation from foamate [21]; (b) surfactant accumulation due to bubble bursting inside dry foam, according to [2].

an increase [16–20]. Simultaneously with the increase of surfactant concentration due to film rupturing, the concentration in the film bulk and Pbs is also growing.

The reason for a surface concentration increase is the shrinkage of the surface area due to ruptured films. When the surface area shrinks, the surfactant amount per unit area Γ is increased, if there are small or no surfactant losses. Khaskova and Kruglyakov [20] revealed the mechanism of surfactant accumulation during film rupturing due to surface area shrinkage (Fig. 3b). However, they considered bubble coalescence in the depth of the foam, in contrast to the assumption here that coalescence within the top layer of bubbles is the most important phenomenon.

Arbuzov and Grebenshchikov [22] noted that film rupturing within the top foam layer should lead to a local surfactant accumulation. However, they did not distinguish accumulation near the foam top from accumulation within the foam as a whole. Thus, a dynamic regime was assumed when the downward surfactant flow prevents its accumulation within the foam top, as it will be considered here.

To the best of our knowledge, first remarks regarding surfactant accumulation near the foam top were made in [23] and [2], however, no clear explanations of the mechanism governing this phenomenon were given. Ruckenstein and co-workers [10] were the first to formulate the problem of surfactant transport in a steady pneumatic foam by emphasizing the importance of surfactant accumulation in the foam top. An equation of surfactant conservation was derived, and an approach for its solution elaborated. Nevertheless, even semi-quantitative results regarding the surfactant accumulation were not derived there. The reason for this deficit is the problem arising in the specification of kinetic parameters. This point will be discussed in more detail below. In Table 1 references are summarised giving remarks regarding surfactant accumulation.

As we can see, the interest in surfactant accumulation within the foam top layer caused by film rupturing significantly increased during the last decade. However,

Table 1.

Selection of references on surfactant accumulation in the foam top layer

Authors	Year	Information
Khristov, Malysa, Exerowa [23]	1984	Application of surfactant accumulation concept to
Exerowa, Kruglyakov [2]	1998	the interpretation of experiments
Bhakta, Ruckenstein [10]	1997	Conservation equation for surfactant transport
Darton, Sun [24]	1999	Short remarks
Krotov, Nekrasov, Rusanov [25]	2002	Short remarks
Neethling, Lee, Grassia [11]	2005	Short remarks

the speculations about the existing phenomena were not supported theoretically, even not by rough estimations.

B. Qualitative Mechanism of Surfactant Accumulation in the Top Bubble Layer of a Foam Caused by External Film Rupture

A schematic picture of the three stages in the rupture of a film in the top foam layer is shown in Fig. 4. Two main mechanisms of thin film rupture are known, one governed by the magnification of capillary waves [26] and the other by the nucleation of vacancies followed by expansion of the hole [27]. No matter what the mechanism is, the formation of droplets after film bursting is possible, and the film surface shrinks (cf. Fig. 4). This film shrinkage promotes surfactant accumulation, while the drop formation can cause a loss of surfactant.

The small droplets formed during film rupture can get involved in an upward stream of air, and in this way part of the surfactant is removed and cannot further contribute to foam stabilisation. However, the droplets can also sediment onto films located below the ruptured film, and donate their surfactant. In fact, the droplet size determines which scenario does occur. With increasing droplet size, the sedimentation velocity increases and if it exceeds the upward air flow velocity, then the droplets will attach to the underlying films and eventually merge with them. Droplet formation and its influence on the surfactant balance is an independent question which is considered below. For the moment we will assume that the surfactant flux caused by droplets formation is negligible.

The ruptured film can be referred to as donor film, because it donates surfactant to the adjacent Pbs and films, while the films to which surfactant is transferred in this way can be called acceptors. The surfactant amount which comes from ruptured films is added to the initial surfactant amount of the acceptor film. Consequently, the surfactant surface concentration in the acceptor increases almost twice (considering adsorption layers far from saturation) because the films are quite similar.

After a donor film bursts, the acceptor becomes the external film. Accordingly, the probability of its subsequent rupture increases and it becomes a donor for the underlying films (acceptors). Any film rupture leads to an increase of the surfactant

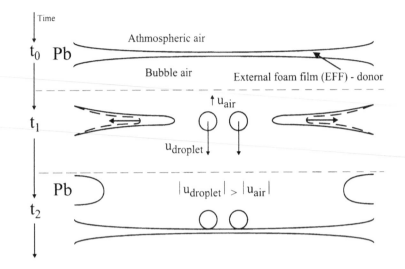

Figure 4. Film surface shrinkage caused by the rupture of a film on top, resulting in an increased surface concentration; formation of droplets during film rupture leads to a loss of surfactant; according to [13].

surface concentration within acceptor films, which becomes much higher than the initial concentration $\Gamma_0 = \Gamma(c_0)$. After n successive ruptures, the surfactant surface concentration in the top layer can increase up to n times. This estimate is very rough because the surfactant balance is much more complicated, and the surfactant losses, if they are large, can suppress the accumulation process.

Surfactant loss arises due to transport of accumulated surfactant through the Pbs. The accumulated surfactant can get involved in a downwards directed water transport through vertical and inclined Pbs. This downwards flow exists even if the water as a whole is immobile, and the water distribution in the foam is steady (Fig. 5). Ruckenstein [10] calls this phenomenon 'steady pneumatic foam', while Exerowa and Kruglyakov [2] use the term 'dynamic foam'. Note, while such foam is steady with respect to the water distribution, it can be still unsteady with respect to the surfactant distribution so that we should consider it quasi-steady.

The water in the Pbs close to their walls moves upwards, because these walls belong to rising bubbles. While water rises near the Pb walls, it consequently moves downwards into the central section of Pb, because the water velocity averaged over the foam cross-section is zero (Fig. 6). The cases of a strong and weak influence of surfactant flow downwards, respectively, on its accumulation within the top layer of bubbles, are illustrated in Fig. 7a and b. The higher the surfactant concentration in the downwards flow is, the higher is the density of the downwards surfactant flux Q_{Pb}, which can be especially large at concentrations close to the CMC. The upwards surfactant flow Q_f is caused mainly by films rising as parts of rising bubbles, i.e., Q_f is proportional to the surface concentration Γ.

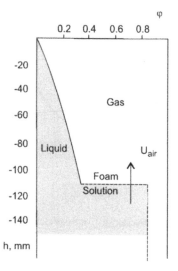

Figure 5. Water distribution in a steady pneumatic foam, according to [15].

If the surfactant amount in the Pbs is comparable with that in films at high surfactant bulk concentrations close to the CMC, then it may occur that:

$$|Q_{Pb}| \approx |Q_f|. \qquad (2)$$

When this condition is valid, the downwards flow prevents a surfactant accumulation in the foam top layer. Accordingly, the proposed model of foam with invariable surfactant concentration in the foam (cf. Fig. 1a) becomes valid. Below the CMC and at sufficiently large values of Γ/c, an opposite condition applies:

$$|Q_{Pb}| \ll |Q_f|. \qquad (3)$$

This condition is sufficient for a surfactant accumulation in the top bubble layer of a foam, because the downwards surfactant flow is too weak to hinder this accumulation. When the surfactant starts to accumulate in the top layer of bubbles, the concentration in the respective Pbs increases monotonously with time. This increase causes an increased surfactant concentration at bubble surfaces located in a lower layer due to the downwards flow in the Pbs (Fig. 7b).

With increasing concentration c the value of Γ/c gradually decreases, due to a further saturation of the adsorption layer, leading to the violation of condition (3) after a certain period of time. When the surfactant fluxes downwards and upwards are balanced, any further surfactant accumulation slows down and the foam comes into a state close to a true steady state. A qualitative characterisation of the processes which cause the surfactant transport from a downwards surfactant flow within the Pbs into the foam films is given schematically in Fig. 8a, b and c.

To illustrate the downwards surfactant flow and surfactant convective diffusion through an upwards stream to the Pb walls, we introduce the surfactant concentration $\bar{c}_d(h)$ averaged over the cross-section of the downwards water flow and

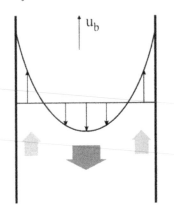

Figure 6. Flow profile of water in a vertical/inclined Pb; according to [13].

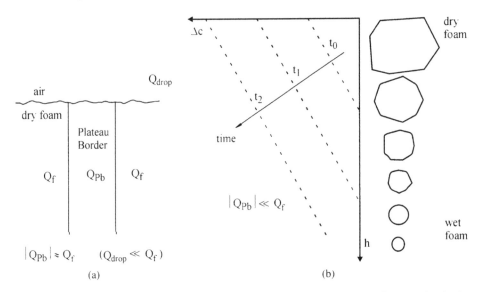

Figure 7. Schematic of the mechanism SAFT-2: (a) surfactant flux upwards and downwards; (b) time evolution of the surfactant distribution in the foam; according to [13].

surfactant concentration $\bar{c}_u(h)$ averaged over the cross-section of liquid upwards flow in inclined Pb (Fig. 8a). Both concentrations increase when approaching the foam top, i.e., with decreasing h. As the downwards flow originates from the foam top with a surfactant concentration $c > c_0$, and the upwards flow from the foam bottom with a surfactant concentration $c < c_0$, we have $\bar{c}_d(h) > \bar{c}_u(h)$ at any h as schematically shown in Fig. 8a.

To illustrate the transport of accumulated surfactant into films at any level h, cylindrical coordinates are introduced, with the axis in the Pb axis, and the radial coordinate r gives the distance to the Pb axis. $r = R$ corresponds to the 'boundary' between Pb and an adjacent film while $r > R$ corresponds to the distance between the point located inside a film and the Pb axis. It is seen from Fig. 8b that the sur-

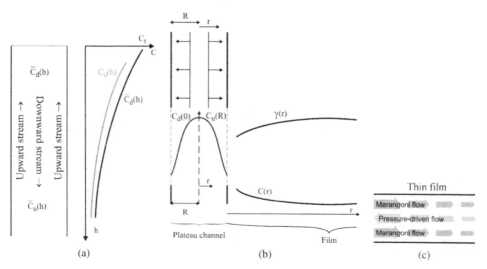

Figure 8. Schematic of the onset of SAFT-2: (a) transport of accumulated surfactant downwards and upwards in the Pb; (b) surfactant distribution in the Pb and in the film; (c) Marangoni flow due to the radial gradient of surfactant generated by SAFT-2; according to [13].

factant concentration is maximum at the axis, $c_d(0)$, and minimum at the Pb wall, $c_u(R)$. However, $c_u(R)$ is higher than that in absence of surfactant accumulation in the foam top, i.e., it is higher than the concentration in the film centre. This is illustrated in Fig. 8b: for $r > R$ the value $c(r)$ decreases with increasing r. Accordingly, the surface tension $\gamma(r)$ near the Pb/film boundary is lower than in the film centre, as also shown in Fig. 8b.

The radial surface tension gradient within the film gives rise to a Marangoni flow which is directed towards higher surface tension, i.e., into the film as shown in Fig. 8c. At the same time, the pressure-driven film drainage has the opposite direction and causes film thinning, i.e., its destabilisation. The Marangoni flow originating from surfactant accumulation in the foam top retards film drainage, and therefore stabilises the film. It is well known that pressure driven drainage by itself causes a radial surfactant gradient and the onset of a Marangoni flow which in turn retards foam drainage.

A special analysis is necessary to determine whether a superposition approximation can be applied to take these two kinds of Marangoni flow and their influence on film drainage into account. For the sake of convenience, the abbreviation SAFT (Surfactant Accumulation in Foam Top) is used to refer to the phenomenon studied here. Presumably, two SAFT modes exist:

SAFT-1: Accumulated surfactant remains mainly within the top layer of bubbles and a rather weak downwards directed flow of accumulated surfactant through the inclined Pbs is required here.

SAFT-2: A not negligible part of surfactant accumulated in the top layer of bubbles penetrates deeply into the foam.

Actually, Figs 7 and 8 illustrate the mode SAFT-2. Also the model proposed by Ruckenstein refers to the case of SAFT-2. In the attempts to analyse SAFT-2 we encounter an essential difficulty, and so far there is no solution to overcome it. While it is rather straightforward to quantify the surfactant transport from the downwards flow through the upwards flow to the Pb walls, it remains unclear what occurs with downwards and upwards flows in the many joints of Pbs. Possibly, the same difficulty prevented Ruckenstein from giving a quantitative analysis of surfactant transport, although he elaborated the approach to the solution of the derived equation for surfactant transport. In the following paragraph, we will focus on the mode SAFT-1.

C. Estimation of Surfactant Loss

Let us assume here a quasi-stationary foam [2, 10] where during the time of rupture of one bubble layer at the top of the foam it simultaneously rises due to the creation of one new bubble layer at the bottom. In such dynamic regime the foam height remains constant, and the liquid in the foam is on average motionless because the downwards flow caused by drainage (in the central part of Pb cross-section) is compensated by the upwards flow of liquid rising with bubbles (close to the Pb walls). Thus, during the time of rupture of the top layer of bubbles the liquid containing an excess of surfactant can penetrate into the foam to the depth approximately equal to the height of one bubble layer. This circumstance can be used to estimate the extent of surfactant "leakage" from external films due to drainage.

Let us consider a top layer of bubbles named Layer 1, and the Layer 2 located immediately below at time t_0 just before the rupture (Fig. 9). These two layers contain almost equal amounts of surfactant M_S^0 and water M_W^0, respectively. When the film in the top layer bursts, Layer 1 donates its amount of water and surfactant to Layer 2. In the time moment t_1, i.e. when Layer 1 disappeared completely, all amount of water and surfactant is donated to Layer 2 which now becomes the new top layer (Fig. 9). When a steady pneumatic foam is considered, the Layer 2 in its new position at the top of foam has to possess the same amount of water M_W^0 as the earlier Layer 1 before it began to burst. On the other hand, Layer 2 initially possessed the same amount of water M_W^0. Thus, after it acquired the same amount of water M_W^0 from the formed Layer 1 it seems to comprise the doubled amount, $2M_W^0$. However, it cannot preserve this amount of water, because it is determined by the requirement of steady water distribution, namely that in its new position at the top, i.e. the amount M_W^0. One has to conclude therefore that, simultaneously with the acceptance of water from Layer 1 it should donate some amount of water to Layer 3. This causes the water loss M_W^0 in Layer 2 in its new position as the top layer.

The maximum loss of accumulated surfactant accompanying the water loss M_W^0 is $(c_0 + \Delta c) \cdot M_W^0$ where Δc is the increase of surfactant concentration within Layer 2. This corresponds to the assumption that the equilibrium between films

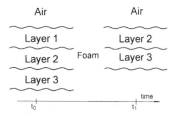

Figure 9. Scheme of the displacement of a bubble layer caused by the collapse of the top foam Layer 1; explanations are given in the text; according to [13].

and Pbs within Layer 2 is established instantly. However, this equilibration takes some time. Accordingly, the concentration in flow through Pbs which determines the surfactant loss is smaller than the maximum possible loss:

$$M_S^{loss} = (c_0 + \alpha \Delta c) \cdot M_W^0, \tag{4}$$

where

$$0 < \alpha < 1. \tag{5}$$

It is now straightforward to describe the surfactant balance in the process of its transport from the donor (bubbles Layer 1) to the acceptor (bubbles Layer 2). The initial amount of surfactant in Layers 1 and 2, namely $2M_S^0$ should be equal to the accumulated amount M_S^{ac} in Layer 2 in its new (top layer) position plus the surfactant losses M_S^{loss}:

$$M_S^{ac} + M_S^{loss} = 2M_S^0. \tag{6}$$

The number of films and Pbs should be accounted for in the expression for total surfactant amount M_S^0. The surfactant amount in the film bulk can be neglected as compared to that at the film surface assuming that

$$\Gamma_0/c_0 \gg h_{cr}. \tag{7}$$

This condition does not impose any limitations because $h_{cr} \sim 10$ nm while the ratio of surface to bulk concentrations generally exceeds 10^{-6} cm. The surfactant amount in Pb surfaces can be disregarded in comparison with that in its bulk. With these simplifications, we obtain

$$M_S^0 = \pi R_b^2 c_0 \left(\frac{2s\Gamma_0}{c_0} + e \frac{R_{Pb}^2}{R_b} \right), \tag{8}$$

where R_b and R_{Pb} are the radii of faces and Pb, respectively, s is the number of films and e is the number of Plateau borders. In the new top position of Layer 2 both the bulk and surface concentrations change; this fact should be accounted for when an expression for M_S^{ac} is derived. In particular, $c_0 + \Delta c$ should be used instead of c_0, and Γ should be replaced by $\Gamma_0 + (d\Gamma/dc) \cdot \Delta c$. This leads to the conservation equation for the surfactant amount (6):

$$2\left(2s\Gamma_0 + ec_0\frac{R_{Pb}^2}{R_b}\right) = 2s\left(\Gamma_0 + \frac{d\Gamma}{dc}\Delta c\right) + e(c_0 + \Delta c)\frac{R_{Pb}^2}{R_b}$$

$$+ e(c_0 + \alpha\Delta c)\frac{R_{Pb}^2}{R_b} \tag{9}$$

and consequently

$$\frac{\Delta c}{c_0} = 2s\frac{\Gamma_0}{c_0}\Big/\left(2s\frac{d\Gamma}{dc} + e\frac{R_{Pb}^2}{R_b}(1+\alpha)\right). \tag{10}$$

For pentagonal dodecahedron $s = 6$, $e = 10$ [2, 28].

Let us consider here three special cases:

$$(1)\quad \frac{\Gamma_0}{c_0} \approx \frac{d\Gamma}{dc} \gg \frac{R_{Pb}^2}{R_b}, \tag{11}$$

then $\frac{\Delta c}{c_0} \approx 1$;

$$(2)\quad \frac{\Gamma_0}{c_0} \gg \frac{R_{Pb}^2}{R_b} \gg \frac{d\Gamma}{dc}, \tag{12}$$

then $\frac{\Delta c}{c_0} = \frac{2s}{e}\frac{\Gamma_0}{c_0}\frac{R_b}{R_{Pb}^2(1+\alpha)}$ is large, but $\Delta\Gamma = \frac{d\Gamma}{dc}\Delta c$ is small;

$$(3)\quad \frac{\Gamma_0}{c_0} \approx \frac{d\Gamma}{dc} \ll \frac{R_{Pb}^2}{R_b}, \tag{13}$$

then $\frac{\Delta c}{c_0} = \frac{2s}{e}\frac{\Gamma_0}{c_0}\frac{R_b}{R_{Pb}^2(1+\alpha)} \ll 1$.

The derived equations lead to important conclusions. In the first case, given by Eq. (11), the surfactant loss small, and therefore the accumulation in the top bubble layer is large. This means that the accumulation mode SAFT-1 is effective. If the surface activity is high, the accumulation attains its maximum value. This is just expected, because the loss is caused by surfactant located within the Pb bulk, and it should be negligibly small when the major part of surfactant is adsorbed. If the surface activity is not high and surfactant is located mainly in the Pbs' bulk, as it is assumed in Eq. (13), the accumulation becomes low, and further decreases with increasing loss (increasing α). As the loss is not negligible, the mode SAFT-2 becomes relevant.

In the second case represented by Eq. (12) the surface activity is high but the adsorption layer is in a state close to saturation, so that the surfactant amount coming from ruptured films increases mainly the bulk solution concentration. Here the surfactant loss can be also large, nevertheless at least half of the initial amount of surfactant in Layer 1 is accumulated within Layer 2. At the same time, in spite of a large increase in bulk concentration, the increase of surface concentration appears relatively small because the surface is almost completely covered. The resulting accumulation mode refers to a mixed mode, because accumulation happens in the top bubble layer and below.

The assumption of simultaneous bursting of all bubbles within the top layer and simultaneous subsequent formation of a new top layer of bubbles is an overestimation. Bubble bursting appears one by one at different places in the top layer. The formation of a new top layer occurs gradually by a replacement of burst bubbles through a new one originally located below. Actually, in any time moment the top layer of bubbles consists of those belonging to the first layer but not yet burst, and of those coming from the second layer.

It seems obvious that the surfactant loss cannot be very high when bubbles burst only one by one, because the physical mechanism which restricts the magnitude of loss is the same as for the bursting of single bubbles. However, the expression for $\Delta c/c_0$ can be different from that derived above, nevertheless predicting approximately the same order of magnitude for the accumulation. Accordingly we generalise the theory now to a more realistic process of bubble bursting one-by-one which can in future lead to the development of a method for the quantification of this phenomenon.

D. Capillary Hydrodynamic Phenomena Following the Rupture of a Single External Foam Film

The rupture of a single external film in the foam results in a sudden local loss of mechanical equilibrium which initiates a sequence of capillary-hydrodynamic processes leading to the establishment of a new mechanical equilibrium. The conditions for a mechanical equilibrium in a capillary system, and in particular in foams, were extensively studied and explained in [7, 29]. To establish a new mechanical equilibrium, a redistribution of liquid between the films and neighbouring Pbs, and of surfactant at the interfaces and in the liquid bulk should happen. Both processes are closely related to each other which makes the analysis of the system behaviour extremely complex.

For the analysis of the top foam layer we are mainly interested in external film rupture. A polyhedral bubble in the top layer of a foam consists of one external film (EF) and several internal films (IF) (see Fig. 10).

After EF ruptures, its surfactant and liquid partially accumulate within the neighbouring EFs and IFs and partially within the Pbs. Let us consider now the sequence of capillary hydrodynamic phenomena following the rupture of a single external foam film.

(i) *Hole formation within an external foam film and its expansion.*

When a film ruptures, a moving toroidal rim is formed around the hole, which accumulates the respective surfactant and water (Fig. 11) [30–33]. The rim arrives to several external Pbs and transfers surfactant and water of the ruptured film to these borders.

(ii) *The neighbouring external Plateau borders transform into "needles".*

When an external film ruptures, the neighbouring external Pbs appear to be supported by only two remaining films (one external and one internal) instead of three.

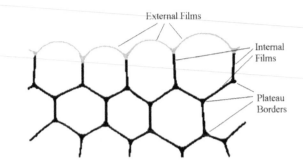

Figure 10. External and internal films in a foam (presented as a 2D foam); according to [13].

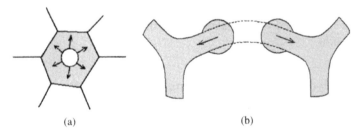

(a)   (b)

Figure 11. Hole formation (a) and expansion (b) in an external foam film; according to [13].

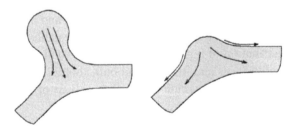

Figure 12. Transformation of external Plateau borders into "needles".

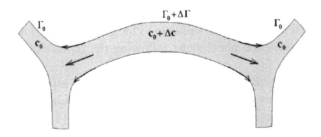

Figure 13. Formation of new external films.

Therefore they transform into "needles" which accept the surfactant and water of the ruptured film (Fig. 12).

(iii) *Formation of new external films.*

Due to the high capillary pressure inside a "needle" and large surface tension gradients, the water and surfactant are transported from the "needles" into the adjacent external and internal films, which merge and form a new external film. A very rapid plug flow establishes a uniform water and surfactant distribution along the newly formed film (Fig. 13).

Films deform to adapt uniform curvature in order to compensate the pressure difference between bubbles and the surrounding air. A fast redistribution of mass results in oscillations of the foam structure near a bubble rupture, which are damped during a rather short time due to dissipative processes in the foam. After establishment of the mechanical equilibrium, thicker external films are formed with a thickness $h_2 > h_{cr}$, where h_{cr} is the critical thickness of rupture.

The surfactant release due to external film rupture leads to increasing surface and bulk concentrations in the adjacent external and internal films. However the rupture of one external film causes only a small increase in surface concentration Γ because of: (i) uniform surfactant distribution within several neighbouring external and internal films, which form the new external films, and (ii) partial surfactant desorption from the interfaces due to the saturation of the surface layer. If the surfactant accumulation in the newly formed films is low, the increase in film thickness may be more important for stabilization.

E. Surfactant Accumulation Due to Multiple Ruptures of External Films

Multiple ruptures in the vicinity of a certain internal film of a top polyhedron increases the surfactant accumulation in comparison with that obtained for a single external film rupture. The illustration of a sequence of film ruptures is given in Fig. 14 for a 2D foam. The rupture of EF2 and further merging of the adjacent internal and external films, namely IF6 with EF4 and IF7 with EF5, leads to the formation of the new external films EF4a and EF5a (Fig. 14b). Their further ruptures unite IF8, IF9 and IF10 into a new EF8a (Fig. 14c). This film receives surfactant from films EF2, EF4a and EF5a.

After all bubbles in the top layer ruptured the newly formed EF8a takes the position of EF2. It accumulates surfactant and liquid from EF2, IF6 and IF7 and half of that from EF4 and EF5 (including their Pbs). Assumed there is no leakage of surfactant due to drainage from external films through internal (vertical and inclined) Pbs, the amount of surfactant in EF8a would be three times larger than that initially in EF2. However, the actual increase in surfactant concentration in the top films is significantly lower, because essential amounts of surfactant are driven into the foam bulk through drainage (Fig. 15).

Right after bubble rupture, part of the surfactant of newly formed external films is transferred to the adjacent external Pbs (Fig. 13). The mechanical equilibrium

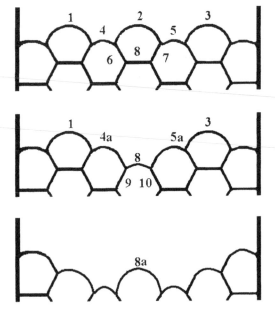

Figure 14. Subsequence of stages of film ruptures: details are given in the text; according to [13].

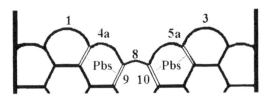

Figure 15. Outflow of surfactant by drainage of liquid from external films (EF4a and EF5a) through internal Plateau borders (Pbs) after the rupture of a film; according to [13].

between the film and the adjacent Pbs is established very rapidly, on timescales of milliseconds. For the present situation, mechanical equilibrium means uniform surfactant distribution at common surfaces of films with the adjacent Pbs. The adsorption equilibrium between the surfaces and bulk of Pbs is established more slowly, within timescales of seconds, which is still much faster than foam drainage. Thus, the excess of surfactant generated by the rupture of bubbles, reaches the external Pbs very fast.

From the external Pbs the surfactant is transferred to adjacent internal Pbs (cf. Fig. 15), which is caused by drainage of the excess liquid from the external Pbs. Because the surfactant concentration within Pbs may increase from c_0 to $c_0 + \Delta c_0$, the amount lost in the internal Pbs may be written as:

$$M_S^{loss} = (c_0 + \Delta c) \cdot \Delta V_W, \qquad (14)$$

where ΔV_W is the amount of excess liquid from EF4a, flowing into the internal Pbs (see the Appendix).

After the rupture of EF2 (before drainage starts) the amount of surfactant and liquid in EF4a with its Pbs is approximately the same as in the films EF4, IF6 plus half of EF2 with their Pbs before rupture. From the respective balance equations one obtains the local increase of surfactant concentration as result of film rupture (see the Appendix):

$$\frac{\Delta c}{c_0} \approx \frac{3\Gamma_0/c_0}{4\frac{d\Gamma}{dc} + 3.5\frac{R_{Pb}^2}{R_b} + 3.5h_{cr}}. \tag{15}$$

Here h_{cr} is the critical film thickness, R_{Pb} is the radius of curvature of the Pb, and R_b is the Pb length (which is of the order of the bubble radius). As discussed above, in most cases the term with h_{cr} is negligible because it is usually small as compared to one of the other two terms. Again three particular cases can be considered:

$$(1) \quad \frac{\Gamma_0}{c_0} \approx \frac{d\Gamma}{dc} \gg \frac{R_{Pb}^2}{R_b}, h_{cr}, \tag{16}$$

then $\frac{\Delta c}{c_0} \approx \frac{3}{4}\frac{\Gamma_0}{c_0}(\frac{d\Gamma}{dc})^{-1} \sim 1$;

$$(2) \quad \frac{\Gamma_0}{c_0} \gg \frac{R_{Pb}^2}{R_b}, \qquad h_{cr} \gg \frac{d\Gamma}{dc}, \tag{17}$$

then $\frac{\Delta c}{c_0} \approx \frac{3\Gamma_0/c_0}{3.5R_{Pb}^2/R_b+3.5h_{cr}}$ is large, but $\Delta\Gamma = \frac{d\Gamma}{dc}\Delta c$ is small;

$$(3) \quad \frac{\Gamma_0}{c_0} \approx \frac{d\Gamma}{dc} \ll \frac{R_{Pb}^2}{R_b}, h_{cr}, \tag{18}$$

then $\frac{\Delta c}{c_0} \approx \frac{3\Gamma_0/c_0}{3.5R_{Pb}^2/R_b+3.5h_{cr}} \ll 1$.

These results are very similar to those considered above, i.e. the local increase in surfactant concentration after film rupture depends on the particular conditions. In the first two cases the surfactant loss from external films due to drainage does not prevent surfactant accumulation within the top foam layer. However in case 2 this does not lead to a significant increase in surface concentration.

An estimation of the surfactant accumulation in 3D foams is more complicated and depends on the particular foam structure. It can be expected, however, that qualitatively the results should be very similar to those provided for 2D foams. The particular foam structure would mainly influence the absolute values but the effect of foam characteristics might remain nearly the same.

The surfactant concentration in the top layer may increase up to the CMC due to multiple accumulation and simultaneously the adsorptions approaches saturation Γ_∞. As the lifetime of films at the foam top τ_f increases with increasing concentration and every subsequent rupture of bubbles in the top layer requires time of the order of τ_f, the increase of concentration retards the rate of concentration increase (Fig. 16). The surfactant accumulation in the top layer practically stops when the CMC is reached, because the downwards surfactant flow increases strongly.

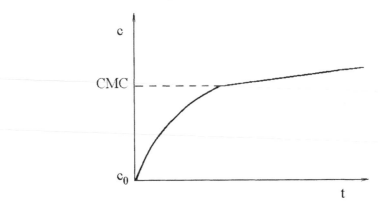

Figure 16. Accumulation of surfactant in the top foam layer.

The downwards drainage along vertical and inclined (internal) channels affects surfactant downwards transport and determines whether the situation for SAFT-2 is fulfilled instead of SAFT-1.

F. SAFT and Surface Rheology

Drainage downwards the internal channels can happen either in form of a Poiseuille-like flow [34–36] or a plug-flow [36, 37]. However, a more detailed analysis shows that for the initial period of the downwards flow a plug-flow pattern has to be considered, because a higher surfactant concentration ($c_1 = c_0 + \Delta c_0$) is inherent in the stream from external into internal channels (Fig. 15). The surfactant concentration decreases in direction downwards the internal channels which causes a Marangoni plug-flow that in turn reduces the surfactant accumulation in the top layer and promotes the SAFT-2 mode. A plug-flow takes place probably only during the initial stage of drainage because the average flow of liquid should remain zero in a stationary column of foam.

G. Conditions for the Onset of SAFT

For the onset of SAFT a dry foam is the necessary initial condition. Dry foams, however, are not always formed, as it is the case, for example, for fatty acid or fatty alcohol solutions [19]. The possibility to concentrate a surfactant in a standard foam fractionation procedure [16–20] indicates the ability of this surfactant to form a dry foam, because foam drying is a necessary step of this technology. Hence, SAFT can occur in solutions of efficient foaming agents. Still, the surfactant concentration has to be sufficiently high for dry foam formation. There is a lower concentration limit below which the foam fractionation method for surfactants cannot be applied [16–20]. This lower concentration limit can be used as a crude estimation for limit of dry foam formation, which in turn leads us to the regularities established by Khaskova and Kruglyakov [20] for the appearance of common black films (CBF) and Newton black films (NBF). Correlations were established between the low concentration limit c_{min} and the concentration c_{bll} which corresponds to the beginning of black spot formation in microscopic films. For example for SDS solutions at 0.1 M NaCl the value of c_{min} is 3.5×10^{-6} M, while c_{bl} is 2×10^{-4} M [20]. Hence, the concentration range for dry foam formation may be estimated as 3.5×10^{-6} to 2×10^{-4} M. Accordingly, this yields a crude estimate for the concentration range of SAFT mode which corresponds to the mechanism of long term foam stabilization at $c < c_{bl}$, according to Kruglyakov and Exerowa [2], with respective experimental evidences given in [23].

For NBF the concentration range is rather narrow: 0.8×10^{-4} to 2×10^{-4} M for SDS in the presence of 0.35 M NaCl. However, it was found by Khaskova and Kruglyakov [20] that c_{min} is rather small as compared to c_{bl}^{N}. It can be concluded that the concentration range for the onset of SAFT can be rather broad under the conditions corresponding to CBF and NBF formation. This conclusion is based on results obtained in [20] for few surfactants only. The problem should be considered in more detail, in particular using the large literature relevant to dry foams.

H. Droplet Formation during Breakage of Foam Films and Its Influence on Surfactant Accumulation

1. Droplet Formation during Film Rupture Caused by Capillary Wave Mechanism

De Vries [38] proposed a capillary wave mechanism for the formation of droplets after film rupture. This model was later extended in [26, 39–43]. The mechanism was developed for the case of monotonic attraction between the two surfaces of a liquid film (van der Waals attraction). Thermally excited capillary waves are always present at film surfaces. With decreasing average film thickness h, the attractive disjoining pressure enhances the amplitude of some modes of fluctuation waves. At a given critical film thickness h_{cr}, corrugations on the two opposite film surfaces can touch each other which then can cause the rupture of the film [42].

In contrast to geophysics, to the best of our knowledge, there are no references about the formation of droplets during film rupture in foam science. Bubble bursting near the water/air interface occurs in oceans and leads to the formation of atmospheric aerosols of different kinds. This phenomenon attracted large attention and is studied over about half a century, as summarised in Table 2. The given data are partially controversial.

The bubble dimension has strong influence on droplet formation as reported by all researchers. The measurement techniques for the determination of droplet size distribution were continuously improved during decades. The studies were performed without any added surfactant, however, natural biosurfactants are typically found in sea water [49]. Blanchard [44] paid attention to the influence of surfactant concentration on droplet formation. He assumed that droplet formation can be at least partly suppressed [50].

Table 2.
History of references to bubble bursting

Reference	d_{dr} [µm]	d_b [mm]
Blanchard (1963) [44]	5–330	0.23–1.60
Blanchard and Syzdek (1988) [45]	0.03–10	1.00–6.30
Resch and Afeti (1992) [46]	0.056–0.8	1.60–5.70
	0.2–5.0	1.60–5.70
Spiel (1997) [47]	10–600	3.95–12.57
Reinke et al. (2001) [48]	1–500	

A bubble can burst under the water/air interface at rather large film thickness. Accordingly, the geophysical studies cited above are relevant to wet foam conditions, while we are essentially interested in dry foams. Nevertheless, the experience in droplet measurement can be used for droplet detection when produced by ruptures in dry foams. Recently the attention to droplet formation during bubble bursting has increased because of the possibility of radionuclide re-entrainment at bubbling water pool surfaces [51].

2. Droplet Formation during Hole Growth

In grey films of thickness $h \leqslant h_{cr}$ black spots can form spontaneously and expand with time [2]. At any time moment a black film of radius $r(t)$ coexists with a thicker grey film at a radius larger than $r(t)$. This scenario does not correspond to bubble bursting. A film bursts when a hole is formed and grows. This occurs when the film is punctured or the surfactant concentration is sufficiently low. Lord Rayleigh [30, 31] was the first to explain the physics of growth of a hole. He assumed that at any instant the liquid from a circular rupture is collected in a rolled-up edge (the rim) which is moving with a kinetic energy equal to the energy of the disappearing surface. Thus:

$$\pi r^2 h \rho_1 v^2 / 2 = 2\gamma \pi r^2 \tag{19}$$

with

$$v = 2\sqrt{\gamma / \rho_1 h}. \tag{20}$$

Note, the unknown quantity $r(t)$ is cancelled out from Eq. (19), the result becomes rather simple and the velocity depends neither on r nor on time.

Ranz [32] observed a moving rim, measured its velocity and film thickness and established agreement with Rayleigh's equation. He observed that the rim is unstable leading to droplet formation and their separation from the rim. Droplet formation was confirmed in systematic investigations by Mysels and co-workers [33, 52, 53] who observed "aureoles" formed by the rim and interpreted it as manifestation of surface concentration gradients. Mysels also proclaims that surface concentration gradients have also a component along the rim being the reason of

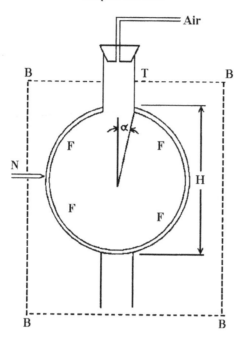

Figure 17. Schematic of the experimental device for the visualisation of hole growth and determination of droplet size distribution; the spherical bubble is punctured; droplets deposit on the internal surface of a cubic-shaped envelope; according to [54].

Table 3.
Droplet formation from rupturing spherical soap films [54]

Initial sphere diameter (from different solutions) (mm)	Number of drops	Drop diameter (µm)	$\frac{\text{Drop area}}{\text{Film area}}$ (%)
80	$80\,000 \pm 20\,000$	50 ± 30	3.2–4.0
40	$13\,000 \pm 2000$	80 ± 30	5.72–6.0
60	$27\,000 \pm 3000$	70 ± 40	3.67–4.0
20	4000 ± 2000	80 ± 30	6.4–9.2
30	680 ± 300	320 ± 150	1.9–2.6

droplet formation. Simultaneously, he rejects the possibility of a Rayleigh instability as reason for droplet formation.

The work of Pandit and Davidson [54] is of special interest, because it contains quantitative information about droplets formed by a moving rim. They punctured a spherical soap film to create a hole and rim (Fig. 17). Their results are summarised in Table 3.

The observed droplet were rather large, however, the films were also rather thick (0.3–0.9 µm). The authors mention that droplet radii decrease with decreasing film thickness. As total droplet surface area was few per cent of the original spherical film, this may be an indication that the surfactant loss due to droplet formation may be small. However, it is not clear whether this regularity persists at a film thickness close to h_{cr}.

The work of Frens and co-workers, lasting over three decades [55–60], was dedicated to black films, the building blocks of dry foams. The formation of droplet from the expanding rim was observed for NBFs [57, 58]. Consequently, surfactant accumulation in the top layer of a dry foam may be complicated by droplet formation. Although no attention was paid to measurements of droplet size, experiments with CBF enable one to conclude that surfactant loss with droplets will be not large. Large loss has to cause large differences in the rim velocity as compared with the Rayleigh equation, which was not observed for CBF.

In contrast to CBF, Frens' measurements with NBF demonstrate a rim velocity lower than those expected from Rayleigh's equation [57, 59]. Although it may be caused by special elastic properties of NBF, as Frens assumes, the elaborated model may not be very exact. If this is true, any influence due to droplet formation cannot be eliminated.

I. Discussion and Conclusions

As mentioned above, the surfactant accumulation in the top bubble layer can be complicated due to the partial removal of accumulated surfactant through its transport with droplets spread out during film rupturing. Assumed droplet formation does not occur, the surfactant accumulation is governed by its downwards transport through vertical or inclined Pbs together with the downwards directed water flow. This flow exists in spite of the fact that the water velocity averaged over any horizontal cross-section of Pbs equals to zero. Near the walls of Pbs the water moves upwards together with rising bubbles, and therefore, a downwards flow should exist near the axis of the Pbs to compensate the upwards flow. These statements hold for steady pneumatic or dynamic foams [2, 9].

With these restrictions we have the possibility for a simple estimation of the accumulated surfactant loss caused by a downwards flow through Pbs of the bubble layer underneath the top layer. This layer accepts water and surfactant released during the rupturing of bubbles in the top layer. Each bubble layer received additional water from top and donates to the bubble layer below it. This requirement arises because the water content in the second layer, which becomes the top bubble layer, has to remain almost the same as it was before this transformation. When the second bubble layer donates part of the accepted water to the layer located below, there is partial loss of accumulated surfactant. Accordingly, Eq. (10) shows that the accumulated surfactant remains mainly within the top layer, if its surface activity is sufficiently high.

The presented approach for estimating the accumulated surfactant loss is based on the simplifying assumption that the difference in water content between top and secondary layer is small. This difference can be easily calculated for the case when there are no large differences in the bubble dimension and amount of Pbs in both bubble layers. To the best of our knowledge, there is no information in literature indicating that this difference can be large. If it is so, the difference in the water content is determined by the difference in the Pbs cross-sections of the two layers, i.e., by the difference in the Pb radii R_{Pb}. This difference is small because there is no physical reason for any large difference in the capillary pressure, and can be approximated by the hydrostatic pressure difference, which in turn is proportional to the difference in the heights of the two layers, i.e., to the bubble diameter $2R_b$. Hence, the following estimation can be proposed:

$$\frac{\Delta R_{Pb}}{R_{Pb}} \sim \frac{2R_b}{H} \ll 1, \tag{21}$$

where H is the height of the foam column.

For the onset of SAFT a minimum concentration is required. Above the CMC the downwards flux of accumulated surfactant through Pbs increases strongly. This leads to a suppression of SAFT. Dry foams are not formed below this minimum required surfactant concentration. Hence the critical concentration for the onset of SAFT should be larger and can be estimated using the results obtained in [20]. It is obvious that a more general approach should be developed.

So far there is no information available about droplet formation during film rupture caused by enhanced capillary waves. Regarding hole formation and expansion, the literature analysis has shown that droplet formation is a universal phenomenon, however, no quantitative data are available about droplet dimensions formed by dry foams, which causes an uncertainty in the prediction of surfactant accumulation.

With increasing droplet dimension, the possibility increases for droplets that will return to the foam films. The drops may sediment on external foam films if their sedimentation velocity exceeds the upwards velocity of air. They may deposit on the external films due to inertial forces, because the initial velocities of droplets are large, while their directions are random.

We can finally draw two main conclusions. At first, surfactant accumulation can be very large if droplet formation does not result in large surfactant loss, provided that the surfactant loss due to drainage of liquid is small. This means that SAFT deserves special attention. At second, due to lack of information about the dimension of droplets formed during rupture of bubbles in dry foams, systematic measurements of the droplet size distribution are needed for further considerations of the described effects.

J. Appendix

Let us consider the balance of liquid and surfactant in the course of rupture of EF2 (Fig. 14a). Just after the rupture of EF2 the newly formed film EF4a receives all surfactant and liquid from films EF4 and IF6 and half of the surfactant and liquid from EF2 (including their Plateau borders). In the 2D foam, as shown in Fig. 14, there are two kinds of Pbs formed either between three different films or between a film and one of the two vertical walls supporting the foam. In the initial state the length of Pb between one of vertical walls and EF2 can be roughly estimated as $3 \cdot R_b$ (where R_b is the average bubble radius approximately equal to the side of a hexagonal cell). The initial amount of liquid in the films EF4, IF6 and half of EF2 with their Pbs can be estimated as

$$V_{W1} = 7C_1 R_{Pb}^2 R_b + 3C_2 R_{Pb}^2 l + (7/2) R_b l h_{cr}, \tag{A1}$$

where $C_1 = 2 - \pi/2 = 0.429$ and $C_2 = \sqrt{3} - \pi/2 = 0.161$ are the numerical coefficients for the respective Pb cross-sections, l is the length of Pbs formed by three different films (i.e. the distance between the vertical walls), and h_{cr} is the critical thickness of film rupture.

Just after the rupture of EF2 the amount of liquid V_{W1} redistributes within the newly formed film EF4a and its Pbs as discussed in Sections 4 and 5. The sudden increase of Pbs cross-section breaks the balance between capillary pressure difference and gravity force what causes a fast drainage of excess liquid from EF4a into internal Pbs ("C" in Fig. 15). The plug flow induced by the surface tension gradient in the channels accelerates this process. Simultaneously the thickness of EF4a decreases from h_2 to a value slightly above h_{cr}. When the local characteristics of foam (Pbs radius and film thickness) return almost to the initial values (before the rupture of EF2), the amount of liquid remaining in EF4a, will be approximately

$$V_{W2} = 4C_1 R_{Pb}^2 R_b + 2C_2 R_{Pb}^2 l + 2 R_b l h_{cr}. \tag{A2}$$

Hence, the amount of excess liquid in EF4a, which is lost due to drainage, is

$$\Delta V_W = V_{W1} - V_{W2} = 3C_1 R_{Pb}^2 R_b + C_2 R_{Pb}^2 l + (3/2) R_b l h_{cr}. \tag{A3}$$

Accordingly, the amount of surfactant, which will be lost due to drainage, can be expressed by Eq. (14). Before the rupture of EF2 the initial amount of surfactant in the films EF4, IF6 and half of EF2 with their Pbs can be written approximately as

$$M_{S0} = 7 R_b l \Gamma_0 + V_{W1} \cdot c_0. \tag{A4}$$

The amount of surfactant in EF4a just after the rupture of EF2 (before drainage) is

$$M_{S1} = 4 R_b l \Gamma_1 + V_{W1} \cdot c_1. \tag{A5}$$

As the amounts of liquid and surfactant do not change during film rupture ($V_{W0} = V_{W1}$, $M_{S0} = M_{S1}$), and assuming $\Gamma_1 = \Gamma_0 + (d\Gamma/dc) \cdot \Delta c$ and $c_1 = c_0 + \Delta c$, we obtain from Eqs (A1), (A4) and (A5)

$$\frac{\Delta c}{c_0} = \frac{3\Gamma_0/c_0}{4\frac{d\Gamma}{dc} + \frac{V_{W1}}{R_b l}} = \frac{3\Gamma_0/c_0}{4\frac{d\Gamma}{dc} + \frac{1}{R_b l}(7C_1 R_{Pb}^2 R_b + 3C_2 R_{Pb}^2 l + (7/2) R_b l h_{cr})}. \tag{A6}$$

Under the assumption $l \approx R_b$ this expression can be simplified to Eq. (15).

K. Acknowledgements

The work was financially supported by a project of the German Space Agency (DLR 50WM0941), COST P21, and the German Science Foundation (DFG Mi418/16-2).

L. References

1. R. Lemlich (Ed.), Adsorptive Bubble Separation Techniques, Academic Press, NY, 1972.
2. D.R. Exerowa and P.M. Kruglyakov, Foam and Foam Films: Theory, Experiment, Application, in "Studies in Interface Science", Vol. 5, D. Möbius and R. Miller (Eds), Elsevier, 1997.
3. R.K. Prud'homme and S.A. Khan (Eds), Foams: Theory, Measurements, and Applications, Surfactant Science Series, Vol. 57, Dekker, 1995.
4. A.J. Wilson (Ed.), Foams: Physics, Chemistry, and Structure, Springer, 1989.
5. V. Bergeron and P. Walstra, Foams, in "Fundamentals of Interface and Colloid Science", Vol. 5, Soft Colloids, J. Lyklema, Elsevier, 2005.
6. V.V. Krotov, Colloid J. USSR, 43 (1981) 33 (English translation).
7. V.V. Krotov and A.I. Rusanov, Physicochemical Hydrodynamics of Capillary Systems, Imperial College Press, London, 1999.
8. K.B. Kann, Capillary Hydrodynamics of Foams, Nauka, Novosibirsk, 1989 (in Russian).
9. E. Ruckenstein and A. Bhakta, Langmuir, 12 (1996) 4134.
10. A. Bhakta and E. Ruckenstein, J. Colloid Interface Sci., 191 (1997) 189; Adv. Colloid Interface Sci., 76 (1997) 1.
11. S.J. Neethling, H.T. Lee and P. Grassia, Colloids Surfaces A, 263 (2005) 184.
12. P. Grassia, S.J. Neethling, C. Cervantes and H.T. Lee, Colloids Surfaces A, 274 (2006) 110.
13. S.S. Dukhin, V.I. Kovalchuk, E.V. Aksenenko and R. Miller, Adv. Colloids Interface Sci., 137 (2008) 45–56.
14. K. Malysa, K. Lunkenheimer, R. Miller and C. Hempt, Colloids Surfaces, 16 (1985) 9.
15. K. Malysa, Adv. Colloid Interface Sci., 40 (1992) 37.
16. A.M. Koganovckij and N.A. Klimenko, Physicochemical Basics of the Extraction of Surfactants from Aqueous Solutions and Disposed Waters, Naukova Dumka, Kyiv, 1978 (in Russian).
17. V.V. Pushkariov and D.I. Trofimov, Physicochemical Basics of the Disposed Waters Cleaning from Surfactants, Khimija, Moscow, 1975, p. 81 (in Russian).
18. A.I. Rusanov, S.V. Levichev and V.T. Zharov, Surface Separation of Substances: Theory and Methods, Khimija, Leningrad, 1981 (in Russian).
19. J.J. Bikerman, Foams: Theory and Industrial Applications, Reinhold Publishing, 1953.
20. T.N. Khaskova and P.M. Kruglyakov, Russ. Chem. Rev., 64 (1995) 235; Kolloid. Zh., 58 (1996) 260 (in Russian).
21. R.C. Darton, S. Supino and K.J. Sweeting, Chemical Engineering and Processing, 43 (2004) 477.
22. K.N. Arbuzov and B.N. Grebenshchikov, Zh. Fiz. Khim., 10 (1937) 32 (in Russian).
23. Khr. Khristov, K. Malysa and D. Exerowa, Colloids Surfaces A, 11 (1984) 39.
24. R.C. Darton and K.H. Sun, Chem. Eng. Research and Design, 77 (1999) 535.
25. V.V. Krotov, A.G. Nekrasov and A.I. Rusanov, Kolloid. Zh., 64 (2002) 793 (in Russian) (English translation: Colloid J., 64 (2002) 716).
26. A. Sheludko, Proc. K. Acad. Wetensch., B65 (1962) 87.
27. B.V. Derjaguin and Y.V. Gutop, Kolloid. Zh., 24 (1962) 431.
28. H.W. Schwarz, Rec. Trav. Chim., 84 (1965) 771.
29. D. Weare and S. Hutzler, The Physics of Foam, Clarendon Press, Oxford, 1999.

30. Lord Rayleigh, Sci. Papers, 3 (1902) 441.
31. Lord Rayleigh, Sci. Papers, 1 (1899) 474.
32. W.E. Ranz, J. Appl. Phys., 30 (1959) 1950.
33. W.R. McEntee and K.J. Mysels, J. Phys. Chem., 73 (1969) 3018.
34. G. Verbist, D. Weaire and A.M. Kraynik, J. Phys.: Condens. Matter, 8 (1996) 3715.
35. D. Weare, S. Hutzler, G. Verbist and E.A.J. Peters, Adv. Chem. Phys., 102 (1997) 315.
36. M. Durand and D. Langevin, Eur. Phys. J., E 7 (2002) 35.
37. S.A. Koehler, S. Hilgenfeldt and H.A. Stone, Phys. Rev. Lett., 82 (1999) 4232; Langmuir, 16 (2000) 6327.
38. A. de Vries, Rec. Trav. Chim. Pays-Bas, 77 (1958) 43.
39. A. Vrij, Disc. Faraday Soc., 42 (1966) 23.
40. I.B. Ivanov, B. Radoev, E. Manev and A. Sheludko, Trans. Faraday Soc., 66 (1970) 1262.
41. I.B. Ivanov and D.S. Dimitrov, Colloid Polym. Sci., 252 (1974) 982.
42. I.B. Ivanov, Pure Appl. Chem., 52 (1980) 1241.
43. P.A. Kralchevsky, K.D. Danov and I.B. Ivanov, in "Foams", R.K. Prudhomme and S.A. Khan (Eds), Marcel Dekker, NY, 1995, p. 1.
44. D.C. Blanchard, Progress in Oceanographie, 1 (1963) 71.
45. D.C. Blanchard and L.D. Syzdek, J. Geophys. Res., 93 (1988) 3649.
46. F. Resch and G. Afeti, J. Geophys. Res., 97 (1992) 3679.
47. D.E. Spiel, J. Geophys. Res., 102 (1997) 1153.
48. N. Reinke, A. Vossnacke, W. Schutz, M.K. Koch and H. Unger, Water, Air and Soil Pollution: Focus, 1 (2001) 333.
49. J. Krägel, A.M. Stortini, N. Degli-Innocenti, G. Loglio and R. Miller, Colloids Surfaces A, 101 (1995) 129.
50. M.P. Paterson and K.T. Spillane, Quart. J. R. Met. Soc. 95 (1969) 526.
51. M.K. Koch, A. Vossnacke, J. Starflinger, W. Schutz and H. Unger, J. Aerosol Sci., 31 (2000) 1015.
52. S. Frankel and K.J. Mysels, J. Phys. Chem., 73 (1969) 3028.
53. K.J. Mysels and J. Stiekelcather, J. Colloid Interface Sci., 35 (1971) 159.
54. A.B. Pandit and J.F. Davidson, J. Fluid Mech., 212 (1990) 11.
55. A.T. Florence and G. Frens, J. Phys. Chem., 76 (1972) 3024.
56. G. Frens, J. Phys. Chem., 78 (1974) 1949.
57. L.J. Evers, S.Y. Shulepov and G. Frens, Faraday Discuss., 104 (1996) 333.
58. L. Evers, S.Y. Shulepov and G. Frens, Phys. Rev. Lett., 79 (1997) 4850.
59. L. Evers, E.J. Nijman and G. Frens, Colloids Surfaces A, 149 (1999) 521.
60. C. Hedreul and G. Frens, Colloids Surfaces A, 186 (2001) 73.

Particle–Bubble Interaction in Flotation

Anh V. Nguyen

School of Chemical Engineering, The University of Queensland, Brisbane,
Queensland 4072, Australia

Contents

A. Introduction

Flotation is a process for separating particles, based on their differences in surface properties, using air bubbles. The flotation process has a long history (over 100 years) of development and widespread applications. One of the earliest applications was in the recovery of sphalerite (zinc sulphide, ZnS) minerals from finely ground tailings of a gravity concentrator at Broken Hill in New South Wales, Australia in 1905. In mineral flotation, the ore is first crushed and finely ground to less than about 200 μm to liberate the minerals, before reagents are added to alter the mineral surface properties. The surface of the wanted minerals is rendered hydrophobic or water-repellent, which allows the particles to attach themselves to air bubbles while leaving the surface of the unwanted particles hydrophilic (water-wetted) and not attachable to air bubbles. After attaching themselves to the air bubbles, the

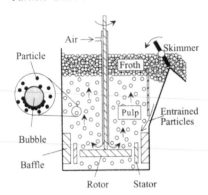

Figure 1. Schematic of flotation where hydrophobic particles are separated by attaching to rising air bubbles to produce a particle-rich froth on the surface of the flotation cell with hydrophilic particles remaining in the suspension to be discharged from the cell. Reproduced with permission from [1].

hydrophobic particles are separated from the suspension by the bubble rise as froth, which forms at the top of the slurry and flows over the lip of the cell (Fig. 1). The hydrophilic particles do not attach to the bubbles and remain in the liquid which is discharged from the bottom of the cell. In this way, the particles are separated, based on their surface hydrophobicities.

Flotation has been used by mineral and chemical engineers for the separation and concentration of aqueous suspensions or solutions of a variety of minerals, coal, precipitates, inorganic waste constituents, effluents, and even microorganisms and proteins. Mineral flotation is by far the most important industrial application of surface chemistry. It is estimated that more than two billion tons of various ores and coal are annually treated by flotation worldwide. This figure, which represents about 85% of ores mined annually, is likely to increase in the future with the depletion of high-grade ore deposits. In addition to mineral processing, flotation has been applied in many diverse industrial areas. It has been adopted for the treatment of industrial wastewater and drinking water, the recycling of plastics and bacteria separation in bioengineering. Flotation has also been applied to remove hydrophobic ink particles from used fibre (de-inking) enabling the fibre to be recycled and re-used in the paper manufacture. In recent years, it has also been used to separate heavy crude oil from tar sands.

Flotation depends on the ability of bubbles to collect particles from the suspension, and carry them to the froth phase and the concentrate launder. The bubble–particle collection determines the attachment of the hydrophobic particles to bubbles by the formation of a finite contact angle at the three-phase gas–liquid–solid contact line. The collection involves a number of bubble–particle interaction subprocesses, which can be divided into (Fig. 2) [1–6]:

   (i) Collision
  (ii) Attachment and
 (iii) Detachment and stability of bubble–particle aggregates.

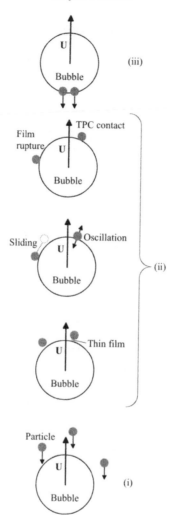

Figure 2. Sub-processes of particle collection by an air bubble as it rises through the pulp: (i) collision, (ii) attachment and (iii) stabilization of particle–bubble aggregates with respect to external stresses.

Collision is the approach of a particle to encounter a bubble which is governed by the fluid mechanics of the particle in the long-range hydrodynamic force field around the bubble, and has been studied extensively [7–16]. The limit of the collision sub-process is determined by the zonal boundary between the long-range hydrodynamic and interfacial force interactions [17]. The inter-surface separation distance at the zonal boundary is of the sub-micrometer order. Once the particle approaches the bubble at a shorter separation distance, the atomic, molecular and surface forces are significant and the attachment sub-process commences, followed by the film thinning and rupture, and the formation of a large three-phase contact (tpc) area. The detachment sub-process can occur during the rise of bubble–particle aggregates to the froth layer and is governed by the capillary force, the particle weight and the detaching forces due to the turbulent stresses. The modelling of the bubble–particle attachment and detachment interactions involves many unsolved complex problems such as hydrophobic attraction and de-wetting hysteresis on physically and chemically heterogeneous surfaces, and is not very well advanced [18, 19].

In this chapter, we will focus on the physics governing bubble–particle collision, attachment and detachment. The attachment of nano- and sub-micron sized particles enhanced by Brownian diffusion will also be included. We will consider here only the case where a bubble is rising in a quiescent liquid or in a laminar flow closely similar to the "plug" flow in a flotation column. The water flow in mechanical flotation cells, where shear and turbulence effects arising from the rotating impeller are important, adds an additional range of complexity. However, the general principles considered here will still apply.

B. Collision

1. Physics of Bubble–Particle Collision Interaction

The collision efficiency is usually derived from a model of the interaction between a particle with radius, R_p, in the approaching stream and a bubble with radius, R_b (Fig. 3). The particle is assumed to obey the Stokes law for resistance. Neglecting the Basset history term, for the equation of motion for the particle the force balance

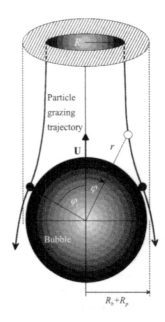

Figure 3. Illustration of the particle grazing trajectories used to define the efficiency of bubble–particle interaction. Particles within the volume enclosed by the grazing trajectories interact with the bubble.

gives [1]

$$\frac{4\pi R_p^3 \rho}{3} \frac{d\vec{V}}{dt} = \frac{4\pi R_p^3 \delta}{3} \frac{d\vec{W}}{dt} - \frac{2\pi R_p^3 \delta}{3} \left\{ \frac{d\vec{V}}{dt} - \frac{d\vec{W}}{dt} \right\}$$
$$- 6\pi \mu R_p (\vec{V} - \vec{W}) + \frac{4\pi R_p^3 (\rho - \delta)}{3} \vec{g}, \tag{1}$$

where V and W are the particle and liquid velocities, respectively, ρ and δ are the particle and liquid densities, t is the reference time, g is the acceleration due to gravity and μ is the liquid viscosity. The arrow describes vectorial variables.

Equation (1) has to be solved to obtain the particle grazing trajectories for predicting the collision efficiency, E_c. Specifically, the radius, R_c, of the grazing trajectories far from the bubble surface must be obtained from Eq. (1) to determine E_c as (cf. Fig. 3)

$$E_c = \left\{ \frac{R_c}{R_b + R_p} \right\}^2. \tag{2}$$

Equation (2) describes the ratio of particle fluxes in interacting with the bubble and in the bubble path. It can be converted into the following expression [1]:

$$E_c = \frac{2}{U + V_s} \int_0^{\varphi_c} [-V_r(\varphi)] \sin \varphi \, d\varphi, \tag{3}$$

where U is the bubble slip (relative to water) velocity, $V_s = 2R_p^2(\rho - \delta)g/(9\mu)$ is the terminal settling velocity for the particle. In Eq. (3), the radial component of the particle velocity, $V_r(\varphi)$, is calculated at one particle radius away from the bubble surface. Otherwise, the radial velocity is a function of both the polar, φ, and radial, r, coordinates. The upper limit, φ_c, for the polar coordinate in the integral in Eq. (3) is the collision angle (also called the angle of tangency) which determines the point of tangency of the particle grazing trajectory with the bubble surface and is defined as $dE_c/d\varphi_c = 0$. It gives [20]

$$V_r(\varphi_c) = 0. \tag{4}$$

Equation (2) has been used to determine E_c when solving Eq. (1) numerically [8, 20–22], while Eq. (3) has been applied when solving Eq. (1) analytically [11, 23]. The determination of R_c also requires the particle motion to be deterministic. This requirement cannot be met in the case of the turbulent particle motion, which is essentially chaotic (non-deterministic). Equation (3) can also be used for the turbulent case.

2. Factors Influencing Collision Efficiency

The determination of R_c is usually carried out by numerical integration of Eq. (1) since the liquid flow field around the bubble is not analytically accessible. The motion equation is also too complicated for developing a full analytical solution.

Nevertheless, a number of approximate analytical solutions have been obtained using various approximations and are summarised in Table 1. A short analysis linking the available predictions for E_c and Eq. (1) is given below.

Table 1.
Available models for the collision efficiency

Author(s)	Collision efficiency, E_c	Notes and symbol definitions
Sutherland [12]	$E_c = 3R$, where $R = R_p/R_b$	Interceptional effect; potential flow. R is the interception number
Gaudin [5]	$E_c = \frac{3}{2}R^2$	Interception effect; Stokes flow
Flint and Howarth [21]	$E_c = \frac{V_s}{V_s+U}$	Gravitational effect
Reay and Ratcliff [22]	$E_c = \frac{V_s}{V_s+U}(1+R)^2 \sin^2 \varphi_c$	Gravitational effect
Anfruns and Kitchener [10]	$E_c = \frac{V_s}{V_s+U}(1+R)^2 + \frac{2U'\psi_c}{V_s+U}$	Combined gravitational and interceptional effects. ψ_c is the stream function ψ calculated at $r = 1 + R$ and $\varphi = \pi/2$
Weber and Paddock [11]	$E_c = \frac{V_s}{U}(1+R)^2 \sin^2 \varphi_c + E_{c,i}$ $E_{c,i} = \frac{3R^2}{2}[1 + \frac{3Re/16}{1+0.249Re^{0.56}}]$ ...(*) $E_{c,i} = R[1 + \frac{2}{1+(37/Re)^{0.85}}]$ ...(**)	Combined gravitational and interceptional effects; intermediate bubble Reynolds number. Equation (*) for bubbles with a fully immobile surface. Equation (**) for bubbles with a fully mobile surface
Yoon and Luttrell [7]	$E_{c,i} = R^2[\frac{3}{2} + \frac{4Re^{0.72}}{15}]$	Interceptional effect; intermediate bubble Reynolds number; fully immobile bubble surface
Dobby and Finch [8]	(1) *Low particle inertia, St < 0.1* $E_c = E_{c,g} + E_{c,i}$ $E_{c,i}$ is given by Weber and Paddock for immobile surface. $E_{c,g}$ is given by Reay and Ratcliff. $\varphi_c = \begin{cases} 78.1 - 7.37 \log Re \\ \quad \text{if } 400 > Re > 20 \\ 85.5 - 12.49 \log Re \\ \quad \text{if } 20 > Re > 1 \\ 85.0 - 2.5 \log Re \\ \quad \text{if } 1 > Re > 0.1 \end{cases}$ (2) *Intermediate particle inertia, Sk > 0.1* $E_c = (E_{c,0})[1.627Re^{0.06}St^{0.54}(V_s/U)^{-0.16}]$ $E_{c,0}$ is given in the low inertial case.	Combined inertial, gravitational and interceptional effects; intermediate bubble Reynolds number; immobile bubble surface; analytical results for low St and numerical results for intermediate St. $St = \frac{Re}{9}\frac{\rho}{\delta}R^2$ is the Stokes number. Range of application of the numerical results for intermediate inertia: $300 > Re > 20$: $0.8 > St$: and $V_s/U < 0.25$

Table 1.
(Continued)

Author(s)	Collision efficiency, E_c	Notes and symbol definitions
Schulze and Plate [9]	$E_c = E_{c,i} + E_{c,g} + E_{c,in}[1 - \frac{E_{c,i}}{(1+R)^2}]$ $E_{c,i} = \frac{U}{V_s+U} 2\psi_c$ $E_{c,g} = \frac{V_s}{U}(1+R)^2 \sin^2 \varphi_c$ $E_{c,in} = \frac{U(1+R)^2}{V_s+U}(\frac{St}{St+a})^b$ For $R \leqslant 1/\xi$: $\psi_c = \frac{4R^2}{2}[1 + \frac{3Re/16}{1+0.249Re^{0.56}}]$	Combined inertial, gravitational and interceptional effects; intermediate bubble Reynolds number; immobile bubble surface; approximate results for both low and intermediate St. Model parameters (a, b and φ_c) are functions of Re. φ_c is given by Dobby and Finch. The surface vorticity is given as $\xi = \frac{3}{2\sin\varphi_c}[1 + \frac{3Re/16}{1+0.249Re^{0.56}}]$
Dukhin *et al.* [13, 25]	$E_c = 3R\sin^2\varphi_c \exp\{-\cos\varphi_c[3K'''(\ln R + 1.8) + (8 - 12\cos\varphi_c + 4\cos^3\varphi_c)/(3\sin^4\varphi_c)]\}$ $\varphi_c = \text{acos}(\sqrt{1+\beta^2} - \beta)$ $\beta = \frac{4R}{3K'''}$ and $K''' = \frac{2(\rho-\delta)UR_p^2}{9\mu R_b}$	Potential flow and bubbles with a mobile surface. Significant effect of centrifugal force on collision. Valid for ultra fine particles. Expression for the collision efficiency is called the Generalised Sutherland Equation (GSE)
Nguyen and Schulze [1]	– Gravitational collision $E_{c,g} = \frac{V_s}{V_s+U}$ – Interceptional collision $E_{c,i} = f(R)\frac{V_s}{V_s+U}[X + Y\cos\varphi_{c,i}]\sin^2\varphi_{c,i}$ $\varphi_{c,i} = \text{acos}\{\frac{\sqrt{X^2+3Y^2}-X}{3Y}\}$ – Inertial collision $E_{c,in} = \frac{(St)^a - b}{(St)^a + c}$ – Turbulent collision $E_{c,t} = 18\sqrt{\frac{3}{15}\frac{\delta}{\rho-\delta}\frac{\lambda_K}{R_p+R_b}}$ – Simultaneous gravitational and interceptional collision $E_{c,gi} = f(R)\frac{V_s}{V_s+U} \times$ $\qquad [X + C + Y\cos\varphi_{c,i}]\sin^2\varphi_{c,i}$ $\varphi_{c,gi} = \text{acos}\{[\sqrt{(X+C)^2+3Y^2} - (X+C)]/(3Y)\}$ $C = \frac{V_s}{U}\frac{1}{f(R)}$ – Overall collision efficiency $E_c = 1 - (1 - E_{c,g})(1 - E_{c,i}) \times$ $\qquad (1 - E_{c,in})(1 - E_{c,t})$	Individual and combined inertial, gravitational and interceptional effects; intermediate bubble Reynolds number; immobile and mobile bubble surface; approximate results for both low and intermediate Stokes number, St. Model parameters (a, b, c, X and Y) are functions of Re and gas volume fraction (as well as the bubble surface mobility) – see [1]. For bubbles with an immobile surface $f(R) = R^2$. For bubbles with a mobile surface $f(R) = R - R^2$. λ_K is the Kolmogorov microscale of turbulence

Table 1.
(Continued)

Author(s)	Collision efficiency, E_c	Notes and symbol definitions
Jiang and Holtham [26]; Reay and Ratcliff [27]; Jameson [2, 28]	$E_c = m(R)^n$	Empirical or computational results. Model parameters m and n are only available for limited conditions but $n > 1$ for cases

The motion of a particle approaching a bubble surface is influenced by a number of factors, including:

- Mass of the particle: inertial effect
- Weight of the particle: gravitational effect, and
- Liquid flow around the bubble: interceptional effect.

These effects are included in the force balance on the particle by Eq. (1) in terms of the steady and unsteady drag, pressure force, and gravitational and inertial forces. As expected, solutions for overall collision behaviour are much more complicated than the expression for each individual effect and the exact solutions are not always attainable. Simpler solutions have been obtained for each of the effects independently. The inertial, gravitational and interceptional collision efficiencies can then be combined to obtain the overall behaviour [6, 9, 24]. However, the overall collision efficiency is not always a simple summation of the effects. In particular, when the collision efficiency is controlled by two or more effects, a complex relationship is needed to predict the behaviour of the collision process [9, 14]. In this case, other factors, such as mobility of the bubble surface, need to be included in the analysis [16].

The mobility of the bubble surface is determined by the surfactants, particles and impurities adsorbed at the bubble surface (Fig. 4) and can be introduced into the evaluation of the collision efficiency through the boundary condition applied to the liquid velocity. The particle velocity V can be modelled if the liquid velocity W is known. Therefore, the next step is focussed on the modelling of the liquid flow field around the bubble. For simplicity only flow at a either fully mobile or immobile surface is considered.

3. Liquid Flow Around a Gas Bubble

The liquid flow around a gas bubble is governed by both the continuity and Navier–Stokes equations. Exact solutions to these equations can be analytically derived when the bubble Reynolds number, $Re = 2R_b U \delta / \mu$, is zero (the Stokes flow) or infinitely large (the potential flow). For bubbles with Reynolds number between these two limits the continuity and Navier–Stokes equations are numerically solved [15].

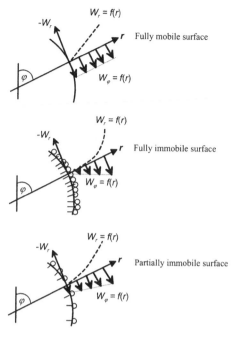

Figure 4. Effect of adsorbed surfactant molecules (line + circle) and other surface contaminants on the liquid velocity components at the bubble surface. A bubble surface in clean (contaminant-free) water is fully mobile (top). A bubble surface fully covered by surfactant and other contaminants is immobile (middle). As it rises, the leading bubble surface becomes partially immobile (bottom) because the surface contaminants are swept to the rear by the shear exerted by the liquid. The tangential liquid velocity as a function of the radial distance, r, measured from the bubble centre is shown by the parallel lines with the arrows. The profiles of the radial, W_r, and tangential, W_φ, components of liquid velocity at the bubble surface are shown by the dashed lines.

The Stokes and potential flows can simply be described in terms of the stream function, ψ, as

$$\psi_{Stokes} = \left\{ r^2 - \frac{3}{2}r + \frac{1}{2r} \right\} \frac{\sin^2 \varphi}{2}, \tag{5}$$

$$\psi_{Potential} = \left\{ r^2 - \frac{1}{r} \right\} \frac{\sin^2 \varphi}{2}, \tag{6}$$

where the radial coordinate, r, is made dimensionless by dividing the dimensional coordinate by the bubble radius and the stream function by $U R_b^2$.

Knowing the stream function, the radial and tangential liquid velocities (made dimensionless by dividing by the bubble slip velocity) can be determined as: $w_r = -\frac{1}{r^2 \sin \varphi} \frac{\partial \psi}{\partial \varphi}$ and $w_\theta = \frac{1}{r \sin \varphi} \frac{\partial \psi}{\partial r}$. The velocity components predicted by the Stokes flow are zero at the bubble surface and the flow corresponds to a water flow at an immobile bubble surface. The potential flow generates the non-zero tangential velocity of water at the bubble surface and represents a flow at a fully mobile bubble

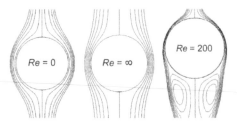

Figure 5. Water streamlines around a bubble at $Re = 0$ (Stokes flow), $Re = \infty$ (potential flow), and $Re = 200$ (numerical results). The aft-and-fore asymmetric streamlines in the right picture are much more complex than those for the idealized cases of the aft-and-fore symmetric Stokes or potential flows, or their linear combination.

surface. The collision efficiency due to interception predicted as a function of ψ_c by Anfruns and Kitchener, and Schulze and Plate in Table 1 is obtained if the equation for w_r is used for V_r in Eq. (3).

Properties of liquid flows around the bubble can be analysed by plotting the water streamlines (i.e. the trajectories of water molecules) passing the bubble — the streamlines are the contours of r *versus* φ at constant values for ψ. Three examples are shown in Fig. 5. The streamlines for both of the flow fields predicted by the Stokes and potential flows are symmetric with respect to the bubble equatorial plane as can be seen from Fig. 5. However, as the bubble Reynolds number increases from zero, the liquid flow field around the bubble loses its symmetry and becomes more and more fore-and-aft asymmetric, with the streamlines being compressed towards the front hemi-bubble surface. A model to predict stream function at intermediate Reynolds number has been derived by a linear combination of the Stokes and potential flows [7]. However, the limitation with the approach is that the fore-and-aft asymmetry of real water flows around a bubble cannot be recovered.

The fore-and-aft asymmetry of fluid flows passing bubbles, spheres, cylinders and other obstructions is well established experimentally [29]. The flow asymmetry is closely related to the boundary layer separation and vortex formation, which usually leads to the decrease in the terminal velocity of bubbles and particles in comparison with the Stokes terminal velocity. Similarly, the fore-and-aft asymmetry of liquid flow near the bubble surface influences both the particle–bubble collision and attachment interactions. The effect of the asymmetry on bubble–particle collision is illustrated in Fig. 6. Since the liquid streamlines are more compressed towards the front of the bubble, the radial liquid velocity changes its direction away from the bubble surface before the liquid passes the bubble equator (this change takes place at the equator in the Stokes flow and potential flow fields). Consequently, the particle colliding area is not the entire front hemi-sphere of the bubble surface, and subsequently, the collision angle, φ_c, is smaller than 90 degrees. Another physics causing the decrease in φ_c is the centrifugal force on low-inertia particles, amplified by non-zero tangential water velocity on a mobile bubble surface as first discovered by Dukhin [25]. For particles with intermediate and high inertia, the effect fore-and-aft asymmetry of the particle trajectories on the collision efficiency decreases. If the

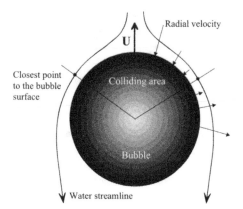

Figure 6. Colliding surface area and direction of the radial velocity of the liquid flow at intermediate Reynolds number for an air bubble.

particle inertia is sufficiently high, the tangency between the particle grazing trajectories and the bubble surface occurs at the bubble equator and the collision angle is equal to 90 degrees. A quantitative analysis is briefly shown in the following.

4. Fore-and-aft Asymmetry of Low and Intermediate Inertia Particle Collision

The liquid velocity at the bubble surface required for determining the particle velocity in Eq. (3) can be approximately derived employing a Taylor series expansion and the results of numerical solution to the Navier–Stokes equations [23, 30]. The final expression for the radial velocity can be described as

$$w_r(r, \varphi) = -\frac{2X \cos \varphi + 3Y \cos^2 \varphi - Y}{2} f(\zeta) + O(\zeta^3), \qquad (7)$$

where $\zeta = r - 1$ is the radial distance from the bubble surface (scaled by dividing by the bubble radius). The symbol $O(\cdots)$ describes the order of magnitude, i.e., the error of the approximation. For the flow at an immobile bubble surface $f(\zeta) = \zeta^2$, while $f(\zeta) = \zeta - \zeta^2$ for the at a mobile bubble surface. In Eq. (7), X and Y are functions of the bubble Reynolds number, the mobility of the bubble surface and the volume fraction of air bubbles in liquid, which are given in [1, 15].

For the collision by interception, the particle velocity is determined by the liquid flow, giving $V_r = U w_r$. The angle of tangency described by (4) is then determined by the φ term in Eq. (7). The solution for φ_c by the collision by interception is given by Nguyen and Schulze in Table 1. The dependence of the collision by interception on the bubble Reynolds number is shown in Fig. 7. For bubbles with a fully immobile surface, the fore-and-aft asymmetry is significant and the angle of tangency decreases with increasing the bubble Reynolds number. For bubbles with a fully mobile surface, the angle of tangency decreases with increasing the bubble Reynolds number, reaching a minimum at $Re = 20$, and then increases back to 90 degrees at the bubble equator. Obviously, the collision by interception is a good approximation for very low-inertia particles which instantaneously follow the wa-

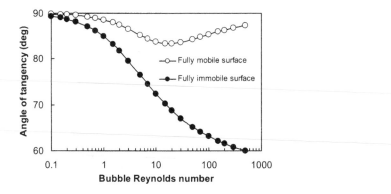

Figure 7. Angle of tangency, φ_c, by interceptional collision as a function of the Reynolds number, *Re*, of single bubbles, described by Nguyen and Schulze's model given in Table 1.

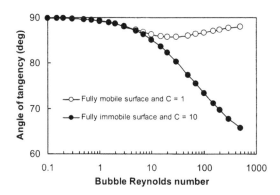

Figure 8. Angle of tangency, φ_c, by combined interceptional and gravitational collision as a function of the Reynolds number, *Re*, of single bubbles and the gravity number, *C*, described by Nguyen and Schulze's model given in Table 1.

ter streamlines around the bubble. If the particle's inertia increases, gravitational and inertial effects become important. Gravitational forces reduce the fore-and-aft asymmetry of bubble–particle collision and increase the angle of tangency towards 90 degrees at the bubble equator (Fig. 8).

Recently, Eq. (1) for the particle motion was numerically solved to validate the Generalised Sutherland Equation (GSE) [20]. The potential flow for water described by Eq. (6) was used in the GSE theory and the numerical computation. The GSE theory with $\beta = 2R/(3K''')$ developed earlier in 1983 [25] is well compared with the numerical results for fine particles (Fig. 9). The earlier GSE theory was later modified to account for the short-range resistance, leading to $\beta = 4R/(3K''')$ [13]. The modified theory predicts the angle of tangency higher than the numerical results (Fig. 9). The GSE analytical solutions predict that the angle of tangency decreases with increasing particle radius, while the numerical solutions only show the decrease for very small (ultrafine) particles. There exists a particle size where the angle of tangency has a minimum. Beyond this minimum, other inertial effects are

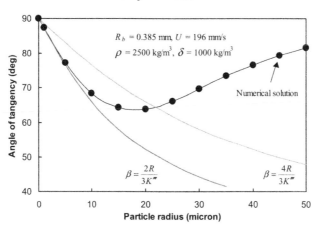

Figure 9. Comparison between the computational results (line and circles) of Eq. (1) and the GSE theory (lines) by Dukhin *et al.* (Table 1) for the angle of tangency, φ_c, *versus* the particle (quartz) radius. Reproduced with permission from [20].

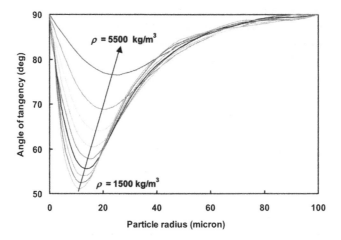

Figure 10. Numerical results of Eq. (1) for the angle of tangency *versus* the particle size and density. The bubble size and rise velocity are the same as those shown in Fig. 9. The particle density increment is 500 kg/m^3.

greater than the centrifugal effect, leading to the increase in the angle of tangency towards 90 degrees when the inertial effects control the bubble–particle collision. The particle density has very strong effect on the angle of tangency and fore-and-aft asymmetry of bubble–particle collision. As shown in Fig. 10, the deviation of the angle of tangency from 90 degrees is significant at low density. Similarly to the results presented in Fig. 9, the angle of tangency decreases from 90 degrees with increasing the particle size, reaching a minimum and then increasing back to 90 degrees. This feature of the computational modelling is beyond the capability of the Generalised Sutherland Theory.

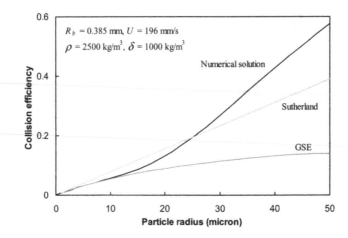

Figure 11. Comparison of the numerical results with the Sutherland Equation and Generalised Sutherland Equation (Table 1) for the collision efficiency. Reproduced with permission from [20].

Comparison between the computational results obtained with Eq. (1) for the collision efficiency and the GSE and original Sutherland equations is shown in Fig. 11. Good agreement between the GSE equation and the computational results is only observed for ultrafine particles. The deviation of the GSE equation from the computational results is significant for non-ultrafine particles. Furthermore, because the angle of tangency has been used for predicting the attachment efficiency [31–34], the significant deviation of the angle of tangency predicted by the GSE theory from the exact solutions shown in Fig. 9 will results in significant differences in modelling and predicting bubble–particle attachment interaction in flotation, in particular in flotation of non-ultrafine particles.

C. Attachment

An intervening liquid film during bubble–particle attachment is formed in which interfacial forces and microhydrodynamic resistance at short inter-surface separation distance become important, governing further stability of the liquid film between the vapour–liquid and solid–liquid interfaces. The net interfacial force between a hydrophobic particle and an air bubble is attractive, causing destabilization of the liquid film. In this case, the liquid film becomes unstable and ruptures, leading to the formation of a three-phase contact (tpc) line and attachment of the bubble. The bubble–particle contact expands further across the particle surface at a certain rate to form a stable wetting perimeter. The three-phase contact expansion initiated during rupture of the film leads to an equilibrium state, governed by the de-wetting dynamics and the properties of the gas–solid, gas–liquid and liquid–solid interfaces. The three steps of bubble–particle attachment are illustrated in Fig. 12. This part of the chapter focuses on unifying interfacial forces and microhydrodynamic resistance to model the attachment interaction. The modelling and measurements of the

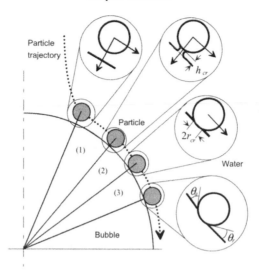

Figure 12. Three steps of bubble–particle attachment: (1) Thinning of the intervening liquid film to critical film thickness, h_{cr}, (2) rupture of the intervening liquid film formed when the bubble and the particle are very close together and formation of the three phase contact nucleus with radius r_{cr}, and (3) the spreading of three-phase contact line from the critical radius to form a stable wetting perimeter with (advancing, θ_a, and receding, θ_r) equilibrium contact angles. Reproduced with permission from [37].

attachment (induction) and contact times and the spreading of the gas–liquid–solid contact line have been reviewed [35]. The interfacial forces, thinning of the intervening liquid film and deformation of the gas–liquid interface during the attachment interaction have also been recently reviewed [36].

1. Extant Models for Attachment Efficiency

The attachment of particles is a function of the contact and attachment times. After colliding with bubble, the particle deforms the bubble surface and then slides along the surface. For attachment to occur the bubble–particle contact time, t_{con}, must not be shorter than the attachment time, t_{at}, i.e., $t_{con} \geqslant t_{at}$. The attachment time is equal to the sum of the times of the bubble–particle attachment steps shown in Fig. 12. The bubble–particle contact time and attachment time were first introduced by Sutherland [12] who developed a model for the contact time based on the particle sliding motion and the potential flow for water described by Eq. (6), leading to

$$V_\varphi = (R_p + R_b)\frac{d\varphi}{dt} = W_\varphi = U\left\{1 + \frac{1}{2}\left(\frac{R_p}{R_b}\right)^2\right\}\sin\varphi. \tag{8}$$

The contact time was obtained by integrating Eq. (8), which was further balanced with the attachment time, giving [12]

$$t_{at} = \frac{2(R_p + R_b)}{U\{1 + (1/2)(R_b/(R_p + R_b))^3\}}\ln\left\{\cot\frac{\varphi_a}{2}\right\}, \tag{9}$$

where φ_a is the polar coordinate at the beginning of the particle-grazing trajectory required for attachment. The attachment efficiency is then determined as

$$E_a = \sin^2 \varphi_a. \tag{10}$$

Eliminating φ_a from Eqs (9) and (10) yields

$$E_a = \text{sech}^2\left\{\left[\frac{1}{2} + \frac{1}{4}\left(\frac{R_b}{R_b + R_p}\right)^3\right]\frac{U t_{at}}{R_b + R_p}\right\}, \tag{11}$$

where the hyperbolic function is defined as $\text{sech}(x) = 2/(e^x + e^{-x})$. In the limit as $R_b \gg R_p$, Eq. (11) can be simplified to the original Sutherland expression given in Table 2.

The Sutherland approach has been extended to include various effects on attachment efficiency such as the bubble Reynolds number, the gas volume fraction, the bubble surface mobility, the fore-and-aft asymmetry of water flow and particle

Table 2.

Available models for the attachment efficiency

Author(s)	Attachment efficiency and time	Notes and symbol definitions
Sutherland [12]	$E_a = \text{sech}^2\{\frac{3U t_{at}}{4R_b}\}$	Potential flow; no inertial and gravitational effects; $R_b \gg R_p$
Yoon and Luttrell [7]	$E_a = \text{sech}^2\{\frac{U t_{at}(45 + 8Re^{0.72})}{30R_b(1 + R_b/R_b)}\}$	Intermediate bubble Reynolds number; fully immobile bubble surface; flow fore-and-aft symmetry; no inertial and gravitational effects; $R_b \gg R_p$
Dobby and Finch [8]	$E_a = \{\frac{\sin\varphi_a}{\sin\varphi_c}\}^2$ $t_{at} = (\varphi_c - \varphi_a)\frac{R_p + R_b}{\overline{W_\varphi + V_s \sin\varphi}}$ $\varphi_c = 9 + 8.1\rho +$ $\quad (0.9 - 0.09\rho)(78.1 - 7.37\log Re)$	Intermediate bubble Reynolds number; immobile bubble surface; flow fore-and-aft asymmetry. Polar angles measured in radians. Averaged variables (overbar) determined from φ_a to φ_c. Liquid velocity determined using numerical data available in the literature
Nguyen and Schulze [1]	Attachment time $t_{at} = \frac{R_p + R_b}{U(1 - B^2)A} \times$ $\ln\{\frac{\tan(\varphi_c/2)}{\tan(\varphi_a/2)}[\frac{\csc(\varphi_c) + B\cot(\varphi_c)}{\csc(\varphi_a) + B\cot(\varphi_a)}]^B\}$ Immobile bubble surface $E_a = \{\frac{\sin\varphi_a}{\sin\varphi_c}\}^2 \frac{X + C + Y\cos\varphi_a}{X + C + Y\cos\varphi_c}$ Mobile bubble surface $E_a = \frac{\sin^2\varphi_a - C_1 X^2 \frac{\cos^3\varphi_a - 3\cos\varphi_a + 2}{3(X + C)}}{\sin^2\varphi_c - C_1 X^2 \frac{\cos^3\varphi_c - 3\cos\varphi_c + 2}{3(X + C)}}$	Intermediate bubble Reynolds number; immobile and mobile bubble surface; flow fore-and-aft asymmetry; $R_b \gg R_p$. Model parameters $(A, B, C, X$ and $Y)$ are functions of Re and gas volume fraction (as well as the bubble surface mobility) — see [1]

motion, and the particle inertia and settling velocity. A summary of the extended models is described in Table 2.

The effect of particle size and bubble with either an immobile or a mobile surface on the attachment efficiency with different attachment times is shown in Fig. 13. For a given particle size, the attachment efficiency increases with deceasing the attachment time. Importantly, attachment efficiency is shown to increase with decreasing particle size, due to the decrease in the tangential velocity and increase in the sliding contact time for smaller particles. Therefore, the equations suggest that small particles become attached more readily than large particles.

For attachment to mobile bubbles the dependence of the attachment efficiency on particle size is not as strong as for the attachment to immobile bubbles. The dependence of attachment efficiency on the bubble size is opposite to the dependence on the particle size, i.e., at a given attachment time, the dependence of attachment efficiency on the bubble size is not significant for immobile bubbles, while the attachment efficiency is shown to increase with decreasing the bubble size.

Unlike the collision sub-process, the attachment sub-process is significantly determined by the interfacial properties of the particle and bubble surfaces. It can be described by attachment time, which is an integral measure of the physics and chemistry of the particle and bubble surfaces, underlying the liquid film drainage and rupture, and the motion of the three-phase contact line for forming a stable contact. The attachment time has been introduced as a useful user-selected parameter in modelling the attachment efficiency (41–43). However, it is difficult to be predicted from the first principles, unifying surface forces, contact angle and electrokinetic properties of the bubble and particle surfaces etc. An alternative development is proposed below.

2. New Development in Modelling Bubble–Particle Attachment and Collection

The divergence-free condition, i.e., $\text{div}(\vec{V}) = 0$ was silently employed to convert Eq. (2) to Eq. (3). The full description of the condition is available in the literature [1] — see page 189. Briefly, under the divergence-free condition, the concentration of particles diverted by a bubble remains constant long its path, allowing the link between the particle fluxes at "infinity" and at the bubble surface. If the divergence-free condition is not met, the theory can be generalised to predict the collection efficiency, E, as

$$E = \frac{2 \int_0^\pi -J_r(\varphi) \sin \varphi \, d\varphi}{(U + V_s)c_\infty},$$
(12)

where J_r is the radial flux of particles calculated at the bubble surface and c_∞ is the particle concentration in the solution (far from the bubble surface). The minus sign occurs in Eq. (12) because the flux is directed towards the bubble surface, while the radial spherical coordinate is directed outwards to the liquid phase, as shown in Fig. 3. The particle flux, $\vec{J}$, can be determined by considering various

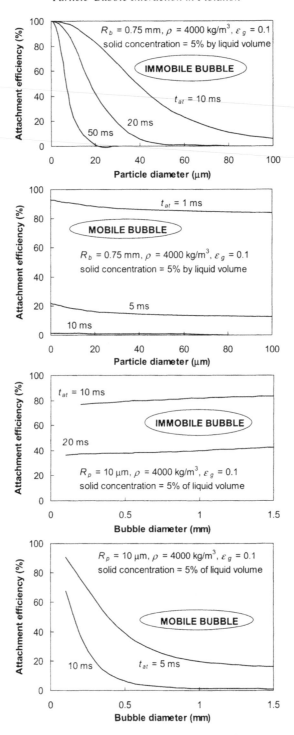

Figure 13. Attachment efficiency *versus* attachment time and particle and bubble diameter as predicted by Schulze and Nguyen (Table 2). ε_g is the gas volume fraction. Reproduced with permission from [1].

particle transport mechanisms, including interception, gravitational settling, inertial deposition and diffusion, giving

$$\vec{J} = c\vec{V} - \mathbf{D} \cdot \text{grad}(c),\tag{13}$$

where c is the particle concentration, $\vec{V}$ is the non-diffusion component of the particle velocity, and $\mathbf{D}$ is the tensor of the particle diffusivity. The first and second terms of Eq. (13) describe the particle fluxes the convection and diffusion processes, respectively. The particle concentration satisfy the conservation of mass, given by

$$\frac{\partial c}{\partial t} + \text{div}(\vec{J}) = 0,\tag{14}$$

where t is the reference time. Equation (14) has to be solved to determine the radial flux in Eq. (12).

The theory presented here has recently been used to predict the collection of nano and submicron particles by flotation [38]. For these particles, the non-diffusion velocity can be determined as

$$\vec{V} = \mathbf{R} \cdot \vec{W} + \mathbf{m} \cdot \vec{F}_{ex},\tag{15}$$

where $\mathbf{R}$ and $\mathbf{m}$ are the tensors of particle resistance and hydrodynamic mobility, respectively, $\vec{W}$ is the water velocity. The external forces, F_{ex}, on the right-hand side of Eq. (15) include gravitational forces, interparticle colloidal forces, and inertial forces arising from the particle and fluid acceleration. Equation (15) is central to the new developments because the introduction of the new terms for the resistance and mobility tensors and the external forces allows the inclusion of the film thickness and colloidal forces in the attachment modelling. The attachment time is not needed here. The modelling also unifies the collision and attachment sub-processes into a single collection model.

There are a few similar flotation theories developed to estimate the flotation recovery of fine particles, including Reay and Ratcliff [22], Collins and Jameson [39] and Ramirez *et al.* [40, 41]. An interesting feature of these predictions is that at a certain particle diameter, below 1 μm, the recovery of fine particles begins to increase with decreasing particle diameter. As the particle size decreases below one micron, its inertia becomes insignificant, and Brownian motion, hydrodynamic interaction and surface force interaction become the driving force for bubble–particle collection. The simplest of the three models, developed by Reay and Ratcliff [22], accounts only for the Brownian motion, while the Collins model [39] accounts for hydrodynamic interactions with a immobile bubble surface and London–van der Waals force, in addition to the Brownian motion. While the Ramirez model [40, 41] accounted for the same effects as the Collins model, the hydrodynamic mobility was calculated as a function of inter-particle distance to take account of the Marangoni effect. This model involved the formulation of the Fokker–Planck equation for the pair-distribution function and was designed to account for bubbles with different mobilities. There are a number of significant problems with each of these models including their neglect of the electrical double layer interaction force and the hydrophobic force in their surface force calculations. The authors also applied the traditional Hamaker constant combining rules in their model, which gave a negative Hamaker constant, indicating that the thin water film between a gas bubble and a solid surface is stable and should never disrupt spontaneously, which is contrary to observation [1]. Below is a new consistent model for particle collection by Brownian diffusion and convection forces, developed based on Eq. (15). Fundamental theories of convection, micro-hydrodynamics and colloidal forces will be incorporated. Flotation experimental results will be used to validate the new model simulations for bubble–particle attachment and collection.

2.1. Microhydrodynamic Resistance and Mobility Tensors

The tensors of diffusivity and mobility are related by the Stokes–Einstein equation: $\mathbf{D} = k_B T \mathbf{m}$, where k_B is the Boltzmann constant and T is the absolute temperature. The tensors are described as

$$\mathbf{m} = m_\infty \begin{bmatrix} \mathbf{i_r i_r} f_1 & 0 \\ 0 & \mathbf{i_\varphi i_\varphi} f_3 \end{bmatrix}, \tag{16}$$

$$R = \begin{bmatrix} \mathbf{i_r i_r} f_2 & 0 \\ 0 & \mathbf{i_\varphi i_\varphi} f_4 \end{bmatrix}, \tag{17}$$

where $m_\infty = 1/(6\pi\mu R_p)$ is the hydrodynamic mobility of particles far from the bubble surface, $\mathbf{i_r}$ and $\mathbf{i_\varphi}$ are the radial and tangential unit vectors, and $\mathbf{i_r i_r}$ and $\mathbf{i_\varphi i_\varphi}$ are the unit dyadics. The microhydrodynamic resistance functions, f_i ($i = 1$ to 4), are equal to unity if the bubble and the particle are far apart. They significantly change with deceasing inter-surface separation distance, $h = r - R_b - R_p$, between the bubble and the particle due to the intervening liquid film, where r is the dimensional radial distance between the particle and bubble centres. The change also depends on the mobility of the bubble surface. The functions can be determined from microhydrodynamics [42–45]. The resistance functions in Eqs (16) and (17) are combinations [38] of the resistance functions numerically determined in [42–45] and reviewed in [1]. The numerical results can conveniently be replaced by simple approximate equations described in Table 3. The approximate equations are

Table 3.
Approximate equations for the hydrodynamic resistance functions

Immobile bubble surface	Mobile bubble surface
$f_1 = \{1 + (\frac{R_p}{h})^{0.915}\}^{-1.093}$	$f_1 = \{1 + (\frac{R_p}{4h})^{0.702}\}^{-1.424}$
$f_2 = \{1 + (\frac{R_p}{3.201h})^{1.256}\}^{-0.796}$	$f_2 = \{1 + (\frac{R_p}{8.109h})^{0.909}\}^{-1.10}$
$f_3 = \{1 + A\{\ln[B(\frac{R_p}{h})^n + 1]\}^p\}^{-q}$	$f_3 = \frac{1.501 + h/R_p}{1.106 + h/R_p}$
$A = 0.498,\ B = 1.207,\ n = 0.986,\ p = 1.027$ and $q = 0.979$	
$f_4 = \{1 + A\{\ln[B(\frac{R_p}{h})^n + 1]\}^p\}^{-q}$	$f_4 = 1$
$A = 0.173,\ B = 0.693,\ n = 1.195,\ p = 1.418$ and $q = 0.641$	

weighted so that they asymptotically reduce to the correct limits at $h \to \infty$ and at $h \to 0$. Some of the numerical constants in Table 1 are slightly different from those reported in [38] because the correlations were re-fitted to the numerical data, giving relative errors within 5%.

2.2. Colloidal Forces

If the inertial forces are neglected, the equation for the external forces on the right-hand side of Eq. (15) gives (Fig. 3)

$$\vec{F}_{\text{ext}} = \frac{4\pi R_{\text{p}}^3 (\rho - \delta)}{3} \vec{g} + (F_{\text{vdW}} + F_{\text{edl}} + F_{\text{nonDLVO}}) \frac{\vec{r}}{r}. \tag{18}$$

The second term on the right-hand side of Eq. (18) describes the three principal colloidal forces, including, the van der Waals, electrical double-layer, and the non-DLVO forces. The first two forces are the key elements of the celebrated Derjaguin–Landau–Verwey–Overbeek (DLVO) theory of colloid stability [46, 47]. The attractive force between hydrophobic surfaces is one of the non-DLVO forces relevant to the bubble–particle collection interaction.

The radial and tangential components of the external forces are obtained from Eq. (18) as

$$F_{\text{r}} = -\frac{4\pi R_{\text{p}}^3 (\rho - \delta) g}{3} \cos\varphi + (F_{\text{vdW}} + F_{\text{edl}} + F_{\text{nonDLVO}}), \tag{19}$$

$$F_{\varphi} = \frac{4\pi R_{\text{p}}^3 \rho g}{3} \sin\varphi. \tag{20}$$

van der Waals Force. This universal force between molecules and atoms has long been known to play an important role in the capture of colloidal particles by surfaces. For the asymmetric bubble–particle systems encountered in flotation, the unretarded van der Waals force is repulsive and was wrongly considered as the driving force for the particle attachment onto the bubble surface [40, 41]. The retarded van der Waals force is influenced by the electrolyte concentration and can be attractive at large separation distances. The prediction of this attractive van der Waals force between a bubble and a particle in flotation systems is only recently available [48] using a combined macroscopic (Hamaker) and the microscopic (Lifshitz) approaches. The most recent expression derived from the combined Hamaker–Lifshitz theory gives [1]

$$F_{\text{vdW}} = -\frac{dE_{\text{vdW}}}{dr}, \tag{21}$$

$$E_{\text{vdW}} = -\frac{A}{6} \left\{ \frac{2R_{\text{p}}R_{\text{b}}}{r^2 - (R_{\text{p}} + R_{\text{b}})^2} + \frac{2R_{\text{p}}R_{\text{b}}}{r^2 - (R_{\text{p}} - R_{\text{b}})^2} + \ln \frac{r^2 - (R_{\text{p}} + R_{\text{b}})^2}{r^2 - (R_{\text{p}} - R_{\text{b}})^2} \right\}. \tag{22}$$

As shown previously, the inter-centre bubble–particle distance, r, is related to the shortest separation distance, h, between the surfaces by $r = h + R_{\text{p}} + R_{\text{b}}$. The Hamaker–Lifshitz function, A, in Eq. (22) is defined by [48]

$$A(\kappa, h) = A^0 (1 + 2\kappa h) e^{-2\kappa h} + A^{\xi}(h), \tag{23}$$

where κ is the Debye constant. The zero-frequency, A^0, and non-zero frequency, A^ξ, terms in Eq. (23) are determined as $A^0 \cong 0.75 k_B T$ and

$$A^\xi(h) = -0.235 \hbar \omega \frac{n_p^2 - 1.887}{n_p^2 - 1} \left\{ \frac{0.588}{[1 + (h/5.59)^q]^{1/q}} - \frac{(n_p^2 + 1.887)^{-1/2}}{[1 + (h/\lambda_p)^q]^{1/q}} \right\}, \quad (24)$$

where $\hbar$ is the Planck constant divided by 2π, n_p is the particle refractive index, $\omega = 2 \times 10^{16}$ rad/s and λ_p is a modified London wavelength accounting for the retardation effect, defined as

$$\lambda_p = \frac{9.499}{\sqrt{n_p^2 + 1.887}}. \quad (25)$$

Electrical Double-Layer Force. The particle and bubble surfaces in aqueous media are charged and have electrical double-layers associated with them. The diffuse layers of the double-layers surrounding the particle and bubble surfaces overlap at close approach, and, as a result, a repulsive or attractive force develops. The determination of the double-layer force depends on the charging mechanism during the overlapping of the diffuse layers. The double-layer interactions at constant surface potential and constant surface charge are usually considered. The actual double-layer interaction occurs between these two limits. The calculation of the double-layer force as a function of the separation distance is recently reviewed by Nguyen *et al.* [49]. For low surface (zeta) potentials (<50 mV), the Hogg–Healy–Fuerstenau approximation for the double-layer interaction at constant surface potentials gives

$$F_{edl} = \varepsilon \varepsilon_0 \kappa \frac{2\pi R_b R_p}{R_b + R_p} \frac{2\psi_b \psi_p \exp(\kappa h) - \psi_b^2 - \psi_p^2}{\exp(2\kappa h) - 1}, \quad (26)$$

where ψ_p and ψ_b are the particle and bubble surface potentials, ε_0 is the dielectric constant of vacuum and $\varepsilon = 80$ for water.

Hydrophobic (non-DLVO) Force. Since this force difficult to theoretically model at present, the empirical correlation fitted with double exponentials has been used, giving

$$F_{nonDLVO} = \frac{R_p R_b}{R_p + R_b} \left[K_1 \exp\left(-\frac{h}{\lambda_1}\right) + K_2 \exp\left(-\frac{h}{\lambda_2}\right) \right], \quad (27)$$

where K_1 and K_2 are the force constants, and λ_1 and λ_2 are the decay lengths of the hydrophobic force. These parameters are determined from force measurements obtained with a surface force apparatus or by atomic force microscopy. The available data for these parameters are reviewed in [1]. The double exponential dependence has no real physical basis — it only describes a difference between DLVO and experimental data for surface forces. Indeed, the double exponential reflects the presence of surface nanobubbles [50–55].

2.3. Water Velocity

The water velocity around the bubble is the key parameter of the convection transport mechanism described by the first term in Eq. (15). Since the mass balance described by Eq. (14) has to be solved for the entire flow field around the bubble, the available close-form solutions for bubbles at intermediate Reynolds numbers like Eq. (7) which are only valid at the bubble surface cannot be used. Therefore, numerical solution to the Navier–Stokes equations must be employed. For simplicity, here we use the Hadamard–Rybczynski theory for small bubbles with small Reynolds numbers and a mobile surface [29] which gives

$$W_r = -U \left\{ 1 - \frac{1}{r} \right\} \cos \varphi, \tag{28}$$

$$W_\varphi = U \left\{ 1 - \frac{1}{2r} \right\} \sin \varphi, \tag{29}$$

where U is the bubble slip velocity and the radial coordinate, r, is dimensionless (made by dividing by the bubble radius).

2.4. Particle Transport Equation in Bubble Spherical Coordinates and Its Solution
At steady state, the time partial derivative in Eq. (14) is zero and the particle transport equation in the bubble spherical coordinates (r, θ) can be written as

$$\frac{1}{r^2} \frac{\partial}{\partial r} (r^2 J_r) + \frac{1}{r \sin \varphi} \frac{\partial}{\partial \varphi} (\sin \varphi J_\varphi) = 0. \tag{30}$$

Since the colloidal forces change rapidly in the limit as $r \to 1$, the exponential stretching of the radial coordinate, $r = \exp(z)$, can be employed to increase the accuracy of the numerical solution of Eq. (30). In terms of the new stretching variable, z, Eq. (30) simplifies to

$$2 J_r + \cot \varphi J_\varphi + \frac{\partial J_r}{\partial z} + \frac{\partial J_\varphi}{\partial \varphi} = 0. \tag{31}$$

After substituting various available equations into Eq. (31) one obtains

$$\frac{\partial^2 c}{\partial z^2} f_1 + \frac{\partial^2 c}{\partial \varphi^2} f_3 + \frac{\partial c}{\partial z} B_z + \frac{\partial c}{\partial \varphi} B_\varphi + c B_c = 0, \tag{32}$$

where parameters B_z, B_φ, and B_c are functions of the particle radial and polar positions, liquid velocity, the bubble and particle terminal velocities, surface forces, the bubble and particle diameters, and the particle diffusivity D_∞.

Equation (32) may be solved using the boundary conditions: $\partial c / \partial r = 0$ at $\varphi = 0$ and $\varphi = 180$ degrees, $c(r \to \infty, \varphi) = c_\infty$ and $c(r \to 0, \varphi) = 0$. The first and second derivatives in Eq. (32) on a non-uniform mesh can be approximated with the central finite difference scheme. The over-relaxation method can be applied to solve the algebraic equations. The initial solutions can be obtained using the parabolic approximation with $\partial^2 c / \partial \varphi^2 = 0$ and the Crank–Nicholson numerical method.

In the numerical computation, the (outer) boundary condition far from the bubble surface (at infinity) must be imposed at some finite distance measured from the bubble surface. The finite distance can numerically be found by trial and error. In this procedure, a number of the finite distances can be selected to solve the governing equation. Then the numerical results for the particle concentration obtained with different values for the distance are compared. The finite distance is determined when no significant difference in the particle concentration distributions is found. A relative error of 0.1% in the relative concentration has been used as the quantitative measure [38]. Typically, the finite distance of $200R_p$ was found to satisfy the requirement in the report [38].

Due to the nature of the molecular interactions, the colloidal forces, in particular the van der Waals and double-layer forces are infinitely large, as the particle approaches the bubble surface. Therefore, the particle transport equation described by Eq. (32) is numerically very stiff and extremely small step sizes in the radial and tangential directions are required to prevent the numerical oscillation and instability, particularly in the region in the vicinity of the bubble surface. It is also noted that the colloidal forces which cause the stiffness problem diminish at large separation distances. Therefore, an extremely fine mesh is not required for the region far from the bubble surface. Consequently, the computational domain can be divided into two regions, an inner and outer region. In the inner region, where the colloidal forces are very strong, an exponentially decreasing (towards the bubble surface) step size in the separation distance h can be used. The largest separation distance of the inner region can be the range of colloidal forces, say 500 nm. In the case of no attractive forces, the shortest separation distance of the inner region can be as small as about 10^{-10} times the particle radius [38]. The step size increases gradually towards the end of the inner region. A uniform step size with relatively large increment in the separation distance can be used for the outer region.

Once the particle concentration distribution around the bubble is determined, the local radial flux, J_r, of particles at the bubble surface can be determined from Eq. (13) and the collection efficiency is obtained by integrating Eq. (12).

Predicted values for the collection efficiency are shown in Fig. 14 for the bubble surface potential from 0 to -50 mV. The constants used in the calculations include the particle density, surface potential and bubble diameter, which are similar to the data obtained experimentally. The empirical hydrophobic decay lengths, λ_1 and λ_2 along with the associated hydrophobic force constants, K_1 and K_2, agree with the measured data using an atomic force microscope [56]. Because both the particle and bubble surfaces are charged with the same negative sign, the double-layer force is repulsive. Greater positive charges lead to decrease in the predicted collection efficiencies. The particle diameter where the minimum collection efficiency occurs decreases from around 100 nm down to 20 nm across the range of bubble potentials. For particles sizes below the minimum, the collection efficiency increases due to the increased Brownian diffusivity.

Comparison between the model and the available experimental results is shown in Fig. 15. The experimental results were obtained from the flotation of three different types of nano and sub-micron silica using 5×10^{-5} M cetyltrimethyl ammonium bromide (CTAB) to control the silica hydrophobicity and 30 ppm Dowfroth (polyglycols) to control the bubble size (of 150 microns diameter on average) [38]. The

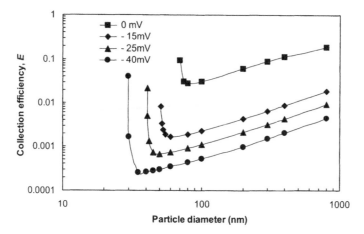

Figure 14. Prediction for the collection efficiency, Eq. (12), *versus* the particle size and the bubble surface potential. The numerical constants used in the computation are bubble diameter (150 μm); particle refractive index (1.54), density (2600 kg/m^3) and surface potential (−35 mV); Debye length (50 nm); and hydrophobic force parameters: $K_1 = -7$ mN/m, $K_2 = -6$ mN/m, $\lambda_1 = 6$ nm and $\lambda_2 = 20$ nm. Reproduced with permission from [38].

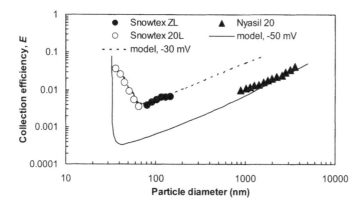

Figure 15. Comparison between model and experimental data for collection efficiency *versus* silica particle diameter. The numerical constants used in the calculation are described in the caption to Fig. 14, except for $K_2 = -20$ mN/m in the simulation with −30 mV for the bubble zeta potential. Reproduced with permission from [38].

zeta potential of the bubble in the CTAB solution was measured as −55 mV, while the model simulation required the bubble zeta potentials of −50 mV and −30 mV. The addition of neutral Dowfroth molecules should take up some of the sites on the bubble surface previously occupied by the CTAB molecules, thereby reducing the overall zeta potential of the bubbles. The collection efficiency of nano silica particle Snowtex 20L increased with decreasing particle diameter. This result is significant as it shows that Brownian motion is the controlling collection mechanism. The predicted particle diameter at minimum collection efficiency reasonably agrees with the experimental data.

Both the modelling and experiment show significant effect of the electrical double-layer and non-DLVO hydrophobic attractive forces on the collection of nano and submicron particles by air bubbles. The theoretical and experimental results also show the collection efficiency to have a minimum at a particle size in the order of 100 nm. The interception and collision mechanisms predominate in flotation of larger particles, while the diffusion and colloidal forces control the collection of particles with a size smaller than the transition size. This new result for flotation of nano and submicron particles is beyond the classical modelling presented in the previous section.

D. Stability and Detachment

The key issue in quantifying bubble–particle aggregate stability is to determine whether or not the adhesive force acting on the attached particle is sufficient large to prevent the particle detaching from the bubble surface under the dynamic forces existing in flotation cells. Significantly, stability and detachment are important for flotation of coarse particles. The focus here is on the force analysis and on the determination of the strength of the particle attachment.

1. Tenacity of Particle Attachment

Adhesive forces acting on the three-phase contact line can be described using Fig. 16. The adhesive forces include:

- Capillary force, F_c,
- Hydrostatic pressure force, F_h, on the area enclosed by the three-phase contact and
- Buoyancy, F_b, of the particle volume immersed in the liquid phase.

The most important detaching forces are the particle weight, F_g, and the turbulent inertial forces, F_t. A description of these forces can be found in the literature [1].

The simple case of attachment of a particle attached to a stationary free surface (a large air bubble) is illustrated in Fig. 16. At equilibrium, the force balance gives

$$F_c + F_h + F_b - F_g = 0. \tag{33}$$

In terms of the polar position, α, of the three-phase contact on the particle surface

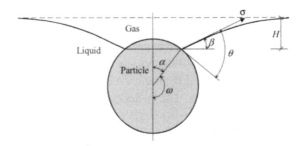

Figure 16. Schematic of a spherical particle attached to a much larger air bubble (a free surface).

Eq. (33) can be rewritten to give

$$2\pi R_p \sigma \sin\alpha \sin(\theta - \alpha) + \pi R_p^2 H \delta g \sin^2\alpha$$
$$- \frac{(2 - 3\cos\alpha + \cos^3\alpha)\pi R_p^3}{3}\delta g = \frac{4\pi R_p^3}{3}\Delta\rho g, \tag{34}$$

where the right-hand side represents the particle weight less its buoyancy, which are independent of the position of the three-phase contact.

The condition for equilibrium of forces on the particle attached to the gas–liquid interface described by Eq. (34) is necessary but not sufficient to determine the strength of the particle–meniscus aggregate since the left hand side changes with α.

Recent numerical solutions to the Young–Laplace equation describing the gas–liquid deformation [57] show that there exists a maximum of the left hand side of Eq. (34) as the position of the three-phase contact varies. This maximum determines the strength of the particle–meniscus aggregate, termed the tenacity, T, of the particle attachment. The particle is detached from the meniscus only if the detaching forces exceed the tenacity. The results of the numerical computation further show that for typical particle size and contact angle encountered in flotation the tenacity of attachment can be approximately described by

$$T = \pi R_p \sigma (1 - \cos\theta)\left\{1 + 0.016\frac{R_p}{L}\right\}, \tag{35}$$

where $L = \sqrt{\sigma/(\delta g)}$ and is the capillary length, which is about 2.7 mm for air–water system.

2. Influence of Bubble Size

The influence of the bubble size on the adhesive force can be included by modifying the pressure force term in Eq. (34) as

$$F_h = \pi R_p^2 \sin^2\alpha\left(\delta g H - \frac{2\sigma}{R_b}\right). \tag{36}$$

The height, H, of the three-phase contact in Eq. (36) is also a function of the bubble radius (see Fig. 17), and can be approximated by

$$H = R_b\left\{1 + \sqrt{1 - \left(\frac{R_p}{R_b}\right)^2 \sin^2\alpha}\right\}. \tag{37}$$

The remaining terms in Eq. (34) are unchanged when the bubble size is considered. Consequently, a similar approach to determining the tenacity of particle attachment can be applied to give [57]

$$T = \pi R_p \sigma(\sqrt{1 + 2G\cos\theta + G^2} - \cos\theta - G) + O\{(R_p/L)^3\}, \tag{38}$$

where G is a function of the particle and bubble radius which is described by

$$G = \frac{R_p}{R_b} - \frac{R_b R_p}{L^2}. \tag{39}$$

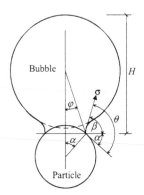

Figure 17. Particle attached to a bubble (not to scale).

As expected, for the case of the particle radius being significantly smaller than the capillary length both Eqs (35) and (38) simplifies into the correct limit described as

$$T = \pi R_p \sigma (1 - \cos\theta) + O\{(R_p/L)^2\}. \tag{40}$$

3. Maximum Size of Floatable Particles

The theoretical developments for the strength of the particle attachment described in Sections 4.1 and 4.2 can be employed to determine the efficiency of bubble–particle stability, as well as in the prediction of the maximum size of floatable particles. For example, to determine the maximum size of floatable particles it is assumed that for mechanical flotation cells turbulence is the main mechanism causing the particle detachment. On this basis, the detaching force, F_{de}, is given as

$$F_{de} = \frac{4\pi R_p^3}{3} \Delta\rho (g + a_m), \tag{41}$$

where g and a_m are the accelerations of gravity and of turbulent eddies, respectively. The condition for the stability of the bubble–particle aggregates can now be described as

$$T \geqslant F_{de}. \tag{42}$$

Inserting Eqs (40) and (41) into Eq. (42) and solving for the particle radius leads to

$$\frac{R_p}{L} \leqslant \sqrt{\frac{3\sigma(1 - \cos\theta)}{4\Delta\rho(g + a_m)}}. \tag{43}$$

For the maximum size, $R_{p\ max}$, of floatable particles, the upper limit of Eq. (43) gives

$$\frac{R_{p\ max}}{L} = \sqrt{\frac{3\sigma(1 - \cos\theta)}{4\Delta\rho(g + a_m)}}. \tag{44}$$

Table 4.
Dependence of the maximum size of floatable sylvinite
particles on the acceleration of turbulent eddies [57]

a_m/g	$R_{p\ max}$ (μm)	
	Eq. (44)	Experimental
4.45	708	700
8.75	529	480
12.10	457	430

Equation (44) exhibits the expected relationship that the maximum particle size is proportional to the contact angle and inversely proportional to the turbulent acceleration. Quantitatively, results are given in Table 1, where the turbulent acceleration, a_m, has been obtained from the experimental data. More accurate prediction could be achieved if Eqs (35) and (38) were used.

4. Efficiency of the Bubble–Particle Aggregate Stability

Knowing the maximum size of floatable particles, the strength and tenacity of the bubble–particle aggregates, the bubble–particle aggregate detachment and stability can be quantitatively described by an efficiency which is directly related to the flotation kinetics [1]. The efficiency of the bubble–particle aggregate stability, E_s, and the detachment efficiency, E_d, are related by the balance equation: $E_s = 1 - E_d$. The detachment efficiency is defined as the ratio of the rate of particles being detached from the bubble surface to the rate of particles being attached, giving [1]

$$E_s = 1 - \exp\left\{1 - \frac{T}{F_{de}}\right\}. \tag{45}$$

The exponential distribution described by Eq. (45) has the property that both the efficiencies E_s and E_d are bounded by 1 and zero.

Since the particle detachment depends on a number of possible mechanisms and external forces, including gravitational forces, tensile stresses, shear stress and bubble vibrations, a number of equations can be established for E_s and are summarised in Table 5. It is noted that the attachment tenacity also depends on the hysteresis of the particle contact angles and, hence, the stability efficiency is a function of the hysteresis.

An example demonstrating how the turbulent energy, the particle hydrophobicity and the particle size influence the efficiency of the bubble–particle aggregate stability is shown in Fig. 18. For small particle radius, the detaching force due to turbulent tensile stress is significantly smaller than the particle tenacity and the efficiency of the bubble–particle aggregate stability is approximately equal to 1. As the particle radius increases, E_s sharply decreases to zero if the turbulent dissipation energy rate is relatively high.

Table 5.
Typical expressions for the efficiency of the bubble–particle aggregate stability

Disruptive force	Stability efficiency	Notes
Tensile stress	$E_s = 1 - \exp\{1 - \dfrac{3\sigma(1-\cos\theta_A)}{4R_p^2(a_m+g)\Delta\rho}\}$ if $\Delta\theta \leqslant \theta_R$ $E_s = 1 - \exp\{1 - \dfrac{3\sigma\sin\theta_R\sin(\Delta\theta)}{2R_p^2(a_m+g)\Delta\rho}\}$ if $\Delta\theta \geqslant \theta_R$ $a_m = \dfrac{2\varepsilon\Delta r/v}{\{30^{3/2}+(\Delta r)^2\varepsilon^{1/2}v^{-3/2}\}^{2/3}}$ $\Delta r = 2(R_b + R_p)$	θ_A is advancing contact angle. θ_R is receding contact angle. $\Delta\theta = \theta_A - \theta_R$ is the contact angle hysteresis. ε is the turbulent energy dissipation. v is the kinetic viscosity
Shear stress	$E_s = 1 - \exp\{1 - \dfrac{\sigma(1-\cos\theta_A)\sin((\Delta\theta)/2)}{0.26\pi R_p\delta\sqrt{\varepsilon v}}\}$ if $\Delta\theta \leqslant \theta_R$ $E_s = 1 - \exp\{1 - \dfrac{\sigma\sin\theta_R\sin(\Delta\theta)\sin((\Delta\theta)/2)}{0.13R_p\pi\delta\sqrt{\varepsilon v}}\}$ if $\Delta\theta \geqslant \theta_R$	

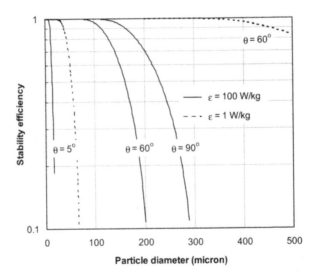

Figure 18. 18 The efficiency of the bubble–particle aggregate stability as a function of the particle size, contact angle θ and turbulent energy dissipation rate ε. Assumptions include disruption by tensile stress and small hysteresis of contact angle: $\Delta\theta \leqslant \theta_R$. Reproduced with permission from [1].

E. Conclusions

In this chapter, bubble–particle interaction in flotation has been reviewed. Three key sub-processes, including collision, attachment and detachment have been considered. The collision efficiency can be determined from the particle approach velocity in the near proximity of the bubble surface. The liquid flow around air bubbles in flotation is characteristically fore-and-aft asymmetric. The particle motion can be fore-and-aft asymmetric, depending on the bubble surface mobility and the particle inertia. The angle of tangency on a fully mobile bubble surface decreases with increasing the particle inertia, reaching a minimum, and then increases back to 90 degrees at the bubble equator. The bubble surface mobility and inertia have been included in the modelling of the collision efficiency.

Detachment and stability can better be determined in terms of the tenacity of the particle attachment, which is the maximum of the adhesive forces and is a function of the particle and bubble size, the surface tension and the particle contact angle. Analysing the tenacity and the simplification of forces acting on particles before detaching can predict the maximum floatable particle size and the stability and detachment efficiency.

Classical models for the attachment efficiency have also been reviewed. The attachment time is the user-selected parameter of the models but is difficult to predict from the first principle at present. An alternative modelling approach incorporating colloidal forces and microhydrodynamic resistance functions due to the liquid film has been analysed. In this new model, there is no need of the attachment time. The new model unifies the collision and attachment sub-processes. A good agreement between the new model and flotation of nano and sub-micron silica particles has been obtained. Both the theoretical modelling and experiments show that there is a particle size of the order of 100 nm, where the collection efficiency has a minimum. Flotation of coarser particles is controlled by hydrodynamics, gravity and inertia. For smaller particles, the particle attachment and flotation are enhanced by Brownian diffusion, and the role of the electrical double-layer force and the hydrophobic surface forces in the particle attachment is critical. The prediction for the attractive forces between hydrophobic surfaces remains a major challenge. Ideally, the experimental data for the direct force measurements between a particle and a gas bubble in the flotation aqueous solutions can be used for the prediction of the particle attachment. Hopefully, the colloid probe technique developed with micro-fabricated atomic force microscope cantilevers will be able to provide the required data for the hydrophobic forces between a particle and a bubble in the near future.

F. Acknowledgement

The author gratefully acknowledges the financial support provided by the Australian Research Council through a Discovery Project grant PD0663688.

G. References

1. A.V. Nguyen and H.J. Schulze, Colloidal science of flotation, Marcel Dekker, New York, 2004, p. 840.
2. G.J. Jameson and S. Nam and M.M. Young, Minerals Sci. Engng., 9 (1977) 103–118.
3. K.L. Sutherland and I.W. Wark, Principles of Flotation, Aust. Inst. Min. Metall., Melbourne, 1955, p. 489.
4. H.J. Schulze, Physico-Chemical Elementary Processes in Flotation, Elsevier, Amsterdam, 1983, p. 320.

5. A.M. Gaudin, Flotation, McGraw-Hill, New York, 1957.

6. J.A. Finch and G.S. Dobby, Column Flotation, Pergamon, Oxford, 1990, p. 180.

7. R.H. Yoon and G.H. Luttrell, Miner. Proc. Extract. Met. Rev., 5 (1989) 101–122.

8. G.S. Dobby and J.A. Finch, Int. J. Miner. Process., 21 (3–4) (1987) 241–60.

9. H.J. Schulze, Miner. Process. Extract. Met. Rev., 5 (1989) 43–76.

10. J.F. Anfruns and J.A. Kitchener, Trans. Inst. Min. Metall. (Sect. C), 86 (1977) 9–15.

11. M.E. Weber and D. Paddock, J. Colloid Interface Sci., 94 (2) (1983) 328–35.

12. K. Sutherland, J. Phys. Chem., 52 (1948) 394–425.

13. Z. Dai, S. Dukhin, D. Fornasiero and J. Ralston, J. Colloid Interface Sci., 197 (2) (1998) 275–292.

14. J. Ralston, S.S. Dukhin and N.A. Mishchuk, Int. J. Miner. Process., 56 (1–4) (1999) 207–256.

15. A.V. Nguyen, Int. J. Miner. Process., 56 (1–4) (1999) 165–205.

16. Z. Dai, D. Fornasiero and J. Ralston, Adv. Colloid Interface Sci., 85 (2–3) (2000) 231–256.

17. B.V. Deryaguin and S.S. Dukhin, Trans. Inst. Min. Metall., 70 (1960–61) 221–233.

18. J. Ralston, D. Fornasiero and R. Hayes, Int. J. Miner. Process., 56 (1–4) (1999) 133–164.

19. R.H. Yoon, Aufbereit.-Tech., 32 (9) (1991) 474–85.

20. M.C. Nguyen, A.V. Nguyen and J.D. Miller, Int. J. Miner. Process., 81 (2006) 141–148.

21. L.R. Flint and W.J. Howarth, Chem. Eng. Sci., 26 (8) (1971) 1155–68.

22. D. Reay and G.A. Ratcliff, Can. J. Chem. Eng., 51 (2) (1973) 178–85.

23. A.V. Nguyen, J. Colloid Interface Sci., 162 (1) (1994) 123–8.

24. J.F. Anfruns and J.A. Kitchener, in "Flotation — A.M. Gaudin Memorial Volume", Vol. 2, M.C. Fuerstenau (Ed.), Am. Soc. Min. Engrs., New York, 1976, pp. 625–637.

25. S.S. Dukhin, Kolloidn. Zh., 45 (2) (1983) 207–18.

26. Z. Jiang and P.N. Holtham, Trans. — Inst. Min. Metall., Sect. C, 95 (1986) C187–C194.

27. D. Reay and G.A. Ratcliff, Can. J. Chem. Eng., 53 (5) (1975) 481–6.

28. N. Ahmed and G.J. Jameson, Int. J. Miner. Process., 14 (3) (1985) 195–215.

29. R. Clift, J.R. Grace and M.E. Weber, Bubbles, Drops and Particles, Academic Press, New York, 1978, p. 380.

30. A.V. Nguyen, Int. J. Miner. Process., 55 (2) (1998) 73–86.

31. Z. Dai, D. Fornasiero and J. Ralston, J. Colloid Interface Sci., 217 (1) (1999) 70–76.

32. B. Pyke, D. Fornasiero and J. Ralston, J. Colloid Interface Sci., 265 (1) (2003) 141–151.

33. J. Duan, D. Fornasiero and J. Ralston, Int. J. Miner. Process., 72 (1–4) (2003) 227–237.

34. J. Ralston, S.S. Dukhin and N.A. Mishchuk, Adv. Colloid Interface Sci., 95 (2–3) (2002) 145–236.

35. A.V. Nguyen, R.J. Pugh and G.J. Jameson, in "Colloidal particles at liquid interfaces", B.P. Binks and T.S. Horozov (Eds), Cambridge University Press, Cambridge, 2006, pp. 238–382.

36. A.V. Nguyen, J. Drelich, M. Colic, J. Nalaskowski and J.D. Miller, in "Encyclopedia of Surface and Colloid Science", P. Somasundaran (Ed.). Marcel Dekker, New York, NY, 2006, in press.

37. A.V. Nguyen, J. Ralston and H.J. Schulze, Int. J. Miner. Process., 53 (4) (1998) 225–249.

38. A.V. Nguyen, P. George and G.J. Jameson, Chem. Eng. Sci., 61 (2006) 2494–2509.

39. G.L. Collins and G.J. Jameson, Chem. Eng. Sci., 32 (3) (1977) 239–46.

40. J.A. Ramirez, R.H. Davis and A.Z. Zinchenko, Int. J. Multiphase Flow, 26 (6) (2000) 891–920.

41. J.A. Ramirez, A. Zinchenko, M. Loewenberg and R.H. Davis, Chem. Eng. Sci., 54 (2) (1999) 149–157.

42. A.V. Nguyen and G.M. Evans, Appl. Math. Model., 31 (2007) 763–769.

43. A.V. Nguyen and G.J. Jameson, Int. J. Multiphase Flow, 31 (4) (2005) 492–513.

44. A.V. Nguyen and G.M. Evans, J. Colloid Interface Sci., 273 (2004) 262–270.

45. V.A. Nguyen and G.M. Evans, Int. J. Multiphase Flow, 28 (2002) 1369–1380.

46. B. Derjaguin and L. Landau, Acta Phys.-chim., 14 (6) (1941) 633–662.

47. E.J.W. Verwey and J.T.G. Overbeek, Theory of the Stability of Lyophobic Colloids, Elsevier, Amsterdam, 1948, p. 218.

48. A.V. Nguyen, G.M. Evans and H.J. Schulze, Int. J. Miner. Process., 61 (3) (2001) 155–169.

49. A.V. Nguyen, G.M. Evans and G.J. Jameson, in "Encyclopedia of Surface and Colloid Science", Vol. 5, A.T. Hubbard (Ed.), Marcel Dekker, New York, 2002.

50. J. Yang, J. Duan, D. Fornasiero and J. Ralston, J. Phys. Chem. B, 107 (25) (2003) 6139–6147.

51. R. Steitz, T. Gutberlet, T. Hauss, B. Kloesgen, R. Krastev, S. Schemmel, A.C. Simonsen and G.H. Findenegg, Langmuir, 19 (6) (2003) 2409–2418.

52. P. Attard, M.P. Moody and J.W.G. Tyrrell, Physica A: Stat. Mech. Appl., 314 (1–4) (2002) 696–705.

53. H.K. Christenson and P.M. Claesson, Adv. Colloid Interface Sci., 91 (3) (2001) 391–436.

54. A.V. Nguyen and G.M. Evans, Exp.Therm. Fluid Sci., 28 (5) (2004) 381–385.

55. A.V. Nguyen, J. Nalaskowski, J.D. Miller and H.-J. Butt, Inter. J. Miner. Process., 72 (1–4) (2003) 215–225.

56. V.S.J. Craig, B.W. Ninham and R.M. Pashley, Langmuir, 15 (4) (1999) 1562–1569.

57. A.V. Nguyen, Inter. J. Min. Process., 68 (2002) 167–182.

Experimental Observation of Drop–Drop Coalescence in Liquid–Liquid Systems: Instrument Design and Features

Giuseppe Loglio [a], **Piero Pandolfini** [a], **Francesca Ravera** [b], **Robert Pugh** [c], **Alexander V. Makievski** [d], **Aliyar Javadi** [e] **and Reinhard Miller** [e]

[a] University of Florence, Department of Organic Chemistry, Via della Lastruccia, 13, I-50019 Sesto Fiorentino (Firenze), Italy
[b] Istituto per l'Energetica e le Interfasi, IENI-CNR , UOS Genova, Via De Marini 6, I-16149 Genoa, Italy
[c] Institute for Surface Chemistry, SE-114 86 Stockholm, Sweden
[d] SINTERFACE Technologies, Volmerstrasse 5-7, D-12489 Berlin, Germany
[e] Max-Planck-Institut für Kolloid- und Grenzflächenforschung, D14476 Golm, Germany

Contents

A. Introduction

The general way of characterising the stability of foams or bubbles is the registration of some integral properties, such as the volume of a foam [1] or the creaming in an emulsion [2]. For liquid film studies, however, no direct correlation between film properties and adsorption layer characteristics on one side, or foam/emulsion stability on the other side do exist. Only few first attempts exist to correlate the properties of these different complexity levels [3].

Bubble and Drop Interfaces
© Koninklijke Brill NV, Leiden, 2011

Since recently there are experimental tools available which allow mimicking the elements in foams or emulsions, i.e. to study the direct interaction between bubbles or drops, respectively. Of course, enormously difficult requirements result in order to control drops or bubbles on a realistic size scale of micrometers. While the dosing of μl and temperature constancy of $\pm 1°C$ is easily managed, such micrometer drops or bubbles would require dosing accuracies of 10^{-9} μl and temperature control of much better than 10^{-3} K. Hence, these tools are typically working with drops or bubbles with diameters of the order of 100 μm. In addition, however, they also allow to look into asymmetric systems like the interaction between a large and a small drop (bubble) and even between a drop and a bubble, as it is a realistic case in foamed emulsions, for example in cosmetics or ice cream [4, 5].

This chapter is dedicated to such type of experimental tools, able to manipulate drops and bubbles on a size scale of 50 to 1000 μm in diameter and describes the design and also some selected protocols for its use in fundamental science but also for very practical systems. The details given further below mainly correspond to the Drop Bubble Micro Manipulator DBMM (SINTERFACE Technologies, Berlin) and represent the current state of the art, while further work on this methodology in direction to smaller drops/bubbles is under way.

B. Importance of Drop–Drop Coalescence

Many multi-phase processes such as flotation, environmental clean-up operations, distillation, tertiary oil recovery and extraction involve the coalescence of air bubbles or oil droplets which are often in the form of foams or emulsions. Also, in many foods such as salad dressing, mayonnaise, and ice cream are in the form of emulsions in which the stability can be related to the performance. Over the past century, foam and emulsion coalescence research has been extensively studied by both the physical chemistry and the chemical engineering community and there has been a continuous effort to understand the nature of the mechanism.

During this period, from investigations mainly using concentrated systems, it has been established that several physicochemical properties (such as interfacial tension, interfacial tension gradients, interfacial shear viscosities, droplet size and droplet approach velocity parameters, etc.) can play important roles in the process. In addition, it has been shown that surface active polar molecules (both in the water phase or oil phase) can cause major changes in the droplet stability enhancing or inhibiting the fusion of droplets.

However, it may be concluded from these studies that the stability of foams and emulsion systems is an extremely intricate phenomenon and in order to reduce the complexity of the process, it is useful as a first step, to study the coalescence of two droplets. In fact, understanding the break-down of the thin aqueous film separating two bubbles or oil droplets is an important precursor to understand stability of a concentrated system. Although the droplet coalescence mechanisms are not fully understood, from fundamental thin film studies to-date several stages have been identified. These have been classified as follows; (a) the initial approach of one droplet toward the other. As this occurs, a hydrodynamic interaction builds-up causing a weak deformation or flattening of the front section of the droplets. Eventually this causes the curvature of the interface to be reverse into a concave lens shaped dimple. Providing the forces acting on the bubble overcome the energy barrier created by the increase in surface energy during deformation then a rapid outflow of liquid from the dimple causes the size of the dimple to decrease. This eventually leads to formation of a plane parallel thin liquid film. (b) Usually, the formation of a thin film between the droplet interfaces leads to drainage which reduces the thickness [6].

In many droplet coalescence studies it is necessary to quantify the stability and this is usually expressed in terms of the induction time which is the time measured from first contact to thin film rupture and binary coalescence. (In the case of the term contact, this refers to the visually observed contact between the drops and since there is always a very thin film of liquid between the drops in contact before coalescence if and when it occurs.) Usually, a significant distribution of the induction times is often observed which may be attributed to the random nature of the disturbances which cause the film to thin and break. Usually, between 20 and 30 experiments are performed under each condition and the average value was taken.

Over the past 20 years, most of the fundamentally studied on thin films have been carried out using the Scheludko cell in which a biconcave thin liquid film is formed in a glass cell by drawing out the liquid from a capillary but the same theoretical approach can be applied to binary droplet experiments. The stability of the thin liquid film can be defined by a disjoining pressure which is dependent on the short range attractive van der Waals forces, steric forces and long range electrostatic double layers repulsive forces. In fact, these are the same forces which can be used to describe the stability of colloidal particles.

In general, the total interaction across the thin film stabilizing the droplets can be defined by the equation;

$$\Pi = \Pi_{vdw} + \Pi_{edl} + \Pi_s, \tag{1}$$

where Π_{vdw} is the van der Waals interaction, Π_{edl} is the electrostatic interaction and Π_s is the steric contribution. For the formation of equilibrium liquid films the following thermodynamic conditions must be satisfied [7].

$$\Pi = P_c > 0 \tag{2}$$

and

$$\left(\frac{d\Pi}{dh}\right) < 0, \tag{3}$$

where P_c is the capillary pressure of the meniscus of the thin liquid film which forces the liquid out of the film and h is the thickness. In order to balance the capillary pressure the disjoining pressure must increase in magnitude as the thickness of the film decreases.

Several different mechanisms have been proposed for the final rupture of a thin aqueous film separated by the droplets but it is generally accepted that rupture occurs due to instabilities arising in the interfacial regions which may be caused by concentration or temperature gradients. These gradients may cause spontaneous convection and surface waves and corrugations. It has been reported that the thin film do not rupture to zero thickness but fractures at finite critical thickness under non-equilibrium conditions (usually between 30 and 50 nm depending on the frequency of the disturbance).

However, it is also important to consider that under dynamic conditions, the external stress on the liquid film which can prevent rupture providing there is surface active material present. Under these circumstances, the Gibbs–Marangoni effect comes into play and can return the film to its original shape. The resistance to surface deformation can be expressed in terms of the surface elasticity which related the increase in surface tension for a unit increase in surface area of the interface. For a liquid film with two interfaces the elasticity of the thin film can be defined by the equation;

$$E = 2\frac{d\gamma}{d\ln A}.$$

(4)

It has also been shown by Christenson and Yaminsky [8] that the elasticity can be defined by;

$$E \propto \left(\frac{d\gamma}{dc}\right)^2,$$

(5)

where c is the concentration of surfactant in the solution and γ is the surface tension. This equation indicates that the elasticity of a film is dependent on the surfactant and the type of surface tension concentration gradient. From fundamental coalescence studies using binary droplets, distinct differences in the dynamic behaviour of binary droplets of can be identified and this type of information can be very important in predicting the performance of concentrated emulsion and foam systems.

C. The Drop/Bubble Micro-Manipulator (DBMM)

The Drop/Bubble Micro-Manipulator (DBMM) is conceived and designed as an additional module, to be adapted and inserted in the optical bench of a standard commercial drop-profile tensiometer, in a similar configuration as the configuration described in recent articles [9–11] for combined profile-analysis and capillary-pressure tensiometers. Actually, the final assembled instrument is essentially a capillary-pressure tensiometer, exhibiting the special feature of two paired measurement cells. The paired cells are closed, contain the liquid-phase forming the drops and keep in opposed positions the active devices, i.e., the piezo-excitation actuator and the pressure sensor. Two capillary tubes connect the closed cell to a single cell, open to the atmosphere, containing the matrix liquid-phase.

The following Section C.1. describes the construction characteristics of a particular DBMM module.

1. Experimental Set-up

The experimental set-up consists of five laboratory-assembled parts: (a) the twin paired measurement cells; (b) the optical system (illuminator, microscope objective, WVBP510 B/W Panasonic CCD video camera with selectable electronic shutter speed); (c) the drop generation system; (d) the computer (Pentium Dual Core 2.5 GHz Intel microprocessor PCI-Personal Computer, with a National Instruments IMAQ PCI-1410 Monochrome Image Acquisition Board and a PCI-6036E Multifunction I/O board), and (e) the vacuum system.

The two closed measurement-cell blocks, made of stainless-steel, hold four principal elements confining an internal cavity: (a) the GE-Druck PDCR-4000 pressure transducer, (b) the model P-843.40 Physik Instrumente piezoelectric actuator, (c) the capillary, which is immersed together with the other twin capillary in an open (28×31-mm^2 inner section) Hellma optical cell and (d) inlet-outlet Hamilton valves.

The structure of the two measurement cells is illustrated in Fig. 1, representing a section of the DBMM module, orthogonal to the light path. As seen in Fig. 1, each cell holds an Eppendorf CustomTip capillary, with a 100-micrometer inner diameter at the tip. The capillaries are mounted at converging directions and show the end part horizontally bent.

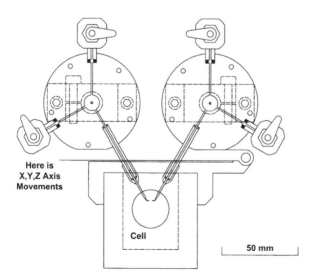

Figure 1. Schematic drawing of the two closed measurement-cell blocks, showing the faced capillaries mounted at converging directions with horizontally-bent tips. The cell inlet and outlet ports, used for filling under vacuum and for coarse displacements of liquid, are closed by Hamilton valves.

Each cell block is supported by a metallic frame which is connected to the other frame by a Physik Instrumente xyz-axis micro-translation stage, allowing the two capillary-tips to face each other at a variable distance.

As also seen in Fig. 1, the interconnected frames are hinged with the basement, providing a possible $90°$-degree-rotation and hence permitting capillary and open-cell substitution.

2. Measurement Methodology and Procedure

The basic measurement methodology, pertaining to the DBMM module, utilises the same established procedures of pressure calibration, optical calibration, drop profile detection and fitting as used in drop profile analysis tensiometry [12, 13], also described in specific aspects in Chapter 2 of this book.

In practice, all instrument functionalities are operated by a dedicated software, according to manual commands (optical calibration and cell cleaning) or to an automatic time-sequence (drop oscillations).

Before the start of each experimental run, two nano-liter-sized drops (with micro-meter diameters) are generated at the tip of the capillaries inside the continuous liquid matrix, by means of a syringe pump. The fine adjustment of the drop size and the drop oscillations are actuated by the piezoelectric device, which allows liquid displacements with a full-scale range of $\Delta V_{pzt} = 188$ nl (at an almost continuous flux, $\Delta V_{pzt} = 6 \times 10^{-3}$ nl/step).

The drop instability phenomenon has been taken into account in the design of the drop hydrodynamic circuit [14]. Thus, filling of the circuit is accomplished under vacuum conditions, preventing air micro-bubbles to be entrapped in the liquid.

For each experimental run, the software controls the size of both drops, following a pre-selected time-line, as shown in the software main panel of Fig. 2.

The software also allows the automatic generation of drop oscillations, in the frequency range from 0.001 up to 4.0 Hz.

For each experiment, the distance between the two drops as well as the values of following specific geometrical and physical quantities are acquired in synchronism for both drops: (a) drop radius, area and volume (b) differential-pressure signal, and (c) temperature.

Single images can be saved by a manual command, each time a particular drop state is worth to be separately examined, such as the unstable (or meta-stable) position of touching drops, as shown in Fig. 3.

Optionally, taking into account the characteristics of the PAL protocol and of the used CCD-camera, a sequence of digital images can be sampled from the video signal and stored in matrix format [11]. The saved matrices are subsequently off-line split into odd and even fields and finally, from further post-processing, consecutive images are reconstructed as a function of time with a time-resolution of 2.0 ms, at 20 ms intervals (see Figs 4 and 5). Moreover, the reconstructed images also allow a quantitative determination of the drop size evolution, of the interdrop relationship and of all the geometrical properties of the two drops (radius, area, volume).

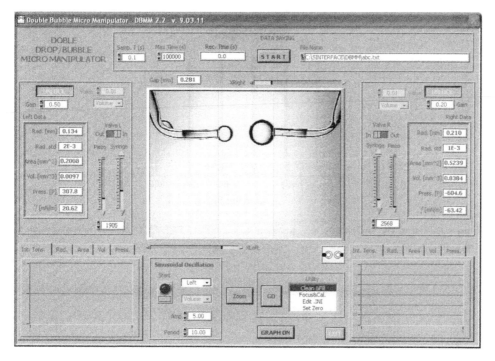

Figure 2. Main panel of the software controlling the size and acquiring the properties of the two drops.

Figure 3. A transient meta-stable state of two touching drops (a liquid bridge connecting the drop is visible).

D. Examples of Application Results

The experiments, illustrated in this section, were performed with the aim to get first a visual documentation of the drop–drop behaviour, and then to show some real application of the given methodology applied to particle covered bubbles.

1. Drop–Drop Interactions as Model Experiments

Figure 6 shows the initial and the final instants of a drop–drop evolution process, in an interval of 12 seconds (a set of 600 pictures). The recorded sequence visualises two distinct drops slowly approaching, becoming into contact and, after a few re-bounds, coalescing into a single drop.

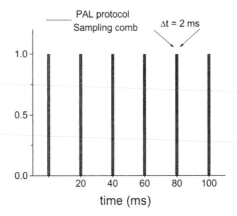

Figure 4. Plot of the sampling comb, pertinent to a standard PAL CCD-camera provided with an electronic shutter.

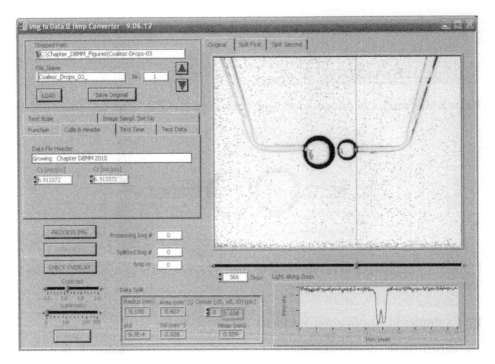

Figure 5. Main panel of the software designed for reconstructing the time-sequences and the drop–drop images and determining the relevant geometrical properties.

The visual sequence can be documented in graphic form by plotting the volume evolution of a single drop, in alternate instants forming a conjunct body with the other drop and rebounding into disjoint drops, and eventually flowing into a single drop (see Fig. 7).

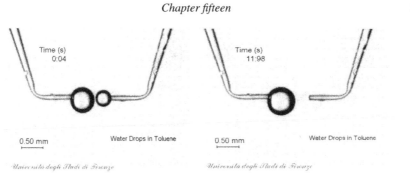

Figure 6. Initial and final states of a drop–drop evolution process. Left panel, volumes of the two drops $V_L = 0.11$ mm^3, $V_R = 0.03$ mm^3. Right panel, volume of the drop $V_C = 0.14$ mm^3.

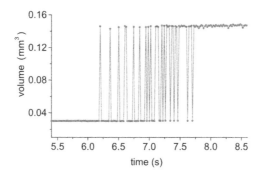

Figure 7. Plot of the volume evolution for a recorded sequence of drop–drop pictures. Between the two extremes of time, as shown in Fig. 6, the volume of a single drop undergoes alternate step variations (forming a conjunct body, rebound drops and a coalesced drop).

Figure 8 illustrates for example a selection of a few sequential video-frames, pertaining to an experimental run with two oscillating drops (initially adjusted at the same size). As seen in Fig. 8, the selected time-interval includes instants of drop–drop contact, the particular instant of drop–drop coalescence and the instants of the coalesced-drop sedimentation.

The unstable drop–drop transient states, visualised by the recorded image sequences, in most cases persist along a prolonged life-time, greater than 20 ms, so that the traditional PAL video-protocol is efficient to provide satisfactory physical information (as demonstrated by the outlined examples of this section).

2. Coalescence Studied Carried with Micron-Sized Particle Coated Bubbles in Flotation

The coalescence of bubbles is an important step in froth flotation which is widely used to separate valuable mineral particles. Initially, the ore is mined and ground in water to produce a suspension of fine particles in a flotation cell. Chemical collectors are then added to the cell which hydrophobize the particles by selective adsorption. Bubbles are generated in the cell resulting in the hydrophobic particles attaching to the bubbles and rising to the surface forming a froth layer. It has been

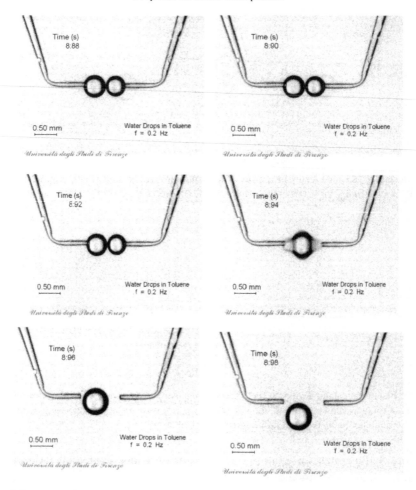

Figure 8. Sequence of video-frames at 20 ms intervals, including the instant of drop–drop coalescence, acquired with a time-resolution of 2 ms.

known for a considerable period of time that the attached particles stabilize the froth and the stability is dependent on the number of attached particles, the particle size, shape, concentration, hydrophobicity and the orientation of particles at the interface [15].

More recently the bubble coalescence studies has been pioneered by Ata and co-workers [16–18] to investigate the stability of two bubbles coated with particles. The experiment is based on producing two bubbles at adjacent capillaries and coating them with glass particles (66 μm) after treatment with a cationic collector CTAB (cetyltrimethyl ammonium bromide). The bubbles were pushed into contact to provoke coalescence. The coalescence process is capture as a typical sequence of pictures taken from the video. In Fig. 9, the results are shown for (a) two uncoated bubbles compared to (b) one coated and one uncoated bubble and (c) two coated bubbles.

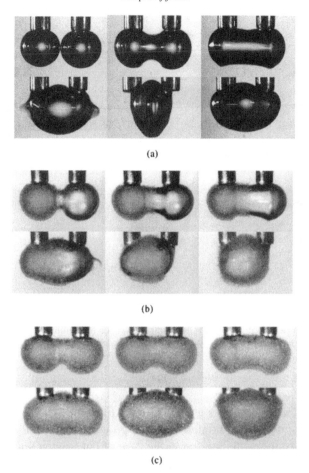

(a)

(b)

(c)

Figure 9. Sequence of coalescence for (a) both bubbles uncoated (b) one coated and one uncoated bubble and (c) both coated bubbles. The time between photographs was 0.5 ms; from [16].

From these studies the stability can be evaluated from the induction time. It was also observed that as the droplets merge the resulting single droplet unit oscillates in various modes until the energy is dissipated to the water phase. These results for air bubbles were found to be rather similar to the sequence obtained for the coalescence of air oil droplets. Essentially, the process begins with the rupture of the thin film which initially separates the two bubbles then the bubbles appear to undergo repeated compression and expansion. This is followed by the formation of a neck between the two droplets which expand, while the remainder of their surfaces is not immediately affected. As the neck expands, a wave is produced along the surface of the newly created single unit and as the wave reaches the farthest point, the droplets bulges slightly increasing in width in the horizontal plane. This bulge then collapses as the wave returns in the reverse direction increasing the vertical height. The process repeats itself in a series of damped oscillatory patterns alternating between

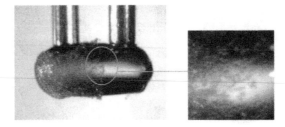

Figure 10. Flow of agglomerated particles from the coated to the uncoated bubble occurring during coalescence; from [17].

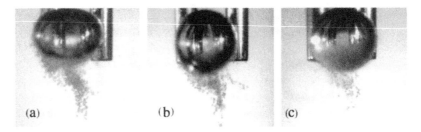

(a)   (b)   (c)

Figure 11. Images of detached particles following coalescence of bubble pairs in carious CTAB concentrations. (a) 2.74×10^{-5} mM. (b) 5.49×10^{-4} mM, (c) 1.1×10^{-3} mM; from [17].

width and height until the droplets eventually settled to an approximate spherical shape.

It may be noted by comparing Fig. 9a with Fig. 9b, that dynamics of coalescence when only one of the bubbles is coated proceeds differently than in the case when both bubbles are coated in that in the case of one coated bubble a large wave moves horizontally from the point of rupture along the surface of the naked bubble causing a bulge. This is probably associated with surface damping effects. Another important observation arising from studies was that in some cases the flow of agglomerated particles rather than individual particles along the interface could be observed between the coated and uncoated bubbles as shown below in Fig. 10.

In addition, from these experiments with coalescence of two bubbles it was found that the process could be vigorous and that the particles residing on the surface of the bubbles could be detached during the merging process as shown in Fig. 11. The degree of detachment was found to be dependent on the hydrophobicity of the particles as controlled by the amount of flotation collector (CTAB).

Another striking observation from the work was that bubbles coated with particles were seen to be more rigid when compared to fluid, deformable bubbles that were produced in CTAB solution alone. Rigid bubbles oscillate less vigorously during the coalescence process, leading to a fewer particles detaching from the surface.

The packing of particles on bubbles is another important aspect of these studies. Nearly spherical glass beads were used to coat the bubbles (see Fig. 12) in a range of CTAB concentration for different coating times and the geometric packing parameters could be approximated to a hexagonal model or square models.

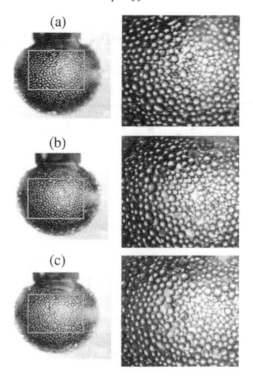

Figure 12. Images of bubbles coated with hydrophobic glass particles for various coating times (a) 30 s, (b) 90 s and (c) 120 s; from [18].

3. Coalescence Studies Carried with Nano-sized Particle Coated Bubbles

Recent works are mostly concerned with the stabilisation of foam using particles below sub-micron, or in nano-size ranges [19]. There is an increasing interest in this field due to the effective stabilisation mechanism offered by small solid particles. Adsorbed colloidal particles have similar properties to surfactant molecules, except that once they are at the interface they are practically completely irreversibly adsorbed and strongly held at the fluid interface. This is because the energy required to remove a small particle from an air/water surface is generally several orders of magnitude greater than the thermal energy, leading the particles to be irreversibly adsorbed, in marked contrast to surfactant molecules that reversibly adsorb and desorb.

We have recently carried out some preliminary experiments on the influence of nanoparticles on coalescence the bubbles. Two bubbles of equal volumes generated at the tips of adjacent capillaries were coated with titanium with mean diameter of 200 nm and then approached each other so that a thin film could be formed between them. The minimum time required for the film to drain to a critical thickness and then rupture (coalescence time) was measured using a CCD camera. The titanium particles were subjected to low-temperature plasma treatment in the presence of a vapour-phase silane coupling agents to make them hydrophobic. Increased treat-

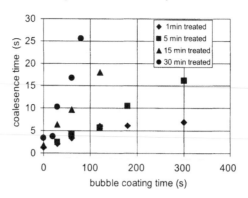

Figure 13. Coalescence time as a function of bubble coating time for various hydrophobization times of titanium particles (achieved by plasma treatment); from [19].

ment time increased the hydrophobicity of the particles. Initial results are shown in Fig. 13 where the coalescence time of bubble pairs is plotted as a function of bubble coating time for various degrees of hydrophobicity. It is clear that the coalescence time or the stability of bubbles increases as the particles become more hydrophobic. The figure also shows that the bubble stability is improved as the surfaces of the bubbles are progressively coated with particles.

4. Coalescence Studies in Commercial Frother Flotation Systems

Gourram-Badri and co-workers [20] have studied the coalescence of air bubbles in chemical frother systems such as MIBC as well as the coalescence of mineralized bubbles using particles of complex sulphide (ZnS, FeS_2 and $CuFeS_2$). The experiments enabled the amount of frother required to reduce the coalescence in the flotation cell. MIBC is used widespread in flotation and is a surfactant with readily adsorbed at the air water interface and reduced the surface tension stabilizing the froth. Using the coalescence apparatus shown in [21] which the bubbles are generated at two capillaries in close proximity it was found possible to evaluate the coalescence process both at the nozzle during bubble rise and in during the bubble rise and finally in the froth.

E. Summary

Coalescence of drops is an important mechanism determining the kinetics of complex disperse systems (emulsions, foams, clouds). Advanced knowledge of the coalescence evolution is mandatory for improving industrial processes and for interpreting natural atmospheric phenomena.

Reliable measurements of drop–drop geometric properties and of the interdrop relationships, together with the inherent visual observation, is a primary source of scientific information for supporting theoretical speculations about emulsion (or foam) stability.

The described hardware-software pair, constituting the experimental set-up for the DBMM module (Drop Bubble Micro-Manipulator), provides fruitful experimental documentation of drop–drop phenomena.

The main feature of the illustrated instrument is the capability of measuring the geometrical properties and the life-time of two drops, after approaching and becoming in contact, and possibly observing the instant of the coalescence event with a millisecond time resolution.

F. Acknowledgements

This work was performed within the framework of "MAP AO-99-052, Fundamental and Applied Studies of Emulsion Stability", FASES project (ESTEC Contract Number 14291/00/NL/SH). The work was also financially supported by projects of the Italian Space Agency (ASI I/002/10/0), the German Space Agency (DLR 50WM0941) and the COST action D43.

G. References

1. D. Langevin, Chem. Phys. Chem., 9 (2008) 510–522.
2. M.M. Robins, Current Opinion in Colloid Interface Sci., 5 (2000) 265.
3. J. Banhart, F. Garcia-Moreno, S. Hutzler, D. Langevin, L. Liggieri, R. Miller, A. Saint-Jalmes and D. Weaire, Europhysics News, 39 (2008) 26–28.
4. T.N. Hunter, R.J. Pugh, G.V. Franks and G.J. Jameson, Adv. Colloid Interface Sci., 137 (2008) 57–81.
5. E. Xinyi, Z.J. Pei and K.A. Schmidt, Food Rev. Intern., 26 (2010) 122–137.
6. A. Scheludko, Adv. Colloid Interface Sci., 1 (1967) 391–464.
7. P.M. Kruglyakov, in "Thin Liquid Films, Fundamentals and Applications", I.B. Ivanov (Ed.), Marcel Dekker, New York, 1988, Chapter 11.
8. H.K. Christenson and V.V. Yaminsky, J. Phys Chem., 99 (1995) 10420.
9. L. Del Gaudio, P. Pandolfini, F. Ravera, J. Krägel, E. Santini, A.V. Makievski, B.A. Noskov, L. Liggieri, R. Miller and G. Loglio, Colloids Surfaces A, 323 (2008) 3–11.
10. S.C. Russev, N. Alexandrov, K.G. Marinova, K.D. Danov, N.D. Denkov, L. Lyutov, V. Vulchev and C. Bilke-Krause, Rev. Sci. Instrum., 79 (2008) 104102–104110.
11. G. Loglio, P. Pandolfini, R. Miller and F. Ravera, Langmuir, 25 (2009) 12780–12786.
12. S. Lahooti, O.I. Del Rio, A.W. Neumann and P. Cheng, Axisymmetric Drop Shape Analysis (ADSA), in "Applied Surface Thermodynamics", A.W. Neumann, J.K. Spelt (Eds), Surfactant Science Series, Vol. 63, Marcel Dekker Inc., New York, 1996, pp. 486–487.
13. G. Loglio, P. Pandolfini, R. Miller, A.V. Makievski, F. Ravera, M. Ferrari and L. Liggieri, Drop and Bubble Shape Analysis as Tool for Dilational Rheology Studies of Interfacial Layers, in "Novel methods to study interfacial layers", D. Möbius and R. Miller (Eds), Studies in Interface Science, Vol. 11, Elsevier, Amsterdam, 2001, pp. 439–485.
14. F. Ravera, G. Loglio and V.I. Kovalchuk, Current Opinion in Colloid Interface Sci., 15 (2010) 217–228.
15. T. Hunter, R.J. Pugh and G.V. Franks, Adv. Colloid Interface Sci., 137 (2008) 57–81.
16. S. Ata, Langmuir, 24 (2008) 6085–6091.
17. S. Ata, J. Colloid and Interface Sci., 338 (2009) 558–565.

18. G. Bournival and S. Ata, Minerals Engineering, 23 (2010) 111–116.
19. G. Bournival, S. Ata and R.J. Pugh, Colloids Surfaces A, 2010, in press.
20. F. Gourram-Badri, P. Conil and G. Morizot, Int. J. Mineral Processing, 51 (1997) 197–208.
21. S. Ata, E.S. Davis, D. Dupin, S.P. Armes and E.J. Wanless, Langmuir, 26 (2010) 7865–7874.

The Tensiograph Platform for Optical Measurement

N.D. McMillan [a], S.R.P. Smith [b], M. O'Neill [c], K. Tiernan [d], D. Morrin [a], P. Pringuet [a],
G. Doyle [a], B. O'Rourke [a], A.C. Bertho [a], J. Hammond [e], D.D.G. McMillan [f],
S. Riedel [c], D. Carbery [a], A. Augousti [g], N. Wüstneck [h], R. Wüstneck [i], F. Colin [j],
P. Hennerbert [j], G. Pottecher [j], D. Kennedy [d] and N. Barnett [k]

[a] Institute of Technology Carlow, Kilkenny Road, Carlow, Ireland
[b] Physics Centre, University of Essex, Wivenhoe Park, Colchester, Essex CO4 3SQ, UK
[c] Carl Stuart Ltd., Whitestown Business Park, Tallaght, Dublin 24
[d] Dublin Institute of Technology, Dublin, Ireland
[e] Starna Scientific Ltd., 52-53 Fowler Road, Hainault, Essex IG6 3UT
[f] TMS Consultants, Pery Square, Limerick, Ireland
[g] Faculty of Science Kingston University, Surrey KT1 2EE, UK
[h] Humboldt University, Berlin, Germany
[i] Max Planck Institute of Colloids and Interfaces, Potsdam, Germany
[j] IRH Environnement, BP 286, 11 bis Rue Gabriel Péri, F-54515 Vandoeuvre-les-Nancy
Cedex France
[k] Ocean Optics Ltd., Ocean Optics BV, Geograaf 24, 6921 EWDuiven, The Netherlands

Contents

Bubble and Drop Interfaces
© Koninklijke Brill NV, Leiden, 2011

A. Introduction

The objective of this chapter is to provide the most comprehensive review and introduction to tensiography i.e. optical measurements on liquid drops. It is recognized that the term "tensiography" can be misleading in that it implies that surface tension plays a major part in determining the results of the measurements. Whilst this may be partly true, it is now clear that optical measurements on liquid drops have a far greater potential than mere surface tension measurement. The tensiograph instrument provides a more universal analysis of the properties of the liquid under test (hereafter LUT).

In nearly every section of this chapter, a facet of tensiography is presented in a theoretical way. Each theoretical section finds itself linked to a practical section based on the experimental work. The two sides of the topic are supported by modelling studies that in all cases qualitatively, but in most cases also quantitatively connect, the theoretical-practical sides of every facet of this study. The work presented is a reasonably satisfactory and balanced study of the technique written by the team that has largely developed optical tensiography from its invention to its present commercial basis. A fundamental symmetry therefore exists in this study, which is now only possible after more than twenty-five years of development of the technique.

The chapter is divided into six sections that move from a general introduction of tensiotraces. The form of the characteristic tensiotrace is introduced with blank (non-absorbing and non-turbid) solutions. The modelling of tensiotraces is described using the Young–Laplace equation. Drop shape and tensiotrace modelling is presented with the discussions of the full theoretical description of tensiotrace measurements that arise from the recent theory of pendant drop scission. The theoretical description of all the tensiotrace measurands taken on the abscissa (equivalent periods or volume measurands depending on which system of tensiotrace representation is used) is presented. The work of providing a comprehensive qualitative descrip-

tion of the photometric tensiotrace measurands has begun with the development
of a ray-tracing modelling approach and the new theory of the RI (refractive index) relationship with the tensiotrace 'rainbow peak'. An analytical theory of drop
spectroscopy is presented, with confirmatory test results of the theory and application studies. Experimental effective path length results are given demonstrating
the excellent predictive capability of modelling. Examples of drop spectroscopy
application studies are reported in pharmaceutical science and water science. The
performance of the drop spectrophotometer is tested against traditional laboratory
instruments. Refractive index measurements of the bulk liquid are reported and surface tension measurement on both non-surface active and surfactant solutions. The
experimental discovery of an empirical relationship between drop period (T_1) and
tensiopeak period (t_3) has also been theoretically modelled, which is a discovery
of some real importance to drop science. To simplify the identification of liquids
it has been recommended that tensiograph calibration standards for fingerprinting
be adopted. This method results in a library of tensiotraces that are presented on
normalised axes. Test samples can then simply be visually fingerprinted from the
recording of a single tensiotrace; modelling has demonstrated that each LUT has its
own unique tensiotrace.

1. The Development of the Tensiograph Instrument

The stalagmometer is an instrument for measuring the size of drops suspended from
a capillary tube, used in the drop-weight method from which the tensiograph was
developed [1]. The stalagmometer/tensiometer is an instrument designed to measure the surface tension of a pendant drop determined by drop volume or drop
weight measurements. The technique began its evolution from an inauspicious beginning of a proposal by a pharmacist Tate [2] in 1864 for making up reliable
preparations of drugs to emerge as a reliable scientific measurement technique
through the work of Rayleigh [3] and Harkins and Brown [4] around the turn of
the nineteenth century.

The tensiograph is the instrument that provides measurements of various properties of a liquid derived from the graphical recording of measurands taken during
the growth of a drop (pendant or sessile) and/or alternatively provides a graphical
fingerprint that can identify or monitor the LUT. This technique originates from the
optical researches of McMillan and co-workers from 1986/7 that was patented and
published in 1992 [5]. The concept was widened to include other modalities through
the collaboration of Augousti with McMillan from 1992. This extended the practical domain of tensiography into capacitive [6] and ultrasonic tensiography [7] and
delivered the first integration of the capacitive and optical modalities/techniques
[8]. Song and co-workers [9] have pushed forward this integration of tensiographic
modalities. The early drop imaging patents of McMillan in 1987 had demonstrated
the intent of this early work to provide a comprehensive drop measurement tool
[10], which would integrate related CCD optical techniques into a single instrument. The work however in this field was led by Neumann to develop what has

become known as ADSA. The ADSA computer program was developed in the University of Toronto's Department of Mechanical and Industrial Engineering. Perhaps the best description of this technique for the purposes of this chapter is the one presented in the 1998 chapter by Neumann and his co-workers [11]. This method employs digital image analysis to detect the edge of a drop and accurately determines a number of surface parameters including the interfacial tension and contact angle. ADSA is capable of determining these parameters with greater accuracy and flexibility than traditional techniques. The ADSA software had been refined so that it facilitates the handling of a wide variety of systems and experiments, including the measurement of interfacial tension, line tension, film tension, surface pressure, surface dilational modulus, surface dilational viscosity, and contact angles for liquid–fluid systems. Naturally optical tensiography wishes to measure itself against this well-established technique.

The first multispectral tensiograph recordings with CCD detectors were made in 1992 by McMillan, O'Mongain *et al.* [12]. The development of this CCD/CMOS technique has been instrumentally advanced by Hao, Qiu and Zhang in 2005 [13]. The original attempts to integrate two modalities, capacitive and optical in the same instrument [8] has led to the more substantial recent work of both Song, Zhang, Qiu, Zhang and Shi [14] and Chen *et al.* [15] with significant integrations of the tensiographic modalities with camera vision technology. Indeed, tentative commercial development of the instrument has been made by SIMTech [16] with their Multispectral Liquid Drop Analyser using photonic signal processing.

On an alternative track, an early indication that tensiography may be used as individual areas of drop analysis rather than an integrated technology came in 1995 with the work of Liu and Dasgupta in Texas for what were essentially spectroscopic applications. They evolved drop analysis into a new and independent technique for gas analysis, using the drop as a windowless optical cell for spectroscopic measurements in a flow system [17] and their system aimed to provide analytical chemistry in a drop [18]. They reported world record sensitivity for detection of chlorine in the auto-sampler they devised based on a drop analyser [19]. They from there proceeded to review drop techniques [20]. Following from this initiative, the significance of the drop analyser has been more recently emphasized with two review papers by Miller and Synovec [21] and Song, Zhang and Qiu [22].

There has been some work towards developing on-line automatic systems. A number of researchers have found the drop volume method for determining surface tension to be time-consuming and so undertook to automate the procedure so as to minimise these difficulties. An automatic drop counter, described by Nikita and Taubman [23], consisting of a drophead, pump, photocell, light source, and an electrical circuit to count the number of drops formed was designed. It was constructed to automatically measure the time between drop detachments at a constant delivery rate to determine drop-volume and using the traditional correction factors of Harkins and Brown a surface tension measurement is made. This apparatus was found to provide sufficient accuracy with short operating time. Liu and Dasgupta

[24] developed an instrument in which light from an LED source passes through a forming drop and is received by an optic fibre, thereby producing a reproducible absorbance signal. Liu and Dasgupta [25] used this system as an automated sensor for gaseous chlorine *via* a droplet of tetramethylbenzidine solution formed at the tip of a tube centred in a cylindrical chamber, through which gaseous chlorine is aspirated. Cardoso and Dasgupta [26] used a fluorometric technique in which a drop containing an absorbing/chromogenic reagent, which is illuminated/excited from within by a transversely mounted optic fibre was used as a sensor for the measurement of low concentrations of atmospheric hydrogen sulphide. Analyte delivery was *via* a teflon delivery tube. Absorbance determination was obtained by simply passing light through the drop by means of an optical fibre and the light intensity transmitted was received by a collector fibre. An automatic system was developed and tested in the 5th Framework EU Project Aqua-STEW [27] for monitoring of solvents on the surface of water and the presence of PAHs and other carcinogenic compounds.

2. *The Optical Tensiograph Instrument and the Tensiotrace*

The optical tensiograph is an Amplitude Modulated Fibre Optic Sensor (AMFOS) instrument that records a tensiotrace. The tensiotrace is an evolutionary signal recorded from the light coupled between a source and collector fibre as the drop moves through its entire life cycle from a remnant drop to a fully evolved drop that finally detaches from the optrode head. Full details of the instrument are given below, but here it may be helpful to explain that the tensiotrace in effect comprises a sequential recording of a series of rainbow-order 2nd, 3rd couplings, etc. each associated with a distinct tensiotrace peak. The instrument has been engineered to deliver measurements of the physical properties of the LUT, namely surface tension/density ratio ($\Gamma = \gamma/\rho$, Shape factor in drops), refractive index (η), colour at a specific wavelength (absorption $= A_\lambda$) and turbidity at a specific wavelength (turbidity $= \tau_\lambda$). The instrument can also of course be used for QA for solid products such as pharmaceutical products if these are dissolved in an appropriate solvent for this purity/fingerprinting assay. It can also be used to sensitively monitor chemical and biochemical kinematics processes.

Tensiography uses optical fibres to illuminate a forming pendant-drop from within (ADSA is the converse of this technique in effect studying the shape of the external form), with source and collector fibres located in a conventional drophead. Figure 1(a) shows a typical concave drophead. The actual problem with such a design is fundamental in that physical damage can be caused due to exposure of the fibres. A patented quartz drophead [28] has been designed that gives the same universal fingerprint capability as this traditional drophead. This new design of drophead is one of the key innovations that facilitates accurate, reproducible and sensitive tensiography.

The resulting optoelectronic signal, which is referred to as a tensiotrace, has a characteristic appearance of one or more peaks as seen in Fig. 1(b). The shape and size of the tensiotrace are governed by (i) the volume of the drop, and hence the

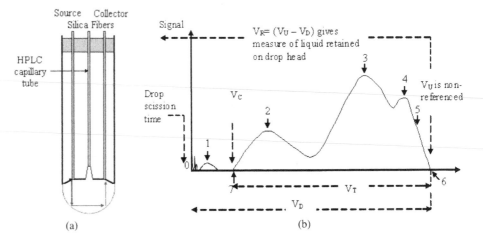

(a)   (b)

Figure 1. Example of a drophead configuration (a); tensiotrace showing the important trace features and the times associated with drop mechanics (b), 0 = separation vibration; 1 = first-order peak (protopeak); 2 = rainbow (second-order peak deuteropeak); 3 = tensiopeak (third-order tritopeak); 4 = shoulder peak (fourth-order tetartopeak); 5 = separation peak (fifth-order pemptopeak); 6 = drop period; 7 = rainbow peak commencement.

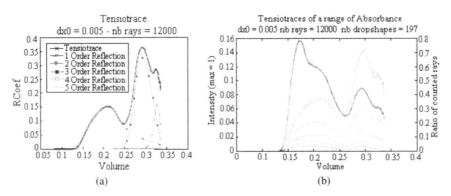

(a)   (b)

Figure 2. Modelling results that show that the rainbow peak is uniquely formed by the 2 reflections rays while the tensiopeak is composed at 80% of 3 reflections rays. 4 and more reflections are only found in the shoulder peak at 48% (a); tensiotraces at absorbance between 0 and 1 (b).

surface tension and density and (ii) the optical properties, refractive index and light absorption of the drop.

The drop period is simply the time taken for the forming drop to detach. The vocabulary of the tensiotrace is important and is given in this diagram, simply because it has been found in practice that all the tensiotrace features are useful measurands. The vocabulary that has been developed for tensiography is designed to mark such measurement differentiation verbally. Modelling shows the complexity of the overlap of reflections as is apparent from the results shown in Fig. 2.

The drophead obviously needs to be placed in an environmentally controlled housing, not least because most of the physical properties that define the tensiotrace

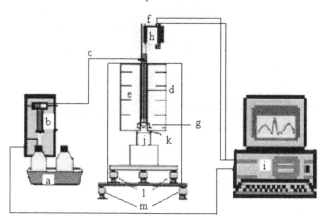

Figure 3. Schematic diagram of drop analyser. (a) Samples for analysis. (b) Stepper-motor-pump. (c) Delivery tube. (d) Drophead. (e) Temperature control (Peltier® control). (f) Light source. (g) Optical eyes. (h) Detector. (i) Computer. (j) Chamber to ensure saturation of atmosphere. (k) Temperature sensors. (l) Auto-levelling drives. (m) Anti-vibration balancing feet.

are temperature dependent. Figure 3 shows a schematic diagram of this instrument design. This is the system [27] successfully used in the 5th Framework Aqua-STEW project for on-line monitoring. The drophead must be levelled to within 0.01° and an active stepper level control system has been devised to achieve this ultra-sensitive 'micro-levelling' of the drophead. This instrumental control feature is not addressed in any of the published drop analyser instrument designs of other workers, but the modelling analysis conducted by the authors has shown this feature is absolutely necessary for reproducible measurement systems. There is also the vital matter of vibration isolation. The system shown uses mechanical spring vibration isolation and this successfully removes all vibrations above 1.5 Hz. On the upper floors of modern buildings this isolation is not sufficient and as with six-figure balances anti-vibration tables, active damping systems or simply sand isolation is required to get the highest quality of measurement. The instrument can always of course be deployed on purpose built weighing stations.

The drop head was fitted into a brass Peltier heater block and maintained usually at 25°C (good stabilisation at 20°C is quite difficult as it is too close to ambient temperature in most laboratories). Measurements were taken with various LED sources and a phototransistor detector. The optoelectronic signal was processed *via* a standard 16-bit resolution PIC microcontroller with the data processed using purpose-written software. The instrument can be fitted with various Ocean Optics hardware accessories such as their pulsed xenon, deuterium CW and high power xenon sources which works with any of their CCD detector systems. Obviously, many other CCD/CMOS systems are now available to record these tensiotraces and the authors used in addition to the Ocean spectrometers the Shamrock 500 Andor Technologies Imaging spectrometer.

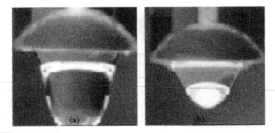

Plate 1. Photograph of a full-formed (a) water (thimble shaped) and (b) 99.8% ethanol (bell shaped) pendant drop on a concave plexiglass drop head, with illumination from the rear.

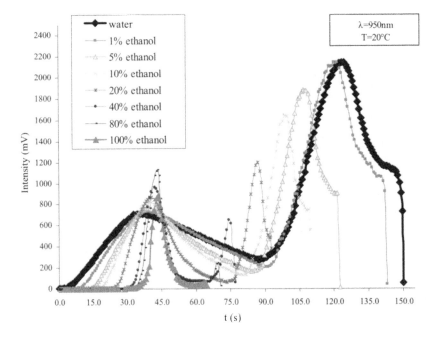

Figure 4. Tensiotraces of mixtures of ethanol and water.

The fundamental and direct method of analysing the properties of a surface in a drop is the ADSA (Axisymmetric drop shape analysis) method. Plate 1 shows the different drop shapes for a water and ethanol drop. The types of shapes for these two extreme liquid types can perhaps be described as thimble and bell shaped drops on 9 mm heads shown in these photographs.

The range of surface tensions in liquids is roughly bracketed by water and ethanol on large dropheads, such as are used in optical tensiograph instruments. Visually, it must be admitted that the variability in these shapes is perhaps rather limited. Whereas, Fig. 4 shows the variation in the form of the tensiotraces for water and 99.8% ethanol at 950 nm is far more radical than the drop shape variations from which the variations here are derived.

The tensiotrace is an amplitude modulation of the temporal signal of light coupled between source and collector fibres in the drophead as the drop evolves in volume from the remnant drop to a fully developed pendant drop. The volume of a drop is given simply by variation as

$$V_D = V_{RD} + qt, \tag{1}$$

here V_D is volume of drop in µl, V_{RD} is the volume of the remnant drop in µl, q is the flow rate in µl s^{-1}, and t is the time in seconds that the liquid has been delivered to effect drop growth.

The rough order of magnitude of these tensiotrace changes from water to solvent tensiotrace is worth noting. The drop period decreases by more than 50%. The rainbow peak rises by about 300% and markedly narrows. The rainbow peak commencement lengthens by a factor of more than 1200%. The tensiopeak reduces by a factor of 1100%. The obvious measurands we have here in these changes are drop period, rainbow peak period, rainbow peak height, rainbow peak commencement, tensiopeak period, tensiopeak height and tensiotrace area. The variations in the tensiotraces provide quantitative information about the liquid under test. Tensiotrace features correlate with the physical properties of the DUT.

3. Instrumental Model and Theory

Tiernan [29] has recently developed the theory that is presented in this section here. From an engineering perspective, the fundamental objective of spectrophotometers is to couple monochromatic light of intensity $P_S(\lambda_t)$ on to a transparent cuvette containing a test sample and to accurately measure, $P_t(\lambda_t)$ the intensity of the light exiting the cuvette. The concentration of the test sample is derived from accurate measurements of $P_S(\lambda_t)$ and $P_t(\lambda_t)$. A key aim of this research is to improve the performance of the tensiograph drop analyser, operating in the mode of a quantitative spectrophotometer, by reducing measurement error. This study will seek to provide an analysis of measurement error for spectrophotometers in general and as such provides the framework for the engineering design process. Five distinct types of error are examined, namely (1) static errors resulting from tolerances on instrumentation construction, (2) cuvette based errors, (3) wideband noise, (4) offsets and (5) drift.

The system model of Fig. 5 arranges the spectrophotometer optical/mechanical/electronic components into five separate categories. The emission spectrum of the light source is represented by $S(\lambda)$. The coupling of light to the container holding the test analyte is represented by $K_S(\lambda_t)$. Incorporated within this are perhaps lens, mirrors, collimator, monochromator and optic fibre. The wavelength selected by the monochromator is λ_t. The light beam directed at the cuvette has intensity

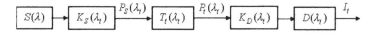

Figure 5. Block schematic for spectrophotometer transmittance.

P_S at wavelength λ_t. The transmittance of the test sample contained within the cuvette is $T(\lambda_t)$ while the light intensity exiting the cuvette is $P_t(\lambda_t)$. This light is coupled to a photo-detector using arrangements of lens, mirror or fibre optic cable having an overall transmission coefficient of $K_D(\lambda_t)$. The photo-detector has spectral sensitivity $D(\lambda_t)$ and produces the electrical signal I_t, which is related to the light intensity coupled to it. This signal may then be measured electronically to determine the analyte transmittance. Its value corresponds to the product of system model transmittances and is given by

$$I_t = S(\lambda)K_S(\lambda_t)T_t(\lambda_t)K_D(\lambda_t)D(\lambda_t). \tag{2}$$

If the spectral characteristics for source, detector and optical couplings were precisely known, it would be a simple matter to measure the transmittance of the test sample and therefore to calculate its concentration by applying the Beer–Lambert law.

Using this approach Tiernan has systematically worked his way through all the instrumental issues. There is no space here to derive these equations but the general results will be presented as they have a major bearing on the engineering of the instrument.

In general there are significant tolerances on the manufacture of optical and electronic components and on their mechanical interfaces. In order to compensate for losses and spectral variations, the transmittance of the test sample is calibrated with respect to a defined reference sample. The importance of instrument calibration is something that is of specific importance to this drop technique. Assuming the time difference between the reference and test sample measurements is not too large such that drift in spectrophotometer components is not significant, then the ratio of test and reference signals is

$$\frac{I_t}{I_0} = \frac{T_t}{T_0}, \tag{3}$$

where subscripts t and 0 represents measurements made on the test and reference sample respectively. Consequently, instrumentation manufacturing tolerances are balanced out and T_t may be calculated from accurate measurements of I_t, I_0 and prior knowledge of the reference sample transmittance T_0.

In discussing a windowless drop spectrometer, comparison with the cuvette-based type of instrument is obviously required, as at this time this is the most widely used standard methodology. Four categories of measurement error associated with cuvettes are examined, i.e., window spacing tolerance, window radiation loss, internal cuvette reflections and manual handling. Unlike manufacturing tolerances these errors are not balanced out through calibration since cuvette based errors differ between test and reference measurements. Assuming $\Delta L_t \ll L_t$, the relative measurement error for sample concentration due to window spacing tolerance is

$$\frac{\Delta C_t}{C_t} \approx -\frac{\Delta L_t}{L_t}, \tag{4}$$

where ΔC_t is the error in measured sample concentration and ΔL_t is the difference between window spacing for the test and reference sample cuvettes. Consequently the spectrophotometer reproducibility is directly limited by the tolerance of the cuvettes that are used. For example, if batch measurements are made using 1% tolerance cuvettes, then the upper limit of spectrophotometer accuracy is also 1%. The minus sign in Equation (4) indicates that the polarity of the measurement error is the opposite of the cuvette path length error, i.e., if the path length is longer than its nominal value ($\Delta L_t > 0$) then this relationship informs us that the estimated sample concentration will be on the low side ($\Delta C_t < 0$).

Ideally the cuvette transmits the entire radiant power incident upon it. In practice some light is absorbed while being transmitted through the cuvette and more is lost through reflection at the air/cuvette boundaries leading to window radiation loss. While the optical properties of high grade glass and quartz cuvettes are manufactured with a high degree of consistency, unless they are thoroughly washed and carefully handled between samples they risk being contaminated resulting in different levels of radiant loss between test and reference cells. Other factors causing dissimilar radiant loss include bubbles in the sample or damp external windows. If we take K_t as the fractional radiation loss in test cuvette and K_0 as the fractional radiation loss in reference cuvette, then it can be shown that the normalised measurement error for sample concentration is

$$\frac{\Delta C_t}{C_t} \approx S_{CL}\Delta K, \tag{5a}$$

$$S_{CL} = \frac{0.43}{A_t}, \tag{5b}$$

where ΔK is the difference in fractional radiant power losses between test and reference cuvette. The error sensitivity function is inversely related to sample absorbance and is given by Eq. (5b). Fresnel losses exist in all cuvettes and if we assume the light incident on the cuvette has normal incidence, then the losses for the air glass boundaries and the liquid–glass boundaries are both a little over 4%. The percentage light energy lost from internal reflections will differ between the test and reference samples should their respective refractive indices not be the same and would therefore contribute to additional measurement error. The normalised measurement error is found to be

$$\frac{\Delta C_t}{C_t} \approx \frac{0.87(R_t - R_0)}{A_t}, \tag{6a}$$

$$\frac{\Delta C_t}{C_t} \approx \frac{0.87\Delta R}{A_t}, \tag{6b}$$

where R_t, R_0 is the cuvette reflectance when holding the test sample and reference samples respectively. The normalised measurement error is given by Eq. (6b). Clearly this error is inversely related to the test sample absorbance, but it is also directly related to the changes in the internal cuvette reflectance. The reflection loss for a standard cuvette ΔR is approximately 0.16 but the error produced is a function of the absorbance measurement A_t. Importantly the reflection losses vary with the concentration of the analyte as the reflectivity at the internal cuvette windows derive from the differences between refractive indices of test and reference samples. Generally, as the concentration of the test sample increases its refractive index deviates further from that of the reference sample. It is theoretically important here to point out that in most assay procedures; the concentration of the analyte is so small that the refractive index variation is negligible. However, there exists a class of applications in which such loss variation would occur and perhaps here the drop cuvette may have a distinct advantage over that of the standard instrument. Based on the Lorentz dispersion model [30], Tiernan derived the relationship of Eq. (7) showing that the difference in refractive index between test and reference solutions is approximately linearly related to the concentration of the test sample, i.e.,

$$\eta_t \approx \eta_0 + \hat{N}^t \frac{G_t}{2\eta_0}, \tag{7}$$

where η_t and η_0 are the refractive indices of the test and reference sample respectively, $\hat{N}^t$ its molar concentration, while G_t is a constant defined by the fundamental properties of the test material. The measurement error is given by

$$\frac{\Delta C_t}{C_t} \approx -\left(\frac{0.87}{A_t}\right)\left(\frac{4\eta_c(\eta_c - \eta_0)}{(\eta_c + \eta_0)^3}\right)\left(\hat{N}^t \frac{G_t}{2\eta_0}\right). \tag{8}$$

According to the first factor in this equation, the error is inversely proportional to concentration while in the last factor it is directly proportional. Consequently, due to counter balancing the normalised error is independent of the test sample concentration but is strongly reliant upon the refractive index differences.

The following numerical example gives an indication of the serious measurement tolerance arising from internal cuvette reflections. Suppose a test is performed using a glass cuvette with $\eta_c = 1.52$ and with a water reference having $\eta_0 = 1.3333$. If a test sample has absorbance $A_t = 0.5$ and η_t of 1.35, then the normalised measurement error is

$$\frac{\Delta C_t}{C_t} \approx -\left(\frac{0.87}{A_t}\right)\frac{dR_t}{d\eta_t}\Delta\eta$$

$$= -\left(\frac{0.87}{0.5}\right)\left(\frac{4 \times 1.52(1.52 - 1.3333)}{(1.52 + 1.3333)^3}\right)(1.35 - 1.3333) = -0.0017$$

or -0.17%.

The manual handling error for a cuvette is really quite severe. The placement of the cuvette in the cuvette holder is subject to small positional errors. Any deviations in position or tilt will affect the beam path-length and shift its direction, thereby adjusting it focus onto the detector. A study was made of the error generated by the manual refilling of the cuvettes. This factor was investigated using the Hitachi UV-vis U-2000 spectrophotometer. A standard error of 1.25% was obtained on a sample of 32 measurements. The test was repeated using a slider attachment for improved mechanical positioning of cuvettes. A standard error of 0.611% was obtained. Obviously, such placement errors do not exist for a drophead system.

The noise issues in optoelectronic circuitry are one that also requires some consideration when comparing the traditional instrument with the drop spectrometer system. Here, a brief review of the traditional noise issues in the classical UV-vis instrument will be given. The spectrophotometer signal will be contaminated with additive random noise. This noise arises from a number of sources with the major contribution being from the photo-detector and associated electronic circuits. The most significant wideband noise signals are thermal noise arising from random motion of electrons in devices, shot noise caused by current flowing in electronic devices and quantization error of electronic data converters. This is a fairly well trod path and textbooks dealing with the subject exist such as Curell [31] are well known. However in the context of spectrophotometers, if the r.m.s. noise level is the same on both test and reference signals, i.e., $\sigma_t = \sigma_0$, then the normalised measurement error may be shown to be equal to the product of the signal-to-noise ratio (SNR) on the reference signal and the error sensitivity function S_{WB}, i.e.,

$$\frac{\Delta C_t}{C_t} = \left(\frac{\sigma_0}{I_0}\right) S_{WB}, \tag{9a}$$

$$S_{WB} = \left(\frac{0.43}{A_t}\right)\left(\frac{\sqrt{1 + T_t^2}}{T_t}\right) \tag{9b}$$

with T_t equal to the transmittance of test sample. As outlined in Eq. (9b), S_{WB} varies with test sample absorbance forming a well-known flat bottomed 'U' shape with a minimum of 2.85 occurring at $A_t = 0.48$. The error sensitivity is however nearly flat at the minimum and test samples with absorbance within the range $0.20 \leqslant A_t \leqslant 0.94$ will have error sensitivity within 3 dB of this minimum. Hence it is best practice to dilute or concentrate (not as easy in practice) samples to have concentrations within this optimal range if issues of additive electronic noise are considered to be problematic. It is also noteworthy that Eq. (9a) and Eq. (9b) indicates that the normalised measurement error is always greater than the SNR of the reference signal.

One important issue in assessing the limitations of drop spectrometers is the DC offset. The spectrophotometer signal will have static offsets added to it. Possible sources for these offsets include stray light in the monochromator, dark current from photo-detector, offset current and voltages in the pre-amplifier and tolerances on level-shifters associated with data conversion circuitry. It can be shown that the normalised measurement error is given by Eq. (10a) with the offset error sensitivity function given in Eq. (10b).

$$\frac{\Delta C_t}{C_t} \approx \left(\frac{K_{OS}}{I_0}\right) S_{OS}, \tag{10a}$$

$$S_{OS} = \frac{0.43}{A_t}\left(\frac{T_t - 1}{T_t}\right). \tag{10b}$$

As indicated in Eq. (10a) the normalised measurement error is directly proportional to the ratio of the DC offset to the reference signal level. The measurement error is also sensitive to the sample absorbance as defined by Eq. (10b). This sensitivity function shows that measurement accuracy improves for lower test sample absorbance, with S_{OS} asymptotically approaching unity as A_t approaches zero. Consequently higher concentration samples will have higher measurement tolerance. For example, a test sample with absorbance of 1.6 will have offset sensitivity of $S_{OS} = -10.4$. If the spectrophotometer had a DC offset error equal to 1% of the reference signal level, then the normalised concentration error is $\Delta C_t/C_t = (0.01)(-10.4) = -0.104$, i.e. -10.4%.

On the other hand, if the sample were diluted by a factor of four giving an absorbance of 0.4, its offset error sensitivity is reduced to $S_{OS} = -1.6$ producing just 1.16% measurement error. It is however possible to calibrate out measurement errors due to offsets by measuring the spectrophotometer output with the light source deactivated. This level should then be subtracted from measured values of I_t and I_0 before calculating sample concentration.

Drift is an issue of perhaps paramount importance to an AMFOS (Amplitude modulated fibre optic system) system, which are subject to major limitations from this error source. The characteristics of the spectrophotometer will change over time due to aging and environmental factors such as temperature variations. There are two forms of drift differing fundamentally in how they affect the accuracy of spectrophotometer measurements. Additive drift is modelled by having a differential static offset applied between test and reference tensiotraces. Multiplicative drift results in different spectrophotometer transfer gains for the test and reference sample measurements.

The main sources of additive drift include photo-detector dark current variations with temperature, stray light variation, drift in offset currents in preamplifier and changes in the voltage of level shifting electronic circuitry. Additive drift results in different values of DC offset applied to the reference and test signals. The normalised measurement error {Eq. (11a)} resulting from additive drift {Eq. (11b)} is

$$\frac{\Delta C_t}{C_t} \approx \left(\frac{\Delta K_{OS}}{I_0}\right)S_{AD} = \left(\frac{\partial K_{OS}}{dt}\right)\left(\frac{\Delta T_S}{I_0}\right)S_{AD}, \tag{11a}$$

$$S_{AD} = -\frac{0.43}{A_t T_t}, \tag{11b}$$

where ΔK_{OS} is the offset difference between test and reference signals, $\frac{\partial K_{OS}}{dt}$ is the rate of offset drift and ΔT_S is the time difference between when measurements of reference and test samples are made.

While it is possible that offset drift may actually reduce measurement error arising from static offsets, it will however increase the error range by the level specified in Eq. (11a). As would be expected, Eq. (12a) indicates that drift induced measurement error increases with both drift rate and longer delays between measurements of test and reference samples. Defined by Eq. (11b), the profile of error sensitivity function S_{AD}, appropriately given this is Tiernan's work, is an inverted Irish currach shaped curve with measurement error increasing for both low and high sample absorbance. The minimum error sensitivity is -2.7, occurring for test samples with absorbance of 0.43. Error sensitivity remains within 3 dBs of the minimum over the absorbance range $0.17 \leqslant A_t \leqslant 0.90$. The measurement error due to additive drift is eliminated with dual-beam type spectrophotometers. These split light into two beams of equal intensity, with one beam passing through the reference and the other through the test sample. Consequently, both test and reference measurements are made simultaneously so ΔT_S equals zero in Eq. (11a). An alternative method would be to cancel the effects of additive drift. This involves measuring the spectrophotometer offsets with the light source turned off just prior to measuring reference and test signals, I_t and I_0. Sample concentration is then calculated with the measured offsets subtracting from I_t and I_0.

The main sources of multiplicative drift include variation in the radiation power of the light source due primarily to changes in its electrical supply voltage, drift in spectral sensitivity of photo-detector, variation in gain of electronic amplifiers and shifts in data converter reference voltage. Multiplicative drift effect is modelled by incorporating additional scaling factors to the spectrophotometer signal. Let G_0 be the overall transfer gain of the spectrophotometer while making the reference measurement and G_t the transfer gain while measuring the test sample.

The normalised measurement error is given by Eq. (12a) and the error sensitivity due to multiplicative drift is given by Eq. (12b).

$$\frac{\Delta C_t}{C_t} \approx \Delta G_N S_{MD} \approx \left(\frac{\partial G_N}{dt} \Delta T_S \right) S_{MD}, \tag{12a}$$

$$S_{MD} = \frac{-0.43}{A_t}. \tag{12b}$$

Equation (12a) shows that measurement error due to multiplicative drift increases in proportion to rate of normalised gain drift and the delay between test and reference measurements. The sensitivity function S_{MD} being inversely related to sample absorbance indicates that measurement error increases dramatically for test samples with low absorbance. Consequently if a spectrophotometer is prone to multiplicative drift it may be necessary to employ more concentrated test samples to minimise measurement error.

In discussing the various pros and cons of drop spectroscopy, obviously the theoretical issues detailed above form the basis of the discussion. An analysis of measurement accuracy for UV-vis spectrophotometers outlined a number of error sources and the process through which they affect accuracy. Three error sources were examined above namely, the instrumentation tolerances, cuvette error and finally, the optical/electronic error. The issues relating to drop spectroscopy are now going to be systematically addressed.

Normalizing the spectrophotometer signal with a reference signal was found to be effective in eliminating measurement errors derived from instrumentation toler-ance.

The drop spectrometer does not obviously use a cuvette. The drophead diameter can be accurately manufactured to a tolerance of better than a micro-metre when manufacturing in quartz. The volume of a drop can be delivered quite routinely to an accuracy of 0.5–1% or better, and then the accuracy of the drop dimensions (r^3 dependence) will be in the range 0.16–0.33%. The accuracy of the drop cuvette dimension is clearly an improved *vis-à-vis* conventional spectrometers because of the quality of the temperature control in the tensiograph ($\pm 0.01°$ reproducibility), which in effect provides an order of magnitude improvement with respect to the dimensions and control of the same, over that of the conventional UV-vis instru-ment. Obviously, the quality of the cuvette has been found to be a limiting factor for the accuracy of spectrophotometer measurements, which has been tackled by flow-through cells and other sample handling innovations, but despite all these at-tempts because of its convenience the cuvette instrument still reigns supreme in laboratories. Key issues are found to tolerances on the spacing between the faces of the cuvette windows as well as micro errors due to placement of the cuvette in the cuvette holder. Differences between the refractive index of test and reference sam-ples also induce measurement error, but these may be calibrated out as explained above.

The fundamental difference between the drop spectrophotometer and standard spectrophotometers self evidently is that no cuvette is required. In effect the drop becomes a natural cuvette formed from the liquid under analysis and consequently the associated cuvette based problems such as manufacturing tolerances, handling, cleaning, and storage are avoided. This analysis is somewhat complicated since the drop dimensions are constantly evolving during analysis, but below a theory of drop spectroscopy is presented and notably this theory shows theoretically there is a considerable measurement advantage with drop measurements in very absorbing liquids over that of conventional instruments. The discussion of this specific issue will be postponed to this section.

The optical/electronic error analysis presented above was split into four cat-egories, (1) wideband noise, (2) static offset error, (3) additive offset drift and (4) multiplicative drift. The normalized measurement error arising from each er-ror source has been defined in this analysis. Each equation is the product of two factors, the first arising from the performance of the optical/electronic circuitry and the second, a sensitivity function that relates the circuit error to the spectropho-tometer measurement error. Figure 6 compares the four error sensitivity functions by plotting them in decibels on a common scale. It is noteworthy that three of the functions show increased sensitivity at low absorbance. This arises because con-centration measurements derived from the equation involves extracting information

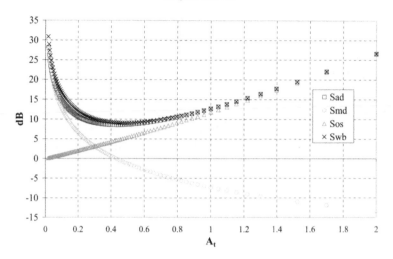

Figure 6. Error sensitivity functions for wideband noise S_{WB}, offset error S_{OS}, additive drift S_{AD} and multiplicative drift S_{MD}.

from the signals I_0 and I_t that are of approximately equal amplitude — a situation that is generally sensitive to signal error. Another set of three sensitivity functions has increased sensitivity at high absorbance. This arises because the test signal I_t has small amplitude and will therefore have relatively large signal-to-error ratio. The offset error sensitivity function S_{OS} falls off at low absorbance values since $I_t \rightarrow I_0$ and for any offset error, K_{OS}

$$\frac{I_0 + K_{OS}}{I_t + K_{OS}} \rightarrow \frac{I_0}{I_t} \quad \text{as } I_t \rightarrow I_0. \tag{13}$$

At high absorbance values the multiplicative drift sensitivity function S_{MD} diminishes because the absolute error $\Delta G_N I_t$ becomes smaller as I_t is reduced. It is noteworthy that there is no general optimum operating point for making measurements, as the minima of the sensitivity functions do not coincide. Consequently it is not possible to recommend a general best operating region for all spectrophotometers since the optimum operating region is dependent upon the relative magnitude of each error source. Typically, the error sources are statistically independent so their variances are additive. This has the important practical result of making the largest error source dominate the concentration measurement error. So, for example, if a spectrophotometer had relatively poor offset error then the best operating region may be for sample absorbances $A_t < 0.3$ while on the other hand if drift from a light source were problematic then a test sample with $A_t > 0.8$ may be better. However, if detailed information about instrumentation optical/electronic error is not known, a reasonable suboptimal operating region is $0.3 \leqslant A_t \leqslant 0.7$.

Another noteworthy feature of the sensitivity functions is that, with the exception of S_{MD} for sample absorbance greater than 0.43, they are always greater than 0 dB. Consequently the measurement error is always greater than the signal-to-error performance of the instrument. The degree of this deterioration is dependent upon the absorbance of the test sample, but it can be quite significant for absorbencies that are remote from optimal sensitivity. For example, suppose a spectrophotometer has wideband SNR of 40 dB and a test sample has measured absorbance of 1.2. According to Fig. 6 the wideband noise sensitivity is 15 dB at this absorbance, so the normalised measurement error is just −25 dB.

The measurement accuracy of the tensiograph has been improved by redesigning its opto-electronic circuitry. The design objective is to increase the value of the first factor in the measurement error expressions defined by Eqs (9a), (10a), (11a) and (12a). i.e., to reduce σ_0/I_0, K_{0S}/I_0, $\Delta K_{0S}/I_0$ and ΔG_N.

B. Drop Shape/Tensiotrace Model

1. Drop Shapes

The following describes briefly the equations used to reconstruct the shape of a hanging drop held up by surface tension. Figure 7 shows a vertical section through the central axis of a pendant drop. The drop has radius r_0 at its top, where it is attached to the drophead. At a point Q on its surface (radius r, height z above the bottom of the drop), there is a pressure difference Δp between the inside and the outside of the drop.

The pressure forces are balanced by surface tension forces given by the classic Young–Laplace (Y–L) equation:

$$\Delta p = p(\text{inside}) - p(\text{outside}) = \gamma\left[\frac{1}{R_\perp} + \frac{1}{R_\parallel}\right], \tag{14}$$

where $R_\perp$ and $R_\parallel$ are respectively the radii of curvature out of plane and in plane at Q, and γ is the surface tension.

The derivation of the equations used in the drop shape model is well known. The equations that are important here as they are the ones used in the computer model using the two characteristic dimensionless parameters β and X_0, defined by

$$\beta = \frac{\rho g r_0^2}{2\gamma}, \qquad X_0 = \frac{r_0}{R_0}. \tag{15}$$

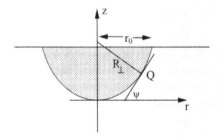

Figure 7. Schematic representation of a hanging drop held up by surface tension.

β is characteristic of the liquid that constitutes the drop, and is the same for all drops formed with the same liquid and the same orifice, whilst the parameter X_0 (the inverse relative radius of curvature at the apex of the drop) changes with drop size. R_0 is the radius of curvature at the apex of the drop.

The volume can be introduced through an auxiliary variable:

$$V = \int dz R^2 \tag{16a}$$

$$dV/ds = R^2 \sin \psi, \tag{16b}$$

where the actual volume is $\pi r_0^3 V$.

Figure 8 shows typical semi-drop shapes obtained for water pendant from a drophead of radius $r_0 = 3.25$ mm, for which $\beta = 0.708$. Identical drop shapes would be obtained for any material with the same value of β; for example, ethanol on a drophead of radius 2.01 mm. The values of X_0 are shown in the legend.

The Y–L equation only gives the condition for equilibrium of the drop surface, and say nothing about how the drop is attached to the device i.e. the drophead, that forms the drop. There is a surface tension associated with the attachment, denoted T_A. Suppose the drop is attached at a horizontal surface, which it meets at an angle α as shown in Fig. 9(a). For equilibrium at the attachment line

$$T \cos \alpha = T_A; \quad \text{i.e. } \cos \alpha = T_A/T. \tag{17}$$

Thus there is in principle a fixed contact angle at the upper horizontal surface. T_A can for some materials be negative, in which case $\alpha > \pi/2$. Normally the liquid "wets" the contacting surface, in which case $\alpha < \pi/2$. Many tensiographic measurements have been made using a drophead comprising a flat disk of radius r_0, with a well-defined edge, to which the drop attaches. Of course, on a microscopic level, this edge cannot be absolutely sharp; we suppose that it is rounded, as in Fig. 9(b). Then it is clear that the drop can maintain a contact angle by attaching

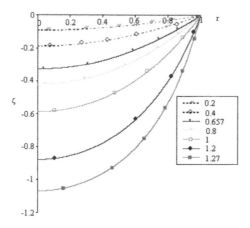

Figure 8. Evolution of a water drop shape on a 3.25 mm radius.

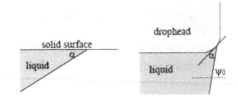

Figure 9. Contact angle geometry.

itself to the edge of the disk at an appropriate point on the rounded edge. This will not usually significantly change the contact radius r_0. ψ_0 is the angle the tangent to the liquid surface makes with the horizontal at the place of attachment, and α the actual contact angle at the rounded edge. Thus ψ_0 is the effective contact angle, subject to the limits: $\psi_0 > \alpha$ for the drop just to attach to the horizontal portion of the disk edge, and $\psi_0 < (\pi/2 + \alpha)$ for the drop just to attach to the vertical section. If $\psi_0 < \alpha$ the drop will shrink on the horizontal portion until the contact angle reaches α, and if $\psi_0 > (\pi/2 + \alpha)$ attachment is not possible.

The requirement that $\psi_0 \geqslant \alpha$ means that there is a minimum size for a drop that attaches to a flat drop-head by wetting the whole of the drop-head. If the drop-head is not flat but is indented (as are some drop-heads used in tensiography) the detailed consideration is different, but there is still a minimum volume V_{min} for a drop that can be supported by the drop-head. In Fig. 8, the third smallest drop shape represents the smallest water drop (volume V_{min}) that can be supported on a flat quartz drop-head of radius $r_0 = 3.25$ mm, where the water/quartz contact angle α is $32.7°$.

At the other end of the scale, there is a maximum drop volume V_{max} that can be obtained by integration of the Y–L equations. At small drop volumes V, an increase in X_0 (inverse apex radius of curvature) always leads to an increase in V, but ultimately there is a maximum drop volume V_{max} beyond which no solution exists for the Y–L equations. The largest drop in Fig. 8 has volume V_{max} for the value $\beta = 0.708$. The situation is a little more complicated for drops with small r_0 because the larger drops here have $\psi_0 > \pi/2$, but there is still a maximum volume V_{max}.

2. Modelling Tensiotraces

A computer programme has been written that simulates the response of the tensiograph instrument using a ray tracing procedure. The passage of a cone of rays from the source fibre (comprising at least 10^6 rays) is followed through the liquid, taking account of reflections at the supported pendant drop surface. The model uses the Fresnel reflection coefficients at the drop surface (which depend on the refractive index), and can also include the effects of absorption. Drop shapes are generated using the Y–L equation. They range between the smallest realisable physical drop that can be supported on the drop-head and the largest drop that can exist, as explained above. The emission from the multimode fibre is modelled by a cone of rays from the source fibre that are followed through the liquid, taking account of reflections at the surface of the drop. If the ray eventually strikes the detector fibre within the acceptance angle of the fibre, its intensity is added to the total.

Typical results are shown in Fig. 2. The total reflectivity (the envelope curve) includes all contributions. The individual curves show the contributions from rays that undergo 2, 3, 4 and 5 reflections at the drop surface (there is no contribution from single reflections); these curves are simple ray counts, and do not include losses on reflection at the drop surface. The results of this analysis are extremely sensitive to the positions of the source and detector fibres, illustrating the need for precision construction of the instrument if reproducible results are to be obtained.

There are two principal peaks in most tensiotraces. The modelling has demonstrated that the first of these is caused mainly by rays that undergo 2 reflections at the drop surface, and is called the "rainbow" peak by analogy with the path of rays that produce a rainbow. The second peak, called the "tensiograph" peak, is produced by rays that undergo 3 reflections, and there are often subsequent peaks of higher order that may or may not be clearly resolved depending upon the exact details of the constructions of the drop-head.

The ray-tracing results reveal the nature of the reflection processes that lead to the rainbow and tensiopeak couplings. The modelling results obtained were from ray tracing calculations made by Smith [32]. It is found that the ray paths for the drop shape that gives the first maximum in the detected intensity with $n = 2$ reflections at the drop surface. It is found that this maximum for $n = 2$ corresponds to the condition that the rays enter and leave the fibres in a symmetric fashion.

3. Scission of Drops

The point at which a slowly growing drop becomes unstable and detaches itself is an issue of obvious theoretical importance. We will refer to the onset of drop separation as the 'scission point'. This to our knowledge is the first time anyone has used this terminology. Tate's law makes the simple assumption that the drop falls off when its weight $W = \rho g V_{\text{ideal}}$ can no longer be supported by the surface tension force $2\pi r \gamma$ acting around the edge of the drop, so

$$V_{\text{ideal}} = 2\pi r \gamma / \rho g \qquad (18)$$

is the idealised maximum volume and the other symbols are defined above in Section B1.

Clearly Tate's law provides an upper limit to the volume of a detaching drop, a limit that is attained only if the drop meets the drop-head vertically at the scission point. V_{max} is an upper limit on the volume at scission for a quasi-static drop — a drop that grows sufficiently slowly that kinetic effects are unimportant. A tensiotrace shows the properties of a drop whose volume grows from that of the remnant drop V_{rem} to the volume at scission V_{sc}. The volume delivered per drop is

$$V_{\text{drop}} = V_{\text{sc}} - V_{\text{rem}} \qquad (19)$$

and it is relatively easy to measure this quantity accurately.

4. Tensiotrace Length and Remnant Drop Size

The problem of obtaining a value for the surface tension directly from measurement of drop volume has been the subject of much empirical investigation in the past. More recently, Basaran and co-workers [33] have made a significant advance on modelling the dynamics of drop scission, using the one-dimensional Navier–Stokes equation for a Newtonian liquid dripping from a capillary. We will briefly review this work. Tate's law, discussed above, assumes that the volume of an idealised drop is given by $\tilde{W} = \rho g \tilde{V}_{ideal} = 2\pi R \gamma$ where $\tilde{W}$ is the weight of the drop of volume $\tilde{V}_{ideal}$, R is the radius of the capillary from which the drop falls and the other quantities have been previously defined above. This equation in dimensionless form becomes $\frac{N_B V_{ideal}}{2\pi} = 1$ where $V_{ideal} = \frac{\tilde{V}_{ideal}}{R^3}$ and the gravitational Bond number $N_B = \frac{\rho g R^2}{\gamma}$. Note that this quantity is the same as the parameter β introduced in Eq. (16), apart from a factor of 2: $N_B = 2\beta$. Harkins and Brown [4] realised that the drop volume that actually falls from the capillary $\tilde{V}_f$ is less than the ideal volume $\tilde{V}_{ideal}$. By a painstaking empirical study of the drop volumes they developed a correction factor through correlating the ratio $\psi(\phi) = \tilde{V}_f/\tilde{V}_{ideal}$ with the 'scaled capillary radius', ϕ, where

$$\phi = \frac{R}{\tilde{V}_f^{1/3}}. \tag{20}$$

Harkins and Brown thereby produced an accurate and reliable method of determining surface tension from drop volume/weight measurements for a liquid of known density dripping from a capillary of radius R, using the formula

$$\gamma = \frac{\rho g \tilde{V}_f}{2\pi R \psi}. \tag{21}$$

Their correction fraction $\psi(\phi)$ was empirical, and was presented in look-up tables as a polynomial of ϕ. The restriction on the range of the correction factor $0.5352 \leqslant \psi \leqslant 0.7208$ was extended by Wilkinson [34] to $0.7208 \leqslant \psi \leqslant 0.9438$. Basaran *et al.* derived an empirical relationship between the Bond number and the scaled capillary radius

$$N_B = 3.60\phi^{2.81} \tag{22}$$

which, when plotted as a linear relationship on a graph of N_B and $\phi^{2.81}$, was shown to fit both Harkins and Brown's results and those of Wilkinson. The relationship was obtained from numerical modelling using the one-dimensional Navier–Stokes equation for a Newtonian liquid dripping from capillary. Basaran *et al.* also investigated viscous and dynamic effects, described using the Weber number (measure of inertia of liquid in relation to the surface tension) $We = \frac{\rho \tilde{Q}^2}{\pi^2 \gamma R^3} = 10^{-7}$ and the Ohnesorge number (measure of viscous and surface tension forces) $Oh = \frac{\mu}{\sqrt{\rho R \gamma}} = 0.01$. Here $\tilde{Q}$ is the volumetric flow rate through the capillary, which is in our system constant and independent of time and $m^3 \ s^{-1}$ (for our system 0.813 $\mu l \ s^{-1}$) and μ is the viscosity of the LUT measured in Pa s. Basaran and his colleagues showed that the Lando and Oakley [35] best fit polynomial

$$\psi = 0.14782 + 0.27896\phi - 0.166\phi^2 \tag{23}$$

does not reduce to the appropriate value of $\psi = 1$ as $\phi \to 0$.

The impressive theoretical development of Basaran is one that shows that at various points along the ψ versus ϕ curve the drop separation is from different processes of drop scission. At low values of ϕ below 0.8 there is not a great deal of elongation. From computation of the drop shapes such elongated drop separation was found to occur at values of ϕ above about 1.0. The processes shown in this study correlate with observations made by the authors, but there is in the case of water a very substantial pull-back from the separated drop onto a concave drop-head, which leads to the production of a satellite drop being released from the drophead after the rebound process in the remnant drop. Such processes do not figure in the drop shape computations of these authors. Basaran *et al.* in their conclusions specifically highlight the advantage of the drop volume method over that of ADSA — "considerably larger experimental errors may be incurred when using other methods including the pendant drop method, which requires painstaking accurate digitisation of experimentally recorded drop profiles". This work also provides a well-established theoretical foundation for drop separation in situations that cannot be described as quasi-equilibrium. This is very useful because vanishingly small flow rates become a problem with highly volatile liquids.

We now apply these considerations to a tensiotrace — see Fig. 1(b). The tensiotrace is plotted as a function of time, which is equivalent to volume as the liquid is delivered at a constant rate. There are two significant points on the trace: the volume V_C at which the signal commences, and the volume V_D at which scission occurs. These volumes are measured using a scale on which the volume at the beginning of the tensiotrace is zero. This zero volume start position does not occur in practice because of the remnant drop. V_D clearly is then the same as Basaran *et al.*'s $\tilde{V}_f$; thus the surface tension can in principle be determined directly from the volume V_D using their prescription.

However, our modelling procedure gives us further information. The commencement volume V_C' is very clearly indicated in the modelling, where the prime ('), means that the volume is referred to an absolute volume scale in which the zero corresponds to no liquid on a flat drop-head. The difference between V_C and V_C' is just the volume of the remnant drop V_R: thus

$$V_C' = V_C + V_R, \tag{24a}$$

$$V_{sc} = V_D + V_R. \tag{24b}$$

Equation (24b) results from the assumption that scission occurs at the calculated volume $V_{sc} = V_{max}$. Thus, there is in principle a method of combining the tensiotrace measurements and modelling to give a method of eliminating the effect of remnant drop volume from the measurements. We are currently investigating the consistency of this approach as compared to the standard empirical drop volume methods.

Considerable amount of work was undertaken by Morrin [36] using water to check and calibrate the performance of the drop analyser. It was shown that the Harkins–Brown correction factor could be used with the drop volume measurement on the tensiograph, generally with better accuracy being obtained with slower pump delivery. This work demonstrated clearly that the drop analyser delivers respectable surface tension measurements that were as accurate and reproducible.

5. Remnant Drop Studies

The only existing tensiograph study of remnant drop relationships are those by Gao and Zeng [37]. They used the vibration of the remnant drop to investigate the size of this drop. This was a method suggested initially by McMillan, Feeney *et al.* [38] and developed then by McMillan, Carbery *et al.* [39]. The real issue of studying the size of the remnant drop has not been advanced much in these studies that were somewhat limited and largely qualitative. The new work reported here is very different and shows for the first time the real power of tensiography in dealing with these vital issues of defining the volume of a remnant drop and thus experimentally advancing the work of Harkins and Brown and supplementing the recent theoretical contribution by Basaran and his co-workers [33].

The rainbow commencement is a powerful tensiotrace measurand for the study of remnant drops volumes. Figure 10(a) shows drawings taken from camera recordings of various remnant drops of various liquids. Figure 10(b) shows the very striking variability in the measured rainbow commencement taken on water and alcohol type of drops, illustrated here by a typical representative of this liquid type 2-methoxyethanol.

The graphical plot of a series of rainbow commencements measured against surface tension measurements for a series of alkanes and alcohols is seen in Fig. 11(a).

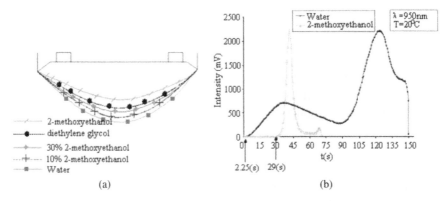

Figure 10. Drawings made from camera images of various remnant drops showing physically the variability for different liquids (a); tensiotraces of water and 2-methoxyethanol showing massive variation in the measured commencement position (b).

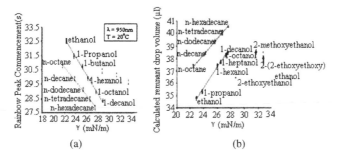

Figure 11. Relationship between rainbow peak commencement and known surface tension values for alkane and alcohols (a); Calculated remnant drop volumes from series of alkanes and alcohols (b).

From a tensiotrace, measurements of the remnant drop can be obtained. The modelling results of the commencement for the primary alcohols with error-bars obtained from running the computer model have been compared. Paradoxically, there are experimental errors in the modelling because of random generation of rays producing variations in the resulting tensiotrace! The conclusions that can be drawn from this preliminary study into the commencement is that there is strong prima facie evidence for a law given relationship here and it would be suggested that an analytical relationship exists that describe this tensiotrace measurand. The commencement measurand is based on what is believed to be an ideal optical detection configuration similar to the Abbé refractometer where the light is totally spatially separated from the region of no light. The drophead has been engineered so that light does not couple from the source to collector fibre for any remnant drop and that in all measurements there is a quiescent period in which there is no light coupling. The rainbow peak then appears at a distinct time in the tensiotrace rising from the zero (base line) to produce a very well defined optical measurement point. The experimental data suggests that the relationship for alcohols might be a first-order dependence, but the modelling study on primary alcohols unequivocally shows that this graph of commencement against surface tension is indeed a second-order curve. The experimental study however does clearly establish that there are some liquids that do not appear to lie on any curve so if an analytical relationship exists here it is rather complex. The second-order curve may perhaps be linearised by using Harkins–Brown correction factors but here all that need be said is that the work is ongoing. The point worth reiterating here is that the drophead was designed to ensure that for all liquid types this measurand could be obtained and that this is an extremely well defined and clean measurement that has minimal experimental error unlike a peak position which is a difficult point to determine and that is most accurately obtained from the first derivative curve of the data set.

　　The question that must be answered really is why does there exist commencement relationships such as shown here in these experimental and modelling studies? The physical relations that describe these results remain still unknown, but the variation discovered here strongly suggests that these are law given. The Harkins–

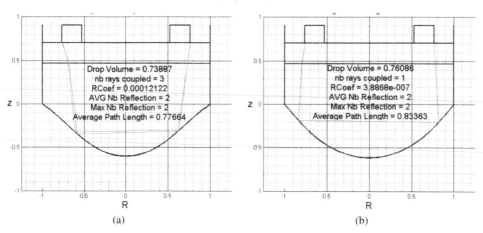

Figure 12. Ethanol drop on an 8.5 mm drophead at the commencement volume (a); water at the commencement volume on an 8.5 mm drophead (b).

Brown correction factors relate to the size of the remnant drops and the commencement develops after some additional volume has been added. The modelling work has been extended somewhat to investigate the form and size of the remnant drop for water and ethanol at the commencement volume. Figure 12(a) and (b) show respectively the commencement positions for each of these liquid types. Visually, the drops are different and this is an important finding. The first optical coupling of photons between the source and collector fibres represents a very well defined optical position. The various trend-lines in Fig. 14 which are accurately predicted by the computer model nevertheless should find some analytical explanation. The explanation must come in part from the fact that the wide emission cone of multimode fibre used in this instrument illuminates the edge of the DUT no matter what the physical property of the drop. It is on this segment of the drop edge that there is a partitioning of rays between those that are non-TIR rays and those that are TIR. The model shows clearly that it is the latter that contribute in a measureable way to the coupled light. More work is obviously required, but given that this commencement point is so beautifully defined experimentally, it is clear that this measurand is an important one giving excellent reproducibly with a resulting small measurement error.

The method of making this computation can be summarised simply. A table of sample alcohols and glycols in this study was constructed. This table included values of both the measured drop volume on detachment and the calculated Tate 'ideal drop' volume. The study revealed that the remnant drop volume calculated from the difference between these quantities does not have a single correlation with the rainbow-peak commencement volume. Figure 11(a) does show some complex relationships of commencement with this calculated remnant drop volume. It is felt there must be some significance that Fig. 11(a) and (b) show a negative and positive slope for the relationships of commencement with respectively surface tension and calculated remnant drop volume for the chemical families; importantly the same basic structure in the graphical relationships exist in both graphs. The same satellite chemicals 2-methoxyethanol, 2-ethoxyethanol and 2-(2-ethoxyethoxy) ethanol do not fall on the primary alcohol plot. This whole matter of commencement deserves further study.

Secondly, there are numerous occasions in which these chemical series reveal linear relationships and many more other situations where distinctive trend lines can be used to identify the family of chemicals. Importantly, Morrin [36] has undertaken a very extensive study of surfactants using this technique that are dealt with in Section 5 of this chapter.

C. Drop Spectroscopy

1. Tensiograph Absorption Theory

The way in which a tensiograph changes with increasing absorption in the LUT is somewhat more subtle than in a standard optical absorption measurement because the optical effective path length is not a fixed parameter of the system. In the standard type of optical absorption measurement, the light passes through a fixed effective path length ℓ. The relative optical intensity due to absorption along a path of length ℓ is

$$R = \exp\{-\alpha'\ell c\}, \tag{25a}$$

$$I = I_0[1 - \exp\{-\alpha_T\ell\}], \tag{25b}$$

here α' is the absorption coefficient of the optical material measured using units of reciprocal length. The effect of turbidity is more complicated but as can be seen from Eq. (25b) it has a similar exponential dependence. Under turbid conditions, the direct ray suffers a loss of intensity over a path ℓ similar to the loss due to absorption, where α_T is the turbidity scattering coefficient. However, the intensity $I_0 \exp\{-\alpha_T\ell\}$ removed from the direct ray is not absorbed, but now contributes to a turbidity-generated optical energy density within the liquid drop. We assume that this energy density is in the form of isotropic radiation, which escapes from the drop at its air surface and also escapes into the drop-head. Some of the latter radiation will find its way into the exit fibre, and will be detected as indirect turbidity-scattered light in addition to the light detected by the direct rays. The calculation of the effect of turbidity is straightforward, given these assumptions, but the details are too lengthy to be specified in this report.

The optical absorbance A is defined by $A = -\log_{10}\{R\}$, so

$$A = \log_{10}(e)\alpha'\ell c = 0.4343\alpha'\ell c = \varepsilon c\ell. \tag{26}$$

This is the well-known Lambert–Beer's law, stating that A is linearly proportional to the effective path length. Here ε is the molar absorptivity and c the molar concentration.

In the tensiograph, the situation is unfortunately more complex. Not only does the effective path length increase as the drop increases in size, but also even for a drop of fixed size there is a range of paths of different lengths for the light going from source fibre to detector fibre. Nevertheless, it is important to distinguish between the tensiograph and UV-vis absorbance. A_T indicates the tensiograph absorbance. The Eq. (27) is consistent with the empirically derived tensiograph absorbance initially defined by McMillan, Finlayson *et al.* [40] and is given by the following equation in terms of measurements taken at a single tensiotrace point. Then:

$$A_T = \log_{10} \frac{\langle V_{0i} \rangle}{\langle V_{tj} \rangle} (0 < A_T), \tag{27}$$

$\langle V_{0i} \rangle$ is the average height of the tensiopeak (data point i) for the blank, $\langle V_{tj} \rangle$ is the average height of the same peak (data point j) for the sample.

The usual situation is for the measurement to be taken at a volume and data point corresponding to peak maxima (tensiopeak or rainbow). It is worth emphasising that the quantity A_T is not the same as absorbance measured in a UV-vis instrument represented simply by A, although it is perhaps in practice almost indistinguishable from this quantity.

In the tensiograph absorbance measurements the solute is usually in such dilute concentrations, in ppm for example, that the shapes of the measured drops are all indistinguishable. Only in the case of water drops with surfactants or alcohols at relatively high concentration in the present study will this condition not hold. The shape of the drop to a very large extent determines the position of the peaks in the tensiotrace with regards to volume, so it follows that if the LUT all have similar physical properties; their peaks are coincident with respect to their volumes. We can see thus mathematically this physical condition translates in Eq. (27) to $i \approx j$. The standard integration time controls of the CCD will enable the measurement accuracy of these height measurements to be improved by the usual experimental averaging procedures. There are two important measurement options. The first practical situation is when the spectroscopic measurement data is taken using the appropriate peak data that is extracted from the series of temporal tensiotrace data. The second situation is when the pump is programmed to stop at the peak maxima and then CCD or CMOS camera averaging is used. These two approaches will be called respectively the dynamic and static quantitative modes.

The drop spectrometer is an instrument that delivers a signal format of a voltage output measured against the volume of a liquid delivered, which of course, relates directly to time of liquid delivery. This voltage is derived from an A/D conversion in a PIC circuit when the instrument is operating *via* LED sources, or alternatively *via* the A/D in a CCD detector circuit. In both cases the voltage represents stored or dynamic charge that derives from the photon flux.

An important study was undertaken using a typical blue-dye tensiotrace (concentration of 1.257 μM) from a set of experimental measurements taken with this chromophore. The recorded tensiotraces (from a series of experimental measurements on a range of concentrations of the liquid aliquot) can in every drop case correspond on the volume or time axis. The same measurement volumes in all drops correspond with the next because the drops are almost identical in shape, all being in essence just water; the concentration of blue-dye in each drop differs but nevertheless in practice the concentrations are infinitesimal. It is found that each point (volume) on the tensiotrace abscissa produces its own unique calibration graph. The resulting calibration graphs are linear but have different slopes. It is found that at different points (drop volumes) in the tensiotrace, there exist different sensitivities and this informs us that the absorption produced by the chromophore is in fact encoding the entire tensiotrace with information on this absorbing species but is complex. In each of these drop volume measurement-positions, the pathlength obviously varies. These quantities determine in the last analysis the form of the modified Beer's law relationship that is found.

This fact raises serious issues about the optimum way we can obtain a measurement of the absorbance of the LUT. The issues surrounding the optimum averaging procedure have been considered in some detail below, but it is here perhaps worth pointing out the obvious. In general, the longer the effective path length (EPL) in the drop, then the greater the calibration sensitivity of the calibration graphs. The calibration can obviously be done in a way that is directly analogous to the standard procedure used in quantitative UV-vis spectrophotometry. Standard aliquot solutions can be used to generate a calibration graph. The CCD or CMOS system of course averages in time as it generates many tensiotraces for the LUT in a second. Averaging is therefore used in the traditional way to determine the result and the standard deviation, which can of course be determined to obtain a measure for the experimental error.

Suppose a beam of light passes through an absorbing material in such a way that there is a range of paths from source to detector. Let the probability that the path length lies in the range ℓ to $\ell + d\ell$ be $P(\ell)\, d\ell$ ℓ, normalised such that $\int P(\ell)\, d\ell = 1$. The EPL (effective path length) can be defined from Eq. (28a), where Eq. (28b) is for the specific case of the tensiograph.

$$\ell_{\text{eff}} = A/\varepsilon c, \tag{28a}$$

$$\ell_{\text{eff}} = A_T/\varepsilon c = \ell_1 - 1.15278\varepsilon c \Delta \ell^2. \tag{28b}$$

It should be noted that ℓ_{eff} and ℓ_1 are different variables.

$$\ell_1 = \int \ell P(\ell)\, d\ell. \tag{29}$$

The overall relative absorption factor is therefore given by the quantity

$$R = \int \exp\{-\alpha' \ell c\} P(\ell)\, d\ell. \tag{30}$$

Clearly the absorbance $A = -\log_{10}\{R\}$ is now not simply equal to $\varepsilon c \ell_1$. One can perform an expansion valid for weak absorption in order to see the size of this effect. To second-order in α,

$$R \approx \int \left\{ 1 - \alpha' \ell c + \frac{1}{2}\alpha'^2 \ell^2 c^2 \right\} P(\ell)\, d\ell = \left[1 - \alpha' \ell_1 c + \frac{1}{2}\alpha'^2 \ell_2^2 c^2 \right], \tag{31}$$

where ℓ_2 is the rms path length defined as $\ell_2^2 = \int \ell_2 P(\ell)\, \mathrm{d}\ell$. Then

$$A_T = 0.4343\alpha'\ell_1 c - 0.21715\alpha'^2 c^2 \Delta\ell^2. \tag{32a}$$

In terms of the molar absorptivity, which is the usual formulation employed in chemistry.

$$A_T = \varepsilon c\ell_1 - 1.15278\varepsilon^2 c^2 \Delta\ell^2. \tag{32b}$$

This result is one of considerable analytical importance to drop spectroscopy and shows that, for small absorption, the optical absorbance falls below the linear Beer's law by an amount proportional to the variance

$$\Delta\ell^2 = (\ell_2^2 - \ell_1^2). \tag{33}$$

If a simple Beer's law applies to a tensiographic spectroscopic measurement and the calibration graph is linear, then mathematically this means $\ell_2 = \ell_1$ and the measurement analysis is considerably simplified.

It is perhaps worth here just giving some more thought to this correction factor and its relation to the form of what we might call the drop spectroscopy Beer's law relationship. Common sense would immediately suggest that if drop shapes of the LUT were essentially unchanged, as is indeed the case in the usual practical working situation, when analysing a set of very dilute concentrations of analyte, then both (ℓ_1) and the variance $(\Delta\ell_2)$ are in essence just a quantifiable function of the absorbance of the test liquid. This 'thought experiment' was tested using the model and Fig. 13(a) and (b) shows the result from modelling 12 000 rays for a water drop on a flat 9 mm drophead. These results require some explanation as they bear on the practical issue of the quantitative determination of concentration using a drop spectrometer.

Figure 13(a) is important as it quantifies the variation in both the EPL and the variance. It is important to say that in water the tensiopeak occurs at a volume of 0.3 on this diagram. We see that as the drop grows in volume the EPL increases

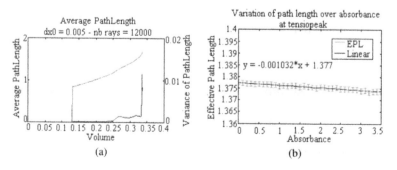

(a)   (b)

Figure 13. Effective path length (EPL) (upper curve) and variance ($\Delta\ell^2$) (lower curve) as functions of drop volume modelled for water on a 9 mm drophead (a); Modelled variation in the effective path length (EPL) and variance in the EPL in a water drop on a 9 mm diameter head for absorbance values 0 to 3 A-units (b).

in an approximately parabolic fashion until close to the drop separation. The EPL varies between 0.8 and 1.8 cm (approximately 2 and 4 times r_0). The variance is extremely small until 3rd order reflections come into operation when a variance of around 0.0015 is obtained. The variance increases in a radical way at this point which is really an anomaly arising from the fact that just at the end of the drop cycle we observe somewhat paradoxically that a very strong 1st order reflection develops. These rays have a shorter path length than rays with 2, 3 or more reflection and thus increase the variance. This is not a problem in practice because the drop separates from the head having reached its maximum volume probably because of the smallest perturbation due to vibrations.

Figure 13(b) shows a graph of ℓ_1 and $\Delta\ell^2$ plotted against absorbance of the LUT in the range $A = 0$ to 3.5 and the average path length is seen in this figure to be given by the relationship:

$$\ell_1 = -0.0132A_T + 1.377. \tag{34}$$

It is clear from this figure that the variance in the path length, which is of course calculated from the model is almost constant and has a value of $\Delta\ell^2 = 0.0021$. Tensiotraces of water and dye solutions have been modeled assuming that the analyte does not change in a measurable way the other physical properties of the water. This is found in practice to be a very accurate modelling assumption. Absorbance values have been modelled for the range 0 to 3.5. The measurement at a peak is always made somewhat problematic because of noise, which affects the value in the most critical way at such a point of inflexion. The path length at the tensiopeak is plotted against absorbance to show that the path length does not change at a given volume.

The graph as theoretically predicted has a slight negative slope as the path length is shortened with absorbance. The cause for this foreshortening of the EPL being simply that attenuation of various path lengths inside the drop is obviously greater for the longer than the shorter paths. It should also be noted that the slope on this graph is only visible due to the zoom factor used. Both the slope and variance shown by the error-bars at each of the modelled absorbance positions are very small.

This result shows that tensiography is capable of measuring absorbance from measurements taken at the tensiopeak position and this to a very good approximation will be a simple Beer's law relationship.

The modelling shows that the EPL and its variance are quantifiable numbers so an exact relationship including the very small analytical correction can be obtained for Eq. 32. It is clear however that the equation we have for drop spectroscopy devolves to one with a slightly circular form, in that the measured tensiograph absorbance appears on either side of the equation. Numerical methods can however be used to solve this second-order equation in concentration.

$$A_T = 0.4343\alpha'(-0.0132A_T + 13.77)c - 0.00448\alpha'^2c^2, \tag{35a}$$

$$A_T = \varepsilon c(-0.0132A_T + 13.77) - 0.00238\varepsilon^2c^2. \tag{35b}$$

The general approach would be then to take ratio measurements of absorbance at corresponding positions (volumes) in the tensiotrace from a measurement and blank drop. There are a series of tensiotraces recorded at each wavelength (λ) and each wavelength in general will present a different tensiotrace to the next if there is an absorbing species present in the LUT. We could perhaps best give the analytical relationship for tensiographic absorbance $A_T(\lambda, V)$ showing the functional dependences for the dynamic quantitative measurement mode. The dependences are given here in italic were λ is the wavelength in nm, V is any selected volume of the drop measured in µl, and A is the absorptivity of the liquid measured using a standard UV-vis measured in A-units. For measurements taken on the growing drop in a dynamic mode then the volume is a continuously changing variable

$$A_T(\lambda, V) = \varepsilon(\lambda)c\ell_1(A, V) - 1.15128\varepsilon(\lambda)^2\Delta\ell^2(A, V)c^2 \tag{36a}$$

but at the tensiopeak position the modelling tells us that

$$A_T(\lambda, V) = \varepsilon(\lambda)c(-0.0132A_T + 13.77) - 0.00242\varepsilon(\lambda)^2c^2. \tag{36b}$$

For the static quantitative mode when the pump is stopped, so as to keep the volume fixed, so that the reflection inside the drop are maximized at the rainbow peak, the volume is now a fixed value and not a variable. Thus in this measurement approach offers a simplified situation:

$$A_T(\lambda) = \varepsilon(\lambda)c\ell_1(A) - 1.151278\varepsilon(\lambda)^2\Delta\ell^2(A)c^2, \tag{37a}$$

$$A_T(\lambda) = \varepsilon(\lambda)c(-0.0132A_T + 13.77) - 0.00242\varepsilon(\lambda)^2c^2. \tag{37b}$$

2. Experimental Study into the Effective Path Length

In order to determine the tensiographic absorbance, it is being assumed that the tensiograph peak maximum is in an equivalent position for each trace. Probably, the first question to ask is about how the EPL varies with the concentration variation of the analyte. Here experimental determination of the EPL has been carried out using three different methods. First of all, by comparing absorbance values obtained by UV-vis spectroscopy and tensiography assuming a simple uncorrected Beer's law approximation to obtain EPL estimates; then by the ray tracing method on camera images of drops; finally by using ray-tracing software developed by Smith.

The tensiograph peak height maximum has been used previously for absorbance measurements [5]. The volume (temporal) position of these peaks may vary slightly since drop volume and drop shape varies slightly between samples. For theoretically well-grounded absorbance reading, the path length of light coupling between the source and detector fibre should be identical between reference and test liquids. Clearly, this is not possible in an absolute way if there are any changes in the shape of the drop arising from differences in the small quantities of say the absorbing species. However, for the majority of liquids, the variation in the path length for the measurement volume (peak height maxima) is absolutely negligible and does not cause any measurable error contribution to the measurement. Some attempts

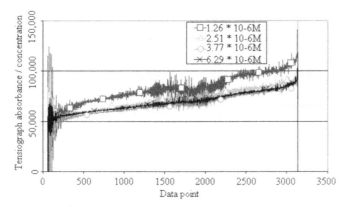

Figure 14. Plot of normalised path length obtained from measurements on blue dye for solutions blue 1.26 μmol, red 2.5 μmol, green 3.77 μmol and yellow 6.29 μmol.

have been made to quantify the changes arising from variations in the LUT that are discussed below.

2.1. Path Length Determination from Absorbance Measurements — Proof of the Theory

A study has been made using dilute solutions of the food colorant dye (FCF blue #1 (C.I./42096)). Solutions of this dye were prepared in de-ionised water in the range 0–6.29 μM. It was assumed that given the dilute nature of these solutions a first-order Beer's law relationship would be obtained. At each point in the tensiotrace, calculations were made using Eq. (27) and from this a calculation of path length was made using Eq. (37). The initial data in the tensiotrace is unreliable as it corresponds to a period of drop oscillation as can be seen from the initial scatter on the data in Fig. 14.

From these calculations the initial drop path length is calculated to be 9 mm, rising thereafter to 12 mm at drop volume corresponding to a time of 90 s (data point 3000). The importance of this result is that we are predicting that the EPL in these drops should be identical as the concentration of food dye is so small that the drop shape should be unchanged. The results here confirm this beautifully, for although the tensiotraces are markedly different the computation produces graphs (Fig. 14) of three sets of data (red, green and yellow) that sit within experimental error one on top of the other. The measurement for the 1.26 μmol solution is however significantly offset from the others three lines. The displacing of this line is interpreted as arising from measurement instrumentation error due to measurements being below the instrument detection limit. This result is taken as a very good experimental proof that first-order Beer's law drop spectroscopy relationships hold for liquids in which there are very minimal variations in drop shape due to the analyte.

Tensiotrace recording of a series of acetaminophen solutions in the range of concentrations 625, 1250, 2500, 5000, 8000 and 11 500 ppb were obtained. Interestingly, in this spectral study there is no need of a separate blank to be used as the acquisition point of a tensiotrace formed at $\lambda_{410\ nm}$ of this pharmaceutical active does not absorb and the sample can thus itself be used to provide the blank reading. In fact, light intensity at each wavelength from $\lambda_{200\ nm}$ to $\lambda_{410\ nm}$ was saved for each temporal acquisition point of tensiotrace formation. The Eq. (27) could be used to calculate tensiograph absorbance at the specified drop volume. From the tensio-spectra, the position of λ_{max} was then measured. From the tensiograph measurement this was found to be 242.5–243.5 ± 0.5 nm for acetaminophen measured in water that is in agreement with Gerhardt [41] and Clarke [42]. This study was repeated for acetaminophen with ethanol used as the solvent with a result 249–250 ± 0.5 nm that is in agreement with the standard values reported for example by Clarke [42]. Referencing against data from one CCD acquisition time number from the reference tensiotrace at 410 nm for the peak maxima, it was shown that the peak positions were in practice identical for all practical purposes for all the test solutions of acetaminophen at the measurement wavelength λ_{max}. Secondly, an estimate of the EPL has been made assuming a simple approximation of an uncorrected Beer's law relationship. The standard molar absorptivity of λ_{max} for acetaminophen was taken to be $3.16 \times 10^4\ 1\ mol^{-1}\ cm^{-1}$ for this study. The result demonstrates that for drop spectroscopy Beer Lambert law is very closely obeyed despite the fact that we know from the above theory that there is in fact a second-order correction needed. The estimated EPL remains 15 ± 0.1 mm for all the test solutions measured against the blank. This result shows an impressive measurable reproducibility for all these solutions in the full range of concentrations of acetaminophen. Furthermore, as can be seen below in the qualitative section, drop spectra of acetaminophen were formed by determining absorbance at each wavelength from data to give spectra, which were equivalent to those recorded on the UV-vis spectrophotometer. This practical test is all that matters fundamentally here; spectra recorded with the tensiograph are visually indistinguishable from those recorded with a standard instrument and most importantly the λ_{max} are the same. The authors have found that in a wide range of test solutions that similar impressive results were obtained lending strong support to the use of drop spectroscopic analysis.

2.2. Practical Issue of Path Length Variations for Drop Spectroscopy

It appears on the basis of practical investigation that variations in water drops containing non-surfactant dye molecules do not present a problem for the application of the ratio-calculation as in Eq. (27). The authors have in many experimental studies undertaken found linear Beer's law type relationships except in two cases. Firstly in measuring UV-vis absorbance and tensiographic absorbance for colour standards at 470 nm in the concentration range 0 HU to 300 HU a non-linear relationship was observed. In the second a visually similar relationship was discovered for measurements on solutions of Blue FCF measured at 660 nm in the concentration range 0 to 6.29 μM shown in Fig. 14. The use of a blank solution measurement (distilled water in most cases, and a low concentration admixtures of methanol was used as a blank for measurements of the PAH compounds as the solvent was required to get these compounds into solution) in most experiments has been used. In both these cases a very good fit was obtained using Eq. (36).

In other experimental studies, a self-referencing situation has been used in which the sample itself is used as the blank. Self-referencing is possible if a CCD detector is employed and there are wavelengths at which the test sample does not absorb. Such a situation must be considered as providing the ideal reference tensiotrace. The drop cuvette used in drop spectroscopy can usually be considered as closely approximating to a constant in the measurement process as the EPL will not vary from drop-to-drop. Linear calibrations have been repeatedly found [36]. For surfactant solutions however, and indeed some other LUT, there will be an appreciable alteration in the shape of a drop with increased concentration of the analyte.

An attempt to give some rough measures on the variation in EPL that will arise in different test liquids has been made through an experimental study. Such information is clearly vital in proscribing strict limits on the range of solutions that can be measured with this technique. Table 1 gives the calculated variation in light throughput in the drop for what is hopefully a fairly representative set of solution types. This table gives measures of throughput for varying ranges of concentration of these liquids. These liquids have been selected as being typical of those used in the general chemical procedures. Variation of average path lengths with concentration for a series of solutions taken from tensiographic measurements applying Eq. (27) was undertaken.

Table 1 gives some guidance for the practical drop spectroscopist into the range of liquid types and associated concentrations that might be used to obtain optimised linear Beer's law calibrations. The variations in each case show the borderline (threshold) at which variations in transmitted light in these solutions occur. For example, if we take the last chemical in the table, namely sodium chloride, then so long as solutions with concentrations below 9.4 mmol are used, then there is no change in light transmission due to concentration variations in this analyte. In such a situation photometry can be undertaken and we can anticipate that there will be a linear Beer's law type relationship.

This threshold condition is the same for both tensiograph peaks, so no variation is observed in either the rainbow or tensiopeak transmission. In practice, if dilutions of the analyte are below the threshold given in this table, then there will be no measurable departure from linearity in the Beer's law relationship. The values at the top of each of these lists of percentage variations is a threshold, below which the quantitative measurement will deteriorate. The threshold of the surfactant SDS is at a low concentration, which is not at all surprising.

It is very well known that water drops change shape radically if there is only the smallest concentration of this surfactant present. It would not be expected that any significant concentration of a surfactant could be present without changing the shape of the drop.

Table 1.
Variation of average pathlengths with concentration for a series of solutions

	Variation in coupled light at rainbow peak	Variation in coupled light at tensiopeak
Glycerol (M)		
9.5 m	100	100
0.19	88	92
0.665	76	90
0.95	66.5	87
Di-sodium hydrogen orthophosphate dodecahydrate (M)		
428 μ	100	100
48.5 m	89	94
0.188	73	89
0.285	64	87
Ammonium Nitrate (M)		
9.5 μ	100	100
475 μ	95	95
9.5 m	93	94
0.19	93	96
Sucrose (M)		
9.5 μ	100	100
475 μ	109	105
9.5 m	110	105
0.19	110	105
Sodium dodecyl sulphate (M)		
95 n	100	100
9.5 μ	102	100
238 μ	80	85
713 μ	66.5	72
Sodium chloride (M)		
9.4 m	100	100
0.24	90	95
0.95	67	87
1.9	49	84

3. Modelling Results

The form of the modelled tensiotraces for increasing absorbance are shown in Fig. 2(b). There is an observed reduction in both the rainbow peak and tensiopeak heights with increasing concentration due to increasing absorbance; the reduction in the latter peak height is greater because of the increased path length in the drop which has a greater volume. These modelling results are discussed below in relation to the experimental results shown in Fig. 15(a).

Tensiotraces of blue dye solutions at 660nm (PEEK drophead)

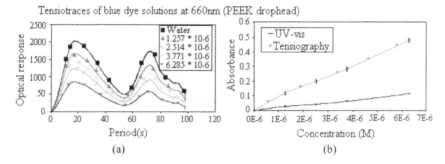

(a)

(b)

Figure 15. Tensiotraces of FCF blue #1 solutions at 660 nm using PEEK drophead (a); comparison of UV-vis absorbance and tensiographic absorbance for Blue FCF at 660 nm in the concentration range 0 to 6.29 µM (b).

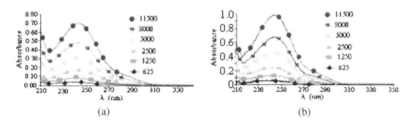

(a)

(b)

Figure 16. Drop spectra for concentrations of acetaminophen in ppb using data obtained at the rainbow (a) and tensiopeak positions (b), respectively. Relative absorbance shown here normalising the tensiopeak value to unity. Note the rainbow absorbance measurement is lower due to shorter EPL at this volume.

The experimental arrangement is such that volume is added with time but of course this linear time-axis corresponds to a linear volume-axis. The drop-period is constant in this sample as the surface tension is constant because the dye is at such low concentrations the bulk property of the measured samples is identical within practical limits. Here it is only necessary to point out, that the modelling assumes a standard linear relationship between the light absorption and increasing path length of the ray. There are many varying coupling ray paths inside the drop between the source and collector fibre.

4. Single Drop Spectroscopy at Two Wavelengths

The ultra-violet spectra of different concentrations of acetaminophen were obtained using a double beam ultra-violet spectrometer at 20°C. These results were compared with the drop spectra obtained with a drop spectrometer in Fig. 16(a) and (b).

The experimental approach here is to obtain measurements during the usual procedure for tensiotrace generation, but the Ocean CCD fibre spectrometer used in these studies has a grating for a wavelength range of 200–410 nm. Using the same acquisition time, at the same stage of drop generation, the tensiograph absorbance of each concentration of acetaminophen in 10% (v/v) ethanol/water was then obtained applying Eq. (27). In these measurements, it is clear that the tensiotrace here

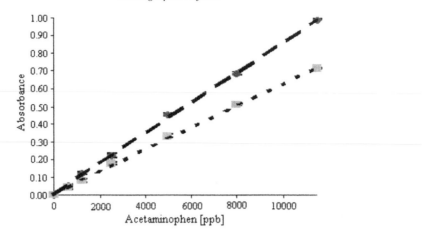

Figure 17. Variation in absorbance of acetaminophen concentrations at $\lambda = 244$ nm using (i) rainbow (lower line) and (ii) tensiopeak (top line) spectroscopy. It might be noted that error bars are shown on this graph, but are too small to be seen.

presents for the same measurement volume of both the reference or test solutions a constant EPL to a very good approximation. Obviously, corresponding times in the CCD measurement must be used to determine the unknown concentration of the acetaminophen. The measurement assumes that here the concentrations of sample do not materially change the shape of the drop and hence the EPL in test and reference tensiotraces correspond to a high degree.

The spectra obtained using drop spectroscopy compare most favourably with the UV spectra obtained spectrophotometrically on standard instruments. The measurements of positions of λ_{max} correspond not only in these studies, but it has been found in numerous other spectra that the measurements of these peak positions can be accurately and reproducibly obtained using drop spectroscopy.

Figure 17 shows the results of measurements taken on the acetaminophen sample, measured in 10% (v/v) ethanol/water. The tensiographic absorbance measurements are obtained from corresponding data acquisition time representing respectively the rainbow and tensiopeak periods from a tensiotrace formed at all wavelengths between 200 and 410 nm. Light intensity of 10% ethanol and acetaminophen concentrations were obtained from an acquisition point from (i) rainbow and (ii) tensiopeak period and both approaches delivered visually similar plots. This qualitative result is interesting and can be extended in its implications. It informs us that any set of corresponding data from the same time in the tensiotrace can be used to produce spectra that will correspond to that obtained by a UV-vis spectrophotometer. The larger volume of the drop and the correspondingly larger EPL for the 'tensiopeak absorbance' with respect to 'rainbow peak absorbance' measurement means in practice that the former produces higher tensiograph absorbance measurements.

The importance of this result is that this graphically really confirms Eq. (37), in that the equation predicts an EPL dependence on the sensitivity of the measurement at various positions in the tensiotrace. The sensitivity (slope) of the graphs is as predicted, with the greater slope coming from measurement point in the drop growth when the EPL is longer. This experimental quantitative result also can be extended in its implications. It means that any data from a set of corresponding tensiotrace times recorded on a set of aliquots could be used for a test solution's quantitative concentration determination. The penalty in using times that do not correspond to the peak maxima is reduced sensitivity of the quantitative measurement. There is one other factor that should be mentioned here. The number of reflections producing the coupling in the drop at the rainbow peak is just two reflections, while the situation in the tensiopeak is more complex with a number of reflection orders being present. It might be argued that the rainbow peak is a better measurement position because of this simpler optical coupling arrangement, and this despite the fact that it usually has lower calibration sensitivity (smaller peak heights). However, in the view of all the authors, such an advantage is only of marginal theoretical value; the real issue is the relative analytical sensitivities of a peak position regardless of how many types of reflection orders there are present.

This graph shows that in this practical measurement situation Eq. (37) can be approximated to a single term, as the second-order correction factor is negligible. These are respectively:

For tensiopeak absorbance $A_{T\ Tensiopeak} = 9 \times 10^{-5} C_t$ (for this fit the correlation coefficient $R = 1$). Here C_t is the concentration of the analyte. For rainbow peak absorbance $A_{T\ Rainbow} = 6 \times 10^{-5} C_t$ ($R = 0.999$). Again C_t is the concentration of the analyte.

Here the tensiograph calibration sensitivities $\varepsilon_T(\lambda, t)$ are respectively 9×10^{-5} and 6×10^{-5}. Quite simply, the tensiopeak sensitivity is 50% larger than the rainbow because of the longer EPL in the drop corresponding to these two tensiotrace-peak positions. The calibration sensitivity indeed here represents a good instrument performance, remembering that the concentration is measured in ppb. In summary, the reason for this improved calibration sensitivity for the tensiopeak measurement is because of the increased EPL in the drop at the tensiopeak measurement volume over that of the corresponding drop volume of the rainbow peak.

It has been found that in nearly all-practical measurement situations that it is possible to ignore the second-order correction factor in Eq. (32).

5. Absorption and Sensitivity Analysis

In this section an attempt will be made to determine the sensitivity of the drop spectrometer instrument. The preparation of colour standards was as follows: 1.246 g of potassium chlorplatinate and 1000 g of cobalt (II) chloride hexahydrate were dissolved in 1 litre of 0.1 M HCl to give a 500 Hazen Units stock solution [43]. The stock solution was diluted into seven aliquots with colour intensity ranging from 25 to 175 Hazen units. A second set of standards was prepared to cover the range of 0 to 500 Hazen units in increments of 100.

Ten replicate absorption readings of each standard were taken at 465 nm. The readings were made using a 1 cm cell in a UV-vis spectrophotometer (Varian Cary$_{50}$). The average absorbance and standard deviation of the readings were calculated.

Solutions were tested with both the UV-vis spectrophotometer and the fibre drop spectrometer. The standard method for colour determination is the Platinum Cobalt (P–C) method [44] in which absorbance is taken at 465 nm. The centre wavelength of the LED used in the primary comparative study of sensitivity was 470 nm and therefore this source provides a good measurement for this P–C method. A comparison between the two methods is shown in Fig. 15.

The gradient of the UV-vis absorbance measurement *versus* concentration is the calibration sensitivity. For the measurements presented here, there is a value for the UV-vis spectrophotometer measurement of 1.10^{-4}. In the case of the drop spectrometer, the gradient is 4.10^{-4}. The latter steeper gradient quantitatively shows that the drop instrument has 4 times the sensitivity of the standard instrument used in these measurement comparisons. Since these results were obtained, instrumental modifications of the drop analyser have further reduced the measurand error and thus further improved the analytical sensitivity (calibration sensitivity/standard deviation of the measurand) of the technique. The analytical sensitivity of the drop analyser is now comparable with that of the classical UV-vis. The tensiograph data in Fig. 15 was accurately fitted using Eq. (37) and this curve-fit passes through all the 3σ-error bars and indeed the fitted line is close to the centre of all the error bars with the exception of the last point. However, the fit-line intersects the error bar on the last data point, showing that the fit is valid. The experimental evidence is not strong enough statistically as only graphical testing here has been conducted to say that the analytical theory of drop spectroscopy, Eq. (37) has been proved definitively. That being said, it is clear that since the theoretical line of the analytical equation plot fits comfortably inside the 3-sigma (range) error-bars that there is very strong evidence at this early stage that the equation holds for all drop measurements. Clearly, further work is required using statistically designed experiments to prove the equation to a 'level-of-confidence'. Figure 15(b) can also be fitted with this second-order equation. The first five data points in the calibration graph are accurately fitted with a first-order Beer's law relationship with EPL of about 10 mm, which is consistent with all experimental and modelling results obtained here. The absorption coefficient here (molar absorptivity is not used in this fitting simply because the concentration of these solutions is given in Hazen Units) is $0.1068\ \mathrm{HU}^{-1}$. The fitting dependence here on this absorptivity is very sensitive indeed, but less so for the variance term, which is found to be 0.0109. The latter value on the Hazen unit plot sits at the bottom of a sharp trough and hence we can conclude that the fitting error increases sharply as we move away from this point. Importantly, the fitting using the Eq. (37) is 13.8 times better than can be achieved with a linear Beer's law fit.

The tensiograph spectrophotometric measurement has been shown to be more sensitive than the traditional UV-visible spectrophotometers in a series of measurements of food colorant dye for example FCF blue#1 (C.I./42096). Solutions of this dye were prepared in de-ionised water in the range 0–6.29 μM and a comparison of calibration graphs for UV-vis and tensiographic absorbance is shown in Fig. 15 below. Again, the sensitivity is much higher for drop spectroscopy (respectively 76877 and 18006) giving lower detection limits for the drop technique. One interesting and unexpected advantage of the drop spectrometer is that it appears that the upper limit of the dynamic range in the Beer's law analysis marked by the departure from linearity for a standard instrumental analysis where the sensitivity decreases, may be higher in a tensiograph. The slope in drop spectrometer calibrations often shows an increase at higher concentrations, inexplicably an increasing sensitivity. Such somewhat idiosyncratic behaviour deserves some more detailed study in the future.

A series of organic pollutants were analysed. These are herbicides, pesticides, PAHs (Polyaromatic hydrocarbons) and other organic molecules (e.g. atrazine, metabenzthiazuron, benzo a pyrene, anthracene, pentachlorophenol), which may be found in polluted waters. The usual method of analysis is *via* liquid chromatography or optical detection. As a consequence of the presence of aromatic rings in these molecules, they strongly absorb in the UV region, but have very little absorbance in the visible range. Those organic molecules do not absorb in the visible, and it was therefore necessary to use a tensiograph instrument with a UV light source coupled with a UV detector. The most appropriate work to report here is for two of these priority organic pollutants that were analysed by tensiography at their wavelength of maximum absorbance. These were anthracene ($C_{14}H_{10}$) and naphthalene ($C_{10}H_8$). They are both PAHs (Polyaromatic hydrocarbons) and both are found in industrial waters.

Although anthracene and naphthalene are not very soluble in water, because of their strongly absorbing properties, even trace amounts of these substances can be detected. The molecules have greater solubility in organic solvents such as alcohols and benzene. Here in the present study, methanol has been used to make up standard solutions. Anthracene samples were prepared by making up water based solutions. The solution was then filtered before being analysed with the tensiograph. The molarity of this sample was calculated from the calibration plot of a range of known concentrations of naphthalene and anthracene standards solutions in methanol, run on UV-vis spectroscopy at their λ_{max}. The tensiographic analysis was carried out at 220 nm and 252 nm and conducted using a pulsed xenon light source and a continuous wave deuterium source. It was found that the pulsed xenon source gave poor repeatability between successive measurements, so the deuterium source was preferred as it is a continuous light source, giving good repeatability between measurements.

Tables 2 and 3 are of some importance to water monitoring applications. The results presented here are part of a 5th Framework project — Aqua-STEW (Surveil-

Table 2.
Summary of detection limits obtained by tensiography for the measurements of naphthalene and anthracene with the pulsed xenon source and continuous wave deuterium source

	Tensiography		IRH
	Rainbow peak height detection limit (M)	Tensiograph peak height detection limit (M)	Detection limit (M) in IRH laboratory
Pulsed Xenon source			
Napthalene in de-ionised water	362×10^{-7}	8.1033×10^{-7}	4.5117×10^{-10}
Napthalene in volvic (mineral) water	1.258×10^{-6}	7.38×10^{-6}	Not measured
Napthalene in tap water	1.3357×10^{-6}	9.629×10^{-7}	Not measured
Continuous wave deuterium source			
Napthalene in de-ionised water	1.33×10^{-7}	1.33×10^{-7}	4.5117×10^{-10}
Anthracene in de-ionised water	4.3×10^{-8}	2.9×10^{-8}	1.683×10^{-10}

Table 3.
Estimation of detection and quantification limits of eleven priority substances all using the High Power Xenon source coupled with the S1024 deep well spectrometer

Substance	Formula	Estimated detection limit (M)	Estimated quantification limit (M)	IRH quantification limit (mgl^{-1})	IRH quantification limit (M)
Napthalene	$C_{10}H_8$	2.43E–09	8.09E–09	2.00E–05	1.55E–10
Anthracene	$C_{14}H_{10}$	2.11E–08	7.04E–08	1.00E–05	5.61E–11
Simazine	$C_7H_{12}N_5Cl$	4.16E–09	1.39E–08	3.00E–04	1.49E–09
Mecoprop	$C_{10}H_{11}O_3Cl$	1.26E–07	4.20E–07	5.00E–05	2.33E–10
Biphenyl	$C_{12}H_{10}$	4.58E–08	1.53E–07	5.00E–05	3.24E–10
Linuron	$C_9H_{10}N_2O_2Cl$	7.47E–09	2.49E–08	5.00E–05	3.24E–10
4-Octylphenol	$C_{14}H_{22}O$	1.86E–08	6.18E–08	1.00E–04	4.85E–10
MCPA	$C_9H_9O_3Cl$	5.64E–08	1.88E–07	1.00E–04	4.99E–10
Atrazine	$C_8H_{14}N_5Cl$	4.94E–08	1.65E–07	3.00E–05	1.39E–10
Isoproturon	$C_{12}H_{18}N_2O$	1.73E–08	5.78E–08	5.00E–05	2.42E–10
Siuron C	$C_9H_{10}N_2OCl$	9.28E–09	3.09E–08	3.00E–05	1.29E–10

lance Techniques for Early Warning [obviously of PAH and other pollution incidents]). Simulated on-line and laboratory measurement programmes were undertaken as part of this project and one of the principal targets for this project was some 20 priority pollutants (hereafter PP), and the conclusions drawn from the tables presented for this study was that tensiography could reach on measuring many of these dangerous molecules at detection limits that made the monitoring worthwhile. Associated data mining techniques were developed to provide the means to measure the target PP in real waters that of course had a changing background of water quality. The full 5th Framework EU Report is with the Commission [27] and a number of papers and reports have flowed subsequently to more fully document the outcomes of this report.

6. Evaluation of the Reproducibility of UV-vis Spectrophotometry

An experimental study was made into the reproducibility of a Shimadzu 460 UV-vis spectrophotometer operated firstly at ambient temperature, and then secondly, with improved operation under temperature control. The ambient temperature study was made to determine the drift in the instrument when operated in a laboratory in which variable quantities of sunlight were incident on the instrument and can therefore be considered a worst-case operational situation. The investigation showed a drift of 12% (approximately 1.7% per hour) over a period of 7 hours. In a second study, a Hitachi U-2000 UV-vis spectrophotometer was tested and a drift of 7.2% (approximately 2.4% per hour) was observed over 3 hours. The instrument was then thermostated and measurements were taken over forty hours. The temperature was initially 27°C, but drifted to 27.5°C over the duration of the test period. The absorbance readings in this time increased continuously and showed an overall 7.8% increase (approximately 0.2% per hour). These measurements, demonstrate the importance of temperature control on a fairly typical laboratory type UV-Vis instrument, showing the very obvious importance of temperature control.

Finally, a study was made of the error generated by the manual refilling of the cuvettes. This factor was investigated using the Hitachi instrument with temperature control. A standard error of 1.25% was obtained on a sample of 32 measurements. The test was repeated using a slider attachment for improved mechanical positioning of cuvettes. A standard error of 0.611% was obtained.

These simple practical tests may explain the improved performance seen for drop spectroscopy over that of the UV-vis cuvette instrument. Obviously, drop delivery is highly reproducible under temperature control (the tensiograph is able to maintain the temperature to within an accuracy of 0.1°C and reproducibility of 0.01°C). Automated sampling gives reproducible drop sizes. The observed drift of the UV-vis spectrophotometers, even when working in a differential dual beam mode, also demonstrated an important operational limitation. For standard UV-vis methods, the measurement procedure takes approximately one hour. Within this time, the drift might be anything from 2.5% without temperature control to 0.2% with control.

7. Summary of Drop Spectroscopy

There perhaps exist some useful practical instrumental advantages of the drop analyser over the standard UV-vis spectrophotometer. The improved analytical reach of drop spectrometers could be important in some research fields and details of the performance of the instrument for the other principal measurands have been tabulated. Some discussion and experimental detail has been provided on the relative sensitivities of the drop and standard UV-vis technique.

Importantly, the new analytical theory of drop spectroscopy has been developed and presented here for the first time in the context of a book. This new theory has been tested against experiment including some detailed enquiries into the average EPL. The best technique for quantitative drop spectrometer measurement has been determined by a statistical theoretical enquiry and the result established by experiment. The performance of the drop spectrometer has been evaluated against a number of traditional types of modern spectrophotometers, with measurements provided by an internationally accredited standards laboratory (Starna). The drop analyser shows some comparable detection limit and sensitivity to the classical technique, for example, in priority pollutants measured in real waters. Experimental evidence of micro-errors in absorbance measurements from traditional cuvette placement has been suggested as one contributory shortcoming in traditional instruments. For whatever reason, it is an important finding that drop spectroscopy generally delivers comparable or improved detection limits to the traditional temperature controlled spectrometers. The drop technique has also been shown to be useful in research pharmaceutical and biotechnological applications. Although not detailed here, the technique has been applied to many other areas of applications.

The fact that the fibre drop head system does not provide a constant EPL is at first sight a disadvantage for all the coupled rays. However, it is well known that there are measurement problems associated with cuvettes. Errors in mechanical construction and collimation mean that a standard pathlength is only really true at a theoretical level in traditional UV-vis spectrophotometry. The EPL for solutions can be determined accurately (providing certain physical characteristics of the liquid are known) for the drop analyser using the experimental or computational methods described in this work. This study has shown that although drop spectroscopy is a departure in this established technique, there are clearly applications where the improved performance and in particular here the advantages of microvolume samples do count heavily. The ease of cleaning a quartz drophead indeed makes automation much simpler than with flow-through cells in traditional instruments. The temperature control of the drop analyser also offers better system reproducibility than can be obtained with cuvettes; it is suggested here from a simple experimental investigation of the standard technology that the repositioning error of the cuvette has shown to be the major performance-limiting factor in this regard for established UV-vis instruments.

D. Refractive Index-Rainbow Peak — Bulk Properties

1. Modelling and Qualitative Description of Tensiotrace

In the first section of this chapter the theory of the basic ray-tracing modelling approach was outlined. Here, modelling results will be used not only to describe the

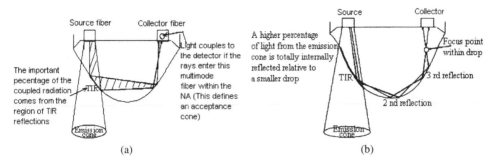

Figure 18. Outline of a water drop from a superimposed CCD camera image, showing 2nd-order reflections. The emission cone is in many situations wider than shown in this diagram and will in circumstances illuminate across the centre of the drop (a); ditto for tensiopeak third order reflection (b).

quantitative results, but also to guide the conceptual description and introduction to tensiography for the bulk properties of the LUT. The broad and comprehensive experimental study of tensiotrace variability presented below requires both a conceptual and theoretical underpinning. Figure 18 shows a pictorial representation of the drop, source and collector fibres and schematically shows some ray paths at the volume corresponding to the first major peak in the tensiotrace known as the rainbow peak.

Figure 2 shows the modelling results derived from using 197 exactly modelled drop shapes with 12 000 rays calculated for a total of 2 364 000 rays. The fibre modelled here is a 1000 micron diameter multimode step-index fibre of numerical aperture 1.47 with core of 1.49 and cladding 1.4433 ($NA = \sqrt{\eta_1^2 - \pi_2^2} = 0.37$ for a fibre working in air with a refractive index of unity. Here η_1 is the refractive index of the core of fibre and η_2 is the refractive index of the cladding of fibre.). The NA is given by $\eta_L \sin\theta_E$ where η_L is the refractive index of the test liquid and into which the exit rays leave the fibre. It can be seen from these relationships that the emission angle θ_E, which is half the angle of the cone of light emanating from the fibre, will narrow with increasing refractive index of the LUT. This emission has an angular intensity variation that can be accurately modelled by a $\cos^2\theta$ intensity dependence. The emission from this end of the fibre is conical emission whose apex is just inside the end of the fibre at a focal position determined by the NA, diameter of fibre and refractive index under test, but is for practical purposes of providing a qualitative description a constant.

It can be readily understood from Fig. 2 that the number of rays coupled between the collector and source fibre has a somewhat complex relationship. Importantly, the ratio of coupling rays and the energy (intensity) the rays deliver to the detector are not simply related and this salient fact requires an explanation. The scale for ratio on the RHS of this diagram, is simply the percentage of rays that couple between source and collector fibres, so we see that in water, the geometry of the fibres is engineered so that the first peak (first-order protopeak) couples 80% of all the rays that emerge from the source fibre. They do not however contribute much to the energy received at the detector. Basically, to a close approximation the optical system here is symmetric and the only light that can propagate back to the detector on the collector fibre must enter this fibre within the same numerical aperture. The angle of emission cone also varies with the refractive index of the LUT and narrows as this refractive index increases. It will be noted that the rays from the emission cone on the LHS of the drop here make various angles of incidence with the drop interface, and the amount of light reflected varies with this angle of incidence and can be calculated directly from the Fresnel reflection coefficients that are determined by the refractive index values and the angle of incidence [45]. These coefficients are simply calculated from the Snell's law angle of incidence θ_L and angle of refraction out of the liquid θ_a and the amount of light reflected is calculated simply from the sum of the parallel and perpendicular polarisations

$$r_{\Leftrightarrow} + r_{\Updownarrow} = \frac{\tan(\theta_L - \theta_A)}{\tan(\theta_L + \theta_A)} + \frac{\sin(\theta_L - \theta_A)}{\sin(\theta_L + \theta_A)}. \tag{38}$$

However above the critical angle of incidence, the light is totally internally reflected (hereafter TIR) and all the light is retained inside the drop. Snell's law gives this angle simply as

$$\theta_c = \sin^{-1}(\eta_A/\eta_L), \tag{39}$$

where η_L is the refractive index of the LUT and η_A is the refractive index of the second optical medium that is usually air. As can be seen from Fig. 18 the coupling between source and collector fibres in the rainbow peak obviously depends on the product of the two reflections at the drop base. This light coupling from one, two, or more reflections required to bring the ray from source to collector fibre shall be referred to the throughput. There is also small Fresnel reflection losses at the end of both fibres to take into account.

Consideration of the TIR light indeed is all that is needed to provide a good qualitative explanation of the characteristic form of the tensiotrace. It must be remembered that an accurate quantitative model is required to back up this conceptual vision of the drop reflection processes and how they explain the form of the tensiotraces. Studies of high-order reflections in pendant drops have been made by Ng, Tse and Lee [46] but these were made by laser light incident on the waist of the drop from outside the drop and are fundamentally different to the successive progression of orders seen by the reflection processes inside the pendant drop in this tensiograph instrument. The first thing perhaps to say about the present modelling reported here is that this is perhaps the first mathematically accurate study of various orders of rainbow type reflections inside a drop from a symmetric fibre vertical geometry. Finite Element Analysis (FEA) studies of rainbows in smaller drops are presently underway in Essex and Tallaght. The second point here, is that the tensiotrace is derived from the complex progression of pendant drop shape by this ray tracing procedure. The tensiotrace can be thought of for a blank (no absorbing liquid) solution simply as being a mapping of the one curve namely the drop shape, to the other, the tensiotrace. The important point theoretically is that these curves contain the same information about the drop shape, but the tensiotrace also is determined by the physics of the optical system inside the drop, namely the type of fibre and the refractive index of the LUT, and for coloured or turbid solutions also the amount of absorption and scattering of the coupled light.

Initially, no light is coupled between the collector and source fibres because of the size of the remnant drop which makes it impossible for light to reflect around the drop base and enter within the acceptance cone of this collection fibre. However, very little of the first-order photons arrive at the detector. Throughput in this first-order coupling is from rays that are weakly reflected (approximately 4% reflection coefficient) off positions close to the bottom of the drop base. The low Fresnel reflection coefficients for first-order reflections arise because the angle of incidence that the first-order rays present to the drop surface are such that most of the energy is refracted out of the drop and thus lost from the coupling process. The rainbow peak commencement is produced as a consequence from the development of second-order reflection, rather than the first-order. This is an optically well defined position and remnant drops have gravitational forces that dominate those of the surface tension force and as a consequence shapes of remnant drops have very little difference from one to the other. The commencement position marks the onset of coupling between the source and collector fibres in the drophead. The broad emission from the 1 mm multimode fibres used means that the rays impinge on a large part of the drop edge below the source fibre. It can be seen from camera studies that the onset of optical coupling occurs at approximately the same volume of the measurement drop. It is very hard in such small drop volumes from camera images to discern any shape differences between liquids of different physical types. Drop shape variations become evident only in larger pendant drops. The commencement position varies we believe in the first instance because coupling occurs in all drops about the same size (volume) of drop, hence the commencement position varies because the volume of remnant drops varies radically from one liquid type to the next. This discussion does not take into account refractive index variations, but perhaps is useful as a first approximation. Because we know from the classic work of Harkins and Brown [4], the remnant drops immediately after separation vary greatly from liquid to liquid, then it is not surprising to find that there is such a radical difference in the volume addition required to increase the remnant drop to a position of first optical coupling, known as the rainbow commencement positions for liquids. The authors have developed a software tool TraceMiner specifically for the accurate determination of the rainbow commencement but it suffices here to say that this is a highly reproducible and useful measurand.

Arising out of the Fresnel net losses from various reflections at the drop surface throughout the complex processes involved in various reflection orders and variabilities of drop shapes, it is found that in water there are serious losses in the rainbow peak (second-order reflection) and less net loss in the tensiopeak and shoulder peak (third and fourth orders). Indeed, it can be seen in Fig. 2 that more coupled ray in the rainbow peak translate into less intensity, than in the corresponding situation for the tensiopeak. As will be seen in the next section this finding is of some real importance. The first-order rays that couple to the detector in a protopeak are not contributing to the tensiotrace at all in any meaningful way, because they are reaching the collector fibre by a reflection of the base of the drop, which only reflects a very small 4% of the light. This is because the rays are not at total internal reflection or are outside of the acceptance angle for this multimode fibre and therefore cannot couple effectively to the detector. The first-order reflections are not contributing to the tensiotrace. The rainbow peak is generated from second-order reflections and it is the maxima in this reflection order that defines the maxima in the intensity graph rainbow peak. The maxima of the tensiopeak for both the straightforward experimentally measured intensity plot and the Excel generated ratio plot have matching positions. The intensity is a product of the number of photons, which is computed from knowledge of the Fresnel intensity and the number of rays. This fact will be seen as being of central importance in the discussion of the physics of the rainbow peak, which lies at the heart of the measurement functionality of this instrument.

2. Critical Angle Dependency

During the rainbow peak stage, the light beam undergoes two internal reflections off the sides of the drop before hitting the drophead base. In effect, any light ray emitted by the source fibre that enters the collector fibre will have changed direction by approximately 180° (rays must have approximate normal incidence if they are to be transmitted by the optic fibre). Equation (40) results from the 180° rotation been executed by two reflections and restricts the sum of the angles of incidence

$$2\theta_b + 2\theta_f \approx 180°, \qquad \theta_b + \theta_f \approx 90°. \tag{40}$$

The critical angle for liquids generally ranges from 49° to 41°. It is impossible for high critical angle liquids such as water (critical angle of about 48.8°) to produce two TIRs subject to the geometric constraint of Eq. (40). It can at best have one TIR with the second reflection located on the steep transition region shown in Fig. 19(a). Fresnel reflection coefficients for water is compared with that for methanol (critical angle of 47.3°) in Fig. 19(a). Both have a similar shape except for the horizontal offset due to their different critical angles.

Typically for an angle of incidence just 0.4° below the critical angle, just 50% of their light energy is reflected while only about 10% of light is reflected at five degrees below the critical angle. This sharp transition at the critical angle is typical for all test liquids. The intensity of a reflected ray will be lower in a water drop than for a drop with higher refractive index such as a methanol drop except of course in the TIR region. This discrepancy is largest for incidence angles in the transition region, which coincidently correspond to the incidence angles found during rainbow peak production. Figure 19(b) plots the transmittance for a double reflection obtained by multiplying the Fresnel coefficients of each reflection, subject to their incidence angles being constrained according to Eq. (40). The methanol plot has a higher peak due to its lower critical angle. It is noteworthy that the location of

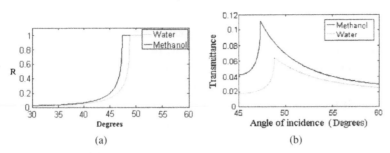

(a)   (b)

Figure 19. Fresnel reflection coefficients for water and methanol as a function of angle of incidence (a); transmittance of ray undergoing two reflections as a function of the angle of incidence of the first reflection (b); the sum of incidence angles is constrained to be 90°.

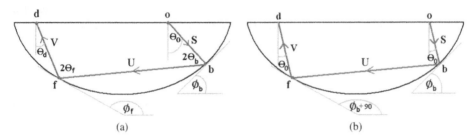

(a)   (b)

Figure 20. 2D model for rainbow peak (a); optimum conditions for rainbow peak (b).

both transmittance peaks occur when the angle of incidence corresponds to the critical angle of the liquid. This point represents the situation where the first reflection is just at the TIR threshold. This allows the second reflection to have a maximum angle of incidence and consequently a maximum reflection coefficient. Ultimately the very high sensitivity of the rainbow peak to changes in the drop refractive index derives from the sharp transition in Fresnel coefficients just below the critical angle.

3. 2D Geometric Model for Rainbow Peak

Figure 20(a) represents an arbitrary symmetrical drop shape where ϕ_b and ϕ_f are the angles of the tangents to the drop surface at locations **b** and **f**. The emission angle that ray **S** forms with the source fibre is θ_0 while the incident angle that ray **V** forms with the collector fibre is θ_d. The angles of incidence for the first reflection (R1) and second reflection (R2) are θ_b and θ_f respectively.

Both fibres are assumed to be normal to the horizontal axis. The relationships between the angles of incidence are not easy to geometrically demonstrate, but may be summarized through the following three equations:

$$\theta_b = \phi_b - \theta_0, \tag{41}$$

$$\theta_f = \phi_f - 2\phi_b + \theta_0, \tag{42}$$

$$\theta_d = 2(\phi_f - \phi_b - 90°) + \theta_0. \tag{43}$$

4. Drop Shape Requirements

The formation of a rainbow peak necessitates that reflections within the drop are TIR and that the angle of incidence of light hitting the collector fibre be within its acceptance angle. In order for the R1 reflection to be TIR, $\theta_b > \theta_C$ where θ_C is the critical angle of the liquid. Therefore from Eq. (41),

$$\phi_b > \theta_C + \theta_0. \tag{44a}$$

Since θ_C is typically greater than 41° for most test liquids, the slope of the RHS of the drop must be larger still in order to develop a TIR. For this reason small drops can never produce a rainbow peak as their sides are not steep enough. As the drop grows in size, its sides become increasingly steeper and eventually condition Eq. (44a) is met. In order for the R2 reflection to be TIR, then $\theta_f > \theta_C$. Therefore from Eq. (42) and taking into account the condition from Eq. (44a) then,

$$\phi_f > \theta_C + 2\phi_b - \theta_0, \tag{44b}$$

$$\phi_f > 3\theta_C + \theta_0. \tag{44c}$$

In practice this would require θ_f to be at least greater than 142° for methanol and 146° for water in order to be TIR. This means in practice that point **f** must reside within the lower left quadrant in Fig. 20(a).

Any ray entering the collector fibre must have angle of incidence θ_d less than the fibres acceptance angle θ_A, if it is to be transmitted. Therefore from Eq. (43),

$$(\phi_f - \phi_b - 90°) < \frac{1}{2}(\theta_A - \theta_0). \tag{45}$$

In practice the RHS of Eq. (48) will equate to a small value since θ_A is approximately 10° for glass fibres and θ_0 is always $\leqslant \theta_A$. Therefore

$$\phi_f \approx \phi_b + 90° \tag{46}$$

i.e., the slope of the drop at the point **f** must therefore be approximately at right angles to the drop slope at point **b**. Under these circumstances, Eq. (43) simplifies to $\theta_d = \theta_0$ so the incidence angle of the ray entering the collector fibre (assuming it is located in the correct position) equals the emission angle from the source fibre. Consequently all emitted rays incident on locations **b** and **f** are guaranteed to be accepted by the collector fibre. However Eq. (43) also indicates that θ_d is very sensitive to small percentage changes in the slope of the drop at locations **b** and **f**. Consequently, regions surrounding the optimal point where rainbow peak production is feasible will be small. Furthermore the range of drop sizes that meet the geometric constraint for rainbow peak production will also be limited with the result that the rainbow peak always forms within a restricted region towards the start of the tensiotrace. Figure 20(b) displays the ideal drop geometry necessary for producing a rainbow peak.

In summary, R1 reflections for liquids with higher critical angle will be constrained to increasingly higher regions on the upper RHS of the drop while their R2 reflections confined to decreasingly lower regions of the LHS. Such liquids would therefore be less likely to also meet condition Eq. (45) which may explain why some low refractive index liquids do not produce a rainbow peak.

5. 3D Interpretation of Rainbow Peak

The 2D geometric model for rainbow peak generation cannot provide a complete analysis for what is essentially a 3D process. Such an analysis requires the detailed computer modelling that has been described above and this accounts fully for the 3D drop geometry and beam trajectory. A 3D perspective for rainbow peak production process is nonetheless provided by Figs 21–23. These figures display the spatial distribution of the angles of incidence for R1 and R2 beam reflections as a colour map for three drop sizes corresponding to the transition from pre-rainbow peak to rainbow peak and post-rainbow peak. In each diagram the location of the tips of the source and collector fibres are indicated by the magenta and black rings respectively. The upper graphic of each figure shows the R1 distribution on the far side of the drop. It also shows the subsequent projection of the R1 reflection on to the opposite side of the drop followed by its projection onto the drophead base. These projections are useful as they show the path taken by the light beam propagated within the drop. The lower graphic shows the R2 incidence angles distribution on the collector fibre side of the drop. Collectively they provide a 3D visualisation

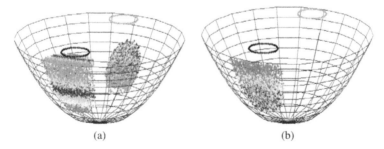

(a)   (b)

Figure 21. Colour map of R1 angles of incidences with projections onto the collector fibre side and drophead base (Pre RBP) (a); colour map of R2 angles of incidences (Pre RBP) (b).

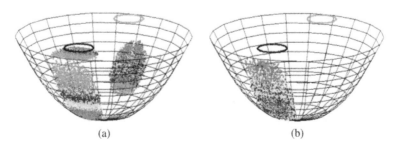

(a)   (b)

Figure 22. Colour map of R1 angles of incidences with projections onto the collector fibre side and drophead base (RBP) (a); colour map of R2 angles of incidences (RBP) (b).

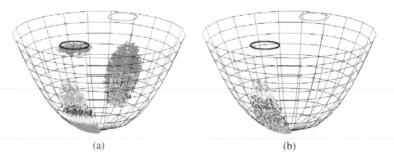

(a)   (b)

Figure 23. Colour map of R1 angles of incidences with projections onto the collector fibre side and drophead base (Post RBP) (a); colour map of R2 angles of incidences (Post RBP) (b).

Table 4.
Colour maps used in the graphics

Angle of incidence	Colour
$\theta_i > 49°$	Red
$49° \geqslant \theta_i > 47°$	Yellow
$47° \geqslant \theta_i > 45°$	Green
$45° \geqslant \theta_i > 43°$	Blue
$43° \geqslant \theta_i > 41°$	Cyan
$41° \geqslant \theta_i > 37°$	Magenta

of the process of rainbow peak production. Table 4 displays the colour map used in the graphics.

Rays with angle of incidence lower than 37° are not displayed since they have relatively low energy and do not substantially affect the amplitude of the rainbow peak. Any ray with angles of incidence above 49° will be a TIR for virtually all test liquids. Figure 21(a) shows that moving upwards along the R1 reflection footprint tends to produce larger angles of incidence, consistent with the increasing slope of the drop. It is apparent that the R1 footprint is composed of a blend of angles, particularly at its upper region due to its close proximity to the source fibre. The colour bands are much more distinctive when projected onto the opposite side of the drop, as the increased path length allows the different ray angles to become spatially separated — a process similar to the formation of natural rainbows. This projection also shows that there is an "image inversion" associated with the propagation of the light between footprints, with the top of the R1 footprint mapping to the bottom of the R2 and *visa versa*. For instance light on the top of the R1 footprint is reflected towards the bottom of the R2 footprint and then subsequently to the far edge of the footprint on the drophead base. Figure 21(b) shows the corresponding spatial distribution of incidence angles for the R2 footprint. Like the R1 distribution this colour map is also composed of a blend since light incident at any point on the R2 footprint arrives from a range of angles.

The top region of the R2 foot print has largest incident angles due to higher slopes at the top of the drop and because incident rays are moving in an upwards direction (from the bottom of the R1 footprint). Moving downwards along the R2 footprint sees the slope of the drop reducing while at the same time the direction of the incident rays become increasingly downwards — the net effect being that the incident angles tend to decrease. Light from the R2 footprint is propagated upwards towards the drophead base with only the front region hitting the collector fibre. Another "image inversion" is apparent with light from the bottom of the R2 footprint propagating to the front of the drophead footprint while light emanating from the upper region of R2 footprint transferring to the back of the drophead footprint. The intensity of the drophead footprint tends to be greatest at the front region because this region is typically associated with both R1 and R2 TIRs. Moving further backwards along the drophead footprint sees the light intensity fall off rapidly as the internal reflection have increasingly smaller angle of incidence and are thus no longer TIR.

High critical angle liquids such as water only produce a TIR for the red regions of the R1/R2 colour maps. Therefore the light energy incident on the drophead is largely supplied by the narrow red band at the front of the drophead footprint Fig. 21(a). The intensity of this band is determined by the blend of incidence angles on the bottom region of the R2 footprint in Fig. 21. This is mostly green with significant blue and yellow components. This suggests that the average angle of incidence of about $46°$ for this region resulting in a reflection coefficient of approximately 0.18. Liquids such as alcohols produce TIRs extending into the yellow bands. Consequently the high energy band at the front of the drophead footprint is widened to include both red and yellow strips. Due to the lower critical angle of alcohols, this high-energy band will also be of higher intensity than that obtained for water. The net effect of this larger and more intense R3 TIR band is that alcohols produce a much bigger rainbow peak than water. Low critical angle liquids such as oils have TIRs extending further to the green and blue bands. This will both broaden and strengthen the high intensity band at the front of the drophead. However, most of this high intensity band would not hit the collector fibre in Fig. 21 so its rainbow peak would tend to be delayed compared to that of water or alcohols. In Fig. 22 the drop has increased in size and its extra weight causes its sides to become steeper. This tends to increase the angles of incidence within the R1 footprint shown in Fig. 22(a), increasing the red region on the top and reducing the magenta region on the bottom. Another consequence of the steeper slope is that the R2 footprint of Fig. 22(b) moves downwards towards the base of the drop. Bearing in mind that rays incident on the top region of the R2 footprint are travelling in an upwards direction, this steeper slope tends to reduce angles of incidence as manifested by sparser red and yellow colours found towards the top of the R2 footprint compared to that displayed in Fig. 21(b). Towards the bottom of this R2 footprint, the incident rays are moving in a downwards direction and because it is located nearer to the drop base the angles of incidence tend to be increased compared to the equivalent region of Fig. 21(b).

The overall effect is to produce a reduced colour gradient along the R2 colour map. The drophead footprint in Fig. 22(a) has moved forward due to increased downward directions of the rays incident on the lower region of the R2 footprint, although, this effect is counterbalanced by the lower drop slopes found towards the drop base. For most liquids, the tensiotrace level resulting from the drop shown in Fig. 22 would be greater than that obtained from the smaller sized drop shown in Fig. 21 since a greater portion of the high intensity R3 footprint is incident on the collector fibre.

For high critical angle liquids such as water, any further increases in drop size would generate a fall off in tensiotrace level since the red band would be advancing beyond the centre of the collector fibre and effectively be shortened. Alcohols with a slightly lower critical angle would however still produce larger trace levels until the yellow band advances beyond the centre of the collector fibre. Tensiotrace levels would still increase for test liquids with lower critical angle, until the green and then the blue bands, etc, moves beyond the fibre centre. Consequently changes in the refractive index of drops not only modulate the amplitude of the rainbow peak but also shift its position in the tensiotrace.

Figure 23 represents the drop size where the conditions necessary to produce a rainbow peak start to unravel. Continuing the process described earlier, as the drop gets bigger its sides become steeper causing the R2 footprint to advance into the base of the drop. The base region inherently has larger curvature so any portion of the light beam falling on it will encounter greater variation of surface slopes and thus will be reflected over a wider range of angles.

This is illustrated in Fig. 23(a) where the red band (which extends furthest towards the drop base on the R2 projection) is no longer focused onto a narrow strip on the drophead. The backwards drift of the drophead footprint results from R2 footprint extending further into the base region where the slope is oriented to produce an increasingly backwards reflection. Further increases in drop size will create an additional internal reflection that will eventually go on to produce the tensiograph peak.

6. Stochastic Analysis

The 3D visualization model is useful for providing a qualitative description of the method of tensiotrace production, but is cumbersome for detailed analysis due to complex drop geometry. A stochastic model based on the probability density distributions of angles of incidences (PDFAI), encapsulates much of the detail described in the 3D visualization model but enables easier analysis. A key benefit of the stochastic analysis is that the radiation intensity of footprints may be evaluated simply as the area under its PDFAI curve. Figure 24(a), (b) and (c) respectively display the PDFAI for the three drops outlined in the previous section. In each diagram, the

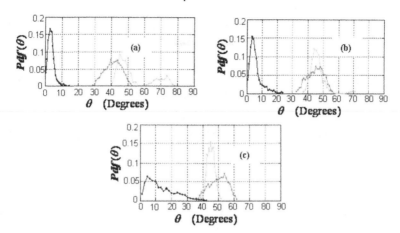

Figure 24. PDFAI (Pre RBP) (a); PDFAI at RBP (b); PDFAI (Post RBP) (c).

PDFAI of the R1 footprint is displayed in red, the R2 PDFAI in green while the R3 PDFAI is shown in blue.

The R1 PDFAI curves are seen to progressively shift to the right in Fig. 24(a)–(c), consistent with the increase in the angles of incidence arising from steeper drop sides. The intensity of the R1 TIR component corresponds to the area under the R1 PDFAI curve for $\theta > \theta_C$. Consequently, aqueous based test solutions produce a relatively low energy TIR component as the R1 PDFAI tails off rapidly for angles above the high forties. The intensity of the R1 reflection in Fig. 24(b) is substantially larger since the R1 PDFAI has shifted to the right. It will be bigger again for the largest drop corresponding to Fig. 24(c) since the R1 PDFAI has shifted further to the right. Test liquids with lower critical angle will always produce a more intense R1 reflection than that obtained from water based solutions since the integration region is correspondingly increased.

The R2 PDFAI in Fig. 24(a) has two peaks. The higher peak is not relevant to rainbow peak analysis and arises from the top part of the R2 footprint that is incident on the base of the drophead. The collector fibre does not transmit this component as the angle of incidence is outside the fibre acceptance angle. Note that this portion of the R2 footprint has not been displayed in Fig. 21 to avoid obscuring the R3 footprint on the drophead base. The upper peak in the R2 PDFAI in Fig. 24(b) is nearly extinguished because as shown in Fig. 22 the R2 footprint has moved downwards and only a very small portion extends onto the drophead. The R2 PDFAI is unimodal in Fig. 24(c) because the R2 footprint is now incident wholly on the side of the drop. The strength of the R2 reflection that contributes to the rainbow peak is defined by the lower regions of the R2 PDFAI only. This is because the rainbow peak is derived from R1 TIR rays multiplied by their corresponding R2 Fresnel reflection coefficient. Figures 21–23 show that the R1 and R2 spatial distributions are mirror images, so the TIR region on the upper parts of the R1 PDFAI are coupled to the lower parts of the R2 PDFAI. Consequently the intensity of the rainbow peak is primarily dependent upon how close the lower part of the R2 PDFAI is to the critical angle of the test liquid.

The R3 PDFAI is evaluated for rays emanating from the R2 footprint that are incident on the drophead. For glass fibres, only rays with incidence angles less than 10° may be transmitted, although this threshold varies slightly with the refractive index of the test liquid. In Fig. 24(a) and (b), most of the R3 PDFAI rays are lower than 10° and would thus be transmitted by the collector fibre if incident upon it. The low angle of incidence derives from the upward direction of rays emanating from the bottom of the R2 footprint as may be deduced from Figs 21 and 22. It is also consistent with Eq. (43), which predicts a low incidence angle on the drophead when the drop shape meets the geometric conditions for rainbow peak production. In Fig. 24(c) the R3 PDFAI is more widely spread out, with less than half its rays having incidence angles below θ_C, so the intensity of light transmitted in the collector fibre is greatly reduced. This increased variance of the R3 PDFAI derives from the lower part of the R2 footprint descending deeper into the drop base producing a more widely spread reflection towards the drophead. This rapid change in the shape of the R3 PDFAI at the rainbow peak is consistent with Eq. 44 where small changes in the slope of the drop where the R1 and R2 reflections take place induce large changes in θ_d. The signature PDFAI features for rainbow peak production consists of R1 and R2 distributions approximately centred on 45° while the R3 PDFAI is concentrated at low incidence angles.

7. Refractive Index Studies

There are a number of very different features to tensiotraces of alcohols and alkanes compared to aqueous solution tensiotraces as can be seen from Fig. 25. The rainbow peaks are extremely narrow; the time taken for the rainbow peak to commence increases over that seen in aqueous solution; and tensiopeak height is very small. These liquids have relatively high refractive indices and as a consequence it is theoretically known that more light is reflected at the surface of the drop, however,

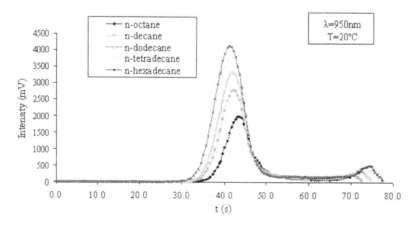

Figure 25. Tensiotraces of a series of alkanes.

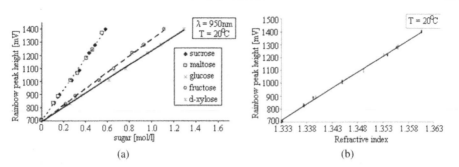

Figure 26. Linear rainbow peak height signal relationship for molar concentrations of various sugars (a); variation of rainbow peak height for all sugars investigated plotted against refractive index (b).

the coupling geometry from source to collector fibre does not lead to a correspondingly greater signal. It is worth noting that the tensiopeak heights are indeed much smaller than the rainbow peak heights. Figure 25 shows an observed increase in rainbow peak height and drop-period of alkanes in the following order: n-octane, n-decane, n-dodecane, n-tetradecane and n-hexadecane. This is to be expected from the tabulated refractive index and surface tension values from literature [47].

The rainbow peak increases in the hexane and alcohol series, similarly arise from an increase in the refractive index of the liquid. For example, an increase in peak heights of primary alcohols in the following order: ethanol, n-propanol, 1-hexanol, 1-heptanol, 1-octanol and 1-decanol [48]. The rainbow peak height is clearly a useful measurand for bulk refractive index determination of the LUT and Fig. 26(a) reveals another set of linear relationships for various sugar solutions over a limited range of concentrations. Figure 26(b) shows a very strong linear relationship of rainbow peak height with refractive index for all these sugars when plotted. It is found that this simple graphical relationship is a universal relationship for sugars. Furthermore, the linearity of results here are excellent with 3σ-error bars shown in (b) but at ±2.8 being too small to show in (a). The strength of the linearity in this rainbow peak height dependence is worth just emphasising once again before leaving this issue.

From the first work of McMillan *et al.* [40] the refractive index dependence of this peak height was discovered with an initial linear variability in this dependence with refractive indices above that of water. At high refractive index values, there is a higher order increase in peak height. The initial work suggested that perhaps there existed a universal trend for rainbow peak-height with refractive index; although not linear.

However, later work has shown in fact that this proposed universal variability is in fact sundered into discrete chemical family trend-relationships, as can be seen in Fig. 27(a) for families of alcohols the primary, secondary and tertiary. Figure 27(b) shows a more detailed experimental trend.

In Fig. 28(a) the variation for butanol mixtures is shown where the 3σ error bars are really too small to see except for two measurements on this plot. The simple

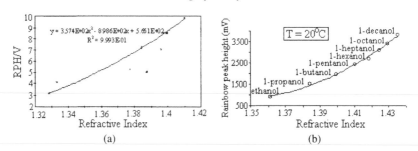

Figure 27. Plot of rainbow peak height for alcohols and water, with trend-line for primary alcohols shown, the secondary alcohols sit beneath this line and the tertiary alcohols below this. Water is included as the reference here and has a refractive index of 1.3333 (a); more detailed analysis of the primary alcohol rainbow peak height relationship with refractive index of the liquids (b).

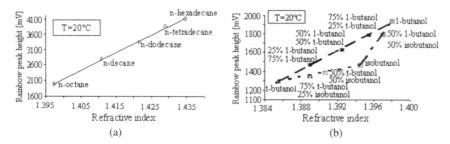

Figure 28. Variation in rainbow peak height with refractive index of a series of alkanes (a); linear graphical relationships for mixtures of butanols when plotted on a rainbow peak height-refractive index graph (b).

conclusion here is that this peak is an exceptionally good measurand and these linear relationships can prove exceedingly powerful in facilitating the analysis of complex mixtures such as beverages that are discussed in the last section of this chapter dealing with tensiograph fingerprinting.

Measurement of surface tensions with the tensiograph has shown that while good agreement is obtained for non-volatile alkanes, for volatile alkanes the agreement is not as good [49]. A first-order increase in refractive index and rainbow peak height of alkanes in the following order: n-octane, n-decane, n-dodecane, n-tetradecane and n-hexadecane is found that is a strong linear trend as shown in Fig. 28(a). Strong linear relationships are surprisingly common in tensiographic analysis which is a very welcome bonus for those who use the technique for analytical applications. Figure 28(b) shows another such relationship for butanol admixtures when plotted against refractive index.

Some improvement in the representation of the refractive index data here can perhaps be obtained from replotting the data. It has been found useful to define the corrected rainbow peak height as the rainbow peak height divided by drop-period. By this artifice, it is discovered an improved trend-line relationship, as evidenced by Fig. 29(a) when compared with Fig. 27.

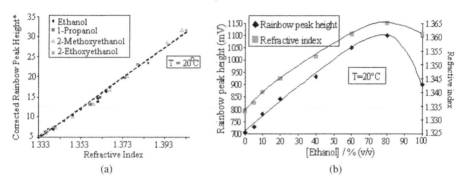

Figure 29. Variation in corrected rainbow peak height (average range values of ±0.16%) with refractive index of ethanol/water, 1-propanol/water, 2-methoxyethanol/water and 2-ethoxyethanol/water (a); variation of both rainbow peak height (H_2) and refractive index plotted against concentration of ethanol in the aqueous ethanol solution (b).

This result is presented here in part to underline the obvious fact that there is a considerable amount of data analysis still needed to be done on tensiotrace measurand relationships and the associated theoretical work to explain what at this point are still many mysterious linear tensiograph relationships. Space is too limited here to investigate the detail of these issues further, but it is perhaps important to point out that the modelling work of Pringuet [50] is directed towards elucidating many of these issues in the next period.

In Fig. 29(b) the very powerful relationship between the rainbow peak height and the refractive index of the LUT is shown in a double ordinate plot of both variables. The value of this graph is that the variability of refractive index of these solutions is anomalous and falls in concentrations above 80% ethanol. This effect makes it an 'acid test' for the study of rainbow peak height (represented by H_2). This relationship was discovered initially by McMillan [40] and led him to name this peak the rainbow peak, since it was shown that from the rainbow peak measurements the refractive index of the LUT could be determined. The second reason for the name is that the peak here is produced by a double reflection inside the pendant drop and is the pendant drop reflection analogue for the double refraction in the water spherical micro-droplet that produces the rainbow.

E. Surface Tension — Drop Period and Tensiopeak Period Relationship

1. Drop Period and Tensiopeak Period Relationships

It has been found in a general and wide-ranging study that there is a linear relationship between the drop period (T_1 measured in seconds) measurand and tensiograph peak period (t_3 measured in seconds). There exists a simple empirical relationship between the two measurands:

$$T_1 = \zeta t_3 \text{ and } \zeta' t_3, \tag{47}$$

where ζ and ζ' are two dimensionless constants known as the 'fundamental tensio-metric constants'. From experiment it was found that two linear relationships exist for what are bifurcated liquid family types. These relationships appear to apply to all liquid types including surprisingly perhaps surfactants. Below, details of these experimental relationships will be given.

This discovery clearly has very major implications and will be the subject of a paper that is devoted exclusively to this and other new relationships that exist between the various tensiotrace measurands [36]. These features can be used to obtain measurements of various physical and chemical properties of liquids. Details of how these measurements are obtained have been recently reported elsewhere [51], based on updating the earlier work of McMillan, Davern *et al.* [52]. The rainbow peak height gives very sensitive measurement possibilities for refractive index of the LUT, the tensiopeak height similarly for colour. Both peaks show sensitivity to both these properties of the LUT, but exhibit very strong specificity, if not completely overwhelming dependence, on the one property (refractive index for rainbow and colour/turbidity for tensiopeak). The design of a drophead for acceptable measurement performance for all liquids, has deliberately sought to maximise these specificities. Independent measurement of both physical quantities are relatively straightforward in all circumstances given calibration data is available. The structure of the separation vibration has been studied with a view to obtaining measurements of surface tension from the frequency of remnant drop vibration and viscosity of the LUT from the damping in this vibration [53].

In order to obtain direct measurement of surface tension from the tensiopeak period, this tensiotrace feature must be corrected for density and drop-generation rate. In an analogous fashion to the treatment of drop-period, the following expression was used and plotted against surface tension:

$$\text{Tensiopeak mass} = \tau_3 = t_3 q \rho, \tag{48}$$

where q is the drop-generation rate (μl s^{-1}) and ρ is the liquid density.

This correction has the effect of converting a time (tensiopeak period) to the mass of the drop at the tensiopeak period.

Experimental work has shown there appears to be a constant relationship between tensiopeak period and drop-volume. As a typical example, the dependence of tensiopeak period on drop-volume of sub-micellar and micellar sodium dodecyl sulphate (SDS), mixtures of 2-methoxyethanol and water and a range of concentrations of CaCl$_2$ is presented in Fig. 30(a). A typical dependence of corrected tensiopeak mass on surface tension of a range of concentrations of sodium dodecyl sulphate may be seen in Fig. 30(b).

The relationship is linear with error bars too small to show for both the micellar and sub-micellar concentrations of SDS. The slope of the graph was found to be 1.19 ± 0.02. A linear plot with the same slope, was also obtained for Triton X-100 and benzalkonium chloride. A typical dependence of corrected tensiopeak

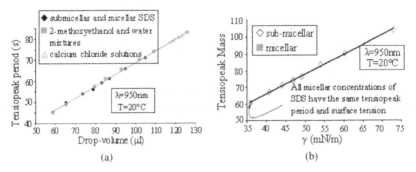

Figure 30. Tensiopeak period relationship with drop volume from various types of typical solutions (a); tensiopeak mass (τ_3) plotted against surface tension for a series of SDS solutions (both sub-micellar and micellar) (b).

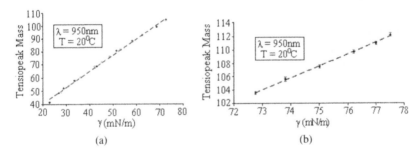

Figure 31. Variation in tensiopeak mass (τ_3) with tensiographic surface tension of mixtures of SDS (i) ethanol and water and (ii) 2-methoxyethanol and water (a); variation in tensiopeak mass with tensiographic surface tension of a range of concentrations of calcium chloride (b).

mass (M_3) on surface tension for a range of concentrations (0–100% (v/v)) of (a) ethanol/water and (b) 2-methoxyethanol/water is presented in Fig. 31(a).

There is an observed linear variation in corrected tensiopeak mass with surface tension of mixtures of ethanol and water, with a slope of 1.20 ± 0.020. The same results have been found for mixtures of (i) ethanol and water and (ii) 2-methoxyethanol and water. There is a negligible difference in the slope of tensiopeak mass *versus* surface tension for both series of alcohols in water as confirmed from the error in the slope. Significantly, this is the same slope that is observed for surfactant solutions, which means that the tensiopeak mass is an effective tensiotrace feature that can be used to determine surface tension of these liquids and solutions. A typical dependence of the tensiopeak mass on surface tension of calcium chloride is presented in Fig. 31(b). All salts analysed sit on the same first-order dependence.

There is again an observed linear empirical variation in tensiopeak mass with surface tension of a range of concentrations of $CaCl_2$ with a slope of 1.75 ± 0.020. Importantly, this dependence has a significantly higher slope than obtained for water, surfactants and mixtures of alcohol and water.

The modelling of this result is perhaps best done in a slightly different way to the above. We can see that there is a relationship between drop period and tensiopeak period and mass and this result is confirmed for a number of liquids. The anomalous results for salts are not included.

This figure shows the relationship between T_1 and t_3 for a range of liquids. It demonstrates through modelling that the experimental result that t_3 and T_1 are linearly related. T_1 and t_3 here are normalised using water as the volumetric standard. The slope is a constant at approximately 0.8. This very strong relationship shows that it is possible to accurately determine the drop period from the tensiopeak period. The surface tension can thus be determined from the tensiopeak period *via* standard tensiometric methods and this means that there is no need for the drop to detach from the head in order to make this measurement of surface tension.

There is indeed no need in tensiography to use Harkins–Brown correction factors, as accurate graphical relationships exist between various tensiograph measurands and surface tension to give a direct measurement of this physical quantity of the LUT. Dynamic studies of processes such as enzyme reactions can thus be carried out on a single drop quantum using pump control. The procedure is to drive the drop volume past the tensiopeak volume in order to determine a surface tension, then to retract the drop below this volume and importantly before the drop separates, and then to continually repeat this measurement of surface tension. This should be a very useful and practical discovery for surface science.

2. Determination of Surface Tension by Tensiography

There are certain expected advantages to the use of tensiography in the determination of surface tension. The system is automated so that a large number of solutions may be rapidly analysed. With the solution pump being isolated from the drophead, there is less danger from vibrations inherent in manual drop-volume manipulations. In the determination of surfactant critical micelle concentration (hereafter the CMC), there is no requirement of a third absorbing species. A possible disadvantage is that slow adsorption processes, such as migration to the surface in surfactant solutions of very low concentration may be 'missed' by the technique.

A number of common surfactant species, one anionic, one cationic and one nonionic, were analysed by tensiography using an LED light-source at $\lambda = 950$ nm and the results presented below. None of the analytes, sodium dodecyl sulphate, benzalkonium chloride or Triton X-100 absorbs at $\lambda = 950$ nm, thus optical absorption effects — which might otherwise diminish peak height — may be ignored. Another important factor in the analysis of these solutions is that refractive index remains constant. Tensiotraces of various concentrations of sub-micellar SDS are shown in Fig. 32.

A water tensiotrace is presented alongside as standard. As expected, tensiotrace drop-period is found to reduce significantly with increased concentration of SDS. Identical behaviour is found for cationic Triton X-100 and benzalkonium chloride. There are a number of unusual features to the tensiotraces, including (i) the dramatic

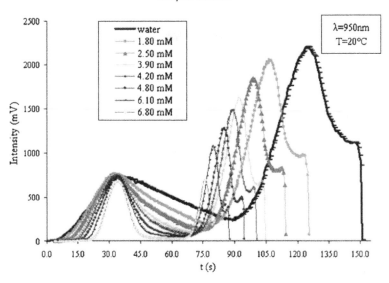

Figure 32. Tensiotraces of sub-micellar sodium dodecyl sulphate.

reduction in tensiopeak height and (ii) the relatively slight reduction in the height of the rainbow peak, which narrows with increasing surfactant concentration. Clearly, the profile of the drop changes little in the early stages of its development but varies significantly just before detachment. A comparison was made of the surface tensions using Harkins–Brown correction factors and the results obtained using the du Noüy ring tensiometer. The surface tension measurements on SDS solutions of 1.80, 2.50, 3.90, 4.20, 4.80, 6.10 and 6.80 mmol/l agreed closely to within the ±0.30% experimental error of both methods. The similarly, acceptable agreement is found in both Triton X-100 and benzalkonium chloride (BAC) sub-micellar solutions.

2.1. Surfactant Behaviour in Micellar Solution

The analysis of micellar surfactant solution by drop-volume gives of course a constant surface tension because of the saturation of the surface by the surfactant molecules above the CMC. An example of tensiotraces for micellar BAC solutions is shown in Fig. 33. It is hopefully useful to study the effect of passing light through the growing drop and following its reflection through tensiotrace development in studies of such molecular solutions.

Micellar SDS, Triton X-100 and BAC have, as expected, constant drop-period. In all these micellar solution, the tensiopeak height is now found to be constant. The measurements using tensiometry agreed with the du Noüy method. There is, however, an increase in rainbow peak height with increasing micellar concentration. This is almost certainly due to an increase in refractive index in these solutions with increasing molecular bulk concentrations, although the saturated surface concentration of course remains unchanged causing a narrowing in the cone of incident light from the LED, with greater transmission to the receiver fibre.

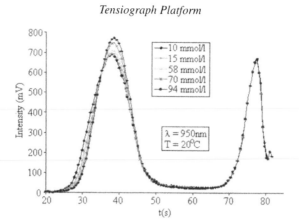

Figure 33. Micellar benzalkonium chloride.

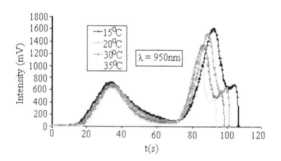

Figure 34. Tensiotraces of 0.015 mmol/l Triton X-100 at different temperatures.

To investigate the capacity of the tensiograph to function as a tensiometer in more complex systems, surfactant solutions at different temperatures and surfactants in the presence of electrolyte were investigated. Figure 34 shows tensiotraces of 0.015 mmol/l Triton X-100 at different temperatures.

As temperature increases and surface tensions decrease, an expected decrease in drop-period is found. There is also a decrease in tensiopeak height with increasing temperature; rainbow peak height also drops, but less dramatically. When correlated with surface tensions obtained by the du Noüy tensiometer, agreement (here with ±0.22% error) is good with those derived from drop-period.

Tensiotraces of sub-micellar SDS in the presence of co-ions are shown in Fig. 35. A reduction in drop period of 3.0 mmol/l SDS with increasing NaCl was noted. There is a dramatic reduction in tensiopeak height which corresponds to a reduction in the number of light rays that are reflected towards the collector fibre; a less pronounced reduction in rainbow peak height is also found with increasing concentration of NaCl. Surface tension results of SDS in 0.01 mmol/l NaCl are consistent with those obtained by du Noüy tensiometry (here within ±0.30% in each method). A similar trend was found where electrolyte was added to sub-micellar BAC.

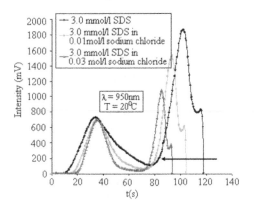

Figure 35. Tensiotraces of SDS in the presence of NaCl.

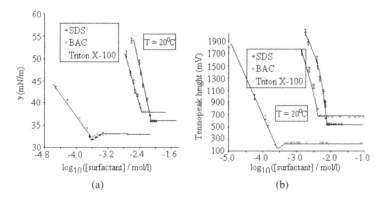

Figure 36. Variation in tensigraphic surface tension with concentration of sodium dodecyl sulphate, benzalkonium chloride and Triton X-100 (a); variation in tensiopeak height with concentration of sodium dodecyl sulphate, benzalkonium chloride and Triton X-100 (b).

2.2. Determination of Critical Micelle Concentration by Tensiographic Surface Tension

Figure 36(a) shows the tensiographic surface tension graphs for the three pure surfactants; sodium dodecyl sulphate (SDS), benzalkonium chloride (BAC) and Triton X-100. The critical micelle concentration (CMC) was determined from these tensiographic graphs and coincided with the value determined from surface tension isotherms by the du Noüy tensiometer.

It is clear from the discussion above that drop-period, optically determined by tensiography, is a satisfactory determinant of surface tension when used with Harkins–Brown correction factors giving CMC values of classic surfactant systems. Other tensiotrace features, principally the tensiopeak and rainbow peak heights, for surface studies will now be analysed beginning with tensiopeak height. The tensiopeak height analysis of both the micellar and sub-micellar tensiotraces of SDS, BAC and Triton X-100 gives the isotherms presented below in Fig. 36(b). The cms's determined from these isotherms agreed with those obtained by the du Noüy tensiometer.

It is important here to note that the plot of tensiopeak height against the log of surfactant concentration exactly mirrors the behaviour in the classical surface tension isotherm (the presence of a minimum in the isotherm of Triton X-100 is a reflection of impurity and is also found in the surface tension isotherm determined by the du Noüy tensiometer). The reflective behaviour of light in the surfactant solution drop therefore shows the same dramatic change in character at the CMC as conductivity, light absorption, light scattering, surface tension and other classical techniques found by previous authors as far back as McBain *et al.* [54] in 1932.

Tensiotraces were analysed of SDS in the presence of an electrolyte and the tensiopeak height graphs in Fig. 37 were used to determine the CMC's. The CMC's are consistent with those by the du Noüy tensiometer. Similarly, an identical result is obtained for benzalkonium chloride in the presence of $CaCl_2$.

Where temperature is varied the technique again allows CMC's to be easily determined. Tensiotraces of SDS and Triton X-100 at different temperatures were analysed using the tensiopeak height again to determine the CMC (Fig. 38(a) and (b)). The CMC's of both SDS and Triton X-100 at temperatures: 15°C, 20°C, 30°C

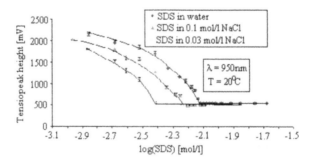

Figure 37. Variation in tensiopeak height with concentration of (i) pure sodium dodecyl sulphate (ii) SDS in 0.01 mol/l NaCl and (iii) SDS in 0.03 mol/l NaCl.

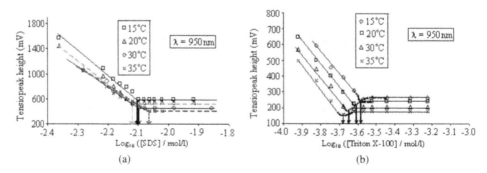

Figure 38. Variation in tensiopeak height with concentration of sodium dodecyl sulphate at different temperatures (a); variation in tensiopeak height with concentration of Triton X-100 at different temperatures (b).

and 35°C, are consistent with those obtained by both the du Noüy tensiometer and the tensiographic drop-period isotherms i.e. SDS is 8.0, 7.6, 7.9 and 8.4 mmol/l for 15, 20, 30 and 35°C. Similarly, Triton X-100 is 0.27, 0.25, 0.22 and 0.20 mmol/l for the same temperatures.

The above results could of course be quite expansive, but given the graphical and tabulated results are presented here in direct comparison with conventional results, it is hopefully just necessary to point out salient facts. It is clear that the tensiotrace has a characteristic appearance, which depends on both surface and optical properties of the LUT. Obviously, the analysis of surfactants tensiopeak has property characteristics, which depend on the CMC of the surfactant.

Rainbow peak height analysis has similarly shown these property characteristics and the CMC can indeed also be obtained from the rainbow peak height measurand. There is a major difference in behaviour of the rainbow peak height above and below the CMC. Above the CMC, the rainbow peak height appears to increase linearly with concentration; below the CMC it decreases gradually with concentration apparently on a smooth curve. It is possible to determine the CMC from rainbow peak height measurements.

When rainbow peak height is plotted against log of concentration for SDS, BAC and Triton X-100, dramatically different behaviours are evident below and above the CMC giving intersection points at the surfactant CMC. In sub-micellar surfactant, the rainbow peak height diminishes with increasing surfactant concentration. At 950 nm, the refractive index remains effectively constant and it is reasonable to conclude that this reduction is due to change in drop curvature leading to light loss. In an attempt to allow for this, the rainbow peak height is now corrected for drop-period across the range of both sub-micellar and micellar solution giving the intercepting linear plots of Fig. 39(a) and (b). The definition of this corrected rainbow peak height (H'_{RP}) is:

$$H'_{RP} = \frac{H_{RP}}{T_1} \tag{49}$$

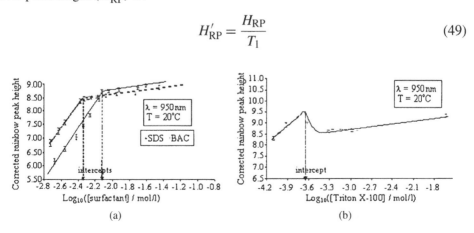

(a)

(b)

Figure 39. Variation in corrected rainbow peak height (average range values of ±0.16%) with concentration of sodium dodecyl sulphate and benzalkonium chloride (a); variation in corrected rainbow peak height (average range values of ±0.16%) with concentration of Triton X-100 (b).

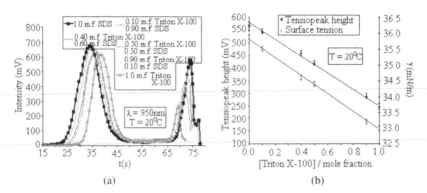

Figure 40. Tensiotraces of varying concentrations of micellar Triton X-100 in SDS (a); variation in (i) tensiopeak height and (ii) surface tension with mole fraction of Triton X-100 in sodium dodecyl sulphate (b).

where H_{RP} is the measured height of the rainbow peak and T_1 is the measured drop period.

Corrected rainbow peak height now increases with increasing sub-micellar concentration of SDS, BAC and Triton X-100. The behaviour of TX-100 is again a little more complex, giving a maximum rather than a simple intercept; this is not unusual given its impure state. The intercepts of these graphs again correspond well to literature values of CMC.

The tensiopeak heights of surfactant mixtures is an important facet of this study and it is found that the tensiotraces of mixed surfactant, show ideal mixing in mixed micellar solutions, i.e. the surface and bulk concentrations are identical. As the mixed surfactants are at micellar concentrations, full saturation of the surface with monomer is expected. It is interesting to use tensiopeak height as indications of preferential adsorption.

The behaviour of mixed micelle systems was investigated by generating tensiotraces (Fig. 40(a)) of varying concentrations of micellar Triton X-100 in SDS. Figure 40(b) shows the variation of micellar concentrations of Triton X-100 in sodium dodecyl sulphate. Results, although in aqueous solution, are presented in mole fraction. There are linear reductions in tensiopeak height and surface tension with increasing mole fraction of Triton X-100 in SDS.

Figure 41(a) shows the generated tensiotraces of varying concentrations of micellar Triton X-100 in BAC. The results, although in aqueous solution, are presented in mole fraction in Fig. 41(b). There are linear reductions in tensiopeak height and surface tension with increasing mole fraction of Triton X-100 in BAC. The tensiopeak height is an accurate indicator of surface adsorption of mixed surfactants.

A linear decrease in surface tension with increasing mole fraction of Triton X-100 in (i) SDS and (ii) BAC was observed. This implies that the surface concentration of each species is the same as its bulk concentration, which implies that there is no preferential surface adsorption in either micellar species.

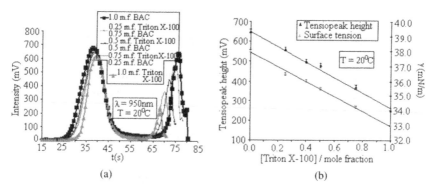

Figure 41. Tensiotraces of varying concentrations of micellar Triton X-100 in BAC (a); variation in (i) tensiopeak height and (ii) surface tension with mole fraction of Triton X-100 in benzalkonium chloride (b).

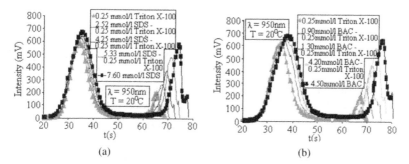

Figure 42. Tensiotraces of varying concentrations of sub-micellar SDS in a constant micellar concentration of Triton X-100 (0.25 mmol/l) (a); tensiotraces of varying concentrations of sub-micellar BAC in a constant micellar concentration of Triton X-100 (0.25 mmol/l) (b).

In summary, it is possible to say that the tensiograph is a very versatile and useful instrument for studying both bulk and dynamic surface properties of liquids. Tensiography gives three important parameters which allow surface properties to be easily and rapidly examined: drop-period, rainbow peak height and tensiopeak height. All three properties are affected by micellization; any one will allow critical micelle concentration to be determined if the analytes are non-absorbing. Surface tension may be determined from drop-period by use of Harkins–Brown correction factors. As a tool in the determination of rapid, accurate surface tension and CMC data tensiography is valuable.

Tensiotraces of varying sub-micellar concentrations (a) of sodium dodecyl sulphate (Fig. 42(a)) and (b) benzalkonium chloride (Fig. 42(b)) in a constant micellar concentration of Triton X-100 (0.25 mmol/l) were examined.

There is an increase in surface tension with increasing concentration of both SDS and BAC in a constant micellar concentration of Triton X-100 (0.25 mmol/l). The experimental studies indicated that there is an increase in the surface concentration of (i) SDS and (ii) BAC at the expense of Triton X-100 with increasing sub-micellar

concentration of ionic surfactant. A sharp transition in SDS and BAC solutions at
the CMC is found as reported above. In the constant micellar Triton X-100 solution
of the other two surfactants, intersecting curves moving away from the respective
CMC points, intersect at the zero surfactant concentration, for the second component. The blank Triton X-100 solution with surface tension determined by the
0.25 mmol/l is the intersecting point.

3. Example of Dynamic Surface Tension Monitoring Applications — Analysis of Protein Surface Adsorption by Tensiography

The aim of this study was to investigate the decline and recovery of the drop period of a BSA sample on 3 polymer substrates (PMMA, PEEK and acetyl nylon)
during adsorption and cleaning processes. Water was used as the solvent, which
was degassed using helium. A series of traces of solvent were run to ensure that a
regular drop period was obtained. A 0.5 mg/ml solution of BSA was prepared using degassed water. This was gradually introduced through the drop-analyser. The
drop-analyser was on automatic mode in that tensiotraces were continually developed, this was carried out in order to gain a comprehensive understanding of protein
adsorption behaviour. Two hundred traces were made throughout. The delivery tube
was removed from the sample into the solvent again, gradually introducing water
into the system with tensiotraces continually developed. In this way the cleaning
process was critically analysed in the same manner as adsorption.

From Fig. 43 it can be seen that the type of polymer to which the protein is
adsorbing has a big effect on the surface tension of the pendant drop. In the case of
both PEEK and Nylon the surface tension profile is quite similar. This is because
both of these materials are hydrophobic. Thus the drop forming on these surfaces
will have a high contact angle and lower surface tension.

In the case of adsorption to PMMA there is a significant difference. The first
difference is that PMMA is a hydrophilic surface, which will mean the drop has
a lower contact angle and higher surface tension. From Fig. 43 it can be seen that

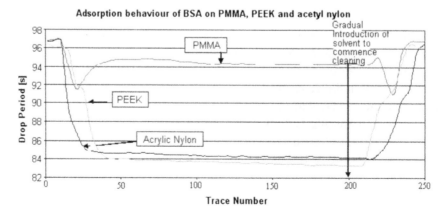

Figure 43. Comparison of BSA adsorption on different polymers.

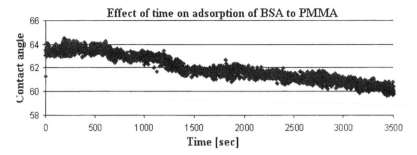

Figure 44. Contact angle of a sessile protein drop of a PMMA substrate.

there is an initial drop in surface tension before a rise to a plateau level. In other experiments on sessile drops of BSA adsorption to PMMA it was found that there was a delay in the protein having an effect on the contact angle of the sessile drop. This can be seen in Fig. 44.

As can be seen above there is no change in the contact angle of the sessile drop for the first 5 minutes. After this time the contact angle starts to gradually drop. This delay can be used to explain the initial drop and subsequent rise in surface tension seen in the tensiography experiments. Initially the protein in the pendant drop migrates to the liquid air interface and thus lowers the surface tension on each drop. As the protein starts to adsorb to the PMMA surface and lower the contact angle of the surface there is a change in the drop shape and a slight rise in the surface tension before it reaches a plateau level and the PMMA surface is covered in an adsorbed protein layer.

F. Calibration Standards for Tensiographic Fingerprinting

The purpose of this section is to establish the methodology for tensiographic fingerprinting and establish this technique for routine quality assurance in a regulated environment. Whilst this is obviously a long-term goal, in general spectroscopic measurement the primary references are often based on fundamental physical constants of nature, for example atomic line spectra for wavelength measurement and in this respect the fundamental nature of the form of the tensiotrace is no different. Ultimately, it is perceived that validation of the technique, as a forensic liquid fingerprinting methodology would be possible. Therefore, with these goals in mind, it is important nevertheless here to establish proper referencing methods for the technique, to assist the path of the evolutionary process.

Reference Materials (RMs) for tensiography need to be liquids, or conceivably solids that could be dissolved up to make a reference solution. Batch certification would provide the essential value assignment and homogeneity data. It would be very desirable for the calibration of the tensiotrace to normalize on both the photometric and volume scales. The tensiotrace would then be a dimensionless graph and the Cartesian coordinates of the various measurands taken from the tensiotrace could be recorded in an archive that would facilitate the simplest of fingerprinting procedures.

The first and most important issue in making these proposals for RMs is the requirements concerning health and safety standards for the handling of these materials. The RMs have to be chemically safe in all regards including: environmental and health effect including physicochemical characterisation and properties of the materials, identification of any explosion risks, hazard information like acute and chronic toxicity, toxicokinetics, ecotoxicity, bioaccumulation, etc., safety with regards to cell and tissue penetration, potential circulatory effects, mutagenicity and genotoxicity should there be any such issues.

Given the absence of established external quality assessment schemes for such a new technique as tensiography, the aim is here to propose RMs that could be made available on a commercial basis. The obvious advantages of samples circulated by external quality assessment schemes are that: the analyst does not have fore-knowledge of the result, a range of concentrations may be presented, there are significant time/cost savings in not having to prepare the materials, the data generated by the use of the material in the laboratory, does not have all the preparation steps as components in the overall uncertainty budget.

The tensiographic RM must have a very accurate and reproducible tensiotrace in a simple and well-established measurement procedure, and must deliver property values suitable for its intended use, up to its expiration date. The property value can therefore be used as a reference value for inter-comparison purposes.

In summary the following criteria should be considered in assessing the fitness of a RM:

(i) they should be chemically safe.

(ii) their property values should be stable for an acceptable time period, under realistic conditions of storage, transport and use.

(iii) they should be sufficiently homogeneous that the property value(s) measured on one portion of the batch shall apply to any other portion within acceptable limits of uncertainty.

(iv) be manufactured in a well-defined and regulated manner.

(v) provide good inter-laboratory correlation.

(vi) generate reliable test data from automatic, semi-automatic and manual equipment.

(vii) the tensiographic property values of the RM should have been established with a precision and accuracy sufficient to meet the needs of the end users and should include spectroscopic, photometric and volume standards.

The solution is happily at hand for the UV-vis region of drop spectroscopy as there exists the full gamut of reference materials in this, the most established field of conventional spectroscopy. These materials are available commercially from a number of suppliers [55]. Well-established techniques can be readily adapted for drop spectroscopy and thereby be validated for tensiographic spectroscopic measurements. Some photometric reference materials have been employed in this present study to demonstrate that the technique is accurate and precise, and provide the essential check on instrument performance on a regular basis to ensure that it is within satisfactory parameters i.e. "under control", and to allow for corrective action to be taken when found to be outside these limits. Their whole range of certified reference materials for absorbance/transmission, wavelength, resolution/bandwidth and stray light could be generally adapted to tensiographic measurement systems. It is possible that given the use of such materials applications of drop spectroscopy could be approved for use in GLP or ISO 9000 controlled environments, and in due course be accredited to ISO/IEC 17025. Therefore, using the above statement, it is conceivable that drop spectroscopic CRMs with a defined traceable path to national or internationally recognised primary materials or procedures could be made available.

By routinely verifying the performance of an instrument it is possible to ensure consistent quality that meets expected performance demands, and as a fingerprint technique tensiography will have to face up to the need to provide not only RMs but CRMs, as more and more market sector laboratories come under enforced regulatory control. For example, it is a technique that will shortly have to face the most demanding field with respect to CRMs, namely forensic science [56] and this gives an excellent if ambitious guide to considerations for the adoption of RMs for tensiography.

The temporal (volume) tensiographic RM is the safest material that can be found, namely triply deionised water. The measured drop period (T_1) from water will provide the calibration measure for the normalised scale. In practice, it will only be the rarest occasion that any drop period in excess of this measured value will be obtained. Such situations do arise only when the second phase is a liquid and does not apply to the regular practical situation for tensiograph fingerprinting.

One photometric standard for tensiography should be the EJ-399-05c1 [57] liquid scintillator anthracene solution supplied by Eljen Technologies. It is a formulation designed for use with capillaries with bore size down to 250 μm and it is therefore ideal for use in tensiography. This product is already in wide use as scintillation standard and therefore has already passed a series of safety checks [58]. The product has a suitably low vapour-pressure of 0.01 mmHg at 25°C and good flashpoint 145°C. While its high flash point and low chemical toxicity make it a relatively safe liquid to handle it is always advisable, however, to exercise reasonable caution when handling organic solvents with regard to personal contact and breathing. Most importantly, EJ-399-05c1 possesses a remarkably high and reproducible refractive index (1.564) and because of its zero absorption below 400 nm and above 520 nm is ideal as a reference standard in tensiography as the photometric standard. It is a somewhat rare liquid among liquid scintillator solvents, which although known for their high refractive indices, are chemically and photochemically unstable and consequently darken over a period of a few months if exposed to air or UV light for even relatively short times. This specific product is excellent for this purpose of a tensiographic RM as it provides an outstanding combination of high scintillation efficiency, high refractive index and excellent chemical stability required to endure the complicated handling associated with encapsulating liquids in capillary arrays. It has a quoted specific gravity of 0.993 and is a liquid of suitable viscosity to be pumped by tensiographic systems. It can be cleaned from the system using standard safe solvents that make it useable in any laboratory.

Anthracene-based references will give the largest photometric signal that can be anticipated and will be used to give a reference signal of 1 on the normalised photometric scale. The actual practical calibration can in most circumstances be achieved simply and economically by measuring triply deionised water for both the temporal and photometric calibration. Clearly, the normalization of the two axes of the tensiotrace would be advantageous in presenting a standardised tensiotrace which will be an issue to be investigated below. Obviously, water will be easy to clean from the system and is absolutely safe. As can be seen from Fig. 45 the coordinates of the water peaks are well known and in a paper [59] the use of water and all other calibration issues are discussed at length.

Although it is not discussed here, it is possible anthracene solutions could serve another purpose as a tensiographic RM. It is possible with CCD or CMOS detectors to record drop fluorescence spectral information and in this case anthracene solutions might offer a suitable standard. However, molecular fluorescence spectroscopy whilst being a sensitive and often selective technique is unlike absorption spectrophotometry because it is not an absolute technique. The instrument thus requires calibration before every series of measurements. This is most easily achieved

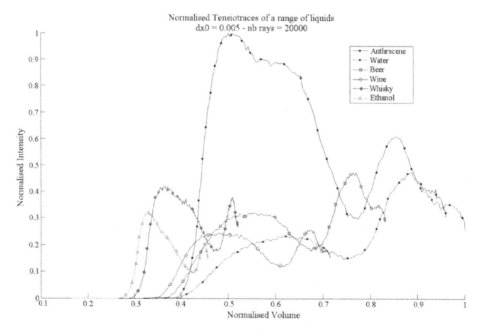

Figure 45. Normalised tensiograph photometric and volume calibration illustrated by anthracene and water.

by using a stable reference material [60], which should absorb and emit at similar wavelengths to the samples of interest. While the use of the general purpose fluorescent reference material set type 6BF (Starna) enables the day to day stability of instruments to be measured it is less obvious how such materials can be adapted to use in the tensiograph, but anthracene (10 µM, 360 nm excitation and 402 nm emission) and naphthalene (60 µM, 290 nm excitation and 330 nm emission) could be employed to provide standards.

The actual reference solution already commercially available for another application is a solution of water in 1-octanol as per the standard. SRM 2890, is certified for water content and is intended for use in calibrating instruments and validating the accuracy of analytical methods. Water concentration values, for RMs 8506a to 8510 are not certified, but represent the "best estimate" of the moisture content determined by NIST, and are intended for use in developing and validating methods for the determination of moisture in oil and similar matrices. The MSDS2890 standard water saturated in 1-octonol provides safe calibration standards that despite not being an absolute reference to an NIST moisture standards are nevertheless 'robust' with regard to their use in tensiography. The vapour pressure of the two liquids is equal and the use of this standard therefore obviates the need to worry about differential evaporation of the two components. Figure 45 shows the modelled tensiotrace of this standard, which the authors are proposing to be the calibration standard for those concerned with the most accurate fingerprint issues. Calibration, validation and control of the tensiographic instrument and its associated reagents is a central requirement here for establishing a Quality Management System (QMS). Here gaining from the work that has been put into establishing procedures for water determination in Karl Fischer titrations, where water standards can be used but are subject to operator error, validation is provided by using ISO 9000 and GLP guidelines with traceability to the NIST reference material SRM 890.

The figure above regroups the tensiotraces of four representative liquids (Beer, Wine, Whisky and Ethanol) and two standards (Anthracene, Water). The tensiotraces are normalised on both axis using Anthracene (photometric standard) and de-ionised water (temporal/volume standard) for respectively the photometric signal normalization and the temporal normalisation.

The enormous variability of tensiotrace forms when compared to the variability of the drop-shapes offers great potential for simple fingerprinting of liquids based on straightforward visual inspection. Each liquid can be uniquely characterised using a few measurands extracted from the tensiotrace when plotted on a normalised-reference scale, namely drop period, rainbow peak period and height, tensiopeak period and height and the rainbow peak commencement. Indeed through long experience of tensiotrace analysis we would claim that every physically distinguishable liquid has a unique tensiotrace and it is suggested that perhaps this signature of the LUT is the best graphical way of representing a liquid that presently exists. Obviously, since refractive index, absorbance/turbidity are wavelength dependent, there exists a family of such tensiotraces all of which it is suggested would be referenced against water (temporal or volume normalisation) and anthracene liquid (photometric normalisation) measured at the wavelength of the standard sodium-D line 589 nm. The computer modelling shows that there is indeed a unique tensiotrace for every physically distinguishable liquid type. This fact can only be deduced from this model and cannot otherwise be proven.

In summary, practical tensiograph differentiation and fingerprinting could become trivial and devolve simply to visual inspection of tensiotraces. For accurate discrimination of liquids look-up tables can be compiled of principle features of tensiotraces based on standardized, normalized and dimensionless measures. The unique fingerprint of a LUT could be sensitively identified based on such universal measures with numerical values referenced to perhaps NIST standards.

G. Conclusion

This wide-ranging review of the present state-of-the-art in tensiography reveals that there now exists a radically improved technological basis of the science. The technique is shown to have exceptionally good quantitative measurement capability. The most developed applications at this time are brewing and water monitoring. The theoretical foundations of tensiography have been fundamentally advanced analytically for spectroscopic work.

A substantial range of measurements have been reported for both bulk and surface measurements that show the technique is now a credible research tool for many fields where a broad characterisation of the LUT is important. One advantage of the technique is that it offers the potential of automating sample handling for a very substantial throughput of samples, which again is perhaps important in bio-applications. Researchers might note that the technique delivers several high quality measurements on the LUT. The tensiotrace can in addition provide a unique fingerprint for the liquid.

The modelling studies have been advanced to the point that very good qualitative agreement is seen in the form of the tensiotrace and measurement trends can be accurately predicted. The instrumental theory for tensiography through the work of the authors now stands adequate comparison with that of the gold standard of UV-visible spectroscopy. Perhaps the most important facet of the work reported here are the range of experiments supported by the modelling that deliver surface measurements without recourse to any of the traditional Harkins–Brown correction factors. The full theoretical quantitative description of the tensiotrace has been provided incorporating into the tensiotrace model the recent theoretical advances of Basaran and co-workers. However, this important theoretical advance is not necessary for the surface measurements made without recourse to correction factors, as it is clear that this tensiotrace provides measurands that in themselves give these uncorrected surface measurements.

It is the hope of the authors that the technique as presented here answers the fundamental issues associated with modern instrumentation for properly established calibration methods. Reference materials have been defined and these will hope-fully enable the technique to be used as a universal liquid fingerprint technique. The very simplicity of the approach recommended for tensiotraces plotted on a nor-malised photometric and temporal (volume) scale should make visual identification of liquids in most broad applications possible such as in medical diagnostics of dis-ease from body fluids. For more delicate and subtle differentiation of liquids such as in the batch analysis of beverages the more careful numerical quantification of peak heights and times, commencements and drop times will prove to be power-ful differentiators. The involvement here of Starna has proved to be particularly important as they manufacture NIST (National Institute for Standards and Tech-nology) UV-vis standards. It is suggested here that through traceable materials for tensiography there now exists a robust technique that can be exploited for many re-search applications. The software toolkit provided with the tensiograph instrument provides the brewers, water monitoring experts and other application people with a considerable range of techniques for qualitative analysis and product fingerprinting. The flexibility of these techniques are such that it is really impossible to prejudge measurement issues and say which measurand from the software toolkit will be the most appropriate for any specific job.

H. References

1. N.D. McMillan, V. Lawlor, M. Baker and S. Smith, From stalagmometry to multianalyser tensiog-raphy: The definition of the instrumental, software and analytical requirements for a new departure in drop analysis, in "Drops and Bubbles in Interfacial Research", Vol. 6, Series Studies in Interface Science, D. Möbius and R. Miller (Eds), Elsevier, 1998, pp. 593–705.

2. T. Tate, Phil. Mag., 27 (1864) 176.

3. Rayleigh, Phil. Mag., 48 (1899) 321.

4. W.D. Harkins and F.E. Brown, J. Am. Chem. Soc., 41 (1919) 499.

5. N.D. McMillan, O. Finlayson, F. Fortune, M. Fingelton, D. Daly, D. Townsend, D.D.G. McMillan and M.J. Dalton, Meas. Sci. Technol., 3 (1992) 746.

6. C.H. Wang, A.T. Augousti, J. Mason and N.D. McMillan, Meas. Sci. Technol., 10 (1999) 19.

7. A. Augousti, J. Mason, H. Morgan and N.D. McMillan, S.J. Prosser and E. Lewis (Eds), IOP Publishing, 2003, pp. 297–302.

8. A.T. Augousti, J. Mason, N.D. McMillan and D. Zhang, IOPP, Bristol, 2001, pp. 197–201.

9. Q. Song, Z.R. Qiu, G.X. Ghang *et al.*, J. Optoelectronics and Adv. Materials, 6 (2004) 393.

10. N.D. McMillan, CCD camera method of analysing the property of liquids, Irish Provisional Patent (1987).

11. P. Chen, D.Y. Kwok, R.M. Prokop, O.I. Rio, S.S. Susnar and A.W. Neumann, Axisymmetric Drop Shape Analysis (ADSA) and Its Applications in "Drops and Bubbles in Interfacial Research 6", D. Möbius and R. Miller (Eds), 1998, p. 61.

12. N.D. Mcmillan, E. O'Mongain, J. Walsh *et al.*, Optical engineering, 33 (1992) 746.

13. G. Hao, Z.R. Qiu and G.X. Zhang, Key Engineering materials, (2005) 295 and 215.

14. Q. Song, G.X. Zhang, Z. Qiu, A. Zhang and Q. Shi, J. Trace and Microprobe Techniques, 32 (2004) 43.

15. H.X. Chen, Z. Qui and G.X. Zhang, Key Engineering Materials, 33 (2005) 705.

16. The Singapore Institute of Manufacturing Technology produced a Multispectral Liquid Drop Analyser in 2003.
17. H. Liu and P.K. Dasgupta, Analytica Chimica Acta, 326 (1996) 13.
18. H. Liu and P.K. Dasgupta, Trends in Analytical Chemistry, 15 (1996) 468.
19. H. Liu and P.K. Dasgupta, Anal. Chem., 67 (1996) 4221.
20. H. Liu and P.K. Dasgupta, Microchemical J., 57 (1997) 127.
21. K.E. Miller and R.E. Synovec, Talanta, 51 (2000) 921.
22. Q. Song, G.X. Zhang, Z.R. Qiu, N.D. McMillan, A.T. Augousti, J. Mason, C.H. Wang and D.F. Zhang, Opto-Electronics Review, 13 (2005) 1.
23. S.A. Nikita and A.B. Taubman, J. Coll. Sci., 23 (1961) 290.
24. H. Liu and P.K. Dasgupta, J. Anal. Chem., 326 (1997) 13.
25. H. Liu and P.K. Dasgupta, J. Anal. Chem., 68 (1996) 187.
26. A.A. Cardoso and P.K. Dasgupta, J. Anal. Chem., 67 (1995) 2562.
27. P. Hennebert and G. Pottecher, 2005. AQUA STEW project (Water Quality Surveillance Techniques for Early Warning by Tensiographic Sensors), Final Report. 5th EU Common Research and Development Programme (Energy, Environment and Sustainable Development), Contract number: EVK1-CT2000-00066.
28. N.D. McMillan, S. Smith and M. Baker, Tensiographic Drophead, 24 August 2006 WO 2006/087390 A1.
29. K. Tiernan, "Drop spectrophotometer design", PhD thesis, Dublin Institute of Technology, Dublin, 2007.
30. M. Marino, A. Carati and L. Galgani, Classical light dispersion theory in a regular lattice, Annals of Physics, 322 (2007) 799.
31. G. Currell, Analytical Instrumentation Performance Characteristics and Quality, Wiley, 2000.
32. S. Smith, Simulations of drop analyser response for water samples, Report to AQUASTEW project, Physics Centre, University of Essex, Colchester, Essex CO4 3SQ, UK, 2004.
33. O. Yildirim, Q. Xu and O.A. Basaran, Physics of Fluids, 17 (2005) 062107–13.
34. M.C. Wilkinson, J. Colloid. Interface Sci., 40 (1972) 14.
35. J.L. Lando and H.T. Oakley, J. Colloid Interface Sci., 25 (1967) 526.
36. D. Morrin, "Experimental studies in tensiography", PhD thesis, Carlow Institute of Technology, Carlow, 2007.
37. L. Gao and L. Zeng, Meas. Sci. Technol., 14 (2003) N50.
38. N.D. McMillan, F. Feeney, M.J.P. Power, S.M. Kinsella, M.P. Kelly, K.W. Thompson and J.P. O'Dea, Instrumentation Science & Technology, 22 (1994) 375–395.
39. D. Carbery, S.M. Riedel and N.D. McMillan, An experimental investigation into the engineering basis of a new fibre optic small volume drop surface analyser, Sensors and their Applications XI, (IOPP, Bristol, 2001), 209.
40. N.D. McMillan, O. Finlayson, F. Fortune, M. Fingelton, D. Daly, D. Townsend, D.D.G. McMillan, M.J. Dalton and C. Cryan, Rev. Sci. Instrum., 63 (1992) 216.
41. C. Gerhardt, Annalen der Chemie, 67 (1853) 149.
42. E.C.G. Clarke, Isolation and Identification of Drugs. Ed. The Pharmaceutical Press, London, 1969.
43. US Standard Methods for the Examination of Water and Wastewater, APHA. 19th Edition (1995).
44. Method 10068, Hach DR/2010 Spectrophotometer Handbook.
45. V.I. Haltrin, Algorithm and code to calculate specular reflections of light from a wavy water surface, Proceedings of the Seventh International Conference on Remote Sensing for Marine and Coastal Environments 20–22 May, 2002, Miami, Florida, USA.
46. P.H. Lee, M.Y. Tse and W.K. Lee, J. Opt. Soc. Am. B, 15 (1998) 2782.

47. J.A. Dean (Ed.), "Lange's Handbook of Chemistry", 15th Ed., McGraw-Hill, 1999.

48. D. Morrin, N.D. McMillan, K. Beverley and B. O'Rourke, Opto Ireland, Proc. SPIE, (2005) 5826.

49. J.A. Dean (Ed.), "Lange's Handbook of Chemistry", 15th Ed., McGraw-Hill, 1999.

50. P. Pringuet, "Tensiotrace modeling", PhD thesis, Carlow Institute of Technology, Carlow, due for submission, 2009.

51. K. Tiernan, D. Kennedy and N.D. McMillan, Materials & Design, 26 (2005) 197–201.

52. N.D. McMillan, P. Davern, V. Lawlor, M. Baker, K. Thompson, J. Hanrahan, M. Davis and J. Harkin, Colloids Surfaces A, 114 (1996) 75–97.

53. D. Carbery, S.M. Riedel and N.D. McMillan, An experimental investigation into the engineering basis of a new fibre optic small volume drop surface analyser, Sensors and their Applications XI, (IOPP, Bristol, 2001), 209–214.

54. J.W. McBain and C.W. Humphreys, J. Phys. Chem., 36 (1932) 300.

55. CRM for UV and fluorescence spectroscopy requirement for laboratories for accreditation for at least one of the internationally recognised standards of GLP, ISO 17025 or ISO 9000. Starna standards provide "evidence of control" that has become an absolute necessity as indeed has the use of Certified Reference Materials with a defined traceable path to national or internationally recognised primary materials or procedures. See for example http://www.starna.co.uk/ukhome/d_ref/f_ref/scientistarticle.html.

56. D.M. Epstien, K.T. Lothridge and W.J. Tilstone, The Use of Certified Reference Materials in Forensic QA 13 INTERPOL Forensic Science Symposium, Lyon, France, October 16–19, 2001.

57. EJ-309 Liquid Scintillator from Southern Scientific Ltd. from Eljen Technology, PO Box 870, 300 Crane Street, Sweetwater TX 79556 USA, Website: www.eljentechnology.com. This product appears to have been changed from the hydrocarbon to a polymeric fluorescence standard for safety reasons in very recent times. The author's work has shown discrepancies in measurement from those shown on data sheet and conclusions given in the text probably need to be revisited.

58. Syracuse Research Corp., NY (United States), Evaluation of the potential carcinogenicity of dibenz(a,h) anthracene. Final report EPA-68-03-3112; EPA-68-03-3182, 1988.

59. A. Augousti, D. Carbery, J. Hammond, N.D. McMillan, R. Miller, D. Morrin, M. O'Neill, B. O'Rourke, P. Pringuet and S.R.P. Smith, Meas. Sci. & Technol., Submitted February 2008.

60. A.K. Gaigalas, L. Li, O. Henderson, R. Vogt, J. Barr, G. Marti, J. Weaver and A. Schwartz, J. Research of the National Institute of Standards and Technology, 106 (2001) 2.

Emulsification with Micro-Structured Membranes and Micro-Engineered Systems

R.M. Boom and C.G.P.H. Schroën

Department of Agrotechnology and Food Science, Food Process Engineering Group,
Wageningen University, Bomenweg 2, 6703 HD Wageningen, The Netherlands

Contents

A. Emulsification with Membranes and Micro-Engineered Systems

Emulsions have been important in almost any field important for mankind: most of our food is based upon an emulsion; many of our products (paint, bitumen, composite plastics) are emulsions, and being able to produce them in all their variations is of the utmost importance.

There is therefore a wide variety of processes used for production of emulsions. They are either based on the use of flow to deform and disrupt larger droplets into smaller ones, or on the alteration of the physicochemical stability of a solution or pre-emulsion. An overview of these methods is given in Table 1.

These processes all have the disadvantage that they require either a large amount of energy input to produce emulsions with small droplet sizes, or they are only applicable in a narrow window of product formulations (phase inversion). In addition, they usually produce polydisperse emulsions. The dissipation of large amounts of

Bubble and Drop Interfaces
© Koninklijke Brill NV, Leiden, 2011

Table 1.

Conventional processes used for emulsification

Class of process	Process	Principle of emulsification	Comments
Rotor–stator systems	Stirred vessel	Shear and elongational flow	– Relatively low intensity of flow field – larger droplets produced
	Colloid mill		– Conical rotator and stator with narrow gap in between – Smaller droplets produced, longer residence time needed
	Toothed mill		– Hollow rotor with periodic gaps (teeth) in a stator similar to colloid mill in performance
Homogenizer	Valve system	Shear and elongational flow Possible contribution of cavitation	– Mixture flows through a valve that is checked with a spring – Flow through a valve gives very intense flow fields and low dynamic pressures that may induce cavitation (see below)
	Nozzle systems		– Similar to valve systems; only here the mixture flows trough a small nozzle (sometimes more than one in sequence)
Ultra-sonica-tion		Cavitation and micro-jets from cavitation	– Cavitation induces strong local flow impinging on individual droplets and breaking them up
Phase inversion	Transitional inversion	Change in interactions between surfactant and dispersed and continuous phases	– Changes in interactions induce formation of a microemulsion system, giving a gradual inversion
	Catastrophic inversion		Changes in interaction induce a sudden inversion, influenced by flow and other parameters

energy during emulsification can be detrimental for fragile components in the system. For the preparation of a double emulsion, one cannot easily use a homogeniser to produce the outer emulsion, since the large energy dissipation would destroy the inner emulsion.

Table 2.

New membrane and microchannel-based emulsification processes discussed in this chapter

Class of process	Process	Principle	Section
Droplet snap-off by localised shear forces	Crossflow membrane emulsification	Formation of individual droplets on pore mouth, snap-off by shear forces of the flow over the membrane	B.1.
	Use of T-shaped micro-junctions	Formation of individual droplets in the junction, snap-off by shear forces of the flow over the junction	B.2.
Interfacial tension induced snap-off	SPG emulsification	Formation of individual droplets in pore mouth, snap-off due to strong non-cylindrical shape and strong interconnectivity of the pores	C.1.
	Use of strongly non-spherical microchannels	Formation of individual droplets in the junction, snap-off by the interfacial tension (interface reduction by transition from deformed droplet into spherical shape)	C.2.
Premix membrane emulsification		Branching and unification of pores inside the membrane split up the dispersed phase into small 'slugs'	D.
Co-flowing systems	Use of Y- and Ψ-shaped micro-junctions	Formation of individual droplets in the junction, snap-off by elongational flow over the droplet (usually by Rayleigh breakup)	E.1.
	Use of porous microchannels	Formation of a liquid thread inside a pore followed by Rayleigh breakup	E.2.

Thus, there is a need for processes that need less energy input and hence are milder, and that produce mono- or narrowly dispersed emulsions. In recent years we have seen the emergence of a new class of processes, which is aimed at these two aspects. These processes do not depend on the creation of an average flow field, but they individually form droplets in or on a microstructured system with microstructural elements. Examples of these are the use of microporous membranes and the use of microchannels that are made with micro-engineering methods. These methods rely on either a very localised flow field, or on utilising the interfacial tension itself as a driving force to snap droplets off.

The geometry of the system often dominates or is at least very important in determining the size of the droplets formed. An overview is given in Table 2.

In the next sections, we will discuss each category separately, on the mechanisms of droplet formation and their differences.

B. Droplet Snap-off Induced by Localized Shear Flow

1. Cross-Flow Membrane Emulsification

Cross-flow membrane emulsification is the oldest form of emulsification with microstructured systems (see Fig. 1). In 1991, Nakashima *et al.* [1] reported on pressing an oil phase though the pores in porous glass membranes made from volcanic ashes (from the Shirazu region in Japan). At the downstream side, an aqueous phase flowed across the membrane. Since these glass membranes are extremely hydrophilic, the oil emerged as drops from the pore mouths, which were taken up by the cross-flowing aqueous phase: an oil-in-water (O/W) emulsion was formed. The same principle has since then also been demonstrated for preparing water-in-oil (W/O) emulsions using a hydrophobic membrane.

The sizes of the droplets obtained are often in the range of 2–10 times the pore diameter [2, 3]. Peng and Williams [3] gave an approach to better estimate the droplet size created with this process, by using a force balance. The droplets, emerging from the pore mouths, are initially still bound to the oil in the pore mouth through the interfacial tension (see Fig. 2), which gives a force of size

$$F_{\text{interface}} = 2\pi R_{\text{p}}\sigma \tag{1}$$

in which σ is the interfacial tension and R_{p} is the radius of the pore mouth. As the emerging droplet extended into the cross-flowing continuous phase, a drag force is exerted on it, given by Stokes' law:

$$F_{\text{drag}} = -1.7(6\pi\eta_{\text{c}}R_{\text{d}}v_{\text{c}}). \tag{2}$$

The cross-flow velocity v_{c} is at half-height of the droplet; η_{c} is the viscosity of the continuous phase. The factor 1.7 stems from the non-uniform flow field around the droplet. To find a value of v_{c}, we may make use of the shear stress, which is related to the gradient in cross-flow velocity (or shear rate) dv_{c}/dz:

$$\tau_{\text{w}} = -\eta_{\text{c}}\frac{dv_{\text{c}}}{dz} \approx -\eta_{\text{c}}\frac{v_{\text{c}}}{R_{\text{d}}}. \tag{3}$$

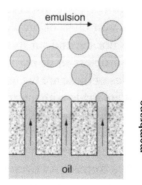

Figure 1. Schematic representation of cross-flow membrane emulsification [1].

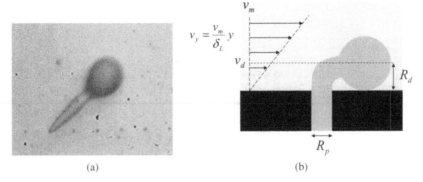

$$v_y = \frac{v_m}{\delta_L} y$$

(a)   (b)

Figure 2. (a) Video microscope image showing a droplet emerging from a small pore (lower left corner), dragged with the cross-flow from the lower left corner to the upper right corner (image from C.J.M. van Rijn and C. Goeting). (b) Schematic overview of the forces acting on the emerging droplet.

This leads to an estimate of the drag force of:

$$F_d = -10.2\pi R_d^2 \tau_w. \tag{4}$$

As the droplet grows, the drag force exerted by the cross-flowing continuous phase, becomes larger and larger. At a certain size, the drag force will balance and then exceed the retaining interfacial tension. The droplet size at which this happens is given by:

$$F_{drag} = F_{interface} \rightarrow \frac{R_d}{R_p} \geqslant \sqrt{\frac{\sigma}{5.1\tau_w R_p}} = \sqrt{\frac{1}{5.1Ca}} \tag{5}$$

in which Ca is the capillary number. One notices here that the droplet size is not proportional to the pore size, as often assumed, but is dependent on the square root of the pore size. Since the droplets are often generated very quickly, the interfacial tension is not a constant but will vary. Using the interfacial tension as a fitting parameter yields a good qualitative description of the droplet size as a function of the wall shear stress (see Fig. 3). The values of the interfacial tensions obtained are [4, 5]:

Surfactant concentration (wt%)	$\sigma_{apparent}$ (mN/m)	$\sigma_{equilibrium}$ (mN/m)
0.05% Tween-20,	18	5.7
0.5% Tween-20	11	5.0
2% SDS	1.7	8.0

Since for SDS the lowest possible interfacial tension would be the equilibrium value; the obtained values cannot be physically realistic. Therefore the force balance model should only be used for qualitative purposes.

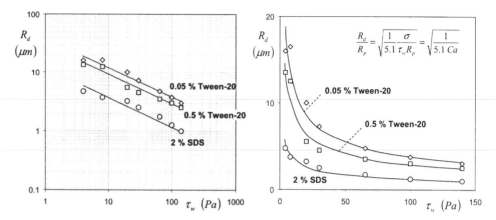

Figure 3. Experiments from Schröder [7–9] (1998 and 1999), fitted by the force balance model by Peng and Williams [3]. The qualitative description of the data is adequate; however the values of σ obtained are not physically realistic. Sunflower oil in water with different surfactants; the membrane used was ceramic with av. pore size 0.4 μm, and with trans-membrane pressure 3 bar.

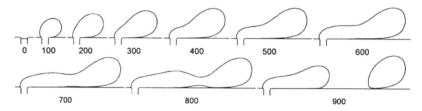

Figure 4. CFD simulations of the shape of the emerging droplet during the formation process (Kelder *et al.* [6]). Modelled after the hexadecane/water/SDS system. The simulations assumed a constant interfacial tension of 5 mN/m. The cross-flow velocity was chosen at 1 m/s.

There are two main reasons why the force balance model is insufficient: the dynamic nature of the interfacial tension (see Fig. 5), and the fact that the droplet is highly non-spherical during the formation process, due to the drag force on the droplet. Kelder *et al.* [6] illustrated the non-spherical form with the help of computational fluid dynamics calculations (see Fig. 4); these figures also show that the break-up process is in fact not always very close to the pore mouth. Therefore, the estimation of the retaining force due to the interfacial tension is not accurate as well.

Droplet Size Distribution

The droplet size distributions obtained with cross-flow membrane emulsification are usually polydisperse, even though the wall shear stress is relatively homogeneous and one would therefore expect monodispersity on the basis of the model described above. However, the droplet formation process is disturbed by a number of effects.

1. Possible coalescence due to slow surfactant dynamics

2. Steric hindrance on the membrane surface

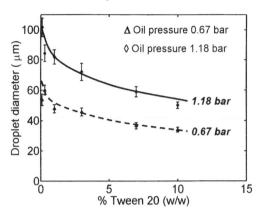

Figure 5. Measured droplet diameter (left hand diagram) with different surfactant concentrations and different transmembrane pressures (from Van der Graaf *et al.* [10]). The dependence of the droplet diameter on the surfactant concentrations shows that the interfacial tension is not at equilibrium.

3. Pressure fluctuations inside the membrane

Coalescence

Polydispersity is often explained by the fact that the surfactants need to diffuse from the bulk of the continuous phase towards the freshly formed interface with the droplets, and adsorb onto the interface. When the interface expansion rate is too fast, the interface remains practically empty of surfactant, implying that the droplets are may not be stabilized during formation. When pores on the membrane surface are too close, the droplets forming are too close and may coalesce into larger droplets (e.g., Christow *et al.* [11]). This leads to polydispersity of the emulsion. For slow stabilisers such as proteins this may indeed be the case; for fast, low molecular weight emulsifiers such as SDS and Tween-20 this does not seem to be the case. Van der Graaf *et al.* [10] estimated the interface expansion rate during membrane emulsification to 17–61 s^{-1} (see Fig. 6).

Comparison with measurements of the dynamic interfacial tension in the hexadecane/water/Tween-20 showed that these expansion rates lead to bare interfaces only with surfactant concentrations below about 0.01%. Since polydispersity has been observed even in systems with far higher low MW surfactant concentrations, it is clear that this mechanism cannot not the only one.

Steric Hindrance

Porous membranes have been developed over the years for filtration (separation) purposes. This process benefits from a high porosity of the membrane, therefore membranes are usually highly porous. Abrahamse *et al.* [12] showed that this can lead to steric interaction between growing droplets, even when they do not coalesce (as illustrated in Fig. 7). A large neighbouring droplet can simply push away another growing droplet. This can yield a combination of very small and very large droplets.

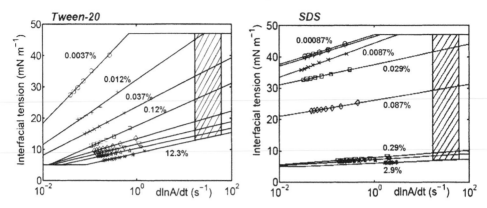

Figure 6. The dynamic interfacial tension of in a hexadecane/water/surfactant system. The grey area is the estimated operation window during membrane emulsification. Left hand graph: Tween-20, right hand graph: SDS as surfactant. From Van der Graaf *et al.*, 2004 [10].

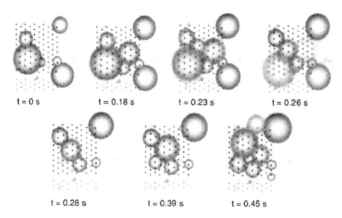

Figure 7. Sequence of snapshots of emulsification: droplets detach due to steric hindrance. From Abrahamse *et al.* [12].

Pressure Fluctuations

Computational fluid dynamics calculations performed by the same authors [13] showed that the pressure inside the pore mouth during the droplet formation process fluctuates greatly. The droplet radius has a minimum in the initial stages of droplet formation ($R_d = R_p$), after which it continually increases. This leads to a steady drop in Laplace pressure. Dynamic effects, due to the very fast inflow of dispersed phase into the droplet enhance this effect. Thus, the pressure in the pore mouth is more than halved during the droplet formation process (see Fig. 8). This sudden pressure drop leads to a redistribution of pressure inside the membrane: the pressures in the membrane near the actively emulsifying pore become lower as well due to the interconnectivity of the pore structure. This leads to coupling of droplet formation between neighbouring pores, which can lead to strong polydispersity.

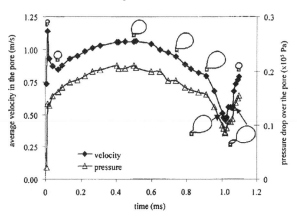

Figure 8. Pressure fluctuations and average velocity of the dispersed phase in a pore, calculated using CFD for a micro-engineered membrane ('microsieve'). From Abrahamse *et al.* [13].

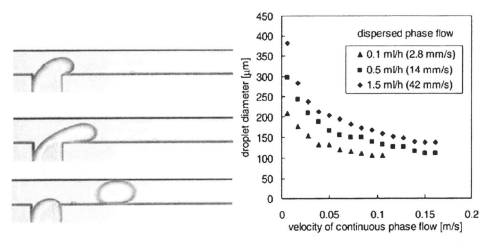

Figure 9. Left hand graph: T-shaped junction used by Van der Graaf and Nisisako *et al.* [14, 15]. Right hand graph: droplet sizes (D_d) relative to the channel dimension (D_m) as function of the flow rate of the continuous phase (from Nisisako *et al.* [15]).

2. *T-Shaped Junctions in Microfluidic Devices*

Akin to cross-flow membrane emulsification, but on a much smaller scale, is the formation of droplets in T-shaped microchannels (see Fig. 9). Two immiscible fluids are brought together through two different microchannels. The phase that wets the microchannel walls will form the continuous phase.

The controllability of the microsystem makes it possible to cascade the process. This makes it possible to produce for example double emulsions or even multiple emulsions with more than one internal phase (see Fig. 10). The size of the channels and the current state of micro-engineering limit this approach to the production of relatively large droplets (~10 μm), however there is no fundamental limit to

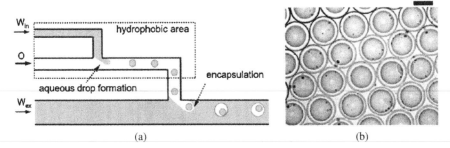

(a) (b)

Figure 10. Cascaded system of T-shaped microchannels for producing multiple emulsions: (a) schematic set-up; (b) double W/O/W emulsion with high internal volume fraction (black bar is 100 μm) (from Nisisako *et al.* [16]).

producing smaller droplets, except the hydrodynamic resistance for flow through the microchannels, which will dictate extremely low fluxes with smaller channels.

Scaling out (massive parallelization) is often thought to be the way to realise larger outputs from microsystems. The fact that two different channels have to come together limits the out-scalability of these devices to perhaps a few hundred parallel channels maximum. Their use may therefore be primarily to analytical and very high-value applications.

Link *et al.* [17] introduced a premix emulsion in the lower end of a T-junction in a microfluidic device (see Fig. 11). They found that the drops of the pre-emulsion divided themselves over the two exiting sidearms, when the flow rate was large enough to enable sufficient elongation of the pre-emulsion droplet at the point of stagnation. This happened when the capillary number Ca ($= \eta v / \sigma$ being the ratio of viscous to interfacial stresses):

$$Ca > \varepsilon_0 (\varepsilon_0^{-2/3} - 1)^2. \tag{6}$$

In this, ε_0 is the initial extension of the droplet in the channels:

$$\varepsilon_0 = \frac{L}{\pi w}, \tag{7}$$

where L is the (extended) length of the droplet and w the width of the droplet in the microchannel. The volumes of the two daughter droplets were proportional to the volume flows in the two side-arms.

C. Droplet Snap-off Induced by Interfacial Tension Effects

1. SPG Membrane Emulsification

Cross-flow emulsification is characterised by the fact that the cross-flowing continuous phase induces droplet detachment. A different flow rate therefore gives different droplet sizes. In some cases, especially at lower flow rates of the to-be-dispersed phase through the membrane pores, one can see an independence of the droplet size

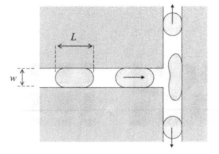

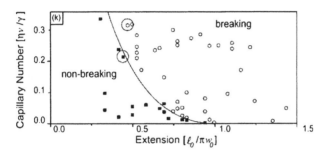

Figure 11. Top drawing: Microfluidic set-up used by Link *et al.* for droplet split-up due to elongation in a T-shaped microchannel. Bottom graph: Conditions under which droplets will or will not break into two droplets. The curve represents the relation discussed in the text. From Link *et al.* [17].

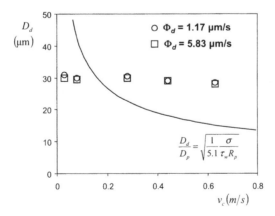

Figure 12. Droplet diameters obtained with different flow rates of the cross-flowing continuous phase, for two different flow rates of the to-be-dispersed phase (as measured by Yasuno *et al.* [18]).

on the cross-flowing rate (see Fig. 12). This has been first observed with Shirazu Porous Glass membranes (SPG), but not with for example ceramic membranes.

With larger flow rates of the dispersed phase, one can see a re-emergence of the dependence on the cross-flowing rate. It is probably that at lower cross-flowing rates, a different mechanism is responsible for droplet detachment. A hypothesis for this is suggested in Fig. 13. The pore structure inside a SPG membrane is highly

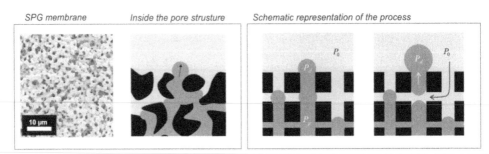

Figure 13. Pore structure of a SGP membrane: micrograph and penetration of the to-be-dispersed phase. The highly interconnected pore structure enables the continuous phase to penetrate into the pore structure, which can displace the dispersed phase and induce droplet detachment (after Boom, 2007).

tortuous and interconnected; much more so than in other types of membranes. When the to-be-dispersed phase is pressed through the pores, which are strongly wetted by the continuous phase, it will first penetrate through those pores that will give the lowest Laplace pressure difference: pores that are slightly larger and may be more cylindrical. The surrounding pores near the downstream side of the membrane will still be filled with the continuous phase. The Laplace pressure difference between the pressure of the dispersed phase $p_{d.pore}$ and the pressure of the continuous phase p_c can be estimated by

$$P_{d.pore} - p_c = \frac{\sigma}{R_p}, \tag{8}$$

R_p being the radius of the pore. As soon as the to-be-dispersed phase emerges at the downstream side, a droplet is formed on top of the membrane. The radius of the growing droplet will soon be larger than the radius of te pore, and the Laplace pressure difference with the continuous phase will drop, according to

$$P_{d.drop} - p_c = \frac{2\sigma}{R_d}. \tag{9}$$

Comparing these two pressure differences yields that as soon as $R_d > 2R_p$, the pressure in the droplet p_{drop} is smaller than inside the pore $p_{d.pore}$. Thus, the to-be-dispersed phase inside the pore can flow from the pore into the droplet, which will further increase the droplet radius, which will further enhance the driving force. Analogous to the effect described earlier with cross-flow membrane emulsification (Abrahamse *et al.* [13]), the pressure inside the pore will decrease as well. Due to the internal Laplace pressure difference between to-be-dispersed phase in the pore and continuous phase in the pore, the continuous phase will now displace the dispersed phase. When this happens near the surface of the membrane, this may result in droplet snap-off. When the pore interconnectivity is too low to allow this, the displacement may not happen: the to-be-dispersed phase will rearrange to adapt to the new pressure distribution. It is therefore logical that this mechanism only occurs in specific membrane morphologies.

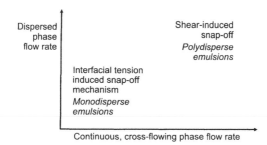

Figure 14. Schematic regime where interfacial tension induced snap-off takes place, and where cross-flow (shear) induced snap-off.

The occurrence of this mechanism has been supported by microscopic observations of droplet formation on a SPG membrane surface (M. Yasuno *et al.* [18]), which indeed shows that this regime is coupled to a situation where only a few pores actively eject droplets, and hence where other pores may still be filled with the continuous phase.

When the flow rates become too high, the continuous phase does not have the time to displace the dispersed phase, while the pressure differences over the pores are too large. It was therefore found [18] that there is a transition at higher flow rates to the shear-induced snap-off mechanism, yielding larger and more polydisperse droplets in general (see Fig. 14).

2. Microchannel Systems

The same mechanism is exploited in microchannels. Nakajima and coworkers (e.g. [18, 19]) explored the use of micro-engineered channels. When a to-be-dispersed phase was pressed through such a channel that was strongly flattened where it opened up into a larger channel, droplets detached spontaneously from the mouth of the channel (see Fig. 15).

The mechanism is basically the same as with the SPG membranes. The flattened mouth of the pore lets the dispersed phase form a flattened droplet between the bottom and top. The Laplace pressure is given by the distance between top and bottom, H_p:

$$P_{d,drop} - p_c = \frac{2\sigma}{H_p}. \tag{10}$$

As soon as this droplet becomes so large that it will bulge out of the channel mouth into the larger channel, it will form a 3-dimensional droplet. In this case this will happen when $R_{drop} > H_p$. Essential is that the flattening of the pore moth is so large, that continuous phase can penetrate deep into the channel. Only in this way can the dispersed flowing out, be displaced by the continuous phase inside the channel. Therefore it has been observed that the channel mouth should be at least three times as wide compared to the height of the channel H_p.

Figure 15. Emulsification process found by Kawakatsu *et al.* [19]. Oil was led through a small channel, which ended on a slit-shaped opening (drawing by Van Dijke *et al.* [20]).

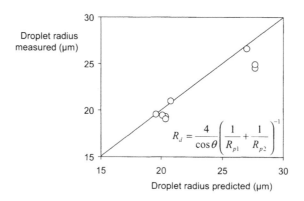

Figure 16. Prediction by Rayner *et al.* [24] of droplet sizes obtained by Kobayashi [22, 23, 25] (2003, 2004 and 2005) with microchannels, featuring varying dimensions, contact angles and interfacial tensions.

Rayner *et al.* (2005) simulated the shape of droplets on a non-cylindrical pore mouth [21], taking into account the complete geometry, and wetting behaviour of the channel walls. By assuming them to be a cylinder with an elliptic cross-section having two characteristic radii, R_{p1} and R_{p2}, they found:

$$R_d = \frac{4}{\cos\theta}\left(\frac{1}{R_{p1}} + \frac{1}{R_{p2}}\right)^{-1}. \tag{11}$$

Comparison with experimental data from Kobayashi [22, 23] shows this relation to be accurate (see Fig. 16).

This mechanism is only valid as long as the continuous phase has the time to displace the dispersed phase inside the microchannel. At higher flow rate of the dispersed phase, one can discern a transition towards a regime where interfacial tension induced auto-snap-off does not take place anymore. The mechanism for snap-off now reverts to shearing of by the cross-flow in the larger channel. Since in many experiments this cross-flow is not well defined, this leads to polydispersity and large droplets. Figure 17 shows this transition. At low flow rates of the dispersed

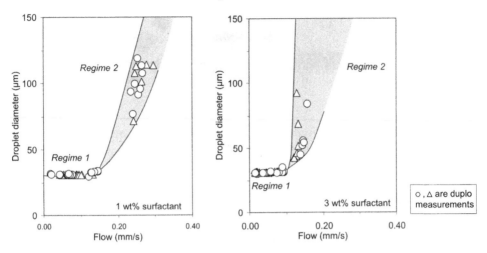

Figure 17. Droplet size as function of the flow rate of oil through the microchannel. At lower flow rates, the droplet size is independent of the flow rate (regime 1), while from a certain critical flow rate, the droplets become larger (regime 2) (Dekkers *et al.* [26]).

phase, the droplet size is only determined through the geometry (and contact angle). When the flow rate is increased one sees a sudden increase of the droplet sizes generated. One can also see that at low concentrations of the surfactant, the dynamic interfacial tension during droplet formation is high, the driving force for snap-off is high and the process can therefore be maintained up to higher flow rates. With a higher surfactant concentration, the transition to larger droplets takes place at a lower flow rate.

D. Premix Membrane Emulsification

In 1998, Suzuki and coworkers [27] introduced a new form of membrane-based emulsification. Instead of letting only the to-be-dispersed phase through the membrane pores, they first prepared a premix emulsion, and pressed this complete emulsion through the membrane pores. The coarse droplets of the premix emulsion were broken up into small droplets having a size in the range of the size of the pores in the membrane (see Fig. 18). Depending on the size of the droplets in the premix emulsion, more than one pass is usually necessary [28].

The flux during the initial pass was reported to be 3–20 m^3/m^{2}h, and in the third pass 6–50 m^3/m^{2}h [29]; see Fig. 19. When one compares this with the typical fluxes obtained with cross-flow membrane emulsification (usually below 0.5 m^3/m^{2}h), it is clear that this process is very efficient. In addition, dispersed phase volume fractions of up to 80% can be used, while the fact that no cross-flow over the membrane is necessary reduces cost and complexity of the process.

The membrane should be wetted by the continuous phase. Thus, starting with an O/W premix, and using a hydrophilic membrane will result in a fine O/W emulsion. However, use of a hydrophobic membrane would result in a fine W/O emulsion,

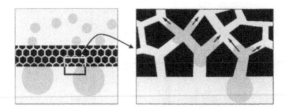

Figure 18. The Suzuki flow-through membrane emulsification. A coarse emulsion is pressed through a porous membrane; at the downstream side, a fine emulsion emerges.

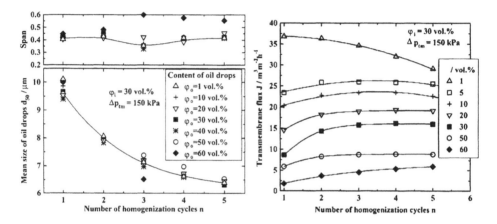

Figure 19. Left hand graph: Droplet sizes as function of the number of passes, for several volume fractions of dispersed phase. Right hand graph: transmembrane fluxes involved (from Vladisavljevic *et al.* [28]). SPG membranes were used with pore size of 10.7 μm.

provided that the surfactant system would permit this (if not, using a hydrophobic membrane would result in separation of the emulsion). Thus, this process can result in phase inversion, depending on the surface properties of the membrane.

Van der Zwan *et al.* [30] investigated the mechanisms playing a role during pre-mix membrane emulsification, using 2D networks of microchannels on a silicon microchip. Since the membrane is wetted by the continuous phase, the droplets of the dispersed phase do not enter the pore easily, since the droplets are larger than the pores. Therefore, the droplets tend to accumulate before and inside the pore network. The volume fraction of the dispersed phase in the membrane is always very high. This necessarily means that the continuous phase moves faster through the pore network than the droplets of dispersed phase, which remain stuck at different places in the structure.

The same authors identified three categories of droplet break-up mechanisms:

– *Snap-off due to localized shear forces.* Inside pores, the same mechanism as with cross-flow membrane emulsification or T-junction microchannels may take place: as the continuous phase flows over a droplet, the resulting shear force can disrupt the droplet into smaller ones.

– *Break-up due to interfacial tension effects* (Rayleigh and Laplace instabilities). Due to geometric variations in pore shape and size, the same mechanism as reported for the SPG membranes and microchannels (see Section C) can take place locally.

– *Break-up due to steric hindrance between droplets.* The very high volume fraction of droplets of the dispersed phase gives rise to steric hindrance. Droplets that flow along with the continuous phase can deform and disrupt other droplets into smaller ones.

The fact that these porous membranes have very many branchings and junctions, means that irrespective of which mechanism plays a role where, a passage through the membrane will always result in the same droplet size.

The most significant backdrop of the process is its sensitivity to depth fouling: since the whole premix, including the surfactant system, has to penetrate through the membrane pores, fouling caused by for example proteins may well block the pore. The process is therefore probably only relevant for applications requiring suitable formulations.

E. Coflowing Systems

Coflowing systems use a flow of the continuous phase in the same direction as the flow of the dispersed phase; this is in contrast to the cross-flow systems (where they are perpendicular to each other; see Section B) and systems in which no flow of the continuous phase is necessary (interfacial tension induced snap-off; see Section C).

1. Microfluidic Flow Focusing

In co-flow emulsification, a droplet emerging from a capillary is surrounded by the continuous phase, which flows and converges towards a small opening (see Fig. 20). The drag of the continuous phase draws a thread of the dispersed phase into the opening. The thread appears to be stable during passage through the opening, and then breaks up.

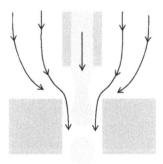

Figure 20. Flow focusing. The system may be built up around a capillary and a cylindrical orifice, or may be built up in 2D with three microchannels.

Coflow microchannel emulsification is often done by using Ψ-shaped microchannels. The middle channel supplies the dispersed phase, while the two outer channels supply the continuous phase. When the three streams merge, the dispersed phase is squeezed into a thread. After the merger of the three streams, a constriction is often applied, leading to 'focusing' of the flow. This strongly reduces the thickness of the thread, accelerates break-up and reduces droplet size [31, 32].

1.1. Dripping Regime

The mechanism of droplet formation depends on the flow rate and the dimensions. At low flow rates, the flow rates are so low that liquids can easily redistribute to accommodate pressure differences; this is called the dripping regime. As the thread is forced through the orifice, it forms a droplet at the exit side of the orifice. At low flow rates, the interface takes shape to form the smallest possible interfacial area inside the orifice (Garstecki *et al.* [33]). At the exit side of the orifice, a droplet is formed, and therefore the interface is close to the walls of the orifice. This results in a build-up of pressure inside the orifice, forcing the dispersed phase to flow out, which results in a sudden collapse of the neck (see Fig. 21).

The droplet size is dependent on the flow rate as:

$$V_d \propto \frac{p_d}{\mu \Phi_c} \tag{12}$$

in which p_d is the pressure of dispersed phase in the central channel, μ is the viscosity of the continuous phase and Φ_c is the flow rate of the continuous phase [33].

At even higher flow rates, inertial effects become important. Bifurcation is observed [34]: the process becomes so fast, that the formation of the next droplet is influenced by the previous one, leading to production of a large and a small droplet. At even higher flow rates, quick further bifurcations are seen, leading through a region of chaos (production of droplets with random sizes), towards a stable region in which three different droplet sizes are produced. This then end up in the jetting regime, where a different break-up mechanism takes place.

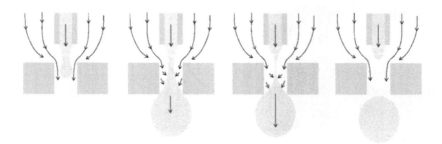

Figure 21. Mechanism of droplet formation at low flow rates (no influence of inertia). The flow of the continuous phase is constricted due to the growth of the droplet at the exit of the orifice. This causes pressure build-up, which squeezes the dispersed phase out of the neck, leading to snap-off.

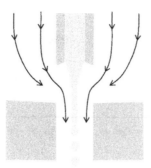

Figure 22. Flow focusing in the jetting regime. The flow rates are so high, that the thread is extended more and becomes much thinner. The resulting droplets are smaller than in the dripping regime.

1.2. Jetting Regime

At higher flow rates, inertial forces become important. The thread of dispersed phase moves so fast that it becomes a jet that is much narrower than the neck encountered in the dripping regime, and can be much longer. The resulting droplets are also significantly smaller (see Fig. 22).

For axisymmetric flow focusing in the jetting regime, quantitative relations have been found to predict the size of bubbles, dispersed in a liquid. Since the equations are insensitive to the fluid properties, they seem to be valid as well for emulsification. Generally, these equations have the form:

$$\frac{D_d}{D_c} = A \left(\frac{\Phi_d}{\Phi_c} \right)^B \tag{13}$$

in which D_d is the diameter of the resulting droplet, and D_c is the diameter of the orifice; Φ_d and Φ_c are the flow rate of the dispersed phase and of the continuous phase, respectively. Gañán-Calvo *et al.* [31] derive based on hydrodynamic reasoning for B a value of 0.37. For high Reynolds numbers, Gañán-Calvo *et al.* [35] use dimensional analysis to find a value for B of 0.4. The value of A is usually found by fitting with experiments; a value of 1.06 to 1.1 was found [36]. Van Hoeve *et al.* [37] combined the hydrodynamics and the dynamics of the Rayleigh–Plateau instabilities to find values for A and B of 1.1 and 0.44, respectively.

1.3. Remarks on Applications

Flow-focusing is a versatile tool for generation of bubbles and drops that should be very monodisperse. In addition, especially in the jetting regime, the droplets can be much smaller than the size of the channels. This is an advantage when working with very narrow channels that clog up easily. Further, the droplet size can be easily set by changing the flow rates and pressures (see above).

The dependency of the droplet size on the flow rates in the system however also makes it difficult to scale on this type of device towards larger scale production. Putting many of these devices in parallel does not easily lead to mondispersity, since flow rates through the channels are dependent on D_c^4 (following Hagen-Poiseuille),

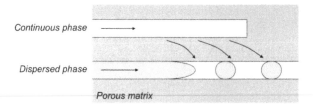

Figure 23. Emulsification by permeation of the continuous phase through a porous matrix (after Geerken *et al.*, 2005).

and since during droplet formation pressure fluctuations may occur. One should therefore envision applications of this device mostly within research or analytical applications, such as high-throughput screening.

2. Porous Channel Based Emulsification

Geerken *et al.* [38] introduced a variant system based on two microchannels in a porous material that was permeable for the continuous phase (see Fig. 23). In one microchannel the to-be-dispersed phase was led; in the other the continuous phase was led. This last channel was a dead-end channel. The continuous phase flowed through the porous matrix into the dispersed phase channel. The dispersed phase was then diluted with the continuous phase. The dispersed phase broke up into droplets that were approximately 1.2 times the diameter of the microchannel.

The channels in this system were still relatively large (300 µm). It is expected that the system will have the same drawback as premix membrane emulsification: depth fouling in the porous material blocking the flow of the continuous phase. However, outscaling of these devices may be easier than with other coflowing devices.

F. Concluding Remarks

The emergence of micro-engineering has finally enabled us to design orifices and systems for a specific mechanism for droplet formation. The process that was started with the pioneering work in the field of membrane emulsification could therefore be naturally extended with the use of micro-devices.

A range of different types of devices has now been realised: T-shaped junctions, building on the experience with membrane emulsification; microchannel systems giving interfacial-tension induced snap-off, analogous to the snap-off in SPG membranes; and coflowing systems, based on the creation of thin threads of the dispersed phase. Each approach has advantages and disadvantages, and one will see applications of each of these devices on a small scale for research and analytical applications in the coming years.

Most of the work done so far is on single microchannels or nozzles, except for research in (micro-engineered) membranes. Combing many thousands or even millions of these devices into scaled-out systems, for large-scale production will be a challenge for the future.

It seems that the field of micro-engineering has opened up a new road in the field of emulsification. The first steps have been taken, and we look forward to the next steps towards practical applications.

G. Acknowledgement

Support by COST Action D43 is gratefully acknowledged.

H. References

1. T. Nakashima, M. Shimizu and M. Kukizaki, Key Eng. Mater., 61/62 (1991) 513.
2. R. Katoh, Y. Asano, A. Furuya, K. Sotoyama and M. Tomita, J. Membrane Sci., 113 (1996) 131.
3. S.J. Peng and R.A.Williams, Trans. Inst. Chem. Eng., 76 (1998) 894.
4. M. Rayner and G. Trägårdh, Desalination, 145 (2002) 165.
5. S. van der Graaf, M.L.J. Steegmans, R.G.M. van der Sman, C.G.P.H. Schroën and R.M. Boom, Colloids Surfaces A, 266 (2005) 106.
6. J.D.H. Kelder, J.J.M. Janssen and R.M. Boom, J. Membrane Sci., 304 (2007) 50.
7. V. Schröder and H. Schubert, Colloids Surfaces A, 152 (1999) 103.
8. V. Schröder and H. Schubert, Emulgieren met Microporösen Membranen. Lebensmittel- und Verpackingstechnik., 43 (1998) 80.
9. V. Schröder, O. Behrend and H. Schubert, J. Colloid Interface Sci., 202 (1998) 334.
10. S. van der Graaf, C.G.P.H. Schroën., R.G.M. van der Sman and R.M. Boom, J. Colloid Interface Sci., 277 (2004) 456.
11. N.C. Christov, D.N. Ganchev, N.D. Vassileva, N.D. Denkov, K.D. Danov and P.A. Kralchevsky, Colloids Surfaces A, 209 (2002) 83.
12. A.J. Abrahamse, A. v.d. Padt and R.M. Boom, J. Membrane Sci., 204 (2002) 125.
13. A.J. Abrahamse, A. v.d. Padt., R.M. Boom and W.B.C de Heij, AIChE J., 47 (2001) 1285.
14. S. van der Graaf, T. Nisisako, C.G.P.H. Schroën, R.G.M. van der Sman and R.M. Boom, Langmuir, 22 (2006) 4144.
15. T. Nisisako, T. Torii and T. Higuchi, Lab Chip, 2 (2002) 24.
16. T. Nisisako, S. Okushima and T. Torii, Soft Matter, 1 (2005) 23.
17. D.R. Link, S.L. Anna, D.A. Weitz and H.A. Stone, Phys. Rev. Lett., 92 (2004) 054503.
18. M. Yasuno, M. Nakajima, S. Iwamoto, T. Maruyama, S. Sugiura, I. Kobayashi, A. Shono and K. Satoh, J. Membrane Sci., 210 (2002) 29.
19. T. Kawakatsu, Y. Kikuchi and M. Nakajima, J. Am. Oil Chem. Soc., 74 (1997) 317.
20. K. v. Dijke, K. Schroën and R.M. Boom, Langmuir, 24 (2008) 10107.
21. M. Rayner, G. Trägårdh, C. Trägårdh and P. Dejmek, J. Colloid Interface Sci., 279(1) (2004) 175–185.
22. I. Kobayashi and M. Nakajima, Eur. J. Lipid. Sci. Technol., 104 (2002) 720.
23. I. Kobayashi, M. Nakajima and S. Mutakata, Colloids Surfaces A, 229 (2003) 33.
24. M. Rayner, Membrane emulsification: modelling interfacial and geometric effects on droplet size, in "Division of Food Engineering Department of Food Technology, Engineering and Nutrition", Thesis, Lund Institute of Technology, Lund, Sweden, 2005.
25. I. Kobayashi, M. Yasuno, S. Iwamoto, A. Shono, K. Satoh and M. Nakajima, Colloids Surfaces A, 207 (2002) 185.
26. K.S. Dekkers, Production of double emulsions by microchannel emulsification, Thesis, 2003, Wageningen University, The Netherlands and Technische Universität Karlsruhe, Germany.

27. K. Suzuki, I. Fujiki and Y. Hagura, Food Sci. Technol. Int. Tokyo, 4 (1998) 164.
28. G.T. Vladisavljević, M. Shimizu and T. Nakashima, J. Membrane Sci., 244 (2004) 97.
29. J. Altenbach-Rehm, K. Suzuki and H. Schubert, Production of O/W emulsions with narrow droplet size distributions by repeated premix membrane emulsification, in "World Conference of Emulsification," Lyon, France, 2002.
30. E. van der Zwan, C.G.P.H. Schroën, K. van Dijke and R.M. Boom, Colloids Surfaces A, 277 (2006) 223.
31. A.M. Ganan-Calvo and J.M. Gordillo, Phys. Rev. Lett., 87 (2001) 274501.
32. Q. Xu and M. Nakajima, Appl. Phys. Lett., 85 (2004) 3726.
33. P. Garstecki, H.A. Stone and G.M. Whitsides, Phys. Rev. Lett., 94 (2005) 164501.
34. P. Garstecki, M.J. Fuerstman and G.M. Whitesides, Phys. Rev. Lett., 94 (2005) 234502.
35. A.M. Gañán-Calvo, Phys. Rev. E., 69 (2004) 27301.
36. S.M. van der Meer, Monodisperse microbubbling — instabilities in coflowing gas–liquid jets, Thesis, 2003, University of Twente, The Netherlands.
37. W. van Hoeve, Monodisperse microbubble formation in microfluidic flow-focusing devices, University of Twente, The Netherlands, 2006.
38. M. Geerken, Emulsification with micro-engineered devices, Thesis, University of Twente, The Netherlands, 2006.

Manipulation of Droplets onto a Planar Interface

T. Gilet, D. Terwagne, N. Vandewalle and S. Dorbolo

GRASP, Département de Physique B5, Université de Liège, B-4000 Liège, Belgium

Contents

A. Introduction

The manipulation of droplets becomes a huge problem as soon as tiny quantities of liquid have to interact with micro-sensors or with another liquid. A broad range of applications can be found in food science, pharmacology, coating, painting and biotechnology among others. Droplets can be used in chemical engineering in order to mix tiny amounts of reactive substances. To achieve such a goal, one has to invent some processes to manipulate a droplet: motion, binary collision, fragmentation, ... Most of efforts were made in the framework of microfluidic [1], i.e. the droplets are driven through close channels using a second liquid (non-miscible with the liquid of the droplet). That technique requires a perfect control of the surface properties: wetting, de-wetting, pinning of the contact line, ... Accurate pumps are needed

Bubble and Drop Interfaces
© Koninklijke Brill NV, Leiden, 2011

to drive the droplet through the network of channels. Finally, the contamination should be prevented by avoiding any contact with a solid element. An alternative technique to the microfluidic channels consists in manipulating individual droplets onto a liquid bath.

Couder *et al.* have shown that a droplet is able to bounce indefinitely on the surface of a liquid pool without coalescing [2]. This effect is basically obtained when the liquid bath is vertically shaken. Moreover, he demonstrated that the droplet moves horizontally due to its interaction with the wave it produces on the liquid surface [3]. This wave-particle interaction reveals some spectacular behaviours including interference patterns [4] and tunnelling effects [5].

In the present work, we describe several operations that can be performed with droplets onto a planar interface, including rebound, division, mixing and size selection. All these operations are done without touching the droplets. By vibrating the bath, a droplet can be stored. The stability of the bouncing droplet is discussed in Section C. Using the phenomenon of partial coalescence, it is possible to decrease the volume of a droplet (Section D). We discuss the best way to mix two droplets (Section E). Finally, by combining the delayed coalescence and the partial coalescence, we show how to have a control on the droplet size (Section F). In the next section, the mechanism related to the delay before coalescence of a droplet with a planar interface is described. This section establishes the theoretical basis of this work.

B. Delayed Coalescence on a Planar Interface at Rest

Let us consider a system composed of two immiscible fluids (at least one of them is a liquid): fluid 1 is heavier than fluid 2 (Fig. 1). A droplet made of fluid 1 is gently laid onto the quasi-planar interface between both fluids. The droplet tends to cross this interface due to gravity: it experiences a coalescence.

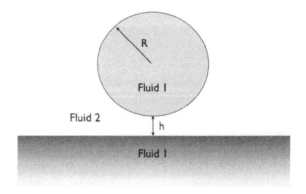

Figure 1. Sketch of a droplet of radius R made of a liquid 1. This droplet is surrounded by a fluid 2 that is not miscible with fluid 1 (may be air). The droplet is laid onto the planar interface between fluid 1 and fluid 2. The coalescence occurs when the interstitial film of fluid 2 (thickness h) breaks.

The merging does not occur immediately. Indeed, the film of fluid 2 that remains between the droplet and its bulk phase has to be drained out. Then, when this film is sufficiently thin, it is broken by attractive forces (Van Der Waals interactions, hydrogen bonds, . . .). The drainage time of the film is influenced by a lot of physical parameters that are discussed here below.

1. Reynolds Model

The first model that describes the drainage before coalescence was proposed by Reynolds [6]. It is based on the lubrication theory: the film of fluid 2 is sufficiently thin compared to the droplet radius R, inertia effects are neglected, and both fluids are Newtonian and incompressible. Reynolds supposes that the film thickness h is constant, and that both interfaces are motionless. This latter hypothesis is approximately satisfied when the dynamical viscosity μ_2 of fluid 2 is much smaller than the viscosity of fluid 1, so the flow inside the film cannot entrain the fluid inside the droplet. In these conditions, the lifetime t_L of the droplet onto the planar interface, due to drainage, is given by:

$$t_L = \frac{9}{16\pi^2} \frac{\mu_2 S^2}{\Delta\rho g R^3 h_c^2},$$
(1)

where $\Delta\rho$ is the difference in density between both fluids, g the gravity acceleration and h_c the critical film thickness for which electrostatic forces become significant. The area of the film S depends on the geometry of the quasi-equilibrium configuration that is reached when the droplet settles on the planar interface.

Usually, S/R^2 is a monotonic function of the Bond number Bo, the ratio between gravity and surface forces (interfacial tension σ):

$$Bo = \frac{\Delta\rho g R^2}{\sigma}.$$
(2)

Moreover, when $Bo \ll 1$, the droplet is approximately spherical onto a planar interface, $S/R^2 \sim 2\pi Bo/3$ and the resulting lifetime is

$$t_L = \frac{1}{4} \frac{\mu_2 \Delta\rho g R^5}{h_c^2 \sigma^2}.$$
(3)

For a millimetric droplet of water in the air laid onto a water surface, a lifetime of about 1 s is found, taking $h_c = 100$ nm [7, 8].

2. Other Influences

The observed lifetime is often smaller than the predicted one. The model of Reynolds has been improved by several authors. For example, more realistic geometrical configurations are detailed in [9]. Another important point is the interfacial mobility: When dynamical viscosities are similar in both fluids, the fluid inside the droplet is entrained by the flow in the film. This entrainment can be reduced when surfactant is added, since it usually rigidifies the interface. An advanced investigation about the influence of interfacial mobility has been made by Ivanov *et al.* [10].

Other effects, such as Marangoni effects, rheological interfacial properties, thermocapillary effects [11], electrocoalescence [12] have been studied in details. The interested reader could refer for example to the review of Neitzel and Dell'Aversana [11].

C. Delayed Coalescence on a Vertically Vibrated Bath

Couder *et al.* [13] have recently discovered that when a droplet is laid on an interface that is vertically vibrated (by using an electromagnetic shaker for example — Fig. 2), it may avoid coalescence. Indeed, when the droplet bounces, the air film is renewed. The oscillation is characterised by the reduced acceleration Γ defined by

$$\Gamma = \frac{A\omega^2}{g}, \tag{4}$$

where A and ω are the amplitude and pulsation of the sinusoidal oscillation; g is the gravity acceleration.

Sustained bouncing is observed when the acceleration is larger than a critical acceleration Γ_C given by [13]

$$\Gamma_C - 1 \propto \omega^2. \tag{5}$$

This criterion allows us to determine whether the droplet bounces or not according to a couple of parameters (A, ω) of the oscillation. Note that this criterion is only valid for high viscosity oil and when the viscosities of both the bath and the droplet are the same. Indeed, we showed that for low viscosity systems, the droplet deformations must be taken into account as eigen modes are excited [14, 15].

Stabilisation by bouncing is limited by the appearance of Faraday waves. This high acceleration limit Γ_F depends on the viscosity of the oil in the bath. Considering a given droplet size and a given viscosity, the bouncing phenomenon may only occur in a delimited region of the (A, ω) diagram, i.e. in between Γ_C and Γ_F (Fig. 3).

A question remains about the stability of the bouncing droplet. Indeed, at low frequencies and for $\Gamma > \Gamma_C(\omega)$, the droplet may bounce for a time longer than

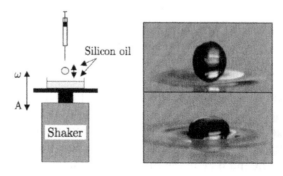

Figure 2. (left) Experimental setup made up of an electromagnetic shaker and a plate on which an oil-filled container is fixed; (right) snapshots of the bouncing droplet.

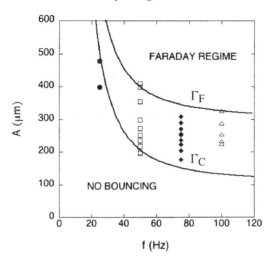

Figure 3. Phase diagram amplitude-frequency for a droplet of 1.5 mm of diameter and made up of 50 cSt silicone oil. Top and bottom curves correspond to the Faraday Γ_F and Couder Γ_C acceleration thresholds respectively (see text). Points represent experimental conditions that have been investigated by Terwagne *et al.* [16] in order to cover at best the region where the droplet can bounce.

a few days. On the other hand, for large frequencies and for $\Gamma > \Gamma_C(\omega)$, the droplet bounces only for a few seconds before coalescing. Let us determine where, in the phase diagram (A, ω), the droplet bounces the longest time.

Due to the large number of parameters: A, ω, R and the viscosity of the oil μ_1, some parameters have been arbitrary fixed. The viscosity and the droplet size are fixed to 50 cSt and 1.5 mm diameter respectively. In these conditions, the area to explore in the A–ω diagram and the investigation points are represented in Fig. 3. For each point, the lifetime has been measured. Moreover, an optical tracking method has been used to determine the position and the deformation of the droplet with respect to the phase of the plate.

1. Experimental Setup

A container is filled with Dow Corning silicone oil (DC200) which viscosity $\mu_1 = 50$ cSt. For each frequency, the acceleration is tuned between Γ_C and Γ_F by increasing the amplitude of the vibration. The droplets are produced by using a syringe and are laid onto the surface of the bath. The lifetime is basically measured using a timer.

The position and the shape of the droplets have been studied by image processing. The motion is recorded from the side with a high-speed camera (1000 fps). A snapshot of an impacting droplet is shown in Fig. 4. During bouncing, the deformation is vertically asymmetric. The evolution of its upper half part is easily determined by a contour detection on the images while its lower part is not accessible at any time of the motion, in particular when the droplet is partly immersed in the bath (Fig. 4).

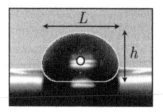

Figure 4. Snapshot of a droplet at its maximum of deformation. The fitted contour of the droplet is represented by a white curve. The height h and the width L are indicated by the arrows while the centre of mass is shown by a white circle.

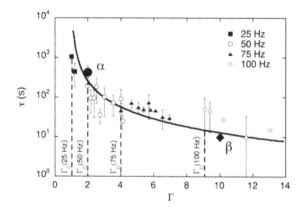

Figure 5. Semi-log plot of the averaged lifetime τ as a function of the reduced acceleration Γ. The symbols correspond to the different frequencies (see legend). The vertical dashed lines represent the Couder threshold Eq. (5) corresponding to the different frequencies. The continuous curve is the fit using Eq. (6). The α and β symbols refer to the cases that have been studied according to their trajectories.

However, since the total volume and the evolution of the upper part of the droplet are known, the lower part may be deduced as soon as a hypothetical shape has been speculated. The lower part shape is modelled by a flat bottom surrounded by a quarter of torus. The shape is fitted to the experimental contour, taking volume conservation into account. An example of this shape (white contour) can be seen in Fig. 4. On the other hand, when the droplet is entirely visible, its shape is oblate. The lower part of the droplet is supposed to behave like a half ellipsoid. Eventually, the trajectory of the centre of mass is deduced as well as the evolution of the aspect ratio $A_r = h/L$, where h and L are respectively the height and the width of the droplet (Fig. 6).

When the amplitude of the oscillation is not sufficiently large, the air film below the droplet is not completely regenerated and the droplet coalesces.

2. *Lifetime of Droplets*

Even beyond Γ_C, we observe that the lifetime of bouncing droplets is finite. This later gives us some information about the stability of bouncing. The lifetime is represented in Fig. 5 with respect to the reduced acceleration Γ. Different symbols correspond to the considered frequency, i.e. circles, squares, diamonds and triangles are for 25, 50, 75 and 100 Hz respectively. The vertical dashed lines represent the Couder threshold $\Gamma_C(\omega)$.

The lifetime is found to decrease with an increase in Γ. The lifetime has been represented in a semi-log plot. A power law has been fitted (continuous curve in Fig. 5), namely

$$\tau \propto \frac{1}{(\Gamma - 1)^{1.4}}. \tag{6}$$

A divergence is expected at $\Gamma = 1$. Lifetimes at 25 Hz are difficult to determine since droplets last for a very long time.

As far as the results corresponding to a given frequency are concerned, the maximum lifetime is found at Γ_C. When the acceleration is increased above the threshold, the droplet deformations increase and destabilize the bouncing. It results in a shorter lifetime.

In the next section, the deformation is finely studied for two different cases: (i) case α: a bouncing droplet characterized by a lifetime of 420 s ($A = 200$ μm and $f = 50$ Hz) and (ii) case β: a bouncing droplet characterized by a lifetime of 10 s ($A = 250$ μm and $f = 100$ Hz).

3. Trajectory and Deformation of Droplets

In Fig. 6, the cases $\alpha(\bullet)$ and $\beta(\blacklozenge)$ are compared: The vertical position of the centre of mass, normalized by A, is plotted on Fig. 6(a) while Fig. 6(b) presents the variations of the aspect ratio with respect to the normalized time $t_r = \omega t / 2\pi$. In both

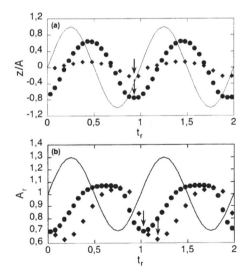

Figure 6. Comparison of (a) the position of the centre of mass and (b) the aspect ratio variation with respect to the reduced time $t_r = \omega t / 2\pi$ for the case α and β (see text). These cases are represented by circles and squares respectively. The continuous curve gives an indication concerning the phase of the plate.

figures, continuous curves represent the position of the plate in arbitrary units in order to visualize the phase of the plate.

In both cases, the motion seems to be perfectly periodic. The normalized position of the droplet is characterized by a larger amplitude in the low frequency case. However, the phase $t_{r,z}$ when the trajectory reaches a minimum is nearly the same in both cases. This particular phase is indicated by an arrow in Fig. 6(a). The trajectory in the case α is more symmetrical than in the case β.

The variation of the aspect ratio has nearly the same amplitude in both cases, but the phase $t_{r,Ar}$ of maximum deformation is observed to be very different. This particular phase is indicated by an arrow in Fig. 6(b). The bouncing droplet is deformed during a smaller laps of time in case α than in the case β. Indeed, the aspect ratio A_r in case α is close to and even higher than unity during a longer reduced time. That means that the time of interaction between the bath and the droplet is smaller than in case β. In that later case, the droplet is continuously deformed by the plate.

In view of these results, a long lifetime is characterized by a large amplitude motion of the centre of mass and a long period without interaction with the bath. Moreover, the phase at which the maximum deformation occurs is of importance. The ideal situation is obtained when $t_{r,z} \sim t_{r,Ar}$.

D. Partial Coalescence

A droplet made of a low-viscosity liquid can experience a partial coalescence when it is gently laid onto a liquid bath [1]: it does not fully empty and a new smaller droplet is formed above the bath interface (Fig. 7). This droplet coalesces partially again and the process repeats until a critical size is reached. Then, the coalescence becomes total and the last droplet fully disappears.

This phenomenon was first reported in 1930 by Mahajan [18], and investigated by Charles and Mason in 1960 [19]. In their experiments, another immiscible liquid was surrounding the droplet (instead of air). No more investigations were made before those of Leblanc [20] in 1993. He was interested in the stability of emulsions. Indeed, partial coalescence considerably slows down the gravity-driven phase separation of two immiscible liquids, for instance when dealing with oil demulsification. Unfortunately, his work remained unpublished. In 2000, Thoroddsen and Takehara [21] studied this phenomenon in more details for an air/water interface. In the same time, partial coalescence with a surfactant addition was described by Pikhitsa and Tsargorodskaya [22]. In 2006, many investigations were conducted in order to fully understand partial coalescence (Blanchette *et al.* [23], Kavehpour *et al.* [24, 25], Feng *et al.* [26, 27]). An experimental study of the influence of both viscosities on the process has been made by Gilet *et al.* [28].

Until now, little is known about micro-flows occurring during coalescence. First, traditional experimental techniques are difficult to be applied at droplet scale. PIV experiments on droplet coalescence were made [29]. Unfortunately in [29], the

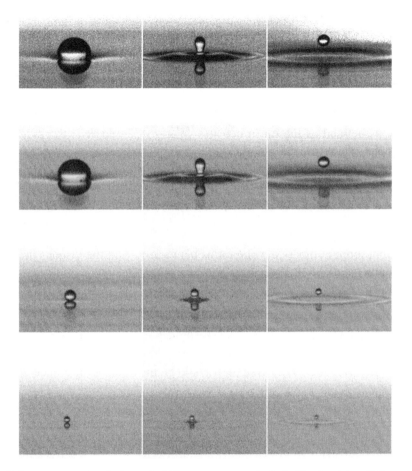

Figure 7. A soapy water droplet is gently laid onto a soapy water bath. The droplet partially coalesces, and a new smaller droplet is formed. This phenomenon occurs several times. When the droplet is sufficiently small, the coalescence becomes total. The arrow on the last snapshot indicates the position of the last micro-droplet before its total coalescence.

droplet was too large to allow partial coalescence. The simplest way to get experimental information about micro-flows is to observe the interface position. From a numerical point of view, the free surface flows with changing topology (such as coalescence, breakdown and pinch-off) are really difficult and expensive to compute. Despite these difficulties, a lot of information has already been collected about micro-flows in coalescence [23, 27].

1. Dimensional Analysis

Charles and Mason [19] attempted to explain the occurrence of partial coalescence by considering the ratio of dynamical viscosities in both fluids. But this only parameter is not sufficient to fully understand partial coalescence mechanisms. As discovered by Leblanc [20], three kinds of forces influences the process dynamics: the reduced gravity, the surface tension and the viscosity in both fluids.

Partial coalescence is usually scaled by the capillary time τ_σ [21, 28, 30], resulting from the balance between interfacial tension and inertia:

$$\tau_\sigma = \sqrt{\frac{\rho_m R^3}{\sigma}}, \tag{7}$$

where ρ_m is defined as the mean density $(\rho_1 + \rho_2)/2$ and R is the radius of the initial droplet. Therefore, the interfacial tension has to be the main force for partial coalescence to occur. In other words, a self-similar process such as partial coalescence is only possible when a single force is dominant (here the surface tension), there is thus no natural length scale related to the balance of two forces.

Reduced gravity is compared to surface tension by means of the Bond number Bo, as mentioned previously. To assess about the influence of viscosities, one can use the Ohnesorge numbers, i.e. the ratios between τ_σ and the viscous times R^2/ν in both fluids

$$Oh_{1.2} = \frac{\nu_{1.2}\sqrt{\rho_m}}{\sqrt{\sigma R}}, \tag{8}$$

where $\nu_{1.2}$ is the kinematic viscosity of the fluid 1 and 2 respectively. According to the π-theorem (Vaschy–Buckingham), a fourth dimensionless number is required in order to describe the whole system. It could be the relative difference of density,

$$\Delta\rho = \frac{\rho_1 - \rho_2}{\rho_1 + \rho_2} \tag{9}$$

linked to the relative difference of inertial effects generated by the interfacial tension in both fluids. Ideally, any dimensionless quantity can be expressed as a function of these dimensionless numbers, especially the ratio Ψ between the daughter droplet radius R_d and the mother droplet radius R

$$\frac{R_d}{R} = \psi(Bo, Oh_1, Oh_2, \Delta\rho). \tag{10}$$

2. The Self-Similar Regime

When $Bo \ll 1$ and $Oh_{1.2} \ll 0.1$, gravity and viscosities have no significant influence on the process [21, 23, 28]. The Ψ function does not depend on Bo and $Oh_{1.2}$ anymore: it becomes a function Ψ_0 that only depends on $\Delta\rho$. In these conditions, the process is self-similar since the densities do not change between two successive partial coalescences. In liquid/liquid systems, the difference of density is small and Ψ_0 is approximately equal to 0.45 [21, 23, 28].

In Fig. 8, the successive stages of the partial coalescence are captured with a high-speed camera. The shape evolution of the interface is relatively constant from one experiment to another [28].

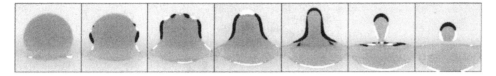

Figure 8. Different stages of a partial coalescence in the self-similar regime. Capillary waves are highlighted by black regions (receding interface) and white regions (progressing interface).

The film rupture occurs where the film is the thinnest. Since the surrounding fluid generates an overpressure (and thus a dimple) at the centre while it is drained outward, the thinnest point is off-centre [9, 31]. When the film is broken, it quickly opens due to high pressure gradients created by surface tension near the opening [32–35].

The emptying droplet takes a column shape, and then experiences a pinch-off that leads to the formation of the daughter droplet. Charles and Mason [19] have suggested that this pinch-off was due to a Rayleigh–Plateau instability. Recently, Blanchette and Bigioni [23] have shown that this hypothesis was wrong. They have solved Navier–Stokes equations for a coalescing water droplet surrounded by air. When the top of the droplet reaches its maximum height, they stop the numerical simulation, set velocities and pressure perturbations to zero, and restart the computation. The coalescence, normally partial, was observed to be total with this flow reset. The pinch-off cannot be due to a Rayleigh–Plateau instability.

During a coalescence, capillary waves are generated at the bottom of the droplet after the film breakdown [20, 21, 29]. A part of them propagates far away, on the planar interface. The other part climbs over the droplet and converges at the top. Those waves are visible in Fig. 8, where the progressing/receding feature of the interface is highlighted. According to Blanchette [23], such a convergence at the top greatly deforms the droplet and delays its coalescence: the horizontal collapse (the pinch-off) can occur before the emptying. Therefore, the pinch-off that enables partial coalescence seems to be mainly due to the capillary waves. The pinch-off usually occurs between $1.4\tau_\sigma$ and $2\tau_\sigma$ after the film breakage. After that, the fluid creates a vortex ring that goes down through the fluid 1 [36–38].

3. Influence of Gravity

For large droplets, gravity is known to be as important as surface tension ($Bo \sim 1$). Therefore, gravity significantly accelerates the emptying of the mother droplet [20, 23]. Moreover, it flattens the initial droplet. When viscosities are negligible ($Oh_{1,2} < 7.5 \cdot 10^{-3}$), the ratio Ψ is monotonically decreasing with Bo (Fig. 9), and it suggests that a critical Bond number Bo_c should exist for which $\Psi = 0$ for $Bo > Bo_c$: coalescence becomes total. Mohamed-Kassim and Longmire [29] have reported total coalescences for Bond numbers equal to 10. However, both Ohnesorge numbers were greater than 0.01, and viscosity played a significant part on the outcome of these coalescences. Until now, no experiment assesses about the existence of Bo_c for negligible viscosities.

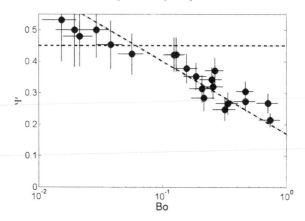

Figure 9. Semi-log plot of the radius ratio Ψ as a function of the Bond number Bo when $Oh_{1,2} < 0.0075$. Dashed lines are guides for the eyes.

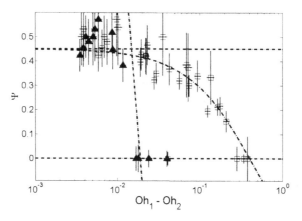

Figure 10. Semi-log plot of Ψ as a function of Oh_1 (triangles) and Oh_2 (squares) influences on Ψ, when $Bo < 0.1$ and the other Ohnesorge number is smaller than $7.5 \cdot 10^{-3}$. Dashed lines are guides for the eyes.

E. Influence of Viscosities

Viscosity forces in both fluids, mainly efficient for small droplets, inhibit the partial coalescence [20, 21, 23, 28]. The ratio Ψ also monotonically decreases with both increasing Ohnesorge numbers (Fig. 10). Critical Ohnesorge numbers $Oh_{1c} \sim 0.02$ and $Oh_{2c} \sim 0.3$ may be defined as the Ohnesorge numbers beyond which coalescence becomes total [20, 21, 23, 28]. Viscosity in fluid 2 leads to a smooth decreasing between the asymptotic regime $\Psi = 0.45$ and the total coalescence $\Psi = 0$. Surprisingly, the behaviour of Ψ with increasing Oh_1 is totally different: the decrease of Ψ is very sharp and premature. The critical Ohnesorge number is 15 times smaller in fluid 1 than in fluid 2.

Capillary waves are damped by viscosity, either inside or outside the droplet. This damping can be seen through the maximum height H_{max} that is reached by the

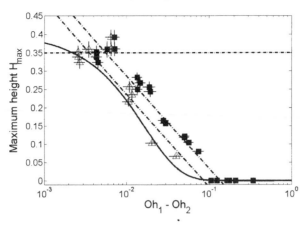

Figure 11. Maximum droplet elevation H_{max} *vs.* Oh_1 (triangles) and Oh_2 (squares). Dashed lines are guides for the eyes. The solid line corresponds to the forecast of the linear waves theory.

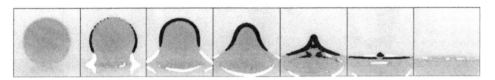

Figure 12. A total coalescence due to $Oh_2 > Oh_{2c}$. Capillary waves are fully damped and no convergence is observed.

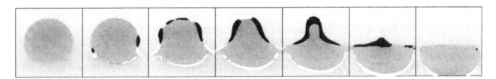

Figure 13. A total coalescence due to $Oh_1 > Oh_{1c}$. Capillary waves are still efficient, convergence is observed, but the vertical collapse is not delayed.

interface during the convergence (Fig. 11). Since the convergence effect is reduced by viscous damping, Blanchette [23] supposed that the capillary wave damping was responsible for the partial to total transition when Ohnesorge numbers are increased (Fig. 12). But it can be seen in Fig. 11 that both viscosities damp the waves in the same way: viscosity in fluid 1 is only 1.5 times more efficient than in fluid 2. This cannot explain the asymmetrical role played by Oh_1 and Oh_2.

Although the delay in vertical collapse due to the convergence of capillary waves is important to get a partial coalescence, it cannot be the only determinant factor. Indeed, as seen in the snapshot of Fig. 13, a coalescence with $Oh_1 > Oh_{1c}$ is total, despite the presence of capillary waves. On the other hand, in Fig. 14, a coalescence with $Oh_2 < Oh_{2c}$ is partial although the capillary waves are fully damped.

Therefore, capillary wave damping cannot be the only viscous mechanism responsible for the partial to total transition. Gilet *et al.* [28] recently proposed another

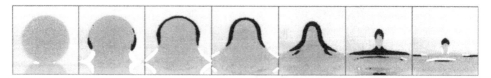

Figure 14. A partial coalescence with $Oh_2 \leqslant Oh_{2c}$, although capillary waves are fully damped and no convergence is observed.

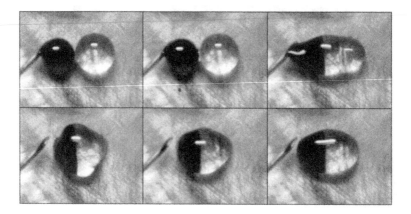

Figure 15. Two droplets with similar sizes coalesce together without mixing.

mechanism that is based on the repartition of the released interfacial energy between both fluids according to their ability to diffuse their kinetic energy. This hypothetical mechanism has still to be checked by PIV experiments.

F. Mixing Droplets Together

When two droplets made from miscible fluids collide with each other, they may coalesce to form a bigger droplet. As in the case of a droplet onto a planar interface, the process is not spontaneous since the film of surrounding fluid between both droplets has to be drained out. The binary coalescence of droplets together is particularly interesting on a microfluidic point of view. If reactive substances are put in each droplet, their coalescence triggers a chemical reaction in the resulting droplet [39]. The reaction kinetics is determined by the ability of both fluids to mix together. As seen in Fig. 15, this ability is really weak when droplets have a similar size: no mixing occurs, except by molecular diffusion. The reaction only occurs at the pseudo-interface between both fluids. The only way to enhance the reaction is to shake the droplet to create mixing flows inside.

The situation becomes interesting when two droplets with a different size coalesce. Indeed, due to a difference of Laplace overpressure, the small droplet fully empties into the large one instead of simply sticking next to it. During the emptying, vorticity is generated near the moving interface [30, 36–38]. As in the coalescence of a single droplet onto a planar interface, this vortex sheet may be able to separate

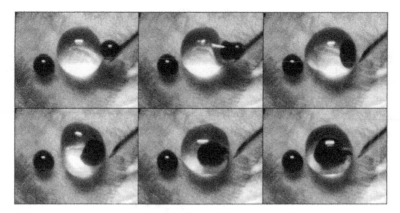

Figure 16. When two droplets with different sizes coalesce with each other, the smallest one empties into the biggest one, and creates a powerful vortex ring that efficiently mixes both fluids.

from the interface and to come deeply into the resulting droplet. The vortex ring completely mixes both liquids, as seen in Fig. 16. Conditions for this separation to occur are detailed in [38] for the planar case, but are not described yet for the binary coalescence. It seems that at least about 25% of relative difference in radius is needed for mixing.

G. Control of the Partial Coalescence with Bouncing

As seen in Section C, droplets are able to bounce indefinitely on a vertically vibrated bath. The air film between the droplet and the bath prevents the droplet from coalescence: it is constantly renewed by the vibration. The vertically vibrated bath is an excellent platform to set up microfluidic experiments. For example, it was recently shown [40] that partial coalescence can take place on this kind of bath under certain conditions. More precisely, a droplet partially coalesces when its viscosity is sufficiently weak (typically less than 5 cSt for a millimetric oil droplet). The Faraday acceleration threshold Γ_F is about 1 g for a forcing frequency of 100 Hz and a viscosity of 3 cSt: much less than the threshold for bouncing Γ_C. The presence of Faraday waves makes the bouncing chaotic and metastable; and this presence cannot be avoided for low-viscosity droplets on a low-viscosity bath. To observe partial coalescence, it is thus more convenient to use a high-viscosity liquid for the bath, in order to significantly increase the Faraday threshold. Typically, a droplet made of 0.65 cSt silicone oil coalesces partially on a 1000 cSt bath vibrating at 100 Hz. In such a system, the coalescing droplet forms a pool at the bath surface. The liquid of this pool cannot immediately mix with the surrounding oil. Consequently, the daughter droplet of the partial coalescence has unavoidably to make its firsts bounces on the pool formed by its own mother droplet. However, this does not significantly affect the bouncing.

As observed by Couder [13], the bouncing threshold Γ_C increases with the droplet radius. For a given acceleration, one expects that small droplets can bounce

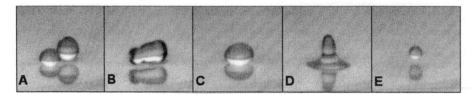

Figure 17. (A) Two droplets are bouncing close to each other on a vertically vibrated bath. (B) Suddenly, they coalesce together. (C) The resulting droplet is too large to bounce indefinitely, so (D) it partially coalesces. (E) The daughter droplet is sufficiently small to bounce.

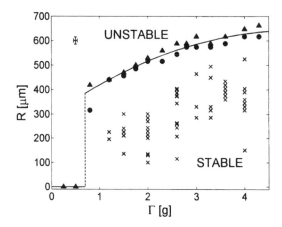

Figure 18. Critical size for bouncing as a function of the reduced vertical acceleration of the bath Γ. Droplets with radii below this size are able to bounce (stable region) while others cannot and have to empty by the way of a partial coalescence.

while large ones cannot. The partial coalescence is then stopped when the daughter droplets are sufficiently small to bounce. The whole system behaves like a low-pass filter that decreases the size of droplets until they reach a suitable size for bouncing. In Fig. 17, (A) two droplets are bouncing side by side. Suddenly, (B) they coalesce together, but (C) the resulting droplet is too large to bounce indefinitely: (D) it coalesces partially. (E) The daughter droplet is sufficiently small to bounce periodically.

The critical size that enables bouncing has been investigated in [40]. It is shown in Fig. 18 as a function of the reduced acceleration of the bath Γ. Since the curve saturates for large Γ, it seems impossible to stabilize larger droplets by increasing the acceleration further.

H. Conclusion

In this work, several methods have been described to manipulate a droplet without touching it. The basic idea is to lay the droplet on a vibrating bath of the same liquid or of a high viscosity liquid. Conditions for bouncing have been described and low frequency vibrations provide longer lifetime. The partial coalescence is a nice way to reduce the size of a droplet without using complex systems. The viscosity limitations have been exposed and discussed. In order to mix droplets, we have shown that the size difference is considered as the most relevant parameter as soon as no additional means are requested. Finally, by combining the bouncing effect and the partial coalescence, the size of the droplets may be automatically selected.

I. Acknowledgements

TG and SD thank FNRS/FRIA for financial support. Part of this work has been supported by Colgate–Palmolive (G. Broze). H. Caps is acknowledged for fruitful discussions. J.P. Lecomte (Dow Corning, Seneffe) is also thanked for the support and for providing silicone oils. This work has been supported by the COST action P21 "Physics of droplets".

J. References

1. H.A. Stone, A.D. Stroock and A. Ajdari, Annu. Rev. Fluid Mech., 36 (2004) 381.
2. N. Vandewalle, D. Terwagne, K. Mulleners, T. Gilet and S. Dorbolo, Phys. Fluids, 18 (2006) 091106.
3. W. Gonzalez-Vinas and J. Salan, Eur. Lett., 41 (1998) 159.
4. Y. Couder and E. Fort, Phys. Rev. Lett., 97 (2006) 154101.
5. A. Eddi, E. Fort, F. Moisy and Y. Couder, Phys. Rev. Lett., 102 (2009) 240401.
6. O. Reynolds, Philos. Trans. R. Soc. London Ser. A, 177 (1886) 157.
7. J. Israelachvili, Intermolecular and Surface Forces, Amsterdam, Academic Press, 2003.
8. S. Dorbolo, E. Reyssat, N. Vandewalle and D. Quéré, Europhys. Lett., 69 (2005) 966.
9. A.F. Jones and S.D.R. Wilson, J. Fluid Mech., 87 (1978) 263.
10. I.B. Ivanov and T.T. Traykov, Int. J. Multiphase Flow, 2 (1976) 397.
11. G.P. Neitzel and P. Dell'Aversana, Annu. Rev. Fluid Mech., 34 (2002) 267.
12. J.S. Eow and M. Ghadiri, Colloids Surf. A, 215 (2003) 101.
13. Y. Couder, E. Fort, A. Boudaoud and C.H. Gautier, Phys. Rev. Lett., 94 (2005) 177801.
14. T. Gilet, D. Terwagne, N. Vandewalle and S. Dorbolo, Phys. Rev. Lett., 100 (2008) 167802 (4 pages).
15. S. Dorbolo, D. Terwagne, N. Vandewalle and T. Gilet, N. J. of Phys., 10 (2008) 113021 (9 pages).
16. D. Terwagne, T. Gilet, N. Vandewalle and S. Dorbolo, Phys. of Fluids, 21 (2009) 054103 (5 pages).
17. D. Terwagne, N. Vandewalle and S. Dorbolo, Phys. Rev. E, 76 (2007) 056311 (6 pages).
18. L.D. Mahajan, Phil. Mag., 10 (1930) 383.
19. G.E. Charles and S.G. Mason, J. Colloid Sci., 15 (1960) 105.
20. Y. Leblanc, Etude des mécanismes de la coalescence partielle d'une goutte à une interface liquide–liquide (Université Paris VII, 1993).
21. S.T. Thoroddsen and K. Takehara, Phys. Fluids, 12 (2000) 1265.
22. P. Pikhitsa and A. Tsargorodskaya, Colloids Surf. A, 167 (2000) 287.
23. F. Blanchette and T.P. Bigioni, Nat. Phys., 2 (2006) 1.
24. E.M. Honey and H.P. Kavehpour, Phys. Rev. E, 73 (2006) 027301.
25. H. Aryafar and H.P. Kavehpour, Phys. Fluids, 18 (2006) 072105.
26. X. Chen, S. Mandre and J.J. Feng, Phys. Fluids, 18 (2006) 051705.

27. P. Yue, C. Zhou and J.J. Feng, Phys. Fluids, 18 (2006) 102102.
28. T. Gilet, K. Mulleners, J.P. Lecomte, N. Vandewalle and S. Dorbolo, Phys. Rev. E, 75 (2007) 036303.
29. Z. Mohamed-Kassim and E.K. Longmire, Phys. Fluids, 16 (2004) 2170.
30. B.S. Dooley, A.E. Warncke, M. Gharib and G. Tryggvason, Exp. Fluids, 22 (1997) 369.
31. S. Hartland, Trans. Inst. Chem. Eng., 45 (1967) T109.
32. J. Eggers, J.R. Lister and H.A. Stone, J. Fluid Mech., 401 (1999) 293.
33. A. Menchaca-Rocha, A. Martinez-Davalos and R. Nunez, Phys. Rev. E, 63 (2001) 046309.
34. M. Wu, T. Cubaud and C.-M. Ho, Phys. Fluids, 16 (2004) L51.
35. S.T. Thoroddsen, K. Takehara and T.G. Etoh, J. Fluid Mech., 527 (2005) 85.
36. A.V. Anilkumar, C.P. Lee and T.G. Wang, Phys. Fluids A, 3 (1991) 2587.
37. P.N. Shankar and M. Kumar, Phys. Fluids, 7 (1995) 737.
38. R.W. Cresswell and B.R. Morton, Phys. Fluids, 7 (1995) 1363.
39. H. Song, D.L. Chen and R.F. Ismagilov, Angew. Chem. Int. Ed., 45 (2006) 7336.
40. T. Gilet, N. Vandewalle and S. Dorbolo, Phys. Rev. E 76 (2007), 035302R (4 pages).

Interfacial Mass Transfer of Growing Drops in Liquid–Liquid Systems

A. Javadi [a,b], **D. Bastani** [a], **M. Taeibi-Rahni** [a], **M. Karbaschi** [a,b] and **R. Miller** [b]

[a] Sharif University of Technology, P.O. box 11365, Teheran, Iran
[b] Max Planck Institute of Colloids and Interfaces, D-14476 Potsdam/Golm (Germany)

Contents

A. Introduction

Regarding technological applications drops and bubbles are dispersed phases of a fluid in a second immiscible fluid phase used for material transportation, mixing, extraction, phase change, etc., with or without chemical reactions. In all these processes the interfacial area between the two phases and the transport properties of solute components in the bulk and at the interface are the major parameters to be discussed. For the integral values of these parameters, elementary knowledge on the single drops or bubbles is essential. Any change in the fluids' properties (chemical composition, density, viscosity, etc.) and solute components, can affect the mentioned parameters significantly. According to the numerous possibilities of different fluids and solute components, it is realistic to speak about a countless number of different liquid systems in fundamental investigations and technological applications. However, all available analytical solutions to theoretical models for describing such systems contain respective simplifications which restrict their applicability. On the other hand, numerical techniques and computational approaches like CFD (computational fluid dynamic) and MD (molecular dynamic) simulations are available but still insufficient for many practical applications.

The development of experimental techniques also is a great challenge for investigations in this filed. It is not easy to penetrate into the bulk of a liquid and catch interfacial information with nanometre scale accuracy (order of molecular size) without a significant perturbation of the bulk and interface. Any experimental tools with minimum perturbations like optical based techniques are also significantly affected by properties of the bulk which are not easy to be separated from the real properties of the interface. The overlap of data analysis problems with the limitations of experimental technique creates many big challenges for researchers in this field. This is why hundred years after the basic studies of Lord Rayleigh and other giants in science who tried to lay down the foundations of fluid dynamics of drops in the 19th century, this subject is still of great interest in the 21st century [1]. The fascination that drops create and their enormous technological usefulness are no doubt some of the reasons why the subject is yet up to date.

One of the most important research areas in this field is the formation of droplets in liquid–liquid systems because of the enormous number of applications such as extraction technology in petrochemical and metallurgical processes, pharmaceutical productions, food processing technologies, particle production, microencapsulation and generation of emulsions. In these systems, when a liquid is injected into a second liquid at low and intermediate velocities (dripping regime), drops are formed, detach and break off from the nozzle or pore. At a velocity above a certain critical value (the jetting velocity), a jet with a certain length is formed and then it breaks up into drops. In this chapter the dripping regime is discussed, mostly. There are generally three distinct stages in the lifetime of a drop: formation, free-fall, and coalescence. Although there are intensive experimental, theoretical and numerical studies on the dynamics of droplet formation in liquid–liquid systems, still many complex problems remain to be solved. Evaluation of the mass transfer during droplet formation is a rather complex process and rather little research has been done on this subject.

In this chapter different approaches and models for evaluation of interfacial mass transfer on growing drops in liquid–liquid systems are discussed. As the interfacial mass transfer rate during drop growth is affected by convective flow and hydrodynamics aspects, we review first the droplet formation and related analytical and modelling approaches. Then different approaches and models for estimating the in-

terfacial mass transfer are discussed. Finally different experimental approaches for
the evaluation of mass transfer during drop formation are presented.

B. Hydrodynamics of Drop Formation

Efforts for understanding and modeling the hydrodynamics of droplet formation on
a nozzle have a long history that can be divided into different levels [2] which are
discussed in this paragraph.

1. Basic Explanation of Drop Formation

Tate [3], and Bashforth and Adams [4] were among the first who presented basic explanations on the formation of drops and proposed possible ideas how to use drops
for the calculation of the surface tension. The theoretical basis of the drop volume
method for measuring the surface tension of a liquid or the interfacial tension between two liquids was founded at the beginning of the last century by Lohnstein [5].
He developed the basic understanding of this method and presented his criticism on
the so-called law of Tate. This law of Tate is actually the basis of the stalagmometer method. Lohnstein made detailed calculations of the volume of detaching and
residual drops as a function of the capillary radius and the capillary constant and
found systematic deviations from the law of Tate. Later Harkins and Humphrey [6]
and Harkins and Brown [7] derived expressions for correlation factors for accurate
calculations of the surface tension from measured drop volumes at negligibly small
flow rates. This target was further developed by others [8, 9], who also considered
additional effects such as viscosity of the continuous phase, nozzle diameter, and
flow rate of the dispersed phase and introduced dimensionless groups such as Weber, Froude and Laplace numbers. For a detailed description and application of the
drop volume tensiometry please see Chapter 6.

2. Development of Two-Stage Models of Droplet Formation

This idea of drop formation in two steps was developed by various authors [10–16].
Rao *et al.* [10] were possibly the first who presented this strategy by referring to the
work of Hayworth and Treybal [8] who did not exactly defined the instance at which
the respective forces act. Thus, they proposed a two-stage approach for incorporating better the hydrodynamic effects during the droplet formation process. In the first
stage (static stage), the drop is assumed to expand until the buoyancy/gravity force
balances the interfacial tension force. At the end of the static stage the drop volume
is given by the equation of Harkins and Brown [7]. During the second stage, when
the drop is detaching from the nozzle, the drop continues to grow.

   A more sophisticated model for drop formation at low velocities in liquid–liquid
systems was proposed by Scheele and Meister [11, 17–19], which reduces some
shortcomings of the model by Rao *et al.* [10] in particular the prediction of drop
volumes smaller than those given by the Harkins and Brown analysis. They developed the understanding of liquid jet and single drop formation regimes on the basis

of experiments with 15 different liquid–liquid systems. The transition from a single drop to the jet regime is explained as follows. At low flow rates drops start to form, grow, neck and break off from the nozzle at regular intervals. Above a certain critical velocity, a jet is formed up to a certain distance from the nozzle where it then breaks up into drops. For even higher nozzle velocities, the jet breaks up into small drops.

Heertjes *et al.* [12, 13] improved the analysis of Scheele and Meister by taking into consideration kinetic and interfacial tension forces, and also defined the applicability range for the Harkins and Brown correction factors which were determined under static (infinitely slow drop formation) conditions. Using a detailed two-stage drop formation model they incorporate essential variables into the second stage, i.e. the forces acting upon a drop, the way in which the dispersed phase enters the drop, the necking of the drop as a function of time, and finally the velocity of necking. This all led to a relationship for the leading-edge velocity, which they compared with experimental data (isobutanol, ethylacetate drops in water–glycerol mixtures) and a good correlation within an accuracy of ca. 15% for all flow rates below the jetting velocity was obtained [13].

Special correlations based on this two-stage drop formation concept have been also by other authors, for example in [15] to obtain correct surface tensions from drop volume experiment. Barhate *et al.* [16] reported that models proposed in literature for predicting the drop volume of aqueous/organic liquid–liquid systems are successful only under restricted conditions. Therefore, they developed a two-stage model starting with the analysis of Rao *et al.* [10] and incorporated the Hadamard–Rybzicynski equation for the drag force with internal circulation. Their results showed some improvement for a number of systems but at higher flow rates, neither model predicted the measured drop volumes satisfactorily.

3. *Computational Fluid Dynamics (CFD) Simulations*

A perfect analysis of the drop formation process should involve the solution of the full Navier–Stokes equation along with the proper boundary conditions. However, the mathematical complexities of the equations, considering the existence of a free boundary two-phase flow, did not allow until recently a solution of this problem. Now numerical simulation methods are available which can provide the solution of the full Navier–Stokes equations [20, 21]. Richards *et al.* [22, 23] is obviously the first who applied direct numerical simulation with the Volume of Fluid (VOF) method to simulate the drop formation in liquid–liquid systems before and after jetting conditions. Good agreement with experiments demonstrated the applicability of direct numerical simulations for these problems [18, 19].

A similar numerical simulation method (SOLA-VOF) along with experimental studies on the analysis of the single drop formation process under pressure pulse condition for toluene drops formed in water were published by Ohta and Suzuki [24]. Zhang [25] presented another numerical VOF/CSF method (Continuum Surface Force) along with experimental studies of the drop formation dynamics of 2-ethyl-1-hexanol/distilled water. This work is one of the best recent numerical/experimental studies on drop formation problem in liquid–liquid systems, based on the solution of the transient Navier–Stokes equation for the axisymmetric free-boundary problem of a Newtonian liquid. Their numerical simulations showed an excellent agreement with the experimental data. The effects of hydrodynamic characteristic on the mass transfer during drop formation in liquid–liquid system performed by Javadi *et al.* [27] is another CFD simulation work with a similar approach which will be discussed further below.

Homma *et al.* [28] reported that the VOF method can deal robustly with jet interfaces, especially for the pinching off or breakup into drops. The pinching off behavior may however depend strongly upon both the resolution of grids and the discretization scheme for the advection of the volume of fraction function (F) which determined the interface position, because F-values are assigned to each cell and the equation is hyperbolic. Although the interfacial tension effects are very important in the liquid–liquid jet issuing problem, the curvature and the normal direction, which are essential to determine the interfacial tension, may be inaccurate because the F-value distribution does not provide the interfacial shape explicitly.

Homma *et al.* [28–30] also simulated the formation of a liquid jet in a second liquid and its breakup into drops using a front tracking/finite difference method. Their simulation results indicated a good agreement with experimental data. They examined how the jet and drop shape dynamics changes with the Reynolds, Weber and Froude numbers and viscosity ratios. They also reported that numerical simulations with the font-tracking method is applicable to jet problems in a wide range of conditions as well as industrial applications.

C. Interfacial Mass Transfer of Growing Drops

Although there are vast experimental, theoretical and numerical studies on the dynamics of droplet formation, the evaluation of the mass transfer during droplet formation is rather complex and there is little research on it. The instabilities associated with mass transfer during drop formation and detachments are well documented experimentally though without significant theoretical explanation [31]. Furthermore, measurements of mass transfer during drop formation involves many operational difficulties because it generally occurs at short times (e.g. 0.2–20 s) and small lengths scales (e.g. 1–20 mm^3) at hardly accessible conditions (tip of submerged nozzle) which creates difficulties for measurement [2]. Consequently, modeling and simulations of this phenomenon are rather complicated due to the variable, flexible, permeable interfacial free boundary. In the present section different analytical solutions and modelling approaches are presented.

1. Analytical Solutions and Modeling Approaches

The history and trend of understanding and modeling of mass transfer during drop formation can be summarized at the following five levels.

1.1. Early Experimental and Basic Studies

Early efforts have been started many years ago for example by Whitman *et al.* [32] who investigated the absorption of CO_2 by water drops in a small column of constant height. They indicated three distinct stages in the life-time of drops or bubbles: formation stage, free fall stage, and coalescence stage. They varied the time of formation and by extrapolation to zero formation time, estimated the amount of gas absorbed during the free fall of the drop. Assuming this to be constant they calculated the absorption during the formation by subtracting this constant value from the total amount of mass transfer. More than 20 years later the concept of three stages was more clearly presented and analyzed experimentally in several investigations [33–37].

The first experimental work for this phenomenon in a liquid–liquid system involved trains of drops reported by Coulson and Skinner [38] who used a different method for measuring mass transfer during drop formation. They formed drops on the tip of a nozzle, suspending them in the continuous phase, and let them disappear back into the same nozzle immediately after formation. They reported that the mass transfer fraction end effects (formation and withdraw) have an amount of about 40% of the total transfer for the 33 cm column and the portion of the formation stage is nearly half of this amount. They reported that extraction efficiency increased slowly with drop formation time (t_f), but for values $t_f > 5$, the increase was very small, and therefore the rate of extraction (mass transfer coefficient) decreased very rapidly. Their results for the mass transfer coefficient in the benzene-benzoic acid–water system showed $K_1 = k_0 \times t_f^{-0.7}$ in comparison with the results from Higbie's penetration theory which indicates $K_1 = 2\sqrt{D/\pi t}$, where k_0 is the constant of model. This comparison demonstrated the effects of the convection during the growth of the free droplet boundary on mass transfer rate in comparison with a pure diffusion interfacial boundary. It will be shown in the next section that the early diffusion-based models generally had this drawback.

1.2. Early General Diffusion Based Models

The transport equation for a component from a droplet (bubble) to a continuous phase can be written as below:

$$\frac{\partial C}{\partial t} + U_r \frac{\partial C}{\partial r} = D\left(\frac{\partial^2 C}{\partial r^2} + \frac{2}{r}\frac{\partial C}{\partial r}\right), \tag{1}$$

$$C = C_0, \qquad t = 0, \qquad r > R, \tag{1a}$$

$$C = C_S, \qquad t > 0, \qquad r = R, \tag{1b}$$

$$C = C_0, \qquad t > 0, \qquad r \to \infty. \tag{1c}$$

Different simplifications, such as neglecting the radial velocity caused by droplet surface growth, assuming a small diffusion depth in comparison with the droplet size, and constant interfacial concentration on the droplet surface, have been applied to reduce the foregoing equation to the following Cartesian 1D unsteady diffusion equation, which is known as Higbie's penetration theory derived for a gas absorption problem at a surface on a deep liquid bulk.

$$\frac{\partial C}{\partial t} = D\frac{\partial^2 C}{\partial x^2}. \tag{2}$$

The analytical solution of this equation using a variable combination technique and considering the foregoing boundary conditions leads to the following result:

$$N_t = 2(\Delta C)\sqrt{\frac{Dt}{\pi}} = K_L t \Delta C \tag{3}$$

in which $K_L = 2\sqrt{D/\pi t}$ is an overall mass transfer coefficient for this problem. Nearly all of the early analytical models for predicting the mass transfer during droplet formation are based on Higbie's penetration theory with different assumption on the hydrodynamics of the process. Most proposed diffusion-based models consider either a surface stretching [39–41] or fresh surface elements [42, 43], or considered both mechanisms [44]. The difference among these models concern the way in which the surface of the growing drop is refreshed. The first mechanism is based on surface stretching where all elements remain at the surface and are stretched as the drop grows. The second mechanism describes the growth of a drop by adding fresh elements to the surface. These and the other approaches have been summarized by Popovich *et al.* [45] in Table 1. In order to facilitate a comparison among different models, they considered a common general function describing the variation of the drop or bubble area with time.

$$A = at^n \tag{4}$$

in which the constants a and n depend on the utilized hydrodynamic models. Combination of this with Eq. (3) the following relationship is obtained for the estimation of the instantaneous mass transfer rate in which the diffusion depth was considered to be negligible in comparison with the droplet size.

$$N_A' = (C_S - C_0)(A(t))\sqrt{\frac{D}{\pi t}}, \tag{5}$$

$$N_A' = (C_S - C_0)k\sqrt{\frac{D}{\pi}}(t)^{[(2n-1)/2]}. \tag{6}$$

Therefore the total amount of mass transfer during droplet formation at time t reads:

$$N_A = \int_0^t N_A' \, dt = \frac{2}{2n+1}k(C_S - C_0)\sqrt{\frac{D}{\pi}}(t)^{[(2n+1)/2]}. \tag{7}$$

This equation is known as Licht solution for the current problem [34]. Heertjes *et al.* [46] assumed that the velocity of diffusion is small in comparison to the velocity of drop growth. Therefore, instead of Eq. (5), the following relation should be applied:

$$N'_A = (C_S - C_0)\sqrt{\frac{D}{\pi}} \int_{A(t=0)}^{A(t=t)} \frac{dA}{\sqrt{t}}. \tag{8}$$

From Eq. (4) $dA = nkt^{n-1} dt$, therefore:

$$N'_A = \frac{2n}{2n-1}(C_S - C_0)k\sqrt{\frac{D}{\pi}} t^{[(2n-1)/2]}, \tag{9}$$

$$N_A = \frac{4n}{4n^2-1}k(C_S - C_0)\sqrt{\frac{D}{\pi}} t^{[(2n+1)/2]}. \tag{10}$$

Groothius and Kramers [42] assumed that any area increase during drop formation produces fresh surface and that there is no mixing between areas of different ages. Thus, if an area $dA(\theta)$ is formed at $t_1 < t < t_1 + dt$, the rate of mass transfer through that area at time t_1 is given by

$$N'_A(t) = (C_S - C_0)\sqrt{\frac{D}{\pi}} \frac{dA(t)}{\sqrt{t_1-t}}. \tag{11}$$

Using equation (4), and assuming the volume of droplet at time t equal to $V = Qt$, in which Q is the flow rate through the nozzle tip,

$$N'_A = (C_S - C_0)\sqrt{\left(\frac{D}{\pi}\right)}(nk) \int_0^{t_1} \frac{t^{n-1} dt}{\sqrt{t_1}\sqrt{1-t/t_1}}. \tag{12}$$

Via the definition of $y = \sqrt{(1 - t/t_1)}$,

$$N'_A = 2n \int_0^1 (1 - y^2)^{n-1} dy \,(C_S - C_0)\sqrt{\frac{D}{\pi}}(k)t_1^{[(2n-1)/2]}, \tag{13}$$

$$N_A = \frac{4n}{2n+1} \int_0^1 (1 - y^2)^{n-1} dy \,(C_S - C_0) \times \sqrt{\left(\frac{D}{\pi}\right)}kt^{[(2n+1)/2]}. \tag{14}$$

The integral part in Eqs (13) and (14) are constant for constant n. Popovich *et al.* [45] indicated that for these models n is a constant for a given assumed area variation with time, so that the rate and cumulative mass transfer for these diffusion-based models can be written in the following general form:

$$N'_A = const\, k(C_S - C_0)\sqrt{\frac{D}{\pi}} t^{[(2n-1)/2]}, \tag{15}$$

$$N_A = const\, k(C_S - C_0)\sqrt{\frac{D}{\pi}} t^{[(2n+1)/2]}. \tag{16}$$

When the drops are assumed to be a sphere and growing at a uniform rate $k = \pi d_f^2 t_f^{-2/3}$, then $A = \pi d_f^2 t_f^{-2/3} t^{2/3}$, and the following relationship for the prediction

Table 1.
Different values of constants for different diffusion-based drop formation mass transfer models [45]

	Assumed mechanism	Constant for N'_A	Constant for N_A
Licht (1953)	The whole area is aged according to the unsteady diffusion theory. Only area variation with time is taken into account	1.0	$6/7 = 0.857$
Groothius (1955)	Any area increase during drop formation produces real "fresh" surface. There is no mixing between areas of different ages	1.77	1.52
Ilkovic (1934, 1938)	The interface movement influences the diffusion layer thickness, stretching it evenly around the sphere	$\sqrt{7/3} = 1.53$	$\sqrt{12/7} = 1.31$
Micheles (1960)	The growing spherical drop advances into the fluid with its centre moving. The diffusion layer is stretched by the moving interface. The largest stretching is at the front of the sphere, the smallest at the rear	$0.495\sqrt{(7 + 6\cos\omega)}$ $\cos\omega = 1$ front $\cos\omega = 0$ equator $\cos\omega = -1$ rear	1.78 1.31 0.495
Heertjes (1954)	The velocity of diffusion is small compared with the velocity of drop growth	4.0	$24/7 = 3.43$

of instantaneous (N'_A) and cumulative (N_A) amount of mass transfer during droplet formation are obtained:

$$N'_A = const(C_S - C_0)\sqrt{D\pi}(d_f^2 t_f^{-2/3} t^{1/6}), \qquad (17)$$

$$N_A = const(C_S - C_0)\sqrt{D\pi}(d_f^2 t_f^{-2/3} t^{7/6}), \qquad (18)$$

where C_S is the saturation concentration, C_0 is the initial concentration, d_f is the final equivalent spherical diameter, t_f is the total drop formation time. Popovich *et al.* [45] used an innovative formation-withdrawal technique for evaluating the drop formation mass transfer, and the predicted values for the constant in Eq. (18) were 1.31 for the surface stretch assumption and 1.52 for the fresh surface elements assumption (see Table 1). They examined the various models *via* Eq. (17) and found that the surface-stretch approach (Licht model) fits their results best (Fig. 1). However, Heertjes and DeNie [44] applied a short column to minimize mass transfer after drop formation and obtained results more consistent with the fresh-surface model calculation.

Although the summarized diffusion-based models are in quite good agreement with applied experimental data, for fast drop formation conditions an internal drop circulation can set in and strongly influences the mass transfer. The existence of significant convection terms and internal circulations are the most important reasons for the failure of the diffusion-based models for fast drop formation studies (Fig. 2).

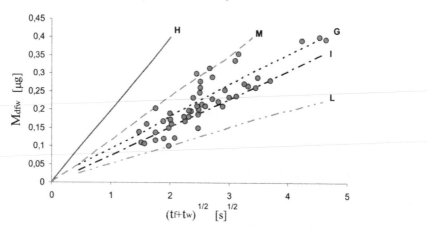

Figure 1. Capability of different diffusion-based drop formation mass transfer models in comparison with results from [45].

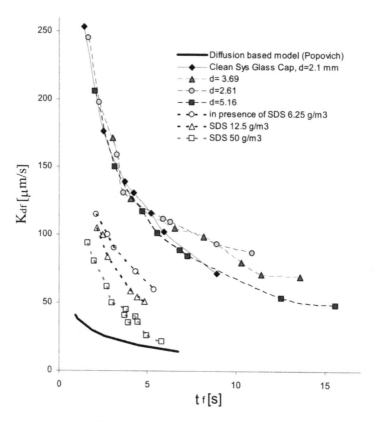

Figure 2. Shortcomings of diffusion-based drop formation mass transfer models (general form by [45]) for prediction of fast formation cases; according to [55].

1.3. Development of Convection Incorporated Models

The need for developing improved drop formation mass transfer models which integrate the convection and internal circulation with molecular diffusion, has been demonstrated in several investigations [41, 46–52, 54, 55]. These efforts can be divided into two main branches: (1) diffusion-based convection incorporated models, (2) dimensional analysis-based convection incorporated models.

The first type of models is precise regarding the analysis based on the transport equation but include simplifications which do not allow an accurate understanding of the fast drop formation conditions. The second groups can be divided into two sub-branches: straightforward dimensional analysis approach and physical-based dimensionless modelling, as discussed in the next sections.

1.4. Diffusion-Based Convection Incorporated Models

One of the first attempts for abandoning the assumption of negligible convective flow in the drop formation mass transfer models is the work by Angelo *et al.* [41], but they kept other simplifications like planar boundary condition and constant driving force (continuous phase concentration). Therefore Eq. (1) was reduced to

$$\frac{\partial C}{\partial t} + U_x \frac{\partial C}{\partial x} = D\left(\frac{\partial^2 C}{\partial x^2}\right). \tag{19}$$

The following result for a constant flow rate was obtained

$$K = (7D/3\pi t_f)^{1/2}, \qquad E_{df} = \frac{36}{\sqrt{21\pi}} \frac{D^{1/2} t_f^{1/2}}{Hd} \tag{20}$$

in which H is the distribution coefficient for equilibrium condition between the dispersed phase and continuous phase, and E_{df} is the fraction of the extraction efficiency during drop formation defined by $E_{df} = (C_{0in} - C_{0out})/(C_{0in} - C_0^*)$, and C_0 is the organic phase solute concentration with the initial concentration in the water $C_0^* = 0.0$.

Walia and Vir [50, 51] have been one of the first investigators who presented an improved diffusion-based model, taking into account the effect of the boundary curvature, convective flow around a forming drop, and the time-dependent concentration change in the continuous phase near the droplet. They assumed that the concentration boundary layer around the drop of thickness x is small in comparison with the radius R, so that $r = R + x$ and a magnitude analysis technique yields

$$\frac{\partial C}{\partial t} - \left(\frac{2x}{3t} - \frac{x^2 t^{-4/3}}{M}\right)\frac{\partial C}{\partial x} = D\left(\frac{\partial^2 C}{\partial x^2} + \frac{2t^{1/3}}{M}\frac{\partial C}{\partial x}\right) \tag{21}$$

with $M = (3Q/4\pi)^{1/3}$. Assuming that the second terms in brackets on both sides of Eq. (21) are small, the solution may be written as $C = C_1 + C_2$, where C_1 is the first approximation, and is obtained by solving the equation

$$\frac{\partial C_1}{\partial t} - \frac{2x}{3t}\frac{\partial C_1}{\partial x} = D\frac{\partial^2 C_1}{\partial x^2} \tag{22}$$

and C_2 is a small correction to this approximation, and was obtained using the following equation:

$$\frac{\partial C_2}{\partial t} - \frac{2x}{3t}\frac{\partial C_2}{\partial x} - D\frac{\partial^2 C_2}{\partial x^2} = 2D\frac{t^{-1/3}}{M}\frac{\partial C_1}{\partial x} - \frac{x^2}{M}t^{-4/3}\frac{\partial C_1}{\partial x}. \tag{23}$$

By substituting $X = xt^{2/3}$, $T = 3/7Dt^{7/3}$, Eq. (22) can be solved as a standard differential equation as below:

$$C_1 = C_0 + (C_0 - C_e)erf(X/2T^{1/2}). \tag{24}$$

For constant equilibrium concentration C_e, C_2 can be obtained from Eq. (24) in which C_1 is inserted into Eq. (23), which gives the following result [50, 51]

$$K = \left(\frac{7D}{3\pi t_f}\right)^{1/2}\left(1 + 2\frac{D^{1/2}t_f^{1/2}}{d}\right),$$

$$E_{df} = \frac{36}{\sqrt{21\pi}}\frac{D^{1/2}t_f^{1/2}}{Hd}\left(1 + \frac{\pi^{1/2}D^{1/2}t_f^{1/2}}{d}\right). \tag{25}$$

While negligible convective flow and planar boundary layer where not assumed for obtaining Eq. (25), a constant concentration within the drop is still assumed. For considering the effect of varying equilibrium concentration due to decreasing concentration within the drop (C'), Walia and Vir used the following material balance for solute over an infinitesimal time dt for zero initial concentration of solute in the continuity condition.

$$QC'dt + Qt\,dC' = QC_0'\,dt - KSC'\,dt/H. \tag{26}$$

Using the value of K from the mentioned equation, and assuming $S = (36\pi Q^2 t^2)^{1/3}$ for a spherical drop growth process at a constant flow rate Q through the nozzle tip,

$$C' + t\frac{dC'}{dt} + \omega(\alpha t^{1/6} + \beta t^{1/3})C' = C_0', \tag{27}$$

where $\omega = (36\pi/QH^3)^{1/3}$, $\alpha = (7D/3\pi)1/2$, $\beta = 2D^{1/2}/(6Q/\pi)^{1/3}$. The following solution was obtained:

$$C' = \sum_{n=0}^{\infty}\sum_{n'=0}^{\infty} a_{n.n'}\alpha^n\beta^{n'}C_0't^{(n/6+n'/3)}, \tag{28}$$

$$E = -1 + \sum_{n=0}^{\infty}\sum_{n'=0}^{\infty} a_{n.n'}\left(\frac{5.2}{H}\frac{D^{1/2}t^{1/2}}{d}\right)^n\left(\frac{10.2}{H}\frac{Dt}{d^2}\right)^{n'}, \tag{29}$$

where

$$a_{n.n'} = -\frac{a_{n-1.n'} + a_{n.n'-1}}{1 + n/6 + n'/3} \quad \text{for } n > 0 \text{ and/or } n' > 0,$$

$$a_{n.n'} = 1 \quad \text{for } n = 0 \text{ and } n' = 0, \tag{30}$$

$$a_{n.n'} = 0 \quad \text{for } n < 0 \text{ and/or } n' < 0.$$

Due to Walia and Vir [50, 51] the Eqs (29) and (30) can be used to predict the fraction of solute extracted by a drop without making any earlier mentioned simplifications. In addition a generalized equation was developed to obtain corrections for other existing equations. For this purpose they indicated that all equations reported in the literature for instantaneous mass transfer coefficient during drop or bubble formation, may, in general, be expressed as:

$$K = k_0 t^{-1/2}. \tag{31}$$

Note, other new models such as the flow expansion approach given in [2, 27] cannot be expressed in this general form.

Equation (27) may be rewritten as

$$C' + t\frac{dC'}{dt} + \omega k_0 C' t^{1/6} = C_0' \tag{32}$$

from which Walia and Vir [50, 51] obtained the following generalization for all cases of constant volume-growth rate of drop formation processes

$$E = \frac{p}{7} - \frac{7p^2}{7 \cdot 8} + \frac{49p^3}{7 \cdot 8 \cdot 9} - \cdots, \tag{33}$$

where $p = 36k_0 t^{1/2}/Hd$.

Their results for low extractions (less than 1% in all cases) in which the effect of changing concentration within the dispersed phase (droplet) and of the planar boundary layer was negligible, showed nearly the same values as expected. Hence, the only equations which proposes values relatively close to experimental data is the new equation by Walia and Vir, in addition to the Groothuis–Kramers and Ilkovic–Baird equations. In contrast, the values obtained by Licht–Pansing and Heertjes–Holve–Talsma equations show quite large deviations (Fig. 3). For larger extraction values, only the series type of solutions of Eq. (33) can explain the experimental results reasonably well (Fig. 4). Note, however, it becomes clear that even after correction the Ilkovic–Baird equation is valid only for sufficiently small values of $D^{1/2}t^{1/2}/d$.

1.5. Dimensional Analysis-Based Convection Incorporated Models

Despite to mentioned capability of the Walia–Vir model [50, 51] for accounting the effects of the convective term on the mass transfer rate during droplet formation; their model was incapable of describing the fast formation conditions required to explain the experimental data of Liang and Slater [55]. Therefore, they developed a circulation/diffusion model which fits the data quite well. For fast drop formation conditions few more investigations are known. In a photographic study, Rajan and Heideger [47] reported a significant effect of internal convection on the measured mass transfer coefficient though the mass transfer resistance acting mainly in the continuous phase. Their results showed that the surface stretch model and fresh surface elements both poorly predict the total mass transfer during drop formation for fast formation rates. Also their results demonstrated a high mass transfer rate at

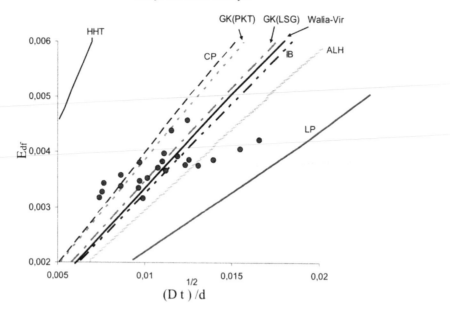

Figure 3. Results of the model by Walia and Vir and comparison with experimental data [50, 51] for low values of extraction fraction (less than 1% in all cases) and results of others models (non-series type). L–P: Licht–Pansing; H–H–T: Heertjes–Holve–Talsma; G–K(PJT): Groothuis–Kramers equation as modified by Popovich–Jeris Trass; G–K(LSG): Groothuis–Kramers for linear surface grow rate; I–B: Ilkovic–Baird; C–P: Calderbank–Patra; A–L–H: Angelo–Lightfoot–Howard.

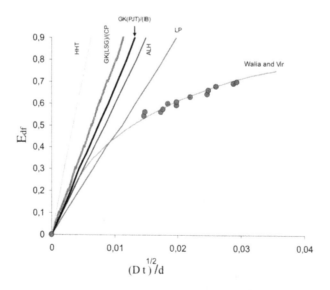

Figure 4. Results of the model by Walia and Vir and comparison with experiments given in [50, 51] for large values of extraction fraction and with results of others' available models (non-series type). L–P: Licht–Pansing; H–H–T: Heertjes–Holve–Talsma; G–K(PJT): Groothuis–Kramers equation as modified by Popovich–Jeris Trass; G–K(LSG): Groothuis–Kramers for linear surface grow rate; I–B: Ilkovic–Baird; C–P: Calderbank–Patra; A–L–H: Angelo–Lightfoot–Howard.

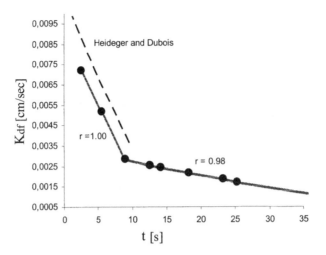

Figure 5. Sharp break in the formation time at about 7.5 s in the correlation of the average mass transfer coefficient with formation time, dispersed phase controlled; according to [52].

the first stages of formation time which is discussed in more detail later. Heideger and Wright [52] examined the mass transfer coefficient over a wide range of formation times such that internal drop convection was important for rapid formation rates and relatively insignificant for slow rates. They reported significant effects of internal convection on the measured mass transfer coefficients for either the continuous or the dispersed phase for drops of 3–5 mm and formation times of 2.5–33 s. The time-averaged mass transfer coefficients as a function of final formation time showed a sharp break in the dependence at a formation time of about 7.5 s (Fig. 5). The authors ascribe this fact to some transition in the circulation occurring within the droplets during formation. Therefore they proposed two distinct time regimes with unique dependencies of the average mass transfer coefficients, resulting from a change in the mechanism of mass transfer. It should be noted that the time of this break up depends on the operational conditions and nozzle tip geometry.

The development of new approaches especially to estimate mass transfer rate for fast drop formation conditions were considered using dimensionless parameters. These models can be divided into two sub-branches: Straightforward dimensional analysis approach and physical-based dimensionless modeling, which are discussed here.

1.6. Straightforward Dimensional Analysis Approach

The model proposed by Skelland and Minhas [48] is perhaps the only presented work using this approach. The following list of variables is used to describe the mass transfer during drop formation.

$$k_{df} = f_n(t_f, d, d_n, u_n, D_d, \gamma, \rho_c, \rho_d, \Delta\rho, \mu_c, \mu_d). \tag{34}$$

A multiple regression analysis was applied to correlate 32 dimensionless variations of equation and proposed the following correlation:

$$N_{MI} = a(N_{Fr})^{b_1}(N_T)^{b_2}(N_{Oh})^{b_3},\tag{35}$$

$$\frac{k_{df}t_f}{d} = 0.0432\left(\frac{u_n^2}{dg}\right)^{0.089}\left(\frac{d^2}{t_f D_d}\right)^{-0.334}\left(\frac{\mu_d}{\sqrt{\rho_d d\gamma}}\right)^{-0.601}\tag{36}$$

in which N_{MI}, N_{Fr}, N_T, N_{Oh} are mass transfer numbers, Froud number, dimensionless time number and Ohnesorge number, respectively. The results of this correlation in comparison with their experimental data for water and Ethyl Acetate, Chlorobenzene and Nujol–CCl₄ dilute systems showed good agreement. From Eq. (36) one can see some new parameters in the model, in particular interfacial tension and viscosity. However, such useful correlation without a reasonable physical basis is not frequently utilized.

1.7. Physical-Based Dimensional Analysis Approach

The diffusion/circulation model developed by Liang and Slater [55] based on their previous experience [54], is one the first drop formation mass transfer models that incorporates convection effects *via* this approach. They developed a practical new model based on the Walia–Vir equation (25) for the prediction of fractional solute extraction during drop formation utilizing the eddy diffusivity part in analogy to the model of Handlos and Baron [56]

$$E_{df} = \beta\frac{(D_{Nf}t_f)^{0.5}}{d_f}\left[1 + \frac{(\pi D_{Nf}t)^{0.5}}{d_f}\right],\tag{37}$$

$$D_{Nf} = b_1(D + D_{EF}).\tag{38}$$

Here D is the molecular diffusion coefficient, b_1 is a proportionality coefficient, and D_{EF} is a time dependent diffusivity associated with internal circulation with the proportionality coefficient b_2. The effects of surfactant can be considered *via* constant b_1 which influences the value of diffusion coefficients. In their model the values of b_1 vary regularly with SDS concentration or interfacial tension, and $b_1 \to 1$ when the surfactant concentration tends to zero. The physical sense of b_1 is that it represents reduction of the effectiveness of surface eddies by the presence of surfactant. Therefore, for some cases without surfactant b_1 was used as a kind of model calibration related to internal circulation effects. A theoretical circulation velocity inside a drop is postulated to be related to the nozzle velocity *via*

$$U_c(\pi d^2/4) = U_N(\pi d_N^2/4)\tag{39}$$

and the pseudo-eddy diffusivity is given by

$$D_E = b_2 U_c d,\tag{40}$$

where b_2 is the proportionality coefficient and U_c is notional circulation velocity. With assuming spherical drop formation, $\pi d^3/6t = U_N\pi d_N^2/4$:

$$D_E = (2/3)^{1/3}b_2 U_N^{2/3}d_N^{4/3}t^{-1/3}.\tag{41}$$

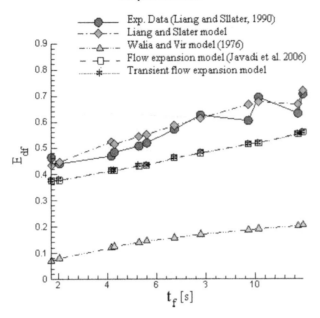

Figure 6. Comparison of different approaches for the prediction of extraction efficiency during drop formation: nozzle diameter 0.406 mm, nozzle velocity 30–195 mm/s, and final formation time 1–13 s; according to [2].

Then they prepared the averaged D_E by integration over the formation time t_f, to give

$$D_{EF} = b_2 d_f^2 / t_f. \tag{42}$$

This is used in Eq. (38) to allow E_{df} to be calculated from Eq. (37) which can be used for the estimation of the fractional extraction E *via* Eq. (33). The nominal circulation time d_f/U_N should be much less than the formation time t_f for the validity of this idea [55]. Comparing the results of their model with their experimental data and others' works showed a noticeable improvement for the prediction of the extraction efficiency when the nozzle Reynolds number was $10 < (\rho_d d_N U_N/\mu_d) < 34$ and $t_f U_N/d_f \gg 1$ (Fig. 6) and showed some shortcomings for $t_f U_N/d_f \approx 1$. However their model showed shortcomings to present a reasonable prediction for the experimental data of Zimmermann *et al.* [52] for acetic acid in toluene (Fig. 7). Their model could describe just the highest values of these experimental data, while the lowest data of this experimental work were close to results of the Walia–Vir model.

2. Flow Expansion Model

This type of models belong to the category of a physical-based dimensional analysis approach to incorporate convection mechanisms, presented first by Javadi *et al.* [2, 27, 57] to cover a wide range of slow and fast drop formations. For the evaluation of the mass transfer during drop formation, in this approach a parameter is defined related to the extent of the convective mixing within and around the growing drop.

For this purpose, it was assumed that the entrance of the dispersed flow into the growing drop from the nozzle is analogous to that from a smaller channel to a larger one (expanding drops). This transfer mechanism has been named "flow expansion". A global time-dependent Reynolds number was defined based on the equivalent of the growing drop diameter as length scale and velocity scale obtained from the flow expansion concept. For definition of these scales, with a constant density and a limited interfacial mass transfer quantity (in comparison with the total mass of growing drop), the following relation was proposed using the conservation of volume of the dispersed phase flowing from the nozzle:

$$\frac{dV_d}{dt} = \frac{d((4/3)\pi r_d^3)}{dt} = \pi R_N^2 U_N, \qquad r_d = \left(\frac{3}{4}U_N R_N^2 t\right)^{1/3}, \tag{43}$$

where, Q, U_N, R_N, r_d are dispersed phase flow rate, nozzle velocity, nozzle cross-section, nozzle radius and growing drop radius, respectively.

Then a velocity scale and consequently, the growing droplet Reynolds number were defined:

$$u_d = \frac{\pi R_N^2 U_N}{\pi r_d^2} = U_N \left(\frac{R_N}{r_d}\right)^2, \tag{44}$$

$$Re_d = \frac{\rho_d U_d D_d}{\mu_d}, \tag{45}$$

where U_d is defined as equivalent mean velocity of flow inside the drop in direction of the drop growth with the viscosity of μ_d. Substituting r_d from Eq. (43) into Eq. (45), the time-dependent growing droplet Reynolds number results:

$$Re_d(t) = 1.1 Re_N \left(\frac{R_N}{t U_N}\right)^{1/3}, \tag{46}$$

where Re_N is the Reynolds number of the fluid flow through the nozzle. It was shown that $R_N/t V_N$ which is a non-dimensional time scale play an important role in the mass transfer rate *via* the related Reynolds number which can be expressed as below with the definition of the nozzle time scale $\tau_N = R_N/U_N$:

$$t^* = t/\tau_N, \tag{47}$$

$$Re_d = 1.1 Re_N (t^*)^{-1/3}. \tag{48}$$

To attain mass transfer coefficient *via* the growing drop Reynolds number, typical semi-empirical mass transfer correlation between Sherwood ($Sh = KL/D$), Schmidt ($Sc = v/D$), and Re numbers are needed, where K and D are the mass transfer and diffusion coefficients, respectively, L is the length scale; and v is the kinematic viscosity. These are available for a constant droplet shape but have to be modified for growing drops in the formation stage. A typical relation for this problem can be written as [58, 59]:

$$Sh = f(Re, Sc) = C_1 Re^{n_1} Sc^{n_2} + C_2, \tag{49}$$

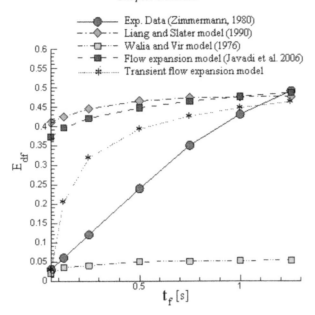

Figure 7. Comparison of different approach for prediction of extraction efficiency during drop formation: nozzle diameter 1 mm, nozzle velocity 53 mm/s, and final formation time 0.1 to 1.3 s; according to [2].

where v in Sc is the kinematics viscosity. The value of $n_2 = 1/3$ has been reported, however for n_1 values of $1/3$ for the Stokes regime, $1/2$ for the boundary layer theory, and $2/3$ for refresh surface assumption or turbulence condition, have been obtained [2, 59]. These correlations can be used for the prediction of mass transfer in continuous or dispersed phases, considering suitable values for the constants. However, these are not applicable during droplet formation, because the mass transfer mechanism changes during the formation time. The flow expansion approach was developed with the purpose of inserting the mentioned time-dependent growing drop Reynolds number from Eq. (48) into Eq. (49), so that:

$$K_{df} = C_1 D Sc^{1/3} Re_N^n R_N^{-1} (t^*)^{(-n-1)/3}, \tag{50}$$

where C_1 is a constant for calibration of the model. Equation (50) shows that the mass transfer coefficient K is a decreasing function of the non-dimensional time t^*. The parameter n can be set constant with values between 0.33–1 related to the extent of the internal convection (turbulence intensity) within droplet. However it can be also set as a function of convective flow. Javadi *et al.* [2] proposed a typical relation $n = n_0 + \Delta n$, in which $\Delta n = \log Re_N^m - 1$ for $Re_N > 10$ as a preliminary attempt with a typical values $n_0 = 0.5$ and $m \approx 1/2$ which can enforce that $0.5 < n < 1$ depends on the nozzle Reynolds number. Similar ideas have been reported by Liang and Slater [55] on the nozzle Reynolds number.

The capability of the model for the prediction of instantaneous mass transfer coefficients was examined with experimental data for a two-component, two-phase

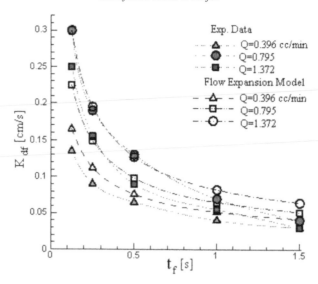

Figure 8. Comparison of the results of the flow expansion model with experimental data from [47] for evaluating the effects of the dispersed phase flow rate on the instantaneous mass transfer coefficients during drop formation; according to [2].

system [47]. An organic liquid with water solubility in the range of 10–25 wt.% was chosen for the dispersed phase; and water as the continuous phase. The organic phase was presaturated with water in order that transfer would occur only in one direction with resistance to mass transfer entirely to the aqueous phase. The results of the flow expansion model show a relatively good agreement with experimental data of [47] as demonstrated in Fig. 8. However for large nozzle diameter (16 Gauge) some shortcomings were observed. Therefore an improvement on the model called "transient flow expansion" was proposed.

Many experimental studies indicate a rapid decrease in the mass transfer coefficients during the early stage of droplet formation, specifically for small nozzle diameters and relatively high dispersed-phase flow rates. Also, a significant difference in the mass transfer rate between small and large diameter nozzles can be observed in the experimental data particularly at the beginning of the drop formation time. It seems that these two facts, which have caused many complications for the prediction of mass transfer during droplet formation in a wide range of flow regimes, can be related to a transition in the flow expansion mechanism [2]. A real process of droplet formation includes generally different stages: (1) early stage of formation, appearance of droplet cap, (2) increasing droplet height with relatively constant diameter equal to the nozzle size, (3) changing the shape to a pear shape, (4) necking, (5) liquid thread is stretched along with rapid rise of the detaching drop, (6) detachment from the nozzle tip, (7) reshape of the remnant liquid at the tip.

For a good drop formation mass transfer model *via* the flow expansion approach, the portion of each stage should properly be considered as different level of mixing,

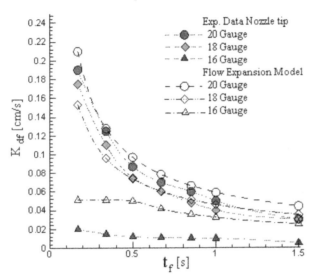

Figure 9. Improvement of the results of the transient flow expansion model in comparison with experimental data from [47] for evaluation of the effects of the nozzle diameter, Gauge No. 20 (outer diameter OD = 0.065 in, inner diameter ID = 0.047 in), 18 (OD = 0.049 in ID = 0.033 in), 16 (OD = 0.035 in, ID = 0.023 in); according to [2].

and internal/external convection terms can occur. In absence of a proper model with the consideration of all drop formation stages, two main periods before and after expansion can be defined. Therefore an initial time for expansion initiation (t_e) higher than $j * t_N$ (where t_N is the nozzle time scale $= R_N/U_N$) is required for the beginning of the droplet expansion with respect to the nozzle size (a time in which the droplet transforms to a pear-shaped form larger than the nozzle diameter). A rough value of $9 < j < 12$ can be estimated, however, many more parameters, in particular the Bond number should be considered. A transient concept can demonstrate that a very small nozzle time scale (t_N), and consequently a small t_e (e.g. less than 0.1 s) can be the main reason for too high values of the mass transfer coefficients. On the contrary, a relatively high value of t_e (e.g. larger than 0.5 s) enforces a restriction of initial high values of the growing droplet Reynolds number resulting in low observed mass transfer coefficients. In other words, it seems the maximum of practical Reynolds numbers of the droplet is given by the nozzle Reynolds number. Thus, for high growing droplet Reynolds numbers (Re_d) an increase in the mass transfer coefficient in the first stages of drop formation is obtained. The results of the transient model show relatively good agreement with experimental data of [47] for all three nozzles size as demonstrated in Fig. 9.

The transient flow expansion model indicates that the Bond number can be an important parameter for the mass transfer rate during droplet formation. The importance of this dimensionless parameter was also mentioned in [48], however, not consequently used. Therefore, the transient flow expansion model is the first mass transfer model that propose a physical mechanism for the effects of the Bond number on the mass transfer rate during the drop formation time.

For the prediction of the total mass transfer Eq. (50) can be integrated over time

$$M_{df} = \int_0^{t_f} K_{df}(t) A(t) \Delta C \, dt, \tag{51}$$

$$M_{df} = C_1 f_c f_{diff} f_{area} \frac{3}{4-n} \frac{Re_N^n T_N^{(n+1)/3} t_f^{(4-n)/3}}{R_N}, \tag{52}$$

where M_{df} is the cumulative mass transfer during droplet formation, C_1 is the constant of the model, $f_c = \Delta C$ is the average concentration gradient factor, $f_{area} = 10.37 U_N^{2/3} R_N^{4/3}$ is the droplet surface growth factor and $f_{diff} = D Sc^{1/3}$ is the mass transfer diffusion factor.

The capabilities of the flow expansion model for the prediction of the total mass transfer [2] was evaluated in comparison with experimental data and theoretical models [50–52, 55] in Figs 6, 7 and 10. Figure 6 shows relatively good agreement between model and experimental data for both, the original and transient model version. In Fig. 7, it becomes clear that the transient flow expansion model (modified) is the only one that can predict the trend of experimental data for the first time adequately. The results of the flow expansion model showed good agreement in comparison with Rajan and Heideger's experimental data [47] for 18 and 20 Gauge

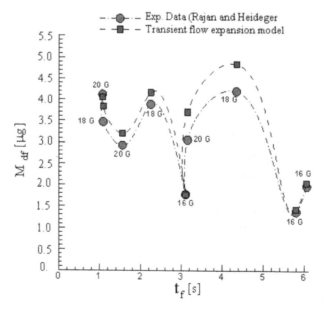

Figure 10. Prediction of the total cumulative mass transfer during drop formation for small, medium and large nozzle sizes (20, 18 and 16 gauge); dispersed flow rate 0.4–1.6 cm^3/min and final formation time 1–6 s; according to [2].

nozzles, dispersed flow rates of 0.4 to 1.6 cm^3/min and final drop formation times of 1 to 4.5 s. A significant difference was only observed for large nozzle size of 16 Gauge, as expected. The transient model shows, however, good agreement with the experimental data for all three size of nozzle, for dispersed flow rates of 0.4 to 1.6 cm^3/min, and the final formation time 1 to 6 s as one can see in Fig. 10.

Rajan and Heideger [47] reported that a square root dependence on time in contrast to relations such as $M_{df} = c(t_f)^n$ proposed by others, cannot describe measured mass transfer data. The least square exponent n was always larger than 0.5, often even larger than 0.9, and showed considerable deviations of about 400% ($0.58 < n < 2.4$) when fitted to experimental data. They concluded that this may be taken as further evidence that unsteady diffusion alone is insufficient to describe mass transfer during drop formation. Nevertheless, Fig. 10 demonstrates that the transient flow expansion model can reproduce the complex variations of the total mass transfer with respect to the final drop formation time in a wide range of nozzle diameter, dispersed phase flow rate, and final formation time. Figure 10 indicates that the trend of the data is too complex to be predicted by a simple relation such as $M_{df} = c(t_f)^n$ or Eq. (18). It is important to estimate the contribution of different model parameters which influence the final drop formation time t_f, i.e. sufficient information and a powerful physical mechanism are required to incorporate the role of various factors [2]. It seems that the simple yet powerful base of the flow expansion model is suitable for this complex task. However, more analytical, experimental and numerical work and also further development of the expansion concept are required to achieve better results. For example, it is expected that the incorporating of a better hydrodynamic droplet formation model such as the two-stage models developed in [10, 11] can improve it. We can also expect that the substitution of a time dependent concentration gradient and the resulting changes in Eqs (50) and (51) can lead to better results for high values of the total interfacial mass transfer (as mentioned in [50, 51]).

The capacity of the flow expansion approach has been compared with other well-known analytical models in Fig. 11. The flow expansion model shows a relatively good agreement (less than 50% error) with experimental data for small and medium nozzle diameters, while other analytical models show differences of up to 500–1000%. The model, however, shows deficiencies for large nozzle size (16 Gauge). The modified transient flow expansion model, on the other hand, shows very good results for this case, as well as for all other nozzle sizes. It also describes well the data for both low and high internal convection, in contrast to the other models.

It should be noted that many other complex hydrodynamic phenomena during growing drop and detachment on a nozzle can affect the mass transfer rate. The importance of the nozzle shape, its material, and wetting condition on the droplet formation process has been discussed in [60]. The impact of Marangoni convection and instability during drop formation is another important issue which can affect the mass transfer rate significantly and should be considered in new models [61–63].

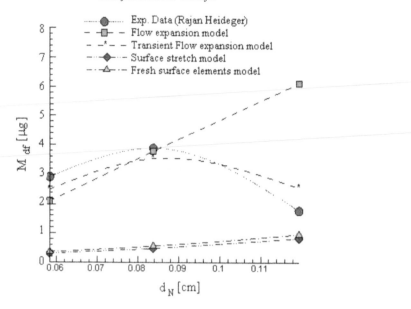

Figure 11. Evaluation of the flow expansion model relative to surface stretch and fresh surface elements models; according to [2].

3. Computational Fluid Dynamic (CFD) Simulations

A review on published numerical work on drop formation in liquid–liquid systems [23–25, 27–29] shows that these investigations have generally evaluated hydrodynamics but without considering the mass transfer. Although a few CFD simulations have studied mass transfer during the rising/falling stage after drop formation [30, 32, 64–66], there have quite limited value for the quantification of the mass transfer during droplet formation. Recently Wegener *et al.* [67] presented a CFD simulation of the transient rise velocity and mass transfer of a single drop with interfacial instability by the using the commercial software STAR-CD. In their work the moving drops were spherical and the concentration inside the drop was uniform at the beginning, however the drop would deform when released from the needle and the solute concentration would change inside the drop. Soleymani *et al.* [68] applied a VOF method for the simulation of drop formation in a solvent extraction system however mass transfer during drop formation was not considered.

To include hydrodynamic aspects into the flow expansion model Javadi *et al.* [27] simulated this phenomenon computationally by an Eulerian volume tracking computational code based on a VOF method [20]. The transient Navier–Stokes equations for the axisymmetric free-boundary problem of a Newtonian liquid was adapted to a vertical dripping and breaking as drops into another immiscible Newtonian fluid (Fig. 12). The results of this simulation and its impact for understanding mass transfer mechanisms and the effects of nozzle diameter and flow rate were discussed. Figures 13 and 14 show the formation of a drop and the creation of the local vorticities and circulating flow.

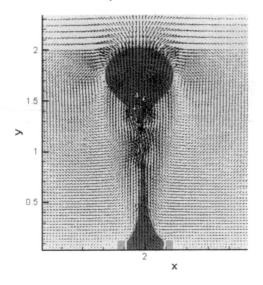

Figure 12. Computational simulation of drop detachment of a 2-ethyl-1-hexanol (2EH) droplet in quiescent water, $R = 0.16$ cm, flow rate $Q = 5$ ml/min; according to [27].

Figure 13. CFD calculated flow field for comparison of the interaction intensity between droplet and second immiscible liquid at $t = 0.04$ s, for small and large nozzle size at a constant flow rate of $Q = 0.795$ ml/min; according to [27].

There are a few computational simulations specifically on mass transfer during drop formation. Lu *et al.* [69] recently presented a numerical simulation based on level set method for simulating solvent extraction features of mass transfer between drops and the surrounding immiscible liquid in axisymmetric cylindrical coordinate. Calculated drop formation times and mass transfer parameters were compared with experimental data in the Methyl Isobutyl Ketone (MIBK)–acetic acid–water solvent extraction system and good agreement was achieved. However, only the Walia and Vir model [50] was considered as reference, although this model is not suitable for fast formation conditions.

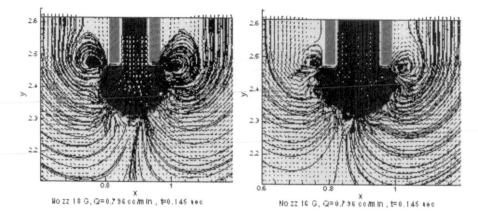

Figure 14. CFD calculated flow field for comparison of the strength and positions of vorticities at the nozzle tip for small and large size nozzle. $t = 0.145$ s, flow rate $Q = 0.795$ ml/min; according to [27].

D. Conclusion

Growing drops in a second immiscible liquid is the basis of flow structure and interfacial transport properties in multiphase system for many fundamental and technological applications. Interfacial mass transfer during drop formation in liquid–liquid systems, after half a century of research, is still a hot research topic currently because of its importance for designing technological processes and development of experimental techniques and devices.

The complexities of this problem for fast dynamic conditions according to the presence of a highly dynamic mobile interface along with diffusional transport, and adsorption from the bulk at the interface, results in a set of conservation equations with not well defined boundary conditions and particular law of transports. The available models developed with several simplifications can be applied for slow growth rates, however, without the presence of surface active agents which add an extra dynamics of adsorption to the interface. For fast expansion rates, even the simple regular systems cannot be analysis easily with the available models.

In addition to the mentioned traditional unsolved problems, nowadays transport phenomena during droplet formation and movement become more and more important in nano- and microfluidics requiring a more accurate analysis. Due to the countless solvents, solutes and respective liquid–liquid interfaces, obviously not a single model and analysis approach can be expected that covers all problems. This chapter can be seen only as an attempt to classify the available models and analysis approaches. The computational fluid dynamic (CFD) simulations based on original macroscopic fluid properties along with numerical methods based on microscopic properties are new efficient tools which help to achieve the mentioned target. However, the present state of the art is yet far from the desired destination. For the validation and improvement of all theoretical approaches, better experimental techniques and setup are required to gain more reliable data.

E. Acknowledgements

The work was financially supported by projects of the European Space Agency (ESA MAP — FASES) and the German Space Agency (DLR 50WM0941).

F. List of Symbols

A droplet/bubble (interfacial) area

C concentration

C_0 initial concentration

C_1 proportionality coefficient in the flow expansion model

D diffusion coefficient

D_E pseudo-eddy diffusivity

D_{EF} time dependent diffusivity associated with internal circulation

E extraction efficiency $(C_{0in} - C_{0out})/(C_{0in} - C_0^*)$

E_{df} extraction efficiency during drop formation

H distribution coefficient between dispersed and continuous phases

K mass transfer coefficient

K_L overall mass transfer coefficient

K_{df} mass transfer coefficient during drop formation

L length scale

M_{df} total mass transfer during droplet formation

N_A total mass transfer during drop formation for surface area A

N' mass transfer rate

F_{Fr} Froud number $= u_n^2/dg$

N_{Oh} Ohnesorge number $= \mu_d/\sqrt{\rho_d d\gamma}$

N_T dimensionless time number $= k_{df}t_f/d$

Q the flow rate through the nozzle tip

R radius

R_N nozzle radius

Re_d Reynolds number of drop

Re_N Reynolds number of nozzle

Sh Sherwood number $(= KL/D)$

Sc Schmidt number $(= \nu/D)$

U velocity

U_c notional circulation velocity

U_d mean velocity of flow inside the drop

U_N fluid velocity in the nozzle tip

V volume

V_d volume of drop

b_1, b_2 proportionality coefficients in Eqs (38) and (41)

d drop diameter

d_f final drop diameter

$f_c = \Delta C$ average concentration gradient factor

$f_{area} = 10.37 U_N^{2/3} R_N^{4/3}$ drop surface growth factor

$f_{diff} = D Sc^{1/3}$ mass transfer diffusion factor

k_0 proportionality coefficient in mass transfer models

r radius of drop (radial direction)

r_d growing drop radius

t time

t_1 time variable

t_e initial time for expansion initiation

t_f total drop formation time

τ_N nozzle time scale $= R_N/U_N$

t^* dimensionless time $= t/\tau_N$

x thickness

γ surface/interfacial tension

ρ_c continuous phase density

ρ_d dispersed phase density

μ_c viscosity of continuous phase

μ_d viscosity of drop phase

ν kinematics viscosity

G. References

1. O.A. Basaran, AIChE J., 48 (2002) 1842.

2. A. Javadi, D. Bastani and M. Taeibi-Rahni, AIChE J., 52 (2006) 895.

3. T.T. Tate, Phil. Mag., 27 (1864) 176.

4. F. Bashforth and H. Adams, An attempt to test theories of capillary action, Cambridge Univ. Press, London, 1883.

5. T. Lohnstein, Ann. Physik, 20 (1906) 237.

6. W.D. Harkins and E.C. Humphrey, J. Am. Chem. Soc., 38 (1916) 240.

7. W.D. Harkins and F.E. Brown, J. Am. Chem. Soc., 38 (1916) 246.

8. C. Hayworth and R. Treybal, Ind. Eng. Chem., 42 (1950) 1174.

9. H.R. Null and H.F. Johnson, AIChE J., 4 (1958) 273.

10. E.V.L.N. Rao, R. Kumar and N.R. Kuloor, Chem. Eng. Sci., 21 (1966) 867.

11. G.F. Scheele and B.J. Meister, AIChE J., 14 (1968) 9.

12. P.M. Heertjes, L.H. De-Nie and H.J. De-Vries, Chem. Eng. Sci., 26 (1971) 441.

13. P.M. Heertjes, L.H. De-Nie and H.J. De-Vries, Chem. Eng. Sci., 26 (1971) 451.

14. L.E.M. Chazal and J.T. Ryan, AIChE J., 17 (1971) 1226.

15. R. Miller and V.B. Fainerman, The drop volume technique, monograph, in "Drops and Bubbles in Interfacial Research", D. Möbius and R. Miller (Eds), Vol. 6, Studies of Interface Science, Elsevier, Amsterdam, 1998, pp. 139–186.

16. R.S. Barhate, N.D. Ganapathi, P. Raghavarao and K.S.M.S. Srinivas, J. Chromatography A, 1023 (2004) 197.

17. B.J. Meister and G.F. Scheele, AIChE J., 13 (1967) 682.

18. B.J. Meister and G.F. Scheele, AIChE J., 15 (1969) 689.

19. B.J. Meister and G.F. Scheele, AIChE J., 15 (1969) 700.

20. B.D. Nichols, C.W. Hirt and R.S. Hotchkiss, "SOLA-VOF: A Solution Algorithm for Transient Fluid Flow with Multiple Free Boundaries", Los Alamos National Laboratory, LA-8355, 1980.

21. S.O. Unverdi and G. Tryggvason, J. Comp. Phys., 100 (1992) 25.

22. J.R. Richards, "Fluid Mechanics of Liquid–Liquid Systems", Ph.D. thesis, University of Delaware, USA, 1994.

23. J.R. Richards, A.N. Beris and A.M. Lenhoff, Phys. Fluid, 7 (1995) 2617.

24. M. Ohta, M. Yamamoto and M. Suzuki, Chem. Eng. Sci., 50 (1999) 2923.

25. X. Zhang, Chem. Eng. Sci., 54 (1999) 1759.

26. D. Zhang and H.A. Stone, Phys. Fluids, 9 (1997) 2234.

27. A. Javadi, D. Bastani, M. Taeibi-Rahni and K. Javadi, "The Effects of Hydrodynamics Characteristics on the Mass Transfer during Droplet Formation Using Computational Approach", ASME International Mechanical Engineering Congress, Paper No: IMECE2006-13283, Chicago, Illinois, USA, November, 2006.

28. S. Homma, G. Tryggvason, J. Koga and S. Matsumoto, "Formation of a Jet in Liquid–Liquid System and Its Breakup into Drops", ASME Fluids Engineering Division Summer Meeting, Washington DC, 1998.

29. S. Homma, J. Koga, S. Matsumoto and G. Tryggvason, "Numerical Investigation of a Laminar Jet Breakup into Drops in Liquid–Liquid Systems", Eighth International Conference on Liquid Atomization and Spray Systems, Pasadena, CA, USA, 2000.

30. S. Homma, J. Koga, S. Matsumoto and G. Tryggvason, Dynamics of mass transfer for an axisymmetric drop, Proc. ICMF2001, 502 (2001).

31. J. Petera and L.R. Weatherley, Chem. Eng. Sci., 56 (2001) 4929.

32. W.G. Whitman, L. Long and W.Y. Wang, Ind. Eng. Chem., 18 (1926) 363.

33. T.K. Sherwood, J.E. Evans and J.V.A. Longor, Trans. Amer. Inst. Chem. Eng., 35 (1939) 597.

34. W. Licht and J.B. Conway, Ind. Eng. Chem., 42 (1950) 1151.

35. B.E. Dixon and A.A.W. Russell, J. Soc. Chem. Ind. Lond., 69 (1950) 284.

36. W. Licht and W.F. Pansing, Ind. Eng. Chem., 45 (1953) 1885.

37. B.E. Dixon and J.E.L. Swallow, J. Appl. Chem., 4 (1954) 86.

38. J.M. Coulson and S.J. Skinner, Chem. Eng. Sci., 1 (1952) 197.

39. D. Ilkovic, J. Chim. Phys. Physicochem. Biol., 35 (1938) 129.

40. H.H. Michels, "The Mechanism of Mass Transfer During. Bubble Growth", PhD thesis, Univ. Delaware, Newark, 1960.

41. J.B. Angelo, E.N. Lightfoot and D.W. Howard, AIchE J., 12 (1966) 751.

42. H. Groothius and H. Kramers, Chem. Eng. Sci., 4 (1955) 17.

43. W.J. Beek and H. Kramers, Chem. Eng. Sci., 17 (1962) 909.

44. P.M. Heertjes and L.H. De-Nie, Chem. Eng. Sci., 21 (1966) 755.

45. A.T. Popovich, R.E. Jervis and O. Trass, Chem. Eng. Sci., 19 (1964) 357.

46. P.M. Heertjes, W.A. Holve and H. Talsma, Chem. Eng. Sci., 3 (1954) 122.

47. S.M. Rajan and W.J. Heideger, AIChE J., 17 (1971) 202.

48. A.H.P. Skelland and S.S. Minhas, AIChE J., 17 (1971) 1316.

49. J.A.C. Humphrey, R.L. Hummel and J.W. Smith, Chem. Eng. Sci., 29 (1974) 1496.

50. D.S. Walia and D. Vir, Chem. Eng. Sci., 31 (1976) 525.

51. D.S. Walia and D. Vir, Chem. Eng. J., 12 (1976) 133.

52. V. Zimmermann, W. Halwachs and K. Schugerl, Chem. Eng. Commun., 7 (1980) 95.

53. W.J. Heideger and M.W. Wright, AIChE J., 32 (1986) 1372.

54. M.J. Slater, M.H. Baird and T.B. Liang, Chem. Eng. Sci., 43 (1988) 223.

55. T.B. Liang and M.J. Slater, Chem. Eng. Sci., 45 (1990) 97.

56. A.E. Handlos and T. Baron, AIChE. J., 3 (1957) 127.

57. A. Javadi and D. Bastani, "Estimation of Mass Transfer during Drop Formation: New Flow Expansion Model", ASME International Mechanical Engineering Congress, Paper No: IMECE2004-62443, California, 2004.

58. L. Steiner, Chem. Eng. Sci., 41 (1986) 1979.

59. J.C. Godfrey and M.J. Slater (Eds.), Liquid–liquid extraction equipment, John Wiley & Sons, Chichester, 1994.

60. C.T. Chen, J.R. Maa, Y.M. Yang and C.H. Chang, Int. Comm. Heat Mass Transfer, 28 (2001) 681.

61. M. Wegener, A.R. Paschedag and M. Kraume, Int. J. Heat Mass Transfer, 52 (2009) 2673.

62. B. Arendt and R. Eggers, Int. J. Heat Mass Transfer, 50 (2007) 2805.

63. A. Javadi, D. Bastani, J. Krägel and R. Miller, Colloids Surfaces A, 347 (2009) 167.

64. W.H. Piarah, A. Paschedag and M. Kraume, AIChH J., 47 (2001) 1701.

65. Z.S. Mao, T. Li and J. Chen, Int. J. Heat Mass Transfer, 44 (2001) 1235.

66. X. Li, Z.S. Mao and W. Fei, Chem. Eng. Sci., 58 (2003) 3793.

67. M. Wegener, T. Eppinger, K. Baumler, M. Kraume, A.R. Paschedag and E. Bansch, Chem. Eng. Sci., 64 (2009) 4835.

68. A. Soleymani, A. Laari and I. Turunen, Chem. Eng. Res. Design, 86 (2008) 731.

69. P. Lu, Z.C. Wang and Z.-S. Mao, Chem. Eng. Sci. (2010) doi:10.1016/j.ces.2010.07.022.

Subject Index

Printed and bound by CPI Group (UK) Ltd, Croydon, CR0 4YY

21/10/2024

01777040-0002